Fluid Mechanics

Fluid Mechanics: An Intermediate Approach helps readers develop a physics-based understanding of complex flows and mathematically model them with accurate boundary conditions for numerical predictions.

The new edition starts with a chapter reviewing key undergraduate concepts in fluid mechanics and thermodynamics, introducing the generalized conservation equation for differential and integral analyses. It concludes with a self-study chapter on computational fluid dynamics (CFD) of turbulent flows, including physics-based postprocessing of 3D CFD results and entropy map generation for accurate interpretation and design applications. This book includes numerous worked examples and end-of-chapter problems for student practice. It also discusses how to numerically model compressible flow over all Mach numbers in a variable-area duct, accounting for friction, heat transfer, rotation, internal choking, and normal shock formation.

This book is intended for graduate mechanical and aerospace engineering students taking courses in fluid mechanics and gas dynamics.

Instructors will be able to utilize a solutions manual for their course.

Fluid Mechanics
An Intermediate Approach

Second Edition

Bijay K. Sultanian

CRC Press is an imprint of the
Taylor & Francis Group, an **informa** business

Designed cover image: Shutterstock

Second edition published 2025
by CRC Press
2385 NW Executive Center Drive, Suite 320, Boca Raton FL 33431

and by CRC Press
4 Park Square, Milton Park, Abingdon, Oxon, OX14 4RN

CRC Press is an imprint of Taylor & Francis Group, LLC

First edition published by CRC Press 2016

Library of Congress Cataloging-in-Publication Data
Names: Sultanian, Bijay K., author.
Title: Fluid mechanics : an intermediate approach / Bijay K. Sultanian.
Description: Second edition. | Boca Raton, FL : CRC Press, 2025. | Includes bibliographical references and index.
Identifiers: LCCN 2024023325 (print) | LCCN 2024023326 (ebook) | ISBN 9781032350790 (hbk) |
ISBN 9781032350806 (pbk) | ISBN 9781003325192 (ebk) | ISBN 9781032883182 (eBook+)
Subjects: LCSH: Fluid mechanics.
Classification: LCC TA357 .S8626 2025 (print) | LCC TA357 (ebook) |
DDC 620.1/06–dc23/eng/20240718
LC record available at https://lccn.loc.gov/2024023325
LC ebook record available at https://lccn.loc.gov/2024023326

ISBN: 978-1-032-35079-0 (hbk)
ISBN: 978-1-032-35080-6 (pbk)
ISBN: 978-1-003-32519-2 (ebk)
ISBN: 978-1-032-88318-2 (eBook+)

DOI: 10.1201/9781003325192

Typeset in Times
by codeMantra

Access the Support Material: routledge.com/9781032350790

Dedicated to my dearest friend Kailash Tibrewal, whose mantra of "joy in giving" continues to inspire me; my wife, Bimla Sultanian; our daughter, Rachna Sultanian, MD; our son-in-law, Shahin Gharib, MD; our son, Dheeraj (Raj) Sultanian, JD, MBA; our daughter-in-law, Heather Benzmiller Sultanian, JD; and our grandchildren: Aarti Sultanian, Soraya Zara Gharib, Shayan Ali Gharib, Loki, and Millie for the privilege of their unconditional love, immensely enriching my life.

Contents

Author

Dr. Bijay (BJ) K. Sultanian, PhD, PE, MBA, ASME Life Fellow, is a recognized international authority in gas turbine heat transfer, aerodynamics, secondary air systems, and computational fluid dynamics (CFD). He is the founder of Takaniki Communications, LLC, providing high-impact technical training programs for corporate engineering teams. He is also a former adjunct professor at the University of Central Florida, where he taught graduate-level courses in turbomachinery and fluid mechanics for 10 years. He has instructed several workshops at ASME Turbo Expos since 2009. He has been an active member of ASME IGTI's Heat Transfer Committee since 1994 and received the ASME IGTI Outstanding Service Award at ASME Turbo Expo 2018, Lillestrom, Norway. He is the author of senior undergraduate and graduate-level textbooks:

1. *Fluid Mechanics: An Intermediate Approach* (2015)
2. *Gas Turbines: Internal Flow Systems Modeling* (Cambridge Aerospace Series #44) (2018)
3. *Logan's Turbomachinery: Flowpath Design and Performance Fundamentals*, Third Edition (2019)
4. *Fluid Mechanics and Turbomachinery: Problems and Solutions* (2021)

During his three decades in the gas turbine industry, Dr. Sultanian has worked in and led technical teams at several organizations, including Allison Gas Turbines (now Rolls-Royce), GE Aircraft Engines (now GE Aviation), GE Power Generation, and Siemens Energy. He has developed several physics-based improvements to legacy heat transfer and fluid systems design methods, including new tools to analyze critical high-temperature gas turbine components with and without rotation. He particularly enjoys training large engineering teams at prominent firms around the globe on cutting-edge technical concepts in engineering design and project management best practices.

During 1971–1981, Dr. Sultanian made several landmark contributions toward the design and development of India's first liquid rocket engine for a surface-to-air missile (Prithvi). He also developed the first numerical heat transfer model of steel ingots for optimal operations of soaking pits in India's steel plants.

Dr. Sultanian is a Life Fellow of the American Society of Mechanical Engineers, a registered Professional Engineer (PE) in Ohio, a GE-certified Six Sigma Green Belt, and an Emeritus Member of Sigma Xi, The Scientific Research Honor Society.

Dr. Sultanian received his BTech and MS in Mechanical Engineering from the Indian Institute of Technology, Kanpur, and the Indian Institute of Technology, Madras, respectively. He received his PhD in Mechanical Engineering from Arizona State University, Tempe, and his MBA from the Lally School of Management and Technology at Rensselaer Polytechnic Institute.

Preface to the Second Edition

Fluid mechanics stands as the captivating core discipline behind modern energy conversion and aerospace propulsion technologies, encompassing rocket engines, gas turbines, steam turbines, hydraulic turbines, and wind turbines, as well as land and marine power plants, compressors, fans, and pumps. Nature's wonders, from tornadoes and hurricanes to Niagara Falls, showcase the elegant principles of fluid mechanics, further demonstrated by mesmerizing marine life and graceful birds like hummingbirds in flight. Despite the apparent complexity of these phenomena, they are governed by three fundamental conservation laws: mass, momentum, and energy. However, despite the simplicity underlying this visually captivating subject, my decade of teaching the first graduate-level course in fluid mechanics at the University of Central Florida revealed that many students found it challenging. This difficulty often stemmed from an insufficient foundation acquired during undergraduate fluid mechanics courses, where the emphasis was often on achieving good grades rather than a deep understanding of the subject matter. Additionally, students tended to rely on equations to solve fluid problems rather than analyzing them from first principles, such as the conservation of mass, momentum, and energy—a physics-first approach.

The primary objectives of Sultanian (2015) were as follows:

1. To comprehensively review and reinforce fundamental concepts of undergraduate fluid mechanics, particularly focusing on control volume analysis of mass, momentum, and energy conservation. This aimed to establish a common baseline for students from diverse educational backgrounds, who may have encountered varied texts and instructional methods.
2. To present intermediate fluid mechanics through a physics-first approach, prioritizing the understanding of physical principles over mathematical formalism. In this approach, mathematics serves as the language of physics, following the conceptual framework rather than dictating it—a departure from the prevalent mathematics-first approach employed in most fluid mechanics courses.

During my time teaching intermediate fluid mechanics at the University of Central Florida (UCF), one homework problem stood out as a memorable experience. The scenario involved a young boy assisting his father with yard work and observing the behavior of water flow from a garden hose. The boy noticed that when he partially covered the hose opening, the water jetted out faster and traveled further. Curious, he asked his father why this happened. The father, believing the flow velocity increased due to the reduced area of the opening, provided an explanation. However, the boy soon realized that although the water jetted out faster, it took longer to fill the bucket. This dilemma prompted the father to reconsider his initial explanation. Despite providing the class with hints and encouraging them to analyze the problem using the Bernoulli equation and ideal flow principles, no student was able to arrive at the correct solution for this incompressible flow problem. This outcome was disappointing, considering the problem's straightforward nature and lack of complexity.

Some of my favorite questions to test how well students understand fluid mechanics remain:

1. What is the physical meaning of Mach number?
2. What happens in a choked flow at $M = 1$?
3. Is there a function of Mach number that remains constant across a normal shock?
4. How does an airfoil develop a lift—not because of the different distance traversed by the fluid particle from the leading edge to the trailing edge of its top (suction) and bottom (pressure) surfaces?
5. Why does total temperature remain constant in a Fanno flow?

6. How does friction accelerate a subsonic Fanno flow—Mach number increases downstream?
7. Just as total temperature remains constant in a Fanno flow, which quantity remains constant in a Rayleigh flow?
8. Which pressure, total or static, remains uniform over the cross-section of a fully developed laminar pipe flow?
9. Which pressure, total or static, and which temperature, total or static, do we measure at the wall of a fully developed pipe flow?
10. For a strongly swirling flow in an abrupt pipe expansion, why does the larger pipe feature on-axis flow recirculation (vortex breakdown)?
11. Why are both $p+0.5\rho V^2$ and $p+\rho V^2$ meaningful physical quantities, and how are they used?
12. What remains constant in a forced vortex, and what remains constant in a free vortex?

Answers to all these questions are available in this edition of the book.

In contrast to the abundance of titles available for undergraduate fluid mechanics, there are relatively few options tailored specifically for first graduate-level courses. However, these books tend to expand significantly with each new edition, often adding 50–100 pages or more, often including chapters beyond the scope of a single semester course at the graduate level. The proposed second edition of Sultanian (2015) aims to buck this trend by streamlining its content, reducing its page count from 558 to approximately 400 pages. This revision will ensure that the text is optimized for a concise, one-semester graduate-level course.

The following changes and enhancements are implemented in the second edition:

1. Chapters 1–4 of Sultanian (2015) are consolidated into the new Chapter 1 ("Review of Undergraduate Fluid Mechanics and Thermodynamics"). This chapter will help students from different universities develop a common understanding of various must-know concepts from their undergraduate courses in fluid mechanics and thermodynamics.
2. Chapters 5, 6, 7, 8, and 10 of Sultanian (2015) are rearranged as Chapters 5, 2, 3, 4, and 6, respectively. Furthermore, the writing in each of these chapters is edited to be as accurate, concise, and simple as possible, but not simpler!
3. Chapter 9 ("Flow Network Modeling") from Sultanian (2015) has been deleted. Upgraded versions of most of the materials in this chapter may be found in Sultanian (2018).
4. This edition features 60 worked examples and over 100 chapter-end problems. Readers may find additional problems and solutions in Sultanian (2021, 2022).
5. In view of Sultanian (2023), Appendix A: Compressible Flow Equations and Tables from Sultanian (2015) is eliminated.
6. Appendices B, C, D, and G of Sultanian (2015) are eliminated.
7. Appendix E: Vorticity and Circulation of Sultanian (2015) appears as Appendix B: Vorticity, Vortex, and Circulation with additional material.
8. Appendix F of Sultanian (2015) appears as revised Appendix A.
9. This edition features a new Appendix C: Euler's Turbomachinery Equation.
10. In this edition, all worked examples are located at the end of each chapter, providing a streamlined approach compared to the first edition, where they are scattered throughout, disrupting the flow of the main text.
11. Wherever possible, the nomenclature of Sultanian (2015) has been simplified, for example, P_s for static pressure has been changed to p, and T_s for static temperature has been changed to T.

REFERENCES

Sultanian, B.K. 2015. *Fluid Mechanics: An Intermediate Approach.* Boca Raton, FL: Taylor & Francis.

Sultanian, B.K. 2018. *Gas Turbines: Internal Flow Systems Modeling* (Cambridge Aerospace Series #44). Cambridge, UK: Cambridge University Press.

Sultanian, B.K. 2019. *Logan's Turbomachinery: Flowpath Design and Performance Fundamentals*, Third Edition. Boca Raton, FL: Taylor & Francis.

Sultanian, B.K. 2021. *Fluid Mechanics and Turbomachinery: Problems and Solutions.* Boca Raton, FL: Taylor & Francis.

Sultanian, B.K. 2022. *Thermal-Fluids Engineering: Problems with Solutions.* USA: Independently published by Kindle Direct Publishing (Amazon.com).

Sultanian, B.K. 2023. *One-Dimensional Compressible Flow: Physics-Based Modeling for Design Engineering.* USA: Independently published by Kindle Direct Publishing (Amazon.com).

Dr. Bijay (BJ) K. Sultanian

Founder & Managing Member, Takaniki Communications, LLC

Former Adjunct Professor, The University of Central Florida

Preface to the First Edition

This book is a unique blend of my passion for fluid mechanics, 40+ years of industry experience, and nearly a decade of teaching graduate-level courses in turbomachinery and fluid mechanics at the UCF. Fluid mechanics can be the most beautiful and most enjoyable area of engineering. Its true beauty lies not in its mathematical descriptions but in the understanding of its underlying physics—the deeper you go, the more beautiful it feels. This is one subject that is vividly displayed in all aspects of nature—a rich source of complex flow visualization. In this book, as a refreshing change, mathematics follows the flow physics, not the other way around.

After over 35 years of industry experience in rocket propulsion, thermal and fluid flow modeling of processes in steel plant engineering, development of physics-based methods and tools for the design of some of the most advanced gas turbines for aircraft propulsion and power generation, and extensive application of three-dimensional (3D) computational fluid dynamics (CFD) technology to the solutions of many design problems, I moved to Orlando, Florida, in 2006 to work for Siemens Energy. The UCF invited me that year to join the faculty as an adjunct professor and teach the graduate course "EML5402—Turbomachinery" in the fall semester. I immediately accepted the assignment. It allowed me to bring my years of industry experience into the classroom to train a new generation of talent for the practical, rather than the theoretical, problems facing today's engineers. The following year, the UCF invited me to teach another graduate course, "EML5713—Intermediate Fluid Mechanics," in the spring semester. The class size in this course grew each year from around 15 in 2007 to around 40 by the fall of 2013. In teaching the second course, I struggled to find an adequate textbook that stressed the fundamental physical harmony of fluid flows as a prerequisite to formulas and problem-solving. Faced with this challenge, I did what any great engineer would do: I designed one that did.

Most undergraduate engineering programs include just a couple of courses on fluid mechanics. The students taking these courses are generally exposed to the concepts of statics and dynamics from prior courses on solid mechanics. As a result, many tend to find the new concepts of fluid mechanics rather overwhelming and nonintuitive. Due to a continued emphasis on mathematical treatment, rather than the physical introduction to the subject, many of these students soon lose sight of the beauty and simplicity of fluid mechanics in favor of simply committing to memory its mathematical equations, including the difficult Navier–Stokes equations (which are nonlinear partial differential equations in three coordinate directions). They soon start treating these undergraduate fluids courses as just another course in advanced mathematics, developing little understanding of the underlying physical concepts. The pressure of examinations and grades in these courses only reinforces this treatment.

As a prerequisite to many advanced graduate courses in the thermofluids stream, many students are required to take some form of graduate-level intermediate fluid mechanics as one of their core courses. Students, weary of their undergraduate experiences, are often immediately frustrated by the prospect of dealing with even more complex mathematics, at times in tensor notation! Although many can recite complex formulas, when asked to articulate the difference between static pressure and total pressure in plain English, and how to compute the latter in incompressible and compressible flows, they cannot. The plain English difference between static temperature and total temperature or the concept of adiabatic wall temperature is surprisingly hard to articulate without fundamental physical understanding. Adding stationary (inertial) and rotating reference frames only amplifies these hidden issues. Overreliance on the Bernoulli equation is a classic demonstration of where a mathematics-first approach fails many engineering students. The concept of choking in one-dimensional flows, for example, can be exceedingly complicated using a mathematics-first approach, but asking students to take a step back to first visualize the phenomenon in nature and understand the physical concept of Mach number can lead to a world of difference. Although many students may be conversant with the equations of Fanno and Rayleigh flows, they find it hard to explain in layman's terms how wall friction accelerates a subsonic airflow in a constant-area duct with or without heating.

Many practicing engineers dealing with fluid flows in various industries are, as a result of the classic academic approach (coupled with dependence on one or more in-house or commercial codes to perform design calculations), surprisingly weak in their intuitive understanding of the physics of these flows. When questioned about the accuracy of their calculations, often an unshaken faith in design tools and their predictions is the answer. For many in the industry, if they, and their competition, have been doing it in a particular way for so many years, it must be right!

With the availability of so many user-friendly commercial 3D CFD codes, today's challenge in the application of 3D CFD technology to design lies not in high-quality and high-fidelity grid generation or in obtaining a fully converged solution with state-of-the-art numerical schemes and turbulence models, but in properly interpreting their computed 3D CFD results for intended design applications. Only a strong foundation in fluid mechanics will help CFD lovers to make good sense of their 3D CFD results.

I strongly believe that this book will benefit not only the graduate and senior undergraduate students of fluid mechanics pursuing their degree programs in the thermofluids stream but also many practicing engineers dealing with thermofluids in the design of commercial and military planes; submarines and cruise ships; automobiles; jet and rocket propulsion; oil and gas pipelines; gas turbines, steam turbines, and generators; pumps and compressors; air-conditioning and refrigeration units; heat exchangers; artificial hearts and valves; dams and irrigation systems; and many other areas of thermofluids engineering.

This book will come to many undergraduate and graduate students, and practicing engineers, as a breath of fresh air. Empowered by the knowledge gained from this book, these students and engineers will develop a unique insight into various fluid flow phenomena and will fall in love with the subject like never before. With focus on a clear understanding of the key fundamental concepts (which are amply illustrated by a number of worked-out real-world examples), this book will help readers develop a variety of problem-solving skills to handle practical fluids engineering problems. For example, they will master the techniques of control volume analysis of mass, linear momentum, angular momentum, and energy in both inertial and noninertial reference frames. They will also develop an unprecedented intuitive understanding of one-dimensional compressible flows, including Fanno flows, Rayleigh flows, isothermal flows, normal shocks, oblique shocks, and isentropic Prandtl–Meyer expansion flows. By eliminating most of the key conceptual gaps for senior undergraduate and graduate-level students, this book will help them transition to learning new topics in advanced graduate courses in thermofluids.

This book also includes two value-added chapters on special topics that reflect the state of the art in design applications of fluid mechanics. Chapter 9 deals with the key details of physics-based modeling of both incompressible and compressible flow networks. These details, notwithstanding the current reality that a number of commercial flow network codes are being used in many industries, are not found in any leading contemporary book on fluid mechanics. Chapter 10 focuses on the applications of CFD technology to practical industrial design problems, often in concert with related flow network solutions, rather than presenting the mathematical details of CFD numerical formulations and solution methods. Additionally, this chapter includes a physics-based methodology to postprocess 3D CFD results to generate section-averaged values for their useful interpretations and design applications.

This book is deemed to be an indispensable companion to all students and practicing engineers engaged in various designs, both new and upgrades, involving fluid flow and heat transfer. Chapters 1–4 form the core course in fluid mechanics. To develop a solid foundation in this subject, all engineers dealing with thermofluids, both in industries and in universities, must master the material presented in the first four chapters.

At universities around the world, at least three distinct graduate or senior undergraduate courses in fluid mechanics can be taught using this textbook. The following syllabi are suggested for these courses; instructors, however, are free to fine-tune these syllabi and reinforce them with their notes and/or additional reference material to meet their specific instructional needs:

1. Intermediate fluid mechanics—a graduate-level core course in fluid mechanics for senior undergraduate and graduate students pursuing their MS and PhD programs in the thermofluids stream (fluid mechanics, heat transfer, propulsion, turbomachinery aerodynamics, combustion, etc.): fluids core course (Chapters 1–4) and selected topics from Chapter 5 to 8.
2. Compressible flows or gas dynamics—a senior-level undergraduate course on compressible flows or gas dynamics: selected topics from fluids core course (Chapters 1–4) and Chapter 5.
3. Industrial fluid mechanics—an elective graduate-level course primarily targeted at practicing engineers in various industries dealing with fluids engineering: fluids core course (Chapters 1–4) and selected topics from Chapters 5, 9, and 10, Appendix G.
   - Mathematical treatment follows flow physics, not the other way around.
   - Over 60 systematically worked-out real-world examples.
   - Over 100 chapter-end problems for which the solutions manual is available to all instructors who adopt this textbook for teaching their university courses in fluid mechanics.
   - Over 250 figures.
   - Clearly explained key thermofluids concepts: rothalpy, stream thrust, impulse pressure, forced/free vortices, windage, vorticity, and circulation.
   - Control volume analysis of linear momentum and angular momentum in both inertial and noninertial reference frames.
   - Enhanced intuitive understanding of internal compressible flows through the use of various flow functions: total- and static-pressure mass flow functions, total- and static-pressure impulse functions, and normal shock function.
   - Physics-based modeling of compressible flow over all Mach numbers in a variable-area duct with friction, heat transfer, and rotation with internal choking and normal shock formation.
   - Compressible and incompressible flow network modeling (Chapter 9).
   - Strengths and weaknesses of state-of-the-art turbulent flow CFD predictions, including physics-based postprocessing of 3D CFD results for design applications (Chapter 10).
   - Compressible flow tables with equations used to generate the tabular values (Appendix I).
   - Closed-form analytical solution of the coupled heat transfer and work transfer in a rotating duct flow (Appendix II).
   - Systematically derived equations to compute pressure and temperature changes in isentropic free and forced vortices (Appendix III).
   - Systematically derived equations to transfer stagnation (total) flow properties from a stationary reference frame to a rotating reference frame and vice versa (Appendix IV).

Dr. Bijay (BJ) K. Sultanian
Founder & Managing Member, Takaniki Communications, LLC
Former Adjunct Professor, The University of Central Florida

Acknowledgements

I am deeply thankful to numerous individuals who have contributed to my journey—my esteemed teachers, close friends, beloved family members, and over 300 students who participated in the graduate-level courses EML5713-Intermediate Fluid Mechanics and EML5402-Turbomachinery, which I had the privilege of teaching at UCF between 2006 and 2016.

Recently, Mr. Sinan Sal and Dr. Erinc Erdem from TUSAS Engine Industries, Inc. (TEI), extended an invitation for me to conduct a one-week short course on gas dynamics for their design engineers at the Eskisehir facility. With more than 65 engineers actively engaged in this design-centric course, I found immense satisfaction in sharing knowledge and engaging in enlightening discussions. I am especially grateful for the thought-provoking questions raised by the participants and for the insightful dialogues with Dr. Erinc Erdem, who graciously hosted the event. The content covered during the course forms the foundation of Chapter 5 (Compressible Flow) in this book.

Finally, I want to express my heartfelt thanks to all readers who graciously take the time to point out any unintentional errors. Your feedback is truly invaluable, and I am deeply grateful to anyone who feels motivated to contribute to the continual exploration and enhancement of the fascinating field of fluid mechanics.

1 Review of Undergraduate Fluid Mechanics and Thermodynamics

1.1 INTRODUCTION

Fluid flows abound in nature and various engineering applications. The continued research across multiple flow phenomena allows for a deeper understanding and more accurate predictions using the state-of-the-art computational fluid dynamics (CFD) technology. This technology enables us to develop advanced commercial and military planes; submarines and cruise ships; automobiles; rocket propulsion; oil and gas pipelines; gas turbines and steam turbines; pumps, compressors, air-conditioning and refrigeration units; heat exchangers; artificial hearts and valves; dams and irrigation systems; and others.

Fluids include both liquids and gases. We precisely define fluid as a substance that deforms continuously under shear stress, no matter how small that shear stress may be. In other words, unlike solids, fluid cannot withstand shear stress unless it is in motion (flowing). Conservation laws of mass, momentum, energy, and entropy, in conjunction with some auxiliary equations, like the equation of state, govern all fluid flows we encounter in nature and various engineering applications. Understanding these laws of fluid mechanics enables us to compute multiple flow properties such as velocity, pressure, and temperature and how they influence drag and lift forces on bodies, including the development of more accurate weather prediction models.

By way of review and reinforcement, this chapter presents material typically covered in one or more undergraduate fluid mechanics courses.

1.2 STREAMLINE

As shown in Figure 1.1a, local velocity vectors in a flow field remain tangent to the streamline at any given instant. Accordingly, no flow can cross a streamline. When bundled together, streamlines form a stream tube. Thus, no flow can enter or exit the surface of a stream tube, which we often use in the analysis of ideal internal flows with impermeable and frictionless walls. While the velocity magnitude may vary along a streamline, its direction is always along the local tangent to the streamline. As shown in Figure 1.1b, relative to the flow over the pressure surface of the airfoil, the flow accelerates along the streamline on the suction surface. Similarly, the flow in a convergent–divergent nozzle, shown in Figure 1.1c, features significant acceleration as it passes through the throat area. As streamlines are instantaneous, making them visible in an unsteady flow is difficult.

1.3 STREAKLINE

A streakline is the locus of fluid particles that have passed through a prescribed point earlier. As shown in Figure 1.2, at $t=0$, particle 1 (cross) has passed through the prescribed point A at the nozzle exit. At $t=1$, particle 2 (solid circle) has passed through A. In the meantime, particle 1 has moved to a new location in the flow. At $t=2$, particle 3 (solid triangle) has passed through A, while particles 1 and 2 have moved to their new locations. Again, at $t=3$, particle 4 (solid square) has passed through the prescribed point A at the nozzle exit, while particles 1, 2, and 3 have moved

DOI: 10.1201/9781003325192-1

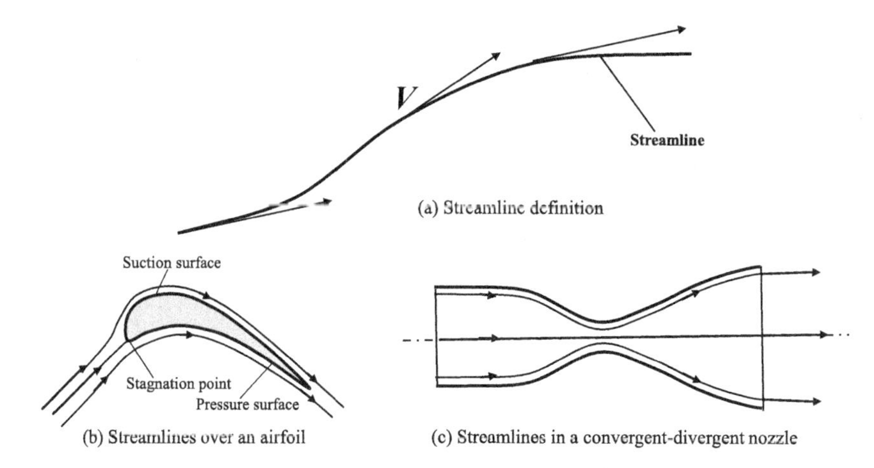

FIGURE 1.1 (a) Streamline definition, (b) streamlines over an airfoil, and (c) streamlines in a convergent–divergent nozzle.

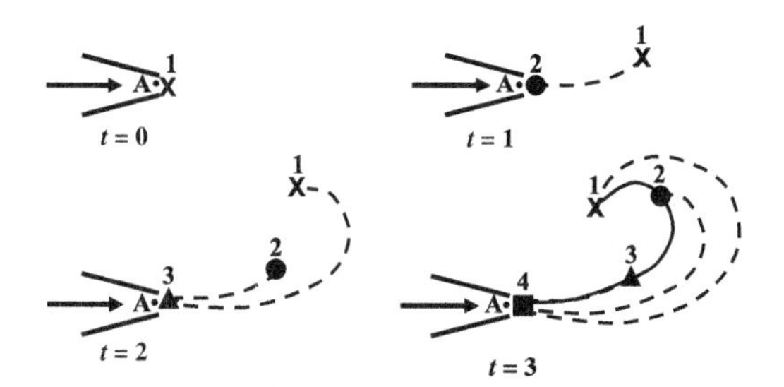

FIGURE 1.2 Streakline.

to their new locations. If we now connect the locations of all four particles, which have all passed through the prescribed point A at the nozzle exit, the resulting solid line shown in the figure is the streakline. Thus, we need the passage of time to generate a streakline.

1.4 PATHLINE

Figure 1.3 shows a streamline and a pathline, an actual path traversed by a given fluid particle. Note that while a streamline is instantaneous, we need time to generate a pathline. Furthermore, note that streamlines, pathlines, and streaklines coincide in a steady flow. These lines reveal the essential properties of a flow field. For example, if the streamlines are curved, we can deduce that the flow must have a gradient of static pressure across them, causing the curvature. Without such a transverse pressure gradient, the streamlines will remain straight. Similarly, in a complex flow, the streamlines clearly show the regions of flow reversal, boundary layer separation, and recirculation.

1.5 STATIC PRESSURE VERSUS TOTAL PRESSURE

The momentum associated with the molecular motions at a point in fluid flow measures the static pressure, which is locally isotropic. A change in the reference frame does not change the static pressure. For example, for the flow in a rotor–stator cavity, the static pressure at a point will be identical to both observers—one sitting on the stator and the other on the rotor.

When we stagnate a flow to rest without any loss in the dynamic pressure (or increase in entropy), the resulting pressure is called the stagnation (total) pressure. In other words, the stagnation pressure

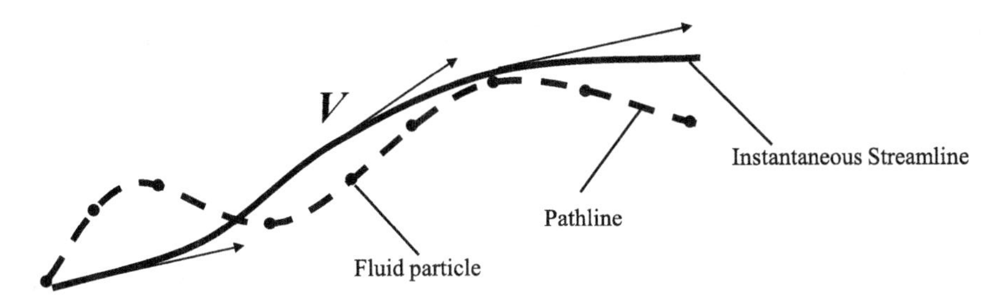

FIGURE 1.3 Streamline and pathline.

is the sum of the static pressure and the dynamic pressure (resulting from the bulk flow velocity). The terms stagnation and total are, therefore, often used interchangeably. Unlike the static pressure, the total pressure is not isotropic. In a flow, we should view the total pressure at a point as a reference quantity obtained through an isentropic stagnation process. Depending on the loss in the stagnation process, the actual total pressure will range from the initial static pressure to the ideal total pressure (zero loss in the dynamic pressure). Note that the contribution of the dynamic pressure to the total pressure makes the total pressure dependent on the chosen reference frame. Furthermore, one must use the static pressure, not the total pressure, to evaluate the pressure-dependent thermophysical properties of a fluid at any point in the flow field.

In an incompressible flow, we use the following equation to calculate the total pressure:

$$p_0 = p + \frac{1}{2}\rho V^2 \tag{1.1}$$

where $\rho V^2/2$ is the local dynamic pressure in the flow.

In a compressible flow (see Chapter 5), however, we calculate the local total pressure using the equation:

$$p_0 = p\left(1 + \frac{\gamma - 1}{2} M^2\right)^{\frac{\gamma}{\gamma-1}} \tag{1.2}$$

where the Mach number $M = V/\sqrt{\gamma RT}$ and $\gamma = c_p/c_v$, which is the ratio of specific heat at constant pressure and constant volume.

1.6 STATIC TEMPERATURE VERSUS TOTAL TEMPERATURE

The average kinetic energy associated with the molecular motions at a fluid flow point measures its static temperature. Like the static pressure, the static temperature remains invariant under a change in the reference frame. We must use the static temperature to evaluate the temperature-dependent thermophysical properties of a fluid at any point in a flow field.

When we bring a fluid flow to rest with no heat transfer (adiabatically), we call the resulting temperature the stagnation (total) temperature. In other words, the stagnation temperature is the sum of the static temperature and the dynamic temperature (resulting from bulk flow velocity) at a point in fluid flow, that is:

$$T_0 = T + \frac{V^2}{2c_p} \tag{1.3}$$

where c_p is the specific heat at constant pressure. In this equation, the total temperature T_0 does not depend on whether the adiabatic stagnation process is reversible or irreversible, or whether the flow is compressible or incompressible. This is, however, not true for the total pressure, whose calculation assumes that the stagnation process is isentropic with no loss in the available dynamic pressure. A stationary thermocouple in a fluid flow will, therefore, always measure the total temperature. Expressing velocity in terms of the Mach number, we can write Equation 1.3 as:

$$T_0 = T + \frac{V^2}{2c_p} = T\left(1 + \frac{\gamma - 1}{2} M^2\right) \tag{1.4}$$

Due to the dependence of the dynamic temperature $V^2/2c_p$ on velocity, the total temperature depends on the reference frame in which we calculate it.

1.7 THE LAGRANGIAN APPROACH VERSUS THE EULERIAN APPROACH

The Lagrangian and Eulerian approaches represent two different viewpoints for fluid flow analysis. The Lagrangian approach uses a control system of an identified parcel of fluid particles, analyzing its exchange of mass, momentum, energy, and so on with the surrounding particles in the flow. This approach in fluid flow is also akin to particle dynamics in solid mechanics. For example, Figure 1.4 shows a few liquid fuel droplets in the complex airflow field in a combustor. One of the design objectives behind using the Lagrangian approach to analyze evaporating fuel droplets is to establish their required residence time in the combustor before the air–fuel mixture is ignited downstream for complete combustion.

Our interest in most engineering applications is the equipment and its performance under various inflows, outflows, and other boundary conditions representing its interactions with the surroundings. Accordingly, we create a suitable control volume around the equipment for the analysis without following any particular set of fluid particles in the flow. This viewpoint makes the Eulerian approach, as depicted in Figure 1.5. The design objective is to estimate the thrust the liquid rocket engine produces under the inflows of fuel and oxidizer and the outflow (exhaust) of combustion products expanding through the convergent–divergent nozzle.

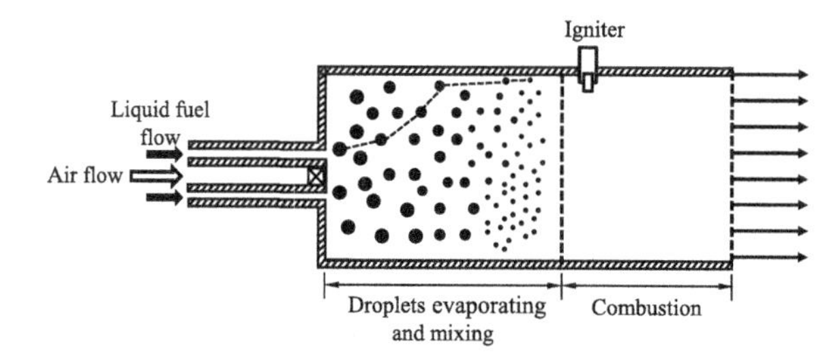

FIGURE 1.4 Evaporating liquid fuel droplets in a combustor airflow—an example of the Lagrangian approach.

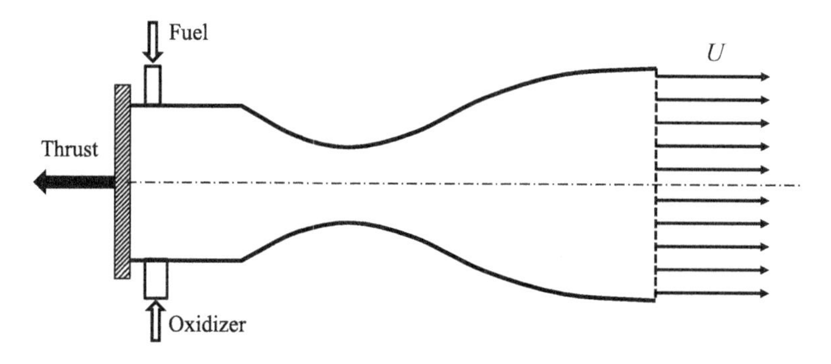

FIGURE 1.5 Thrust generated by the flow through a liquid rocket engine—an example of the Eulerian approach.

1.8 GENERALIZED CONSERVATION EQUATION FOR A SYSTEM

In Newtonian fluid mechanics, conservation laws are written in an inertial reference frame for a system of identified mass—the Lagrangian viewpoint in which we follow the system as it moves from time t to time $t+\Delta t$. We will designate the volume of the system as the material volume (mv) and its surface, which moves with flow velocity V, as the material surface (ms). We can write the substantial (total or material) derivative of an extensive property Φ_{sys} of a system in a flow field as:

$$\frac{D\Phi_{sys}}{Dt} = \Pi_s + \Pi_b = \iiint\limits_{\mathcal{V}_{mv}(t)} (\pi_s + \pi_b)\, d\mathcal{V} \tag{1.5}$$

where Φ_{sys} represents mass, linear momentum, or energy of the system and Π_s and Π_b are the change agents working, respectively, on the material surface and on the material volume, and they can be expressed by the following volume integrals:

$$\Pi_s = \iiint\limits_{\mathcal{V}_{mv}(t)} \pi_s\, d\mathcal{V}$$

and

$$\Pi_b = \iiint\limits_{\mathcal{V}_{mv}(t)} \pi_b\, d\mathcal{V}$$

where π_s and π_b, respectively, are surface and body forces per unit volume.

In Equation 1.5, Φ_{sys} is the instantaneous value of the extensive property Φ for the entire system. Denoting the corresponding intensive (per unit mass) property by ϕ, we obtain:

$$\frac{D\Phi_{sys}}{Dt} = \frac{d}{dt} \iiint\limits_{\mathcal{V}_{mv}(t)} \rho\phi\, d\mathcal{V}$$

We can alternately write Equation 1.5 as:

$$\frac{d}{dt} \iiint\limits_{\mathcal{V}_{mv}(t)} \rho\phi\, d\mathcal{V} = \iiint\limits_{\mathcal{V}_{mv}(t)} (\pi_s + \pi_b)\, d\mathcal{V} \tag{1.6}$$

1.8.1 Differential Conservation Equation: Microanalysis

For an arbitrary system in a flow field, using the Leibniz rule (see Bird, Stewart, and Lightfoot, 2001), we write:

$$\frac{d}{dt} \iiint\limits_{\mathcal{V}_{mv}(t)} \rho\phi\, d\mathcal{V} = \iiint\limits_{\mathcal{V}_{mv}(t)} \frac{\partial(\rho\phi)}{\partial t} d\mathcal{V} + \iint\limits_{S_{ms}(t)} \phi\rho \boldsymbol{V} \bullet \boldsymbol{n}\, dS \tag{1.7}$$

Using the Gauss divergence theorem, we can replace the surface integral in this equation by the volume integral, giving:

$$\frac{d}{dt} \iiint\limits_{\mathcal{V}_{mv}(t)} \rho\phi\, d\mathcal{V} = \iiint\limits_{\mathcal{V}_{mv}(t)} \left[\frac{\partial(\rho\phi)}{\partial t} + \nabla \bullet (\rho\phi \boldsymbol{V})\right] d\mathcal{V} \tag{1.8}$$

whose substitution in Equation 1.6 yields:

$$\iiint_{\forall_{mv}(t)} \left[\frac{\partial(\rho\phi)}{\partial t} + \nabla \bullet (\rho\phi \boldsymbol{V}) \right] d\forall = \iiint_{\forall_{mv}(t)} (\pi_s + \pi_b) d\forall$$

As this equation is valid for any arbitrary material volume, we can equate the integrands on both sides, giving the general differential equation governing the point-wise variation of ϕ in a flow field as:

$$\frac{\partial(\rho\phi)}{\partial t} + \nabla \bullet (\rho\phi \boldsymbol{V}) = \pi_s + \pi_b \tag{1.9}$$

1.8.2 Integral Conservation Equation: Macroanalysis

In most engineering applications, we use a control volume analysis that represents the Eulerian viewpoint. Let us consider the most general case of a deformable control volume having volume $\forall_{cv}(t)$ and surface area $S_{cs}(t)$. For this control volume, using the Leibniz rule, we can write:

$$\frac{d}{dt} \iiint_{\forall_{cv}(t)} \rho\phi d\forall = \iiint_{\forall_{cv}(t)} \frac{\partial(\rho\phi)}{\partial t} d\forall + \iint_{S_{cs}(t)} \phi\rho \boldsymbol{V}_{cs} \bullet \boldsymbol{n} dA \tag{1.10}$$

where V_{cs} is the control surface velocity. If the material volume and surface coincide with the control volume and surface at an instant t such that $\forall_{cs}(t) = \forall_{cv}(t)$ and $S_{cs}(t) = S_{cv}(t)$, we can write:

$$\iiint_{\forall_{mv}(t)} \frac{\partial(\rho\phi)}{\partial t} d\forall = \iiint_{\forall_{cv}(t)} \frac{\partial(\rho\phi)}{\partial t} d\forall = \frac{d}{dt} \iiint_{\forall_{cv}(t)} \rho\phi d\forall - \iint_{S_{cs}(t)} \phi\rho \boldsymbol{V}_{cs} \bullet \boldsymbol{n} dA$$

whose substitution in Equation 1.7 along with:

$$\iint_{S_{ms}(t)} \phi\rho \boldsymbol{V} \bullet \boldsymbol{n} dA = \iint_{S_{cs}(t)} \phi\rho \boldsymbol{V} \bullet \boldsymbol{n} dA$$

yields:

$$\frac{d}{dt} \iiint_{\forall_{mv}(t)} \rho\phi d\forall = \frac{d}{dt} \iiint_{\forall_{cv}(t)} \rho\phi d\forall - \iint_{S_{cs}(t)} \phi\rho \boldsymbol{V}_{cs} \bullet \boldsymbol{n} dA + \iint_{S_{cs}(t)} \phi\rho \boldsymbol{V} \bullet \boldsymbol{n} dA$$

which, when combined with Equation 1.5, becomes:

$$\Pi_s + \Pi_b = \frac{d}{dt} \iiint_{\forall_{cv}(t)} \rho\phi d\forall + \iint_{S_{cs}(t)} \phi\rho \boldsymbol{V}_{rel} \bullet \boldsymbol{n} dA \tag{1.11}$$

where $\boldsymbol{V}_{rel} = \boldsymbol{V} - \boldsymbol{V}_{cs}$. This equation is identical to the Reynolds transport theorem (RTT) for a general deformable control volume (see Hansen, 1979, and Potter and Foss, 1975). In this equation, the first term on the right-hand side represents the time rate of change of Φ within the control volume—because the volume integral is in general a function of time only, we use d/dt instead of $\partial/\partial t$ —and the second term represents the net outflow (total outflow minus total inflow) of Φ through the deformable control surface moving with velocity $\boldsymbol{V}_{cs}$.

1.9 MASS CONSERVATION

1.9.1 Differential Continuity Equation

As no agent can change the mass of a system, for mass conservation, we substitute $\phi = 1$ and $\pi_s = \pi_b = 0$ in Equation 1.9, giving:

$$\frac{\partial \rho}{\partial t} + \nabla \bullet (\rho V) = 0 \tag{1.12}$$

For a steady compressible flow, this equation reduces to $\nabla \bullet (\rho V) = 0$. We can also write this equation as:

$$\frac{\partial \rho}{\partial t} + V \bullet \nabla \rho + \rho \nabla \bullet V = 0$$

$$\frac{\mathrm{D}\rho}{\mathrm{D}t} + \rho \nabla \bullet V = 0 \tag{1.13}$$

For both steady and unsteady incompressible flows, this equation reduces to $\nabla \bullet V = 0$, which states that the velocity field of any incompressible flow must be divergence free.

1.9.2 Integral Continuity Equation

Again substituting $\phi = 1$ and $\Pi_s = \Pi_b = 0$ in Equation 1.11, we obtain:

$$\frac{\mathrm{d}}{\mathrm{d}t} \iiint\limits_{\forall_{cv}(t)} \rho \phi \mathrm{d}\forall + \iint\limits_{S_{cs}(t)} \phi \rho V_{rel} \bullet n \mathrm{d}S = 0 \tag{1.14}$$

which, for a control volume with instantaneous fluid mass m_{cv} and multiple inflows and outflows, we can write as:

$$\frac{\mathrm{d}m_{cv}}{\mathrm{d}t} + \sum_{j=1}^{N_{outlets}} \dot{m}_j - \sum_{i=1}^{N_{inlets}} \dot{m}_i = 0 \tag{1.15}$$

where N_{inlets} and $N_{outlets}$ are the total numbers of inlets and outlets, respectively, without regard to the coordinate direction, stating that the imbalance between mass inflows and outflows equals the rate of accumulation or reduction of mass in the control volume. In a steady flow, the sum of all mass inflows equals the sum of all mass outflows over the entire control volume. Hence, we can write:

$$\sum_{i=1}^{N_{inlets}} \dot{m}_i = \sum_{j=1}^{N_{outlets}} \dot{m}_j \tag{1.16}$$

1.10 FORCE AND LINEAR MOMENTUM BALANCE: LINEAR MOMENTUM EQUATION

1.10.1 Differential Linear Momentum Equation

For the linear momentum of fluid having mass m, we write $\Phi = mV$ or $\phi = V$, whose substitution in Equation 1.9 yields:

$$\frac{\partial(\rho \boldsymbol{V})}{\partial t} + \nabla \bullet (\rho \boldsymbol{V}\boldsymbol{V}) = \pi_s + \pi_b \tag{1.17}$$

where π_s is the surface force per unit volume and π_b is the body force per unit volume. We derive these forces later in Chapter 3 (The Navier–Stokes Equations). We can write the left-hand side of this equation as:

$$\frac{\partial(\rho \boldsymbol{V})}{\partial t} + \nabla \bullet (\rho \boldsymbol{V}\boldsymbol{V}) = \boldsymbol{V}\left[\frac{\partial \rho}{\partial t} + \nabla \bullet (\rho \boldsymbol{V})\right] + \rho\left[\frac{\partial \boldsymbol{V}}{\partial t} + \boldsymbol{V}(\boldsymbol{V} \bullet \nabla)\right]$$

Using the continuity equation (Equation 1.13), the expression within the first set of brackets on the right-hand side of this equation becomes zero. Writing the expression within the second set of brackets as $\rho \, \mathrm{D}\boldsymbol{V}/\mathrm{D}t$, we obtain:

$$\frac{\partial(\rho \boldsymbol{V})}{\partial t} + \nabla \bullet (\rho \boldsymbol{V}\boldsymbol{V}) = \rho \frac{\mathrm{D}\boldsymbol{V}}{\mathrm{D}t} \tag{1.18}$$

Thus, Equation 1.17 becomes:

$$\rho \frac{\mathrm{D}\boldsymbol{V}}{\mathrm{D}t} = \pi_s + \pi_b \tag{1.19}$$

which is the differential form of the linear momentum equation.

1.10.2 Integral Linear Momentum Equation

With $\Pi_s = \boldsymbol{F}_s$ and $\Pi_b = \boldsymbol{F}_b$ for surface forces and body forces, respectively, and $\phi = \boldsymbol{V}$ for the linear momentum $\boldsymbol{M}$, Equation 1.11 becomes:

$$\boldsymbol{F}_s + \boldsymbol{F}_b = \frac{\mathrm{d}}{\mathrm{d}t} \iiint\limits_{\mathcal{V}_{cv}(t)} \rho \boldsymbol{V} \mathrm{d}\mathcal{V} + \iint\limits_{S_{cs}(t)} \boldsymbol{V}(\rho \boldsymbol{V}_{rel} \bullet \boldsymbol{n}) \mathrm{d}A \tag{1.20}$$

which is the integral form of the linear momentum equation for a general deformable control volume. Surface forces include pressure forces from static pressure distributions at control volume inlets, outlets, and walls, and wall shear forces locally parallel to the control surface. In addition, we consider here the body force due to gravity (weight).

As the forces acting in one coordinate direction will not change the linear momentum in the other orthogonal coordinate direction, with no loss of generality, let us only consider the x component of Equation 1.20 in the Cartesian coordinate system attached to the control volume, giving:

$$F_{sx} + F_{bx} = \frac{\mathrm{d}}{\mathrm{d}t} \iiint\limits_{\mathcal{V}_{cv}(t)} \rho V_x \mathrm{d}\mathcal{V} + \iint\limits_{S_{cs}(t)} V_x(\rho \boldsymbol{V}_{rel} \bullet \mathrm{d}\boldsymbol{A}) \tag{1.21}$$

If the flow properties are uniform at each inlet and outlet of the control volume, we can replace the surface integral on the right-hand side of Equation 1.21 by algebraic summations, giving:

$$F_{sx} + F_{bx} = \frac{\mathrm{d}M_{cvx}}{\mathrm{d}t} + \sum_{j=1}^{N_{outlets}} (\dot{m}V_x)_j - \sum_{i=1}^{N_{inlets}} (\dot{m}V_x)_i \tag{1.22}$$

where M_{cvx} is the total instantaneous x-momentum within the control volume. In this equation, identifying the mass velocity—the velocity that produces the mass flow rate—which is always positive, dramatically simplifies the evaluation of the momentum flow rate with the correct sign at each inlet and outlet of a control volume. The linear momentum flow rate at each inlet or outlet is positive if V_x is in the positive x direction; otherwise, it is negative. Sultanian (2015) presents the handling of nonuniform flow properties at control volume inlets and outlets and the linear momentum equation in a noninertial (accelerating or decelerating) reference frame.

1.11 TORQUE AND ANGULAR MOMENTUM BALANCE: INTEGRAL ANGULAR MOMENTUM EQUATION

The angular momentum equation does not represent a new conservation law; it is simply a moment of the linear momentum equation. With its axial coordinate direction aligned with the axis of rotation, a cylindrical coordinate system, shown in Figure 1.6, is a natural choice for all turbomachinery flows. This figure shows that the tangential direction is perpendicular to the meridional plane formed by the axial and radial directions. At any point in the flow, V_x, V_r, and V_θ are, respectively, the axial, radial, and tangential velocity components of the absolute velocity $\boldsymbol{V}$. Similarly, W_x, W_r, and W_θ are, respectively, the axial, radial, and tangential velocity components of the corresponding relative velocity $\boldsymbol{W}$.

The angular momentum vector at point A in the flow about the turbomachinery axis of rotation is the cross product of the radial vector $\boldsymbol{r}$ and the absolute velocity vector $\boldsymbol{V}$. As the radial velocity V_r is collinear with the radial vector, its contribution to angular momentum is zero. The axial velocity V_x contributes to the angular momentum only in the tangential direction and not in the axial direction.

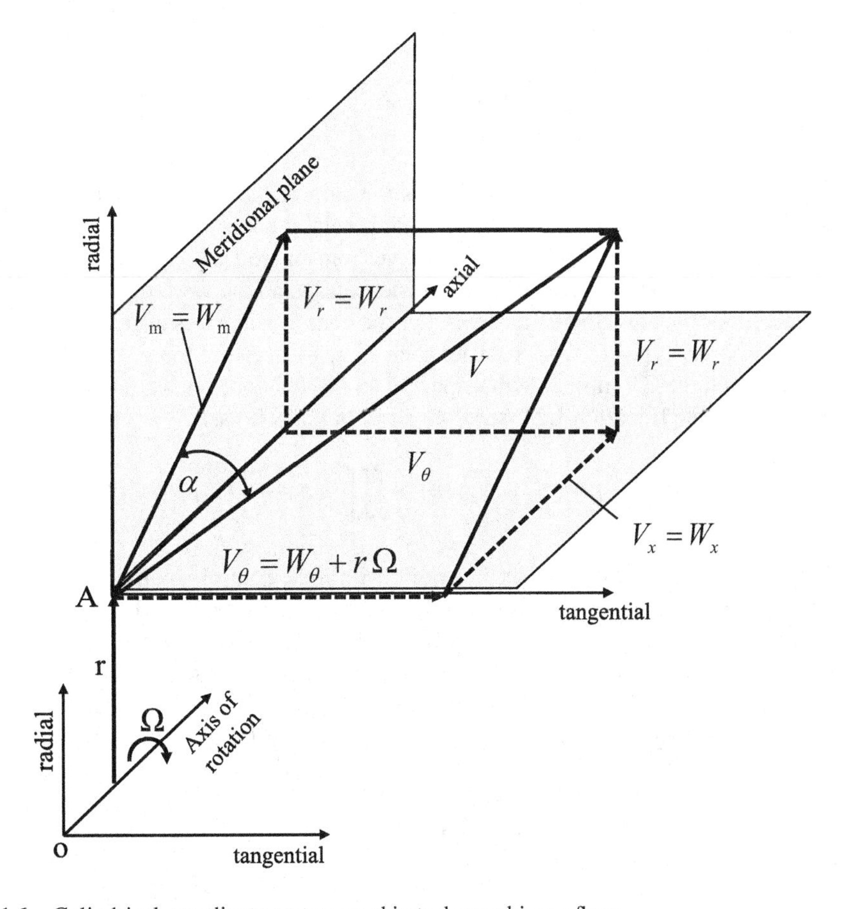

FIGURE 1.6 Cylindrical coordinate system used in turbomachinery flows.

Thus, only the tangential velocity V_θ contributes to the axial angular momentum, giving its flow rate $\dot{H}_x = \dot{m}\, r V_\theta$. In word form, the angular momentum equation reads:

(Total torque from surface forces—e.g., pressure force and shear force)
+(Total torque from body forces—e.g., gravity)
= (Total angular momentum flow rate leaving the control volume)
– (Total angular momentum flow rate entering the control volume)

In equation form, we can write:

$$\Gamma_{sx} + \Gamma_{bx} = \sum_{N_{\text{outlets}}} \dot{H}_x - \sum_{N_{\text{inlets}}} \dot{H}_x \tag{1.23}$$

Sultanian (2015) provided detailed coverage of the integral angular momentum equation in inertial and noninertial reference frames.

1.12 ENERGY CONSERVATION: THE FIRST LAW OF THERMODYNAMICS

1.12.1 Integral Energy Equation

The first law of thermodynamics forms the basis for the energy equation, which states that heat transfer and work transfer change the total energy E of a system, neglecting the transfer of other forms of energy such as chemical, electrical, and electromagnetic. By replacing ϕ with e in Equation 1.11, we obtain the energy equation for a general control volume as:

$$\dot{Q} + \dot{W} = \frac{\mathrm{d}}{\mathrm{d}t} \iiint_{\text{cv}} \rho e \,\mathrm{d}\forall + \iint_{\text{cs}} e(\rho \boldsymbol{V}_{\text{rel}} \bullet \mathrm{d}\boldsymbol{A}) \tag{1.24}$$

where the specific (per unit mass) total energy e is the sum of the specific internal energy u, the specific external kinetic energy $V^2/2$, and the specific potential energy gz due to gravity where we measure z upwardly positive and downwardly negative from an arbitrary datum plane. In this equation, we have used the convention that the heat transfer rate $\dot{Q}$ and the work transfer rate $\dot{W}$ into the control volume are positive, and their transfer out of the control volume is negative.

By including different types of work transfer such as $\dot{W}_p$ by pressure force, $\dot{W}_{\text{shear}}$ by shear force, $\dot{W}_{\text{inertial}}$ (associated with the CV motion with respect to an inertial coordinate system), and $\dot{W}_{\text{shaft}}$ (rate of shaft work), we rewrite Equation 1.24 (see Potter and Foss, 1975) as:

$$\dot{Q} + \dot{W}_p + \dot{W}_{\text{shear}} + \dot{W}_{\text{inertial}} + \dot{W}_{\text{shaft}} = \frac{\mathrm{d}}{\mathrm{d}t} \iiint_{\text{cv}} \rho e \,\mathrm{d}\forall + \iint_{\text{cs}} e(\rho \boldsymbol{V}_{\text{rel}} \bullet \mathrm{d}\boldsymbol{A}) \tag{1.25}$$

Note that the static pressure being isotropic at a point within the control volume does not contribute to $\dot{W}_p$, which occurs only at the control volume inlets and outlets. At an inlet, because $\boldsymbol{V}_{\text{rel}}$ and $\boldsymbol{A}$ are in opposite directions, $\rho \boldsymbol{V}_{\text{rel}} \bullet \mathrm{d}\boldsymbol{A}$ is negative, and the pressure force does work on the flow entering the control volume—considered as positive by our convention. At an outlet, because $\boldsymbol{V}_{\text{rel}}$ and $\boldsymbol{A}$ are in the same direction, $\rho \boldsymbol{V}_{\text{rel}} \bullet \mathrm{d}\boldsymbol{A}$ is positive, and the pressure force opposes the flow exiting the control volume—considered as negative by our convention. Accordingly, for all inlets, we write:

$$\left(\dot{W}_p\right)_{\text{inlets}} = -\left\{ \iint_{\text{cs}} \frac{p}{\rho} \left(\rho \boldsymbol{V}_{\text{rel}} \bullet \mathrm{d}\boldsymbol{A}\right) \right\}_{\text{inlets}}$$

and for all outlets, we write:

$$\left(\dot{W}_p\right)_{\text{outlets}} = -\left\{\iint_{\text{cs}} \frac{p}{\rho}\left(\rho \mathbf{V}_{\text{rel}} \bullet \text{d}\mathbf{A}\right)\right\}_{\text{outlets}}$$

Combining the preceding two equations, we express $\dot{W}_p$ by the following surface integral over all inlets and outlets:

$$\dot{W}_p = -\iint_{\text{cs}} \frac{p}{\rho}\left(\rho \mathbf{V}_{\text{rel}} \bullet \text{d}\mathbf{A}\right) \tag{1.26}$$

where p/ρ is called the specific flow work. Transferring $\dot{W}_p$ to the right-hand side in Equation 1.25 yields:

$$\dot{Q} + \dot{W}_{\text{shear}} + \dot{W}_{\text{inertial}} + \dot{W}_{\text{shaft}} = \frac{\text{d}}{\text{d}t}\iiint_{\text{cv}} \rho e \text{d}V\!\!\!/ + \iint_{\text{cs}}\left(u + \frac{p}{\rho} + \frac{V^2}{2} + gz\right)(\rho \mathbf{V}_{\text{rel}} \bullet \text{d}\mathbf{A}) \tag{1.27}$$

where we can replace $u + p/\rho$ by the specific enthalpy h, giving:

$$\dot{Q} + \dot{W}_{\text{shear}} + \dot{W}_{\text{inertial}} + \dot{W}_{\text{shaft}} = \frac{\text{d}}{\text{d}t}\iiint_{\text{cv}} \rho e \text{d}V\!\!\!/ + \iint_{\text{cs}}\left(h + \frac{V^2}{2} + gz\right)(\rho \mathbf{V}_{\text{rel}} \bullet \text{d}\mathbf{A}) \tag{1.28}$$

If the flow properties are uniform at control volume inlets and outlets, we can replace the surface integral in Equation 1.28 by algebraic summations, giving:

$$\dot{Q} + \dot{W}_{\text{shear}} + \dot{W}_{\text{inertial}} + \dot{W}_{\text{shaft}} = \frac{\text{d}E_{\text{cv}}}{\text{d}t} + \tilde{I}_{\text{outlets}} - \tilde{I}_{\text{inlets}} \tag{1.29}$$

where

$$\tilde{I}_{\text{inlets}} = \sum_{i=1}^{N_{\text{inlets}}}\left[\left(h + V^2/2 + gz\right)\dot{m}\right]_i$$

and

$$\tilde{I}_{\text{outlets}} = \sum_{j=1}^{N_{\text{outlets}}}\left[\left(h + V^2/2 + gz\right)\dot{m}\right]_j$$

where we can combine h and $V^2/2$ into the specific total enthalpy $h_0 = h + V^2/2$. If the velocity profile is nonuniform at any inlet or outlet, we replace $V^2/2$ in these equations by $\alpha\bar{V}^2/2$, where $\bar{V}$ is the average velocity and α is the kinetic energy correction factor (see Sultanian, 2015). For a turbulent flow, we have $\alpha \approx 1$.

1.12.2 Mechanical Energy Equation

The mechanical energy equation (MEE) is a generalized Bernoulli equation—also called the extended Bernoulli equation (see, e.g., Sultanian, 2015). Unlike the original Bernoulli equation, valid for an inviscid flow along a streamline or between any two points in a potential (irrotational) flow, MEE has widespread applications in hydraulic engineering. Before deriving MEE from

Equation 1.29, let us look at alternate interpretations of the total pressure p_0 and the specific potential energy gz for incompressible flows, in which we can express the total pressure p_0 as:

$$p_0 = p + \frac{1}{2}\rho V^2 = \rho\left(\frac{p}{\rho} + \frac{V^2}{2}\right) \tag{1.30}$$

In this equation, the first term within parentheses is the specific flow work, and the second term is the specific kinetic energy. We can interpret p_0 as the total mechanical energy per unit volume.

In a gravitational force field of constant g, the work done per unit mass equals gz, which between any two points is positive for a net downward displacement and negative for a net upward displacement. Note that ρgz becomes work done per unit volume. Consider a vertical free jet of water under negligible air resistance and constant ambient pressure p_{amb}. The static pressure throughout the jet is p_{amb}. Between any two sections, which are apart by z, the total pressure decreases by ρgz for an upward-moving jet and increases by this amount for a downward-moving jet. Accordingly, to satisfy the continuity equation, the jet area will increase going up and decrease coming down. How would you describe the properties of water flow in a frictionless constant area vertical duct?

For an adiabatic steady incompressible flow of constant discharge between sections 1 and 2 of uniform properties, we can write Equation 1.29 in terms of total pressure and its changes as:

$$p_{02} = p_{01} + \sum_{\text{gain}} \Delta p_0 - \sum_{\text{loss}} \Delta p_0 - \rho g(z_2 - z_1) \tag{1.31}$$

where z_1 and z_2 are measured from a fixed datum plane, just as in the original Bernoulli equation. In this equation, $\sum_{\text{gain}} \Delta p_0$ includes all increases in total pressure except due to gravity, for example, by pump work, and $\sum_{\text{loss}} \Delta p_0$ includes all decreases in total pressure except due to gravity, for example, due to turbine work, friction, and entropy generation by other means—e.g., the mixing loss.

Knowing total pressures p_{01} and p_{02}, we can compute the hydraulic power P_{hyd} needed between sections 1 and 2 by the equation:

$$P_{hyd} = Q(p_{02} - p_{01}) \tag{1.32}$$

where Q represents the volumetric flow rate between these sections.

1.13 THE SECOND LAW OF THERMODYNAMICS: THE ENTROPY CONSTRAINT

The first law of thermodynamics embodies energy conservation. However, we can convert energy from one form to another. Available empirical evidence suggests that, while we can ultimately convert all the work into heat (internal energy), the reverse is not true—the second law of thermodynamics restricts the energy conversion process. For example, a reversible work transfer is a mere idealization due to friction in practical applications.

One of the tenets of the second law of thermodynamics is that heat transfer always occurs from higher to lower temperatures. According to Fourier's law of heat conduction, we write $\dot{q} = -k\nabla T$, where the negative sign indicates that heat conduction takes place in the direction of a negative temperature gradient. Other examples obeying the entropy constraint are the Fanno and Rayleigh lines, which are mathematically continuous but do not allow one to go from the subsonic branch to the supersonic branch, or vice versa, due to the maximum entropy constraint at $M = 1$. Due to the entropy constraint, a normal shock must have a supersonic flow upstream and a subsonic flow downstream of the shock, regardless of the value of the upstream Mach number. We discuss these flows in Chapter 5.

1.13.1 The Concept of Entropy

In a fluid system, with different amounts of heat transfer and work transfer, we can go from state 1 to state 2 via different paths, either reversible (best possible) or irreversible. Based on empirical data, for reversible paths A and B connecting states 1 and 2, we find that:

$$\left[\int_1^2 \frac{\delta q_{\text{rev}}}{T}\right]_{\text{A}} = \left[\int_1^2 \frac{\delta q_{\text{rev}}}{T}\right]_{\text{B}} \tag{1.33}$$

This equation suggests the existence of a state property, which we call entropy s, giving:

$$(s_2 - s_1) = \left[\int_1^2 \frac{\delta q_{\text{rev}}}{T}\right]_{\text{A}} = \left[\int_1^2 \frac{\delta q_{\text{rev}}}{T}\right]_{\text{B}} \tag{1.34}$$

which is true for any reversible process connecting states 1 and 2. For an irreversible path C connecting states 1 and 2, we must have:

$$\left[\int_1^2 \frac{\delta q_{\text{irrev}}}{T}\right]_{\text{C}} < (s_2 - s_1) \tag{1.35}$$

because some of the entropy increase along path C occurs from irreversible work transfer—entropy being state property, both reversible and irreversible processes connecting states 1 and 2 will have the same change in entropy, that is, $(s_2 - s_1)_{\text{A}} = (s_2 - s_1)_{\text{B}} = (s_2 - s_1)_{\text{C}}$. The concept of entropy delineates the feasible flow solutions from those that are not physically realizable even if they satisfy the remaining conservation equations. Note that any irreversibility in the flow system will increase its entropy. In addition, the net entropy increases when we heat the flow and decreases when we cool it.

1.13.2 Computation of Entropy Change

Combining the first and second laws of thermodynamics, we can write (see, e.g., Reynolds and Colonna, 2018):

$$\delta q_{\text{rev}} = \mathrm{d}h - \frac{\mathrm{d}p}{\rho}$$

which using Equation 1.34 becomes:

$$T\mathrm{d}s = \mathrm{d}h - \frac{\mathrm{d}p}{\rho} \tag{1.36}$$

being true for both reversible and irreversible processes connecting any two states. For a calorically perfect gas, we can write this equation as:

$$\mathrm{d}s = c_p \frac{\mathrm{d}T}{T} - \frac{\mathrm{d}p}{\rho T} = c_p \frac{\mathrm{d}T}{T} - \left(\frac{p}{\rho T}\right)\frac{\mathrm{d}p}{p}$$

which with the equation of state becomes:

$$\mathrm{d}s = c_p \frac{\mathrm{d}T}{T} - R\frac{\mathrm{d}p}{p} \tag{1.37}$$

Integrating this equation between states 1 and 2 yields:

$$\int_1^2 ds = c_p \int_1^2 \frac{dT}{T} - R\int_1^2 \frac{dp}{p}$$

$$s_2 - s_1 = c_p \ln\left(\frac{T_2}{T_1}\right) - R\ln\left(\frac{p_2}{p_1}\right) \tag{1.38}$$

which, for an isentropic (adiabatic and reversible) process, yields:

$$c_p \ln\left(\frac{T_2}{T_1}\right) - R\ln\left(\frac{p_2}{p_1}\right) = 0 \tag{1.39}$$

from which we obtain the isentropic relations:

$$\frac{p_2}{p_1} = \left(\frac{T_2}{T_1}\right)^{\frac{c_p}{R}} = \left(\frac{T_2}{T_1}\right)^{\frac{\gamma}{\gamma-1}} \tag{1.41}$$

As the process of stagnation is assumed to be isentropic, we can write Equation 1.38 in terms of total pressures and total temperatures as:

$$s_2 - s_1 = c_p \ln\left(\frac{T_{02}}{T_{01}}\right) - R\ln\left(\frac{p_{02}}{p_{01}}\right) \tag{1.42}$$

which for an isentropic process connecting states 1 and 2 yields:

$$\frac{p_{02}}{p_{01}} = \left(\frac{T_{02}}{T_{01}}\right)^{\frac{c_p}{R}} = \left(\frac{T_{02}}{T_{01}}\right)^{\frac{\gamma}{\gamma-1}} \tag{1.43}$$

At sections 1 and 2 of an adiabatic flow in a stationary duct with wall friction, we have $T_{02} = T_{01}$; as a result, Equation 1.42 reduces to:

$$s_2 - s_1 = -R\ln\left(\frac{p_{02}}{p_{01}}\right)$$

$$\frac{p_{02}}{p_{01}} = e^{-\frac{s_2 - s_1}{R}} \tag{1.44}$$

The wall friction always increases entropy downstream of an adiabatic pipe flow. Hence, according to Equation 1.44, the total pressure must always decrease in the flow direction.

1.14 CONCLUDING REMARKS

By way of review and reinforcement, we have introduced in this chapter the concepts of streamlines, streaklines, and pathlines to describe various fluid flows found in nature and engineering. Understanding the difference between static and total pressures and static and total temperatures, presented here, along with the equations to compute them in incompressible and compressible flows, is vital to the physics-based modeling, implementation of boundary conditions, and interpretation of calculated results in thermal-fluids design engineering. Appendix B presents additional kinematic quantities such as vortex, vorticity, and circulation.

A unique feature of this chapter is the generalized conservation equation, which is valid for a time-dependent, deformable control volume. This equation quickly leads to differential (micro-analysis) and integral (microanalysis) equations of mass, linear momentum, and energy, typically derived in undergraduate texts using a fixed control volume and the RTT. Readers may want to compare the two approaches for an in-depth understanding of both.

Interpreting total pressure as total mechanical energy per unit volume simplifies the MEE and its application to handle a one-dimensional incompressible flow as in hydraulic engineering.

This chapter's linear and angular momentum equation presentation is limited to an inertial reference frame. For the development in a noninertial reference frame, interested readers may refer to their undergraduate fluid mechanics texts, including the first edition of this book by Sultanian (2015), which, for the angular momentum equation in a noninertial reference frame rotating at constant angular speed, provides a mathematical deduction showing how the centrifugal and Coriolis terms that appear in the noninertial rotor reference frame disappear in Euler's turbomachinery equation, which uses absolute velocities.

To develop their skills in solving various fluid flow problems, readers may want to review numerous problems with solutions presented in Sultanian (2021, 2022).

WORKED EXAMPLES

Example 1.1

For a steady two-dimensional incompressible flow, the velocity field is represented by $V_x = Ax$, $V_y = -Ay$, and $V_z = 0$. Derive an equation for the streamlines in this flow field.

Solution

Because streamlines have the same slope as the velocity vector at all points, we can write:

$$\left(\frac{dy}{dx}\right)_{\text{streamline}} = \frac{V_y}{V_x} = \frac{-Ay}{Ax} = -\frac{y}{x}$$

By separating the variables in this equation and integrating the resulting expression, we obtain:

$$\ln y = -\ln x + \tilde{C}$$

$$xy = C$$

which represents the streamlines that are a family of rectangular hyperbolas shown in Figure 1.7. For each value of C, we obtain a new streamline. The streamlines with $C = 0$ correspond to the x and y coordinate axes. The figure shows them along with the one with $C = C_1$ as solid boundaries of a corner flow.

Example 1.2

A team of graduate students carried out accurate measurements of the x component of velocity and temperature in a two-dimensional steady flow on a flat plate, as shown in Figure 1.8. They correlated these measurements to $V_x = V_0 x^2$, where V_0 is a constant. Noting that the flow velocity (V_y) is zero along the x axis, they calculated the distribution of y component of velocity from their understanding of incompressible fluid flows. Looking at the data and calculations, the research adviser pointed out that the calculated y component of velocity is 33% higher everywhere. She asked the students to recalculate it considering the temperature variation, which caused the density variation given by $\rho = \rho_0 xy$ (where ρ_0 is a constant) in the flow field, both x and y being

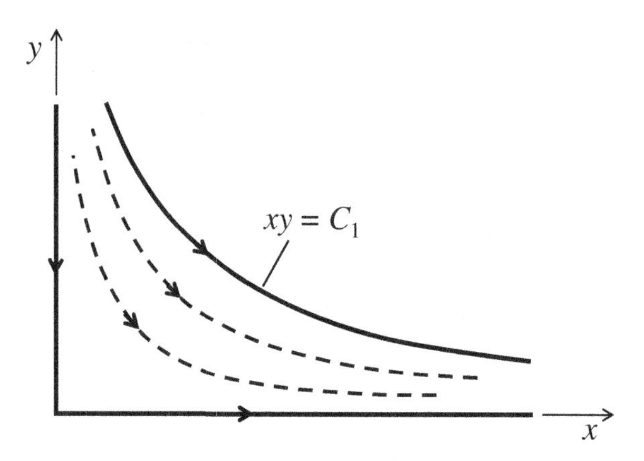

FIGURE 1.7 Streamlines (Example 1.1).

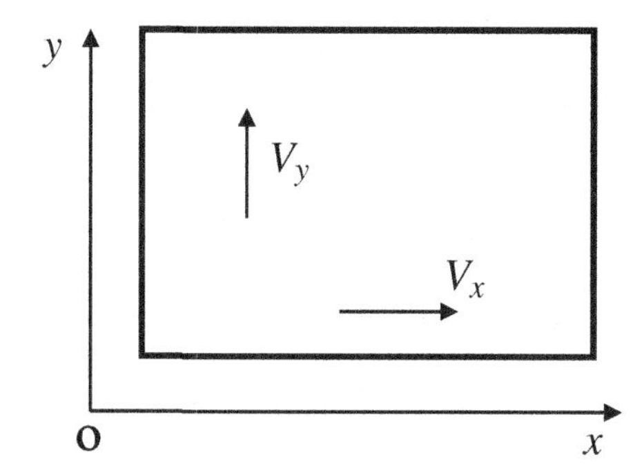

FIGURE 1.8 Two-dimensional flow field (Example 1.2).

dimensionless coordinates. Verify the error estimated by the research adviser in the y component of velocity computed by the graduate students.

Solution

Initially, the graduate students assumed an incompressible flow field with constant density to calculate the y component of velocity ($\tilde{V}_y$) using the following continuity equation:

$$\frac{\partial V_x}{\partial x} + \frac{\partial \tilde{V}_y}{\partial y} = 0$$

giving:

$$\frac{\partial \tilde{V}_y}{\partial y} = -\frac{\partial V_x}{\partial x} = -\frac{\partial \left(V_0 x^2 \right)}{\partial x} = -2V_0 x$$

whose integration along with the boundary condition $\tilde{V}_y = 0$ at $y = 0$ yields:

$$\tilde{V}_y = -2V_0 xy$$

Considering the variation of density in the flow field due to the nonuniform temperature distribution, the continuity equation becomes:

$$\frac{\partial \left(\rho V_x \right)}{\partial x} + \frac{\partial \left(\rho V_y \right)}{\partial y} = 0$$

giving:

$$\frac{\partial(\rho V_y)}{\partial y} = -\frac{\partial(\rho V_x)}{\partial x} = -\frac{\partial(\rho_0 xy V_0 x^2)}{\partial x} = -3\rho_0 V_0 x^2 y$$

where V_y represents the y component of velocity for the actual flow with density variation. Integrating this equation along with $V_y = 0$ for $y = 0$ yields:

$$\rho V_y = -\frac{3}{2}\rho_0 V_0 x^2 y^2$$

$$V_y = -\frac{3}{2}V_0 xy$$

We can calculate the percentage error in the initial calculation of the y component of velocity as:

$$\text{Error} = \frac{(\tilde{V}_y - V_y)}{V_y} \times 100 = 33.3\% \text{ (The research adviser is right!)}$$

Example 1.3

For the flow field of Example 1.1, find the components of local acceleration.

Solution

Because the flow is steady, we only have the acceleration of transport (convective acceleration) in this flow with $V_x = Ax, V_y = -Ay$, and $V_z = 0$. Using Equation 1.13, we write:

$$a_x = V_x \frac{\partial V_x}{\partial x} + V_y \frac{\partial V_x}{\partial y} + V_z \frac{\partial V_x}{\partial z}$$

$$a_x = (Ax)(A) + (-Ay)(0) + (0)(0) = A^2 x$$

$$a_y = V_x \frac{\partial V_y}{\partial x} + V_y \frac{\partial V_y}{\partial y} + V_z \frac{\partial V_y}{\partial z}$$

$$a_y = (Ax)(0) + (-Ay)(-A) + (0)(0) = A^2 y$$

$$a_z = 0$$

For this two-dimensional flow, the results show that the acceleration in the x direction varies linearly with x and that in the y direction varies linearly with y.

Example 1.4

The velocity distribution in an incompressible flow is given by $\boldsymbol{V} = 10x^3 y\hat{\boldsymbol{i}} - 15x^2y^2\hat{\boldsymbol{j}} + 10t\hat{\boldsymbol{k}}$ m/s. First, verify that the given velocity distribution satisfies the continuity equation everywhere in the flow and then determine the velocity and acceleration of a particle at $x = 1$ m, $y = 2$ m, $z = 1$ m, and $t = 1$ s.

Solution

For the given velocity distribution of an incompressible flow:

$$\boldsymbol{V} = 10x^3 y\hat{\boldsymbol{i}} - 15x^2y^2\hat{\boldsymbol{j}} + 10t\hat{\boldsymbol{k}} \text{ m/s.}$$

the continuity equation $\frac{\partial V_x}{\partial x}+\frac{\partial V_y}{\partial y}+\frac{\partial V_z}{\partial z}=0$ yields:

$$\frac{\partial(10x^3y)}{\partial x}+\frac{\partial(-15x^2y^2)}{\partial y}+\frac{\partial(10t)}{\partial z}=30x^2y-30x^2y=0$$

which shows that the given velocity distribution is mass-conserving.

Particle velocity at the required position and time:

$$\boldsymbol{V}(1,2,1,1)=10(1)^3(2)\hat{\boldsymbol{i}}-15(1)^2(2)^2\hat{\boldsymbol{j}}+10(1)\hat{\boldsymbol{k}}\ \text{m/s}$$

$$\boldsymbol{V}(1,2,1,1)=20\,\hat{\boldsymbol{i}}-60\hat{\boldsymbol{j}}+10\,\hat{\boldsymbol{k}}\ \text{m/s.}$$

Particle acceleration at the required position and time:

$$a_x=\frac{\partial V_x}{\partial t}+V_x\frac{\partial V_x}{\partial x}+V_y\frac{\partial V_x}{\partial y}+V_z\frac{\partial V_x}{\partial z}$$

$$a_x=\frac{\partial(10x^3y)}{\partial t}+(10x^3y)\frac{\partial(10x^3y)}{\partial x}+(-15x^2y^2)\frac{\partial(10x^3y)}{\partial y}+(10t)\frac{\partial(10x^3y)}{\partial z}$$

$$a_x=0+(10x^3y)(30x^2y)+(-15x^2y^2)(10x^3)+0$$

$$a_x=0+300x^5y^2-150x^5y^2=150x^5y^2$$

$$a_x(1,2,1,1)=150\times(1)^5(2)^2=600\ \text{m/s}^2$$

$$a_y=\frac{\partial V_y}{\partial t}+V_x\frac{\partial V_y}{\partial x}+V_y\frac{\partial V_y}{\partial y}+V_z\frac{\partial V_y}{\partial z}$$

$$a_y=\frac{\partial(-15x^2y^2)}{\partial t}+(10x^3y)\frac{\partial(-15x^2y^2)}{\partial x}+(-15x^2y^2)\frac{\partial(-15x^2y^2)}{\partial y}+(10t)\frac{\partial(-15x^2y^2)}{\partial z}$$

$$a_y(1,2,1,1)=0+(10x^3y)(-30xy^2)+(-15x^2y^2)(-30x^2y)+0$$

$$a_y=-300x^4y^3+450x^4y^3=150x^4y^3$$

$$a_y(1,2,1,1)=150\times(1)^4(2)^3=1200\ \text{m}/\text{s}^2$$

$$a_z=\frac{\partial V_z}{\partial t}+V_x\frac{\partial V_z}{\partial x}+V_y\frac{\partial V_z}{\partial y}+V_z\frac{\partial V_z}{\partial z}$$

$$a_z(1,2,1,1) = \frac{\partial(10t)}{\partial t} + (10x^3y)\frac{\partial(10t)}{\partial x} + (-15x^2y^2)\frac{\partial(10t)}{\partial y} + (10t)\frac{\partial(10t)}{\partial z} = 10 \text{ m/s}^2$$

$$\boldsymbol{a}(1,2,1,1) = 600\hat{\boldsymbol{i}} + 1200\hat{\boldsymbol{j}} + 10\hat{\boldsymbol{k}} \text{ m/s}^2$$

The results for the acceleration components show that both a_x and a_y are functions of x and y only and not of z. The acceleration in the z direction remains constant throughout the flow field.

Example 1.5: Incompressible Laminar Flow in a Circular Pipe

Figure 1.9 shows an incompressible laminar flow in a circular pipe of constant radius R ($D = 2R$; $A = \pi R^2$). The flow enters the pipe at section 1 at a uniform velocity V and exits at section 3 with a fully developed parabolic velocity profile given by:

$$v(r) = V_{\max}\left[1 - \left(\frac{r}{R}\right)^2\right]$$

Over the entrance length L_e, the wall boundary layer grows all around the pipe until it reaches the pipe axis at section 2, beyond which the flow is considered fully developed (no change in velocity profile with axial distance). In the potential flow outside the boundary layer, the velocity continuously increases and reaches the maximum value on the pipe axis at Section 2. We can apply the Bernoulli equation between any two points on the streamline in this region along the pipe axis. The following key features of this pipe flow are worth noting:

1. From mass conservation (continuity equation) under steady state, the mass flow rate at each section remains constant.
2. For an incompressible flow in this constant area pipe, the average flow velocity also remains constant.
3. The static pressure is radially uniform at each cross section.
4. Due to the nonuniform velocity profile, except at the inlet, the total pressure is radially nonuniform at each cross section.
5. Although the wall shear stress varies in the entrance region, it remains constant in the fully developed flow region beyond section 2.

From the condition of equal mass flow rate at each section in the pipe, we can write:

$$\rho V \pi R^2 = \int_0^R \rho V_{\max}\left[1 - \left(\frac{r}{R}\right)^2\right](2\pi r)\mathrm{d}r$$

Using $\hat{r} = r/R$, $V = 2V_{\max}\int_0^1 (1-\hat{r}^2)\hat{r}\,\mathrm{d}\hat{r} = 2V_{\max}\int_0^1 (\hat{r}-\hat{r}^3)\mathrm{d}\hat{r} = \frac{V_{\max}}{2}$

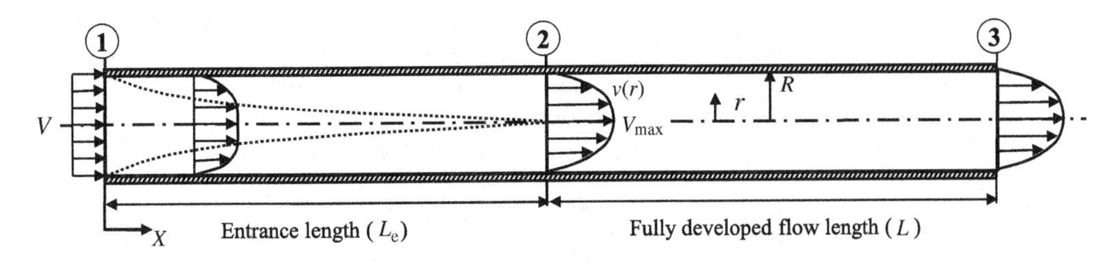

FIGURE 1.9 Incompressible laminar flow in a circular pipe (Example 1.5)

$$V_{max} = 2V$$

which shows that, for the fully developed parabolic velocity profile, the maximum velocity V_{max} at the pipe axis is twice the average velocity V.

We define the momentum correction factor β and the kinetic energy correction factor α as follows:

$$\beta = \frac{\text{Actual momentum flux}}{\text{Momentum flux of the average velocity}} = \frac{\iint_A (v\rho v)\,\mathrm{d}A}{\rho V^2 A} = \frac{\iint_A v^2\,\mathrm{d}A}{V^2 A}$$

$$\alpha = \frac{\text{Actual kinetic energy flux}}{\text{Kinetic energy flux of the average velocity}} = \frac{\iint_A (v^2\rho v)\,\mathrm{d}A}{\rho V^3 A} = \frac{\iint_A v^3\,\mathrm{d}A}{V^3 A}$$

For an axisymmetric parabolic velocity profile, we compute $\beta = 4/3$ and $\alpha = 2$. Note that these correction factors are greater than unity. They allow us to perform integral control volume analysis using algebraic equations based on average velocities. For example, we can obtain the momentum inflow rate or outflow rate of a nonuniform velocity distribution by multiplying the momentum flow rate based on the average velocity by the momentum correction factor β of the nonuniform velocity profile. Similarly, we can obtain the inflow rate or outflow rate of the kinetic energy by multiplying the flow rate of the kinetic energy based on the average velocity by the kinetic energy correction factor α of the given nonuniform velocity profile.

We can evaluate the constant wall shear stress in the fully developed flow as:

$$\tau_w = -\left(\mu \frac{\mathrm{d}v(r)}{\mathrm{d}r}\right)_{r=R}$$

The negative sign in this expression for the wall shear stress indicates that it is equal and opposite to the shear stress acting on the fluid surface in contact. For the given parabolic velocity profile, we write:

$$\tau_w = \frac{2\mu V_{max}}{R} = \frac{8\mu V}{D}$$

which we can express in terms of the shear coefficient, also known as the Fanning friction factor, as follows:

$$C_f = \frac{\tau_w}{\frac{1}{2}\rho V^2} = \frac{16\,\mu}{\rho V D} = \frac{16}{Re}$$

where the Reynolds number $Re = \frac{\rho V D}{\mu}$. Note that the dynamic pressure used in the definition of the shear coefficient C_f corresponds to the average flow velocity V.

Force–momentum balance on the control volume: fully developed flow region. In the fully developed flow region, the axial momentum of the flow entering the control volume at section 2 equals that leaving the control volume at section 3. The force–momentum balance on a control volume in this region, shown in Figure 1.10, reduces to a balance between the net pressure force and the total shear force acting on the control volume.

Thus, we obtain:

$$(p_2 - p_3)\frac{\pi D^2}{4} = \tau_w \pi D L$$

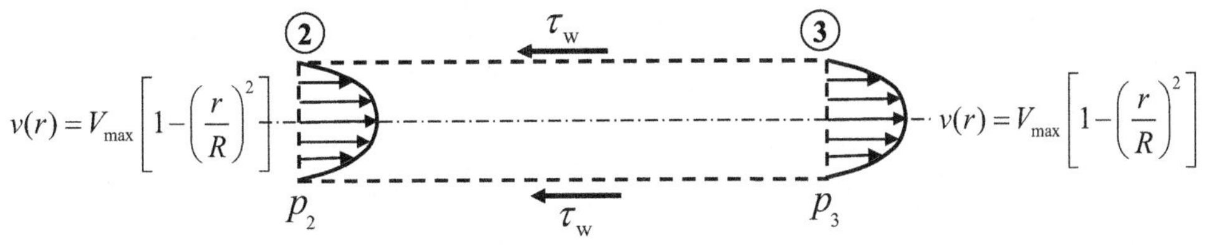

FIGURE 1.10 Control volume showing force–momentum balance in the fully developed flow region (Example 1.5).

$$(p_2 - p_3) = 4\tau_w\left(\frac{L}{D}\right)$$

$$(p_2 - p_3) = 4C_f\left(\frac{L}{D}\right)\left(\frac{1}{2}\rho V^2\right) = f\left(\frac{L}{D}\right)\left(\frac{1}{2}\rho V^2\right)$$

where f is called the Moody or Darcy friction factor, and it is four times the shear coefficient C_f, also called the Fanning friction factor. One must be careful when using the friction factor or the shear coefficient when both are denoted by the same symbol f.

We can compute total pressures at section 3 and section 4 as follows:

$$p_{02} = p_2 + \alpha\frac{1}{2}\rho V^2 = p_2 + \rho V^2$$

$$p_{03} = p_3 + \alpha\frac{1}{2}\rho V^2 = p_3 + \rho V^2$$

In the preceding two equations, we have used the kinetic energy correction factor α to account for parabolic velocity profiles. Thus, in a fully developed pipe flow with constant dynamic pressure, the change in total pressure between any two sections equals the change in static pressure at these sections (i.e., $p_{02} - p_{03} = p_2 - p_3$).

Force–momentum balance on the control volume: entrance region flow. Over the pipe entrance length from section 1 to section 2, both the momentum and kinetic energy flux of the developing velocity profile increase axially. The wall shear stress is at its maximum value near section 1 and reaches the value for the parabolic velocity profile at section 2.

Applying the Bernoulli equation between sections 1 and 2 along the pipe axis, we can write:

$$\frac{p_1}{\rho} + \frac{1}{2}V^2 = \frac{p_2}{\rho} + \frac{1}{2}V_{\max}^2 = \frac{p_2}{\rho} + 2V^2$$

$$(p_1 - p_2) = 2\rho V^2 - \frac{1}{2}\rho V^2 = \frac{3}{2}\rho V^2$$

A force–momentum balance on the control volume shown in Figure 1.11 yields:

$$p_1 A - p_2 A - F_{sh} = \beta(\rho VA)V - (\rho VA)V$$

$$p_1 - p_2 - \frac{F_{sh}}{A} = \rho V^2(\beta - 1) = \rho V^2\left(\frac{4}{3} - 1\right)$$

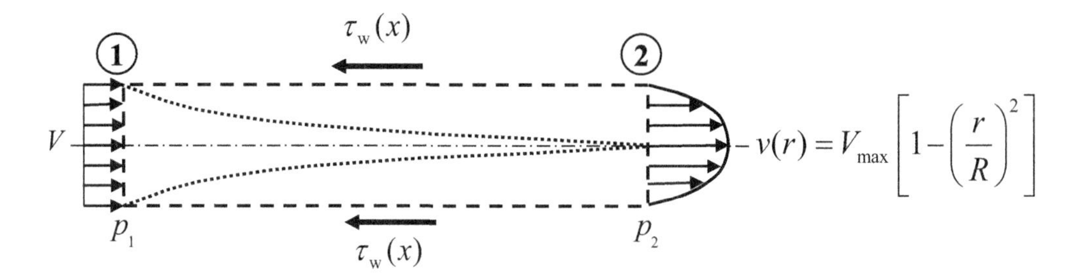

FIGURE 1.11 Control volume showing force–momentum balance in the developing flow region over the pipe entrance length (Example 1.5).

$$\left(p_1 \quad p_2\right) \quad \frac{F_{sh}}{A} = \frac{1}{3}\rho V^2$$

where F_{sh} is the total shear force acting on the control volume. Substituting $\left(p_1 - p_2\right) = \frac{3}{2}\rho V^2$, obtained above from the Bernoulli equation, we obtain:

$$\frac{F_{sh}}{A} = \frac{3}{2}\rho V^2 - \frac{1}{3}\rho V^2 = \frac{7}{6}\rho V^2$$

The total change in static pressure over the entrance length consists of two parts: one to balance the wall shear stress $\left(\Delta p_{sh}\right)$ and the other to increase the flow momentum $\left(\Delta p_{mom}\right)$. Thus, we can write:

$$\left(p_1 - p_2\right) = \Delta p_{sh} + \Delta p_{mom} = \frac{3}{2}\rho V^2$$

where

$$\Delta p_{sh} = \frac{F_{sh}}{A} = \frac{7}{6}\rho V^2$$

giving:

$$\Delta p_{mom} = \frac{3}{2}\rho V^2 - \frac{7}{6}\rho V^2 = \frac{1}{3}\rho V^2$$

At section 1 of Figure 1.11, we compute the total pressure as:

$$p_{01} = p_1 + \frac{1}{2}\rho V^2$$

Similarly, at section 2, we obtain:

$$p_{02} = p_2 + \alpha \frac{1}{2}\rho V^2 = p_2 + \rho V^2$$

Thus, we obtain the change in total pressure in the entrance region as:

$$\left(p_{01} - p_{02}\right) = \left(p_1 - p_2\right) - \frac{1}{2}\rho V^2$$

$$\left(p_{01} - p_{02}\right) = \frac{3}{2}\rho V^2 - \frac{1}{2}\rho V^2 = \rho V^2$$

which is not equal to the change in static pressure in this region. If we were to replace the entrance pipe length with an equivalent fully developed pipe flow length, L_{eq}, we must compute it to yield the same change in static pressure and not in total pressure. The computed equivalent length in this way will ensure that the new pipe with the fully developed flow right from the pipe inlet will satisfy the momentum equation of the original pipe with its entrance length. Thus:

$$(p_1 - p_2) = f\left(\frac{L_{eq}}{D}\right)\left(\frac{1}{2}\rho V^2\right) = \frac{3}{2}\rho V^2$$

Substituting f in the above equation for a fully developed laminar pipe flow, we obtain:

$$\frac{L_{eq}}{D} = \frac{3}{f} = \frac{3}{64} Re = 0.0469 Re$$

For $Re = 2134$, the above equation yields $L_{eq} / D \approx 100$, which is an order of magnitude higher than the value we expect for a turbulent pipe flow.

Example 1.6: Incompressible Flow of Air in an Ejector Pump

Figure 1.12 shows an ejector pump with an incompressible flow ($M < 0.3$) of air. High-velocity annular jet with mass flow rate $\dot{m}_2$ induces the central mass flow rate $\dot{m}_1$ from the quiescent air at the ambient pressure p_{amb}. The flow area for the annular jet at section 3 is A_2 and that for the induced central flow is A_1. The constant air density is ρ.

The mixing loss in total pressure occurs between sections 3 and 4. The flow exits the ejector pump at section 4 at a uniform velocity. In this example, we are asked to determine $\zeta = \dot{m}_1 / \dot{m}_2$ in terms of other design parameters, neglecting any loss due to wall friction.

Solution

Before carrying out control volume analyses in this example with the appropriate boundary conditions, let us understand the following flow physics:

1. Between sections 1 and 2, the flow in the central duct (assumed frictionless) with a bell-mouth entry can be considered a potential flow with the Bernoulli constant $= p_{amb}$. Thus, the ambient pressure at the ejector inlet will be the total pressure at section 1.
2. The velocity distribution at section 2 is nonuniform; the velocity in the annular flow through A_2 is much higher than that in the induced central flow through A_1. As the streamlines are straight, the assumption of uniform static pressure at this section is more appropriate than that of a uniform total pressure.

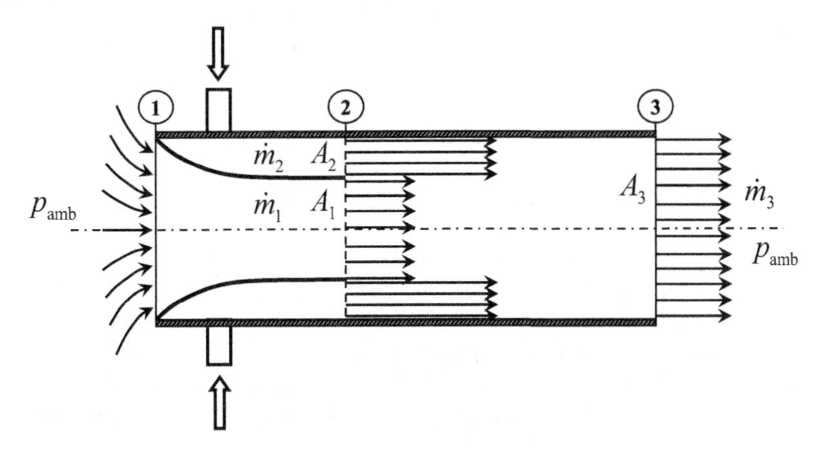

FIGURE 1.12 Incompressible flow in an ejector pump (Example 1.6)

(3) All quantities become uniform at section 3. At this section, however, the exit static pressure, and not the total pressure, must be equal to the ambient pressure p_{amb}, which becomes the static pressure boundary condition ($p_1 = p_{amb}$) at this section.

From the geometry shown in Figure 1.12 and the mass conservation between sections 2 and 3, we obtain the following relations:

$$A_3 = A_1 + A_2; \dot{m}_3 = \dot{m}_1 + \dot{m}_2; V_1 = \frac{\dot{m}_1}{\rho A_1}; V_2 = \frac{\dot{m}_2}{\rho A_2}; \text{ and } V_3 = \frac{\dot{m}_3}{\rho A_3}$$

Using the Bernoulli equation along the ejector axis over sections 1 and 2, we can write:

$$p_{amb} = p_2 + \frac{1}{2}\rho V_1^2 = p_2 + \frac{\dot{m}_1^2}{2\rho A_1^2}$$

$$\left(p_{amb} - p_2\right) = \frac{\dot{m}_1^2}{2\rho A_1^2}$$

The force–momentum balance over the control volume between sections 2 and 3 yields the following equation:

$$(A_1 + A_2)(p_{amb} - p_2) = \frac{\dot{m}_1^2}{\rho A_1} + \frac{\dot{m}_2^2}{\rho A_2} - \frac{(\dot{m}_1 + \dot{m}_2)^2}{\rho(A_1 + A_2)}$$

Substituting $(p_{amb} - p_2)$ from the previous relation obtained from the Bernoulli equation, we can write:

$$\left(A_1 + A_2\right)\frac{\dot{m}_1^2}{2\rho A_1^2} = \frac{\dot{m}_1^2}{\rho A_1} + \frac{\dot{m}_2^2}{\rho A_2} - \frac{\left(\dot{m}_1 + \dot{m}_2\right)^2}{\rho\left(A_1 + A_2\right)}$$

$$\frac{\dot{m}_1^2}{2A_1^2} = \frac{\dot{m}_1^2}{A_1\left(A_1 + A_2\right)} + \frac{\dot{m}_2^2}{A_2\left(A_1 + A_2\right)} - \frac{\left(\dot{m}_1 + \dot{m}_2\right)^2}{\left(A_1 + A_2\right)^2}$$

$$\frac{\zeta^2}{2} = \frac{\beta\zeta^2}{(1+\beta)} + \frac{\beta^2}{(1+\beta)} - \frac{\beta^2(1+\zeta)^2}{(1+\beta)^2}$$

where $\xi = \dot{m}_1/\dot{m}_2$ and $\beta = A_1 / A_2$. The above equation further reduces to the following quadratic equation in ζ:

$$\left(1+\beta^2\right)\zeta^2 + 4\beta^2\zeta - 2\beta^3 = 0$$

We obtain a physically meaningful root of the above equation as:

$$\zeta = \frac{2\beta}{1+\beta^2}\left(\sqrt{0.5\beta^3 + \beta^2 + 0.5\beta} - \beta\right) = \frac{2\beta}{1+\beta^2}\left[(1+\beta)\sqrt{\frac{\beta}{2}} - \beta\right]$$

which shows that the mass flow rate ratio (ζ) in this incompressible flow ejector depends only on the flow area ratio (β). Furthermore, Figure 1.13 shows that ζ increases monotonically with β.

In this example, in the ejector mixing region bounded by section 2 and section 3, we can show that the static pressure increases and the average total pressure, computed by adding mass-averaged dynamic pressure to the uniform static pressure at each section, actually decreases.

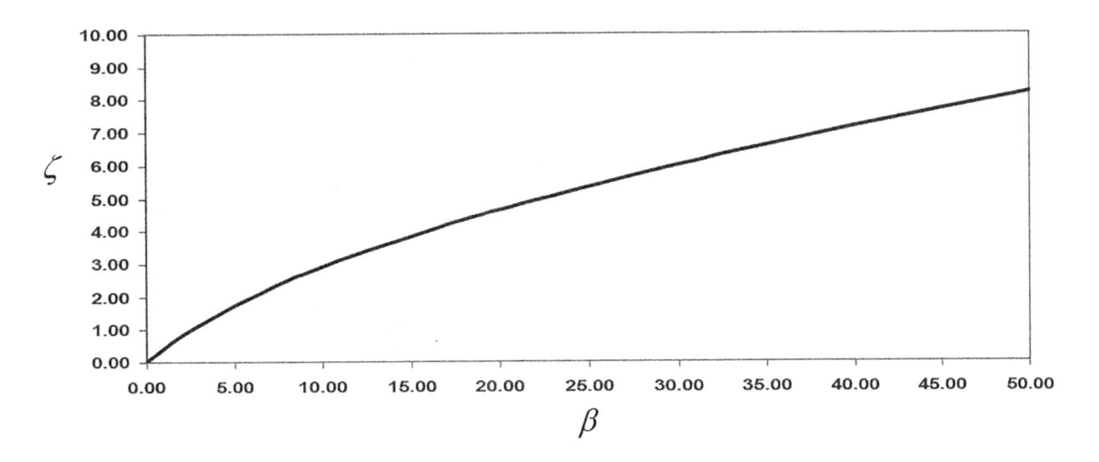

FIGURE 1.13 Variation of mass flow rate ratio ζ with area ratio β (Example 1.6).

Example 1.7

Figure 1.14 shows one-dimensional steady fluid flow at a mass flow rate of $\dot{m}$ in a circular duct of constant diameter D and length L. The pipe wall is at a uniform temperature T_w, which is higher than the fluid total temperature at the pipe inlet. Assuming a constant fluid specific heat c_p and heat transfer coefficient h between the fluid and the pipe wall, derive an expression for the variation of fluid total temperature from the pipe inlet to the outlet.

Solution

Applying steady flow energy balance on the control volume EFGH in Figure 1.14 yields:

$$\dot{m}c_p\left(T_0 + \frac{\mathrm{d}T_0}{\mathrm{d}x}\Delta x\right) - \dot{m}c_pT_0 = (\pi D\Delta x)h(T_w - T_0)$$

$$\frac{\mathrm{d}T_0}{\mathrm{d}x} = \frac{\pi Dh}{\dot{m}c_p}(T_w - T_0)$$

$$\frac{\mathrm{d}(T_w - T_0)}{(T_w - T_0)} = \left(\frac{\pi DhL}{\dot{m}c_p}\right)\mathrm{d}(x/L)$$

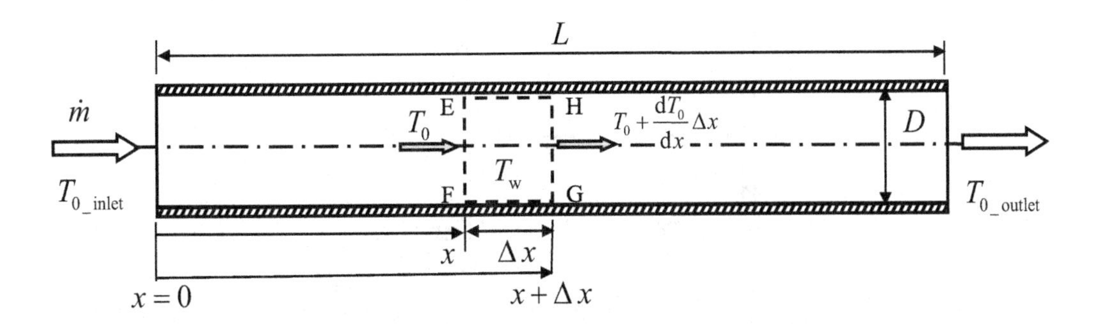

FIGURE 1.14 Heat transfer into fluid flow in a circular pipe at a uniform wall temperature (Example 1.7).

Integrating the preceding ordinary differential equation from the pipe inlet to a distance x along the pipe yields the following expression for the fluid total temperature at x:

$$T_0 = T_w - \left(T_w - T_{0\_inlet}\right)e^{-\eta x/L}$$

where $\eta = \pi DhL/\left(\dot{m}c_p\right)$. This solution shows that the variation of fluid total temperature in the flow through an isothermal pipe is not linear. Instead, from the pipe inlet to the outlet, the difference between the wall temperature and the fluid total temperature decays exponentially with η, also known as the number of transfer units (NTUs) in the literature on heat exchangers. In agreement with the second law of thermodynamics, this solution also ensures that the fluid total temperature will never exceed the pipe wall temperature, regardless of how long the pipe is. Suppose we use the fluid static temperature as the reference temperature to drive the convective heat transfer between the pipe wall and the fluid. In that case, we run the risk of the fluid total temperature at the pipe outlet exceeding the wall temperature, which will be physically unrealistic. How can an isothermal pipe wall heat the fluid to a temperature higher than its temperature? This solution yields the following fluid total temperature at the outlet $(x = L)$:

$$T_{0\_outlet} = T_w - \left(T_w - T_{0\_inlet}\right)e^{-\eta}$$

Example 1.8

As shown in Figure 1.15, water flows at a constant volumetric flow rate of Q_{in} into a large tank of cross-sectional area A_t. The cross-sectional area of the drainpipe near the bottom of the tank is A_p. The drainpipe starts flowing when the water level in the tank exceeds h_o. Derive the differential equation that governs the water level h in the tank as a function of time and hence find the maximum water level the tank will have under steady state.

Solution

With constant density, applying Equation 1.15 to this example yields:

$$A_t \frac{dh}{dt} = Q_{in} - Q_{out} = Q_{in} - A_p\sqrt{2g(h - h_o)}$$

$$A_t \frac{dh}{dt} + A_p\sqrt{2g(h - h_o)} = Q_{in}$$

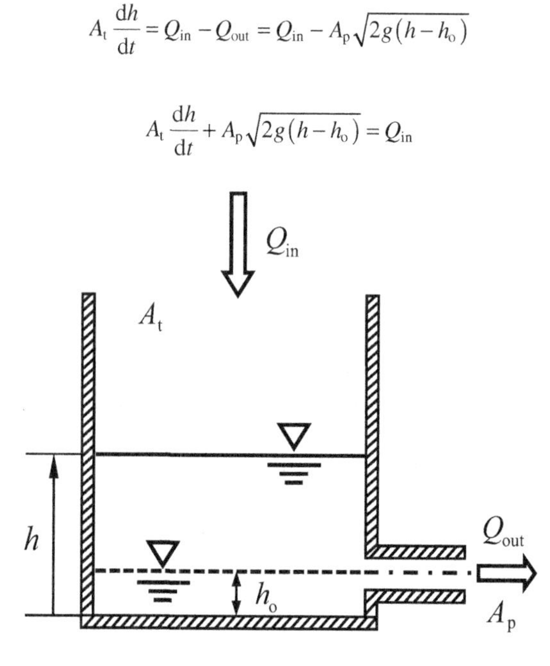

FIGURE 1.15 Water level in a tank with inflow and outflow (Example 1.8).

where the outflow $Q_{out} = A_p\sqrt{2g(h-h_o)}$, obtained using the Bernoulli equation, depends on the hydrostatic pressure in the tank. Under steady state, the time-dependent term vanishes, giving $Q_{out} = Q_{in}$, and the water level in the tank attains its maximum value h_{max} given by:

$$A_p\sqrt{2g(h_{max}-h_o)} = Q_{in}$$

$$h_{max} = h_o + \frac{Q_{in}^2}{2gA_p^2}$$

Example 1.9

Figure 1.16a shows a control volume of a streamtube having a steady uniform flow at its inlet and outlet. Find an expression for the net force acting on the fluid control volume.

Solution

A streamtube consists of a bundle of streamlines; as a result, no flow crosses the lateral surface of a streamtube. The steady mass flow rate at every section of the streamtube from the inlet to the outlet remains constant. Assuming uniform velocities at both the inlet and the outlet and applying the steady form of Equation 1.20 on the entire streamtube, we obtain:

$$\sum \boldsymbol{F} = \dot{m}\boldsymbol{V}_2 - \dot{m}\boldsymbol{V}_1$$

which is shown as a vector diagram in Figure 1.16b.

Example 1.10

Figure 1.17 shows a cylinder on a flat plate in an incompressible, two-dimensional flow with uniform velocity U in the x direction. The flow leaving the face CD of the control volume has a part-linear and part-sinusoidal velocity profile given by the following equations:

Linear: $u = \dfrac{Uy}{2a}$ for $0 \le y \le a$

Sinusoidal: $u = \dfrac{U}{2}\left(1 + \sin\dfrac{\pi(y-a)}{2a}\right)$ for $a \le y \le 2a$

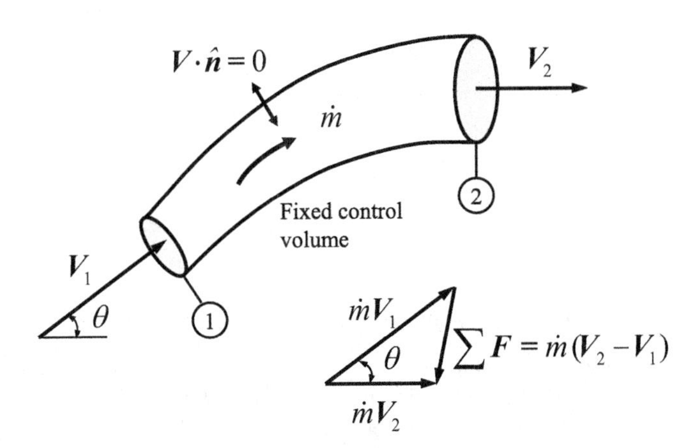

FIGURE 1.16 (a) A streamtube control volume with one inflow and one outflow (Example 1.9) and (b) vector solution.

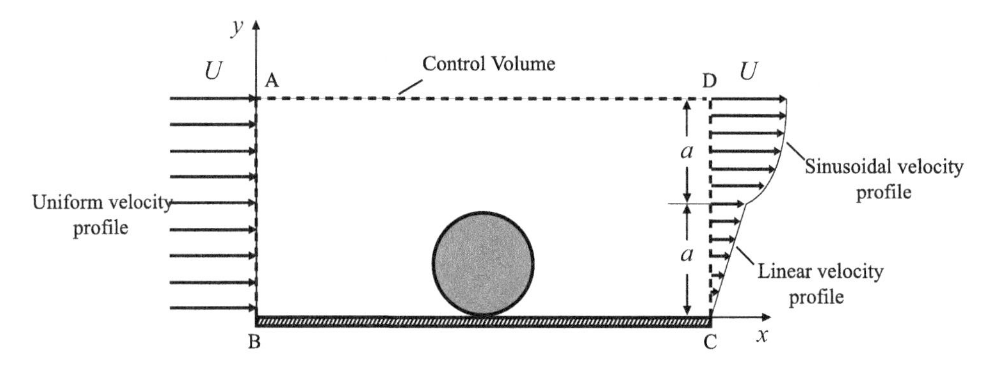

FIGURE 1.17 A cylinder on a flat plate in a two-dimensional flow (Example 1.10).

Calculate the drag force acting on the unit length of the cylinder in combination with the flat plate within the control volume ABCD. The fluid density is ρ. Note that the figure does not show the mass flow rate through AD with a nonuniform y velocity distribution.

Solution

In this example, if the drag force acting on both the cylinder and the flat plate is F_D, the corresponding force acting on the fluid control volume ABCD must be $-F_D$.

CONTINUITY EQUATION

Mass flow rate entering the control volume through AB: $\dot{m}_{AB} = 2a\rho U$

Mass flow rate exiting the control volume through CD: $\dot{m}_{CD} = \dot{m}_{CD\_linear} + \dot{m}_{CD\_sinusoidal}$

$$\dot{m}_{CD\_linear} = \int_0^a \rho\left(\frac{Uy}{2a}\right)dy = \rho\frac{U}{2a}\left[\frac{y^2}{2}\right]_0^a = \frac{1}{4}a\rho U$$

$$\dot{m}_{CD\_sinusoidal} = \int_a^{2a} \rho\frac{U}{2}\left(1+\sin\frac{\pi(y-a)}{2a}\right)dy = \rho\frac{U}{2}\left[y-\frac{2a}{\pi}\cos\frac{\pi(y-a)}{2a}\right]_a^{2a}$$

$$\dot{m}_{CD\_sinusoidal} = \frac{\rho U}{2}\left(a+\frac{2a}{\pi}\right) = \left(\frac{1}{2}+\frac{1}{\pi}\right)a\rho U$$

$$\dot{m}_{CD} = \dot{m}_{CD\_linear} + \dot{m}_{CD\_sinusoidal} = \frac{1}{4}a\rho U + \left(\frac{1}{2}+\frac{1}{\pi}\right)\rho U a = \left(\frac{3}{4}+\frac{1}{\pi}\right)a\rho U$$

From the continuity equation, we obtain the mass flow rate through AD as:

$$\dot{m}_{AD} = \dot{m}_{AB} - \dot{m}_{CD} = 2\,a\rho U - \left(\frac{3}{4}+\frac{1}{\pi}\right)a\rho U = \left(\frac{5}{4}-\frac{1}{\pi}\right)a\rho U$$

MOMENTUM EQUATION IN THE *X* DIRECTION

According to Equation 1.22, the total force acting on the fluid control volume under steady state must equal the net outflow of x-momentum from the control volume.

The x-momentum flow rate entering the control volume through AB: $\dot{M}_{AB} = 2a\rho U^2$

The x-momentum flow rate exiting the control volume through AD: $\dot{M}_{AD} = \left(\frac{5}{4}-\frac{1}{\pi}\right)a\rho U^2$

Note that in evaluating the *x*-momentum flow rate through AD, we have assumed that AD corresponds to the edge of the boundary layer and coincides with the free stream. As a result, all fluid particles exiting through AD have their *x* component of velocity equal to U.

The *x*-momentum flow rate exiting the control volume through CD:

$$\dot{M}_{\mathrm{CD}} = \dot{M}_{\mathrm{CD\_linear}} + \dot{M}_{\mathrm{CD\_sinusoidal}}$$

$$\dot{M}_{\mathrm{CD\_linear}} = \int_0^a \rho u^2 \mathrm{d}y = \rho \int_0^a \left(\frac{Uy}{2a}\right)^2 \mathrm{d}y = \rho \frac{U^2}{4a^2}\left[\frac{y^3}{3}\right]_0^a = \frac{1}{12} a\rho U^2$$

$$\dot{M}_{\mathrm{CD\_sinusoidal}} = \int_a^{2a} \rho \left[\frac{U}{2}\left(1+\sin\frac{\pi(y-a)}{2a}\right)\right]^2 \mathrm{d}y = \frac{\rho U^2}{4}\int_a^{2a}\left(1+\sin\frac{\pi(y-a)}{2a}\right)^2 \mathrm{d}y$$

With the substitution $\zeta = \pi(y-a)/2a$, we simplify this integral as:

$$\dot{M}_{\mathrm{CD\_sinusoidal}} = \frac{a\rho U^2}{2\pi}\int_0^{\pi/2}(1+\sin\zeta)^2 \mathrm{d}\zeta = \frac{a\rho U^2}{2\pi}\int_0^{\pi/2}\left(1+2\sin\zeta+\sin^2\zeta\right)\mathrm{d}\zeta$$

$$= \frac{a\rho U^2}{2\pi}\int_0^{\pi/2}\left(1+2\sin\zeta+\frac{1}{2}-\frac{\cos 2\zeta}{2}\right)\mathrm{d}\zeta$$

$$= \frac{a\rho U^2}{2\pi}\int_0^{\pi/2}\left(\frac{3}{2}+2\sin\zeta-\frac{\cos 2\zeta}{2}\right)\mathrm{d}\zeta$$

$$= \frac{a\rho U^2}{2\pi}\left[\frac{3}{2}\zeta-2\cos\zeta-\frac{\sin 2\zeta}{4}\right]_0^{\pi/2}$$

$$\dot{M}_{\mathrm{CD\_sinusoidal}} = \left(\frac{3}{8}+\frac{1}{\pi}\right) a\rho U^2$$

The force–momentum balance on the fluid control volume finally yields:

$$-F_{\mathrm{D}} = \dot{M}_{\mathrm{AD}} + \dot{M}_{\mathrm{CD\_linear}} + \dot{M}_{\mathrm{CD\_sinusoidal}} - \dot{M}_{\mathrm{AB}}$$

$$-F_{\mathrm{D}} = \left(\frac{5}{4}-\frac{1}{\pi}\right) a\rho U^2 + \frac{1}{12} a\rho U^2 + \left(\frac{3}{8}+\frac{1}{\pi}\right) a\rho U^2 - 2a\rho U^2$$

$$F_{\mathrm{D}} = \frac{7}{24} a\delta U^2$$

which is an exact solution to the problem, showing that around one-third of the incoming momentum flow rate overcomes the drag due to the cylinder and flat plate per unit length along the cylinder axis. This example clearly shows the power of the integral control volume analysis to yield the overall drag force easily. If we had used a two-dimensional CFD analysis for this problem, simulating the boundary conditions along AD and CD would have been challenging. After obtaining the CFD results, we would need to integrate the *x* component of the shear force and pressure force along the wetted wall of the cylinder and the shear force on the flat plate. Despite all this hard work in performing the CFD analysis, the result will still be approximate, which we must validate against the integral control volume analysis result.

Example 1.11

Figure 1.18 shows an incompressible flow in a sudden expansion pipe flow system with pipe diameters D_1 and D_2. For a turbulent flow in each pipe, assume uniform velocity and static pressure in sections 1 and 2. Find the static and total pressure changes between sections 1 and 2 of this flow system. Neglect any shear stress on the downstream pipe wall.

Solution

This example is of significant practical importance, found in many engineering applications. Although somewhat inefficient, this flow geometry is a dump diffuser often used in limited spaces.

CONTINUITY EQUATION

The mass flow rate entering the control volume through section 1: $\dot{m}_1 = A_1\rho V_1$

The mass flow rate exiting the control volume through section 2: $\dot{m}_2 = A_2\rho V_2$ where $A_1 = \pi D_1^2/4$ and $A_2 = \pi D_2^2/4$.

Equating $\dot{m}_1$ and $\dot{m}_2$ for a steady flow yields:

$$\frac{A_2}{A_1} = \frac{V_1}{V_2}$$

MOMENTUM EQUATION

To apply the momentum equation to the control volume shown in Figure 1.18, we make the critical assumption that the static pressure p_1 at the exit of the smaller pipe is uniform over the entire section 1, including the annulus area between the two pipes.

The axial momentum flow rate entering the control volume through section 1: $\dot{M}_1 = \dot{m}_1 V_1$

The axial momentum flow rate exiting the control volume through section 2: $\dot{M}_2 = \dot{m}_2 V_2$

The force–momentum balance on the control volume yields:

$$(p_1 - p_2)A_2 = \dot{M}_2 - \dot{M}_1 = \dot{m}_1(V_2 - V_1) = \rho A_1 V_1 (V_2 - V_1)$$

$$(p_1 - p_2) = \rho\left(\frac{A_1}{A_2}\right)V_1^2\left(\frac{V_2}{V_1} - 1\right)$$

$$(p_2 - p_1) = \rho V_1^2\left(\frac{A_1}{A_2}\right)\left(1 - \frac{A_1}{A_2}\right)$$

$$C_p = \frac{(p_2 - p_1)}{\frac{1}{2}\rho V_1^2} = 2\left(\frac{A_1}{A_2}\right)\left(1 - \frac{A_1}{A_2}\right)$$

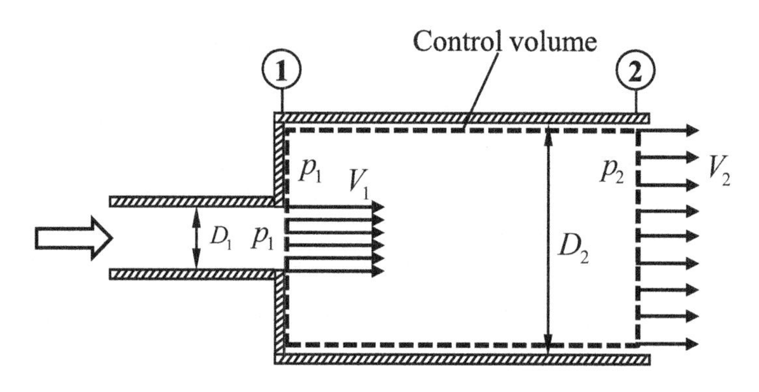

FIGURE 1.18 Sudden expansion pipe flow with uniform properties at the inlet and the outlet (Example 1.11).

When using the sudden expansion pipe flow geometry as a dump diffuser, we call C_p the static pressure rise (recovery) coefficient, which reaches a maximum value of 0.5 when the area ratio (A_1/A_2) is 0.5.

We calculate the change in total pressure between sections 1 and 2 as follows:

Total pressure at section 1: $p_{01} = p_1 + \frac{1}{2}\rho V_1^2$

Total pressure at section 2: $p_{02} = p_2 + \frac{1}{2}\rho V_2^2$

$$P_{t1} - P_{t2} = p_1 + \frac{1}{2}\rho V_1^2 - p_2 - \frac{1}{2}\rho V_2^2 = \frac{1}{2}\rho V_1^2 - \frac{1}{2}\rho V_2^2 - (p_2 - p_1)$$

Substituting $(p_2 - p_1)$ from the result obtained earlier yields:

$$p_{01} - p_{02} = \frac{1}{2}\rho V_1^2 - \frac{1}{2}\rho V_2^2 - \rho V_1^2\left(\frac{A_1}{A_2}\right)\left(1 - \frac{A_1}{A_2}\right)$$

which, with the substitution of V_2 / V_1 for A_1 / A_2, reduces to:

$$p_{01} - p_{02} = \frac{1}{2}\rho(V_1 - V_2)^2$$

which shows that the loss in total pressure in a sudden expansion pipe flow equals the dynamic pressure associated with the difference of average velocities in the two pipes. We can further express this result in terms of the loss coefficient K defined as:

$$K = \frac{(p_{01} - p_{02})}{\frac{1}{2}\rho V_1^2} = \left(1 - \frac{V_2}{V_1}\right)^2 = \left(1 - \frac{A_1}{A_2}\right)^2$$

which yields $K = 1$ when the downstream pipe diameter is very large compared to the diameter of the upstream pipe, implying a complete loss of the dynamic pressure of the incoming flow wholly in the downstream pipe as if it were a plenum.

Example 1.12

Figure 1.19a shows a water jet of velocity V_j deflected by a curved plate mounted on a cart, which is held stationary by the application of a braking force F_b. The jet leaves the plate at an angle θ with the x axis. The cross-sectional area A_j of the jet remains constant over the curved plate. Find the braking force F_b (a) when the cart is stationary and (b) when the cart is moving in the x direction at a constant velocity of U_c, where $U_c < V_j$. The density of water is ρ.

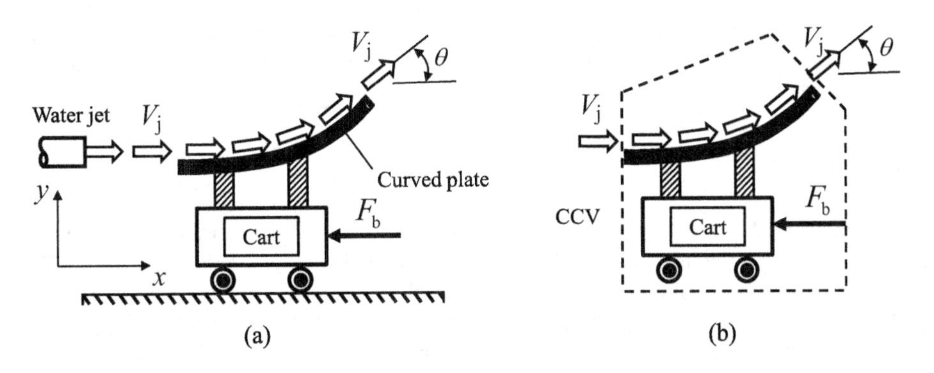

FIGURE 1.19 Water jet deflected by a curved plate mounted on a cart (Example 1.12).

Solution

Figure 1.19b shows the combined control volume (CCV) used in the present analysis. The mass flow rate entering and exiting the control volume remains constant, satisfying the continuity equation. The static ambient pressure around the CCV is uniform, so it applies no net force on the control volume.

We write the force–momentum balance on the CCV in the x direction as follows:

a. Stationary Cart

Because the water density and the jet area remain constant over the curved plate, the velocity of the jet exiting the CCV must be V_j from the continuity equation.

The jet mass flow rate at the CCV inlet and exit: $\dot{m}_{(a)} = A_j \rho V_j$

The rate of x-momentum outflow: $\dot{M}_{out(a)} = \dot{m}_{(a)} V_j \cos\theta$

The rate of x-momentum inflow: $\dot{M}_{in(a)} = \dot{m}_{(a)} V_j$

The total body and surface forces acting on the CCV: $\sum F_x = -F_{b(a)}$

The force–momentum balance on the CCV yields:

$$-F_{b(a)} = \dot{m}_{(a)} V_j \cos\theta - \dot{m}_{(a)} V_j = \dot{m}_{(a)} V_j (\cos\theta - 1) = -\dot{m}_{(a)} V_j (1 - \cos\theta)$$

$$F_{b(a)} = \dot{m}_{(a)} V_j (1 - \cos\theta)$$

b. Cart Moving at a Constant Velocity U_c

In this case, we attach the coordinate system to the moving cart and measure the jet inflow and outflow velocities relative to the cart:

$$W_j = V_j - U_c$$

$$\dot{m}_{(b)} = A_j \rho W_j$$

The rate of x-momentum outflow: $\dot{M}_{out(b)} = \dot{m}_{(b)} W_j \cos\theta$

The rate of x-momentum inflow: $\dot{M}_{in(b)} = \dot{m}_{(b)} W_j$

The total body and surface forces acting on the CCV: $\sum F_x = -F_{b(b)}$

The force–momentum balance on the CCV yields:

$$-F_{b(b)} = \dot{m}_{(b)} W_j \cos\theta - \dot{m}_{(b)} W_j = \dot{m}_{(b)} W_j (\cos\theta - 1) = -\dot{m}_{(b)} W_j (1 - \cos\theta)$$

$$F_{b(b)} = \dot{m}_{(b)} W_j (1 - \cos\theta)$$

For the same incoming water jet and the curved plate with turning angle θ, the braking force required to keep the cart stationary will be higher than to keep the cart moving at a constant velocity U_c. In each case, the braking force becomes zero when the turning angle is zero. The braking force increases as the turning angle rises up to $\theta = 180^\circ$, where each braking force will be twice its value at $\theta = 90^\circ$.

Example 1.13

As shown in Figure 1.20, water flows steadily through a vertical pipe of diameter D. A fully developed laminar flow enters the control volume at section 1 with a parabolic velocity profile. At the control volume outlet, the flow transitions to a fully developed turbulent flow with the 1/7th power-law profile. In the given expression for each velocity profile, R represents the pipe radius. The total shear force F_f and the gravitational body force W_g act on the control volume, which has length L. For a given mass flow rate of $\dot{m}$ through the pipe, the measured static pressures at

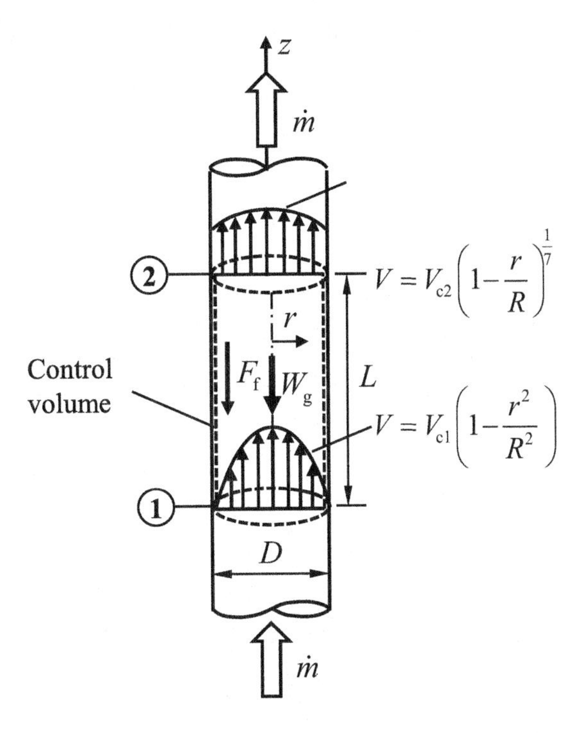

FIGURE 1.20 Water flow through a vertical pipe with nonuniform velocities (Example 1.13).

sections 1 and 2 are p_1 and p_2. Find the total shear force F_f in terms of other quantities given in the problem.

Solution

Because the flow is steady, the mass flow rate at each section of the pipe remains constant, and so does the average velocity (incompressible flow in a constant area pipe), which we calculate as:

$$\overline{V} = \frac{\dot{m}}{\rho A} = \frac{4\dot{m}}{\rho \pi D^2}$$

where ρ is the density of water and A is the flow area.

MOMENTUM EQUATION ALONG THE PIPE AXIS

The rate of z-momentum inflow at section 1: $\dot{M}_1 = \beta_1 \dot{m}\overline{V} = \frac{4}{3}\dot{m}\overline{V}$

The rate of z-momentum outflow at section 2: $\dot{M}_2 = \beta_2 \dot{m}\overline{V} = 1.020\,\dot{m}\overline{V}$

The total surface and body forces acting on the fluid control volume:

$$\sum (F_s + F_b) = (p_1 - p_2)A - F_f - W_g = (p_1 - p_2)A - F_f - \rho g A L$$

The force–momentum balance on the fluid control volume yields:

$$(p_1 - p_2)A - F_f - AL\rho g = 1.02\,\dot{m}\overline{V} - 1.333\,\dot{m}\overline{V}$$

$$F_f = (p_1 - p_2)A - AL\rho g + 0.313\dot{m}\overline{V}$$

Note that the shear force F_f acting on the fluid control volume shown in Figure 1.20 must be positive in magnitude. Because the rate of z-momentum outflow at section 2 is lower than its rate of inflow at section 1, in the absence of friction and gravitational body force, we will have static pressure recovery in the duct.

Example 1.14

Figure 1.21 shows an incompressible flow in a sudden expansion pipe system with pipe diameters D_1 and D_2. The fully developed flow in the smaller pipe has a parabolic velocity profile, characteristic of a laminar flow. The flow exiting the larger pipe is turbulent with a uniform velocity. Assuming uniform static pressure in sections 1 and 2, find the static and total pressure changes between sections 1 and 2 of this flow system. Neglect any shear stress on the downstream pipe wall.

Solution

This example is like Example 1.11 except that the velocity profile entering the larger pipe is non-uniform (parabolic), given by:

$$V_1 = 2\overline{V}_1\left[1-\left(\frac{2r}{D_1}\right)^2\right]$$

CONTINUITY EQUATION

Mass flow rate entering the control volume through section 1: $\dot{m}_1 = A_1\rho\overline{V}_1$

Mass flow rate exiting the control volume through section 2: $\dot{m}_2 = A_2\rho V_2$

where $A_1 = \pi D_1^2/4$ and $A_2 = \pi D_2^2/4$.

Equating $\dot{m}_1$ and $\dot{m}_2$ for this steady flow yields:

$$\frac{A_2}{A_1} = \frac{\overline{V}_1}{V_2}$$

MOMENTUM EQUATION

To apply the momentum equation to the control volume shown in Figure 1.21, we assume that the static pressure p_1 at the exit of the smaller pipe is uniform over the entire section 1, including the annulus area between the two pipes.

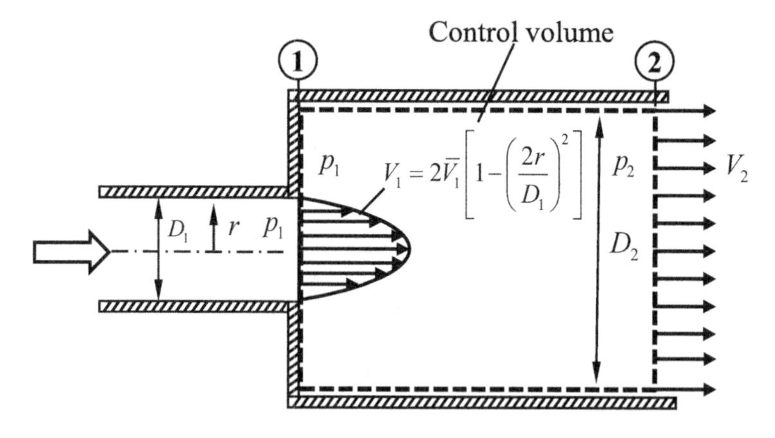

FIGURE 1.21 Sudden expansion pipe flow with a parabolic velocity profile at the inlet and uniform velocity at the outlet (Example 1.14).

We can write the axial momentum flow rate entering the control volume through section 1 as:

$$\dot{M}_1 = \beta_1 \dot{m}_1 \bar{V}_1 = \frac{4}{3}\dot{m}_1 \bar{V}_1$$

Similarly, we write the axial momentum flow rate exiting the control volume through section 2 as:

$$\dot{M}_2 = \dot{m}_2 V_2 = \dot{m}_1 V_2$$

The force–momentum balance on the control volume yields:

$$(p_1 - p_2)A_2 = \dot{M}_2 - \dot{M}_1 = \dot{m}_1\left(V_2 - \frac{4}{3}\bar{V}_1\right) = \rho A_1 \bar{V}_1\left(V_2 - \frac{4}{3}\bar{V}_1\right)$$

$$(p_2 - p_1) = \rho \bar{V}_1^2\left(\frac{A_1}{A_2}\right)\left(\frac{4}{3} - \frac{A_1}{A_2}\right)$$

$$C_p = \frac{(p_2 - p_1)}{\frac{1}{2}\rho \bar{V}_1^2} = 2\left(\frac{A_1}{A_2}\right)\left(\frac{4}{3} - \frac{A_1}{A_2}\right)$$

where we define the pressure rise (recovery) coefficient C_p in terms of the dynamic pressure based on the average inlet velocity $\bar{V}_1$. Note that the C_p reaches a maximum value of 8/9 when the area ratio (A_1/A_2) is 2/3. Compared to the case of uniform inlet velocity in Example 1.11, a part of the recovery in static pressure in this example happens when the nonuniform (parabolic) velocity profile with its higher momentum content at the inlet becomes a uniform one with lower momentum at the exit.

We calculate the change in total pressure between sections 1 and 2 as follows:

Total pressure at section 1:

$$\bar{p}_{01} = p_1 + \frac{1}{2}\alpha\rho \bar{V}_1^2 = p_1 + \frac{1}{2}2\rho \bar{V}_1^2 = p_1 + \rho \bar{V}_1^2$$

where α is the kinetic energy correction factor and equals 2 for the parabolic velocity profile in a circular pipe.

Total pressure at section 2:

$$p_{02} = p_2 + \frac{1}{2}\rho V_2^2$$

$$\bar{p}_{01} - p_{02} = p_1 + \rho \bar{V}_1^2 - p_2 - \frac{1}{2}\rho V_2^2 = \rho \bar{V}_1^2 - \frac{1}{2}\rho V_2^2 - (p_2 - p_1)$$

Substituting $(p_2 - p_1)$ from the result obtained earlier yields:

$$\bar{p}_{01} - p_{02} = \rho \bar{V}_1^2 - \frac{1}{2}\rho V_2^2 - \rho \bar{V}_1^2\left(\frac{A_1}{A_2}\right)\left(\frac{4}{3} - \frac{A_1}{A_2}\right)$$

With the substitution of $V_2 / \bar{V}_1$ for A_1 / A_2, this equation reduces to:

$$\bar{p}_{01} - p_{02} = \frac{1}{2}\rho\left(\bar{V}_1 - V_2\right)^2 + \frac{1}{2}\rho \bar{V}_1^2\left(1 - \frac{2}{3}\frac{V_2}{\bar{V}_1}\right)$$

Compared to the result obtained in Example 3.4 for the loss in total pressure with uniform inlet velocity, this result for the parabolic inlet velocity profile shows additional loss given by the second term on the right-hand side. We can further express this result in terms of the loss coefficient K as:

$$K = \frac{(\bar{p}_{01} - p_{02})}{\frac{1}{2}\rho\bar{V}_1^2} = 2 - \frac{8}{3}\left(\frac{A_1}{A_2}\right) + \left(\frac{A_1}{A_2}\right)^2$$

Again, when the downstream pipe diameter is much larger than the diameter of the upstream pipe, the loss coefficient equals 2. This means that the dynamic pressure associated with the parabolic velocity profile of the incoming flow is completely lost in the downstream pipe, acting as a plenum.

Example 1.15

Figure 1.22 shows an airplane propeller moving to the left with a constant velocity V_p against a head wind having a velocity V_w. Away from the propeller, the ambient air pressure is p_{amb}. In the coordinate system attached to the airplane propeller, the air enters the streamtube, which encompasses the propeller, at section 1 at a relative velocity of $V_{rel1} = V_w + V_p$ and exits at section 4 with the relative velocity V_{rel4}. In terms of the quantities shown in the figure, develop the expressions for (a) propeller thrust, (b) propeller power output, (c) airflow velocity through the propeller, and (d) propeller efficiency. Assume a steady incompressible flow of air with density ρ.

Solution

In this example, the incompressible flow of air through the streamtube remains constant at $\dot{m} = A\rho V_{rel}$, satisfying the continuity equation between sections 1 and 4. As there is no loss in the streamtube between sections 1 and 2, we can write:

$$p_{amb} + \frac{1}{2}\rho V_{rel1}^2 = p_2 + \frac{1}{2}\rho V_{rel}^2$$

Similarly, between sections 3 and 4, we write:

$$p_{amb} + \frac{1}{2}\rho V_{rel4}^2 = p_3 + \frac{1}{2}\rho V_{rel}^2$$

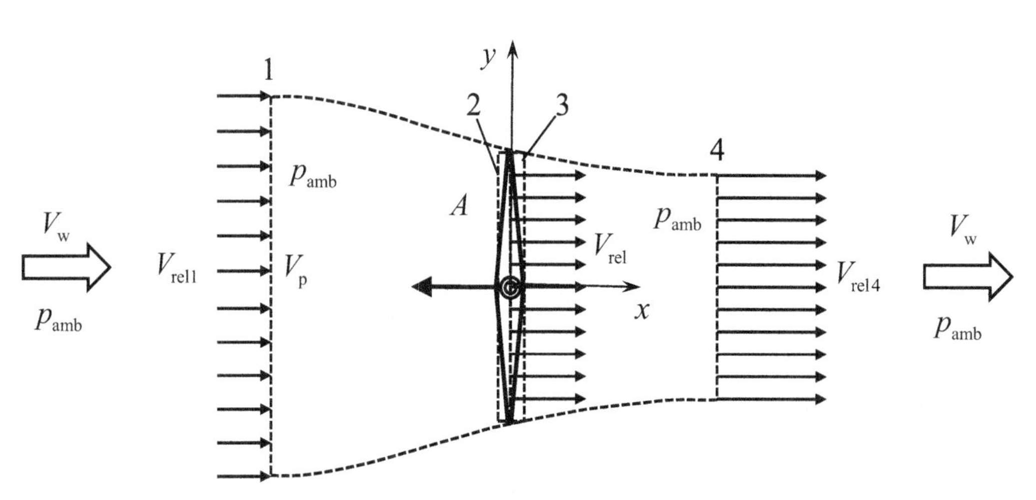

FIGURE 1.22 Propeller of an airplane flying at a constant velocity V_p (Example 1.15).

The two preceding equations yield:

$$p_3 - p_2 = \frac{1}{2}\rho V_{\text{rel}4}^2 - \frac{1}{2}\rho V_{\text{rel}1}^2$$

a. Propeller Thrust

We can evaluate the propeller thrust in two ways: first, using the force–momentum balance on the control volume between sections 1 and 4, and second, using the force–momentum balance on the control volume between sections 2 and 3. Using both of these options, we obtain:

$$F = \dot{m}\left(V_{\text{rel}4} - V_{\text{rel}1}\right) = \rho A V_{\text{rel}}\left(V_{\text{rel}4} - V_{\text{rel}1}\right) = A\left(p_3 - p_2\right)$$

b. Propeller Power Output

Propeller power$=FV_{\text{p}} = \dot{m}\left(V_{\text{rel}4} - V_{\text{rel}1}\right)V_{\text{p}} = \rho A V_{\text{rel}}\left(V_{\text{rel}4} - V_{\text{rel}1}\right)V_{\text{p}}$

c. Airflow Velocity through the Propeller

From the expression for the propeller thrust, we can deduce the following:

$$p_3 - p_2 = \rho V_{\text{rel}}\left(V_{\text{rel}4} - V_{\text{rel}1}\right)$$

For the total pressure remaining constant between sections 1 and 2 and between sections 3 and 4, we write:

$$p_3 - p_2 = \frac{1}{2}\rho V_{\text{rel}4}^2 - \frac{1}{2}\rho V_{\text{rel}}^2$$

The preceding two equations yield:

$$p_3 - p_2 = \rho V_{\text{rel}}\left(V_{\text{rel}4} - V_{\text{rel}1}\right) = \frac{1}{2}\rho V_{\text{rel}4}^2 - \frac{1}{2}\rho V_{\text{rel}1}^2$$

$$V_{\text{rel}} = \frac{V_{\text{rel}4} + V_{\text{rel}1}}{2}$$

which shows that the flow velocity through the propeller is the average of the velocities far ahead and far behind the propeller where the ambient pressure prevails.

d. Propeller Efficiency (η)

$$\eta = \frac{\text{Power output}}{\text{Power input}}$$

$$\eta = \frac{\dot{m}\left(V_{\text{rel}4} - V_{\text{rel}1}\right)V_{\text{p}}}{\dot{m}\frac{1}{2}\left(V_{\text{rel}4}^2 - V_{\text{rel}1}^2\right)} = \frac{V_{\text{p}}}{\frac{1}{2}\left(V_{\text{rel}4} + V_{\text{rel}1}\right)} = \frac{V_{\text{p}}}{V_{\text{rel}}}$$

If we know the air mass flow rate through the propeller, the flow velocity V_{rel} is easily determined as $V_{\text{rel}} = \dot{m}/(\rho A)$, giving:

$$\eta = \frac{V_{\text{p}}}{V_{\text{rel}}} = \frac{\rho A V_{\text{p}}}{\dot{m}}$$

which shows propeller efficiency is the ratio of propeller velocity to airflow velocity through the propeller. As the propeller efficiency is limited to 100% for an ideal propeller, its velocity is always less than that of the air flowing through it.

Example 1.16

Figure 1.23 shows a radially outward isentropic incompressible flow in a variable area rotating duct. For the given mass flow rate and other parameters shown in the figure, find the change in static pressure and total pressure (in the rotating reference frame) between section 1 (inlet) and section 2 (outlet). Neglect any shear stress on the duct wall and assume all properties are uniform over the cross section. The fluid density is ρ.

Solution

In this example, the incompressible flow through the rotating duct is isentropic (zero heat transfer with no frictional loss). As a result, the total pressure without rotation remains constant over the duct. The effect of rotation as a body force is to alter the static pressure distribution to balance itself over the duct control volume.

CONTINUITY EQUATION

Mass flow rate at section 1: $\dot{m} = A_1 \rho V_1$

Mass flow rate at section 2: $\dot{m} = A_2 \rho V_2$

Mass conservation (equal mass flow rate at the inlet and the outlet) yields:

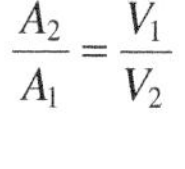

$$\frac{A_2}{A_1} = \frac{V_1}{V_2}$$

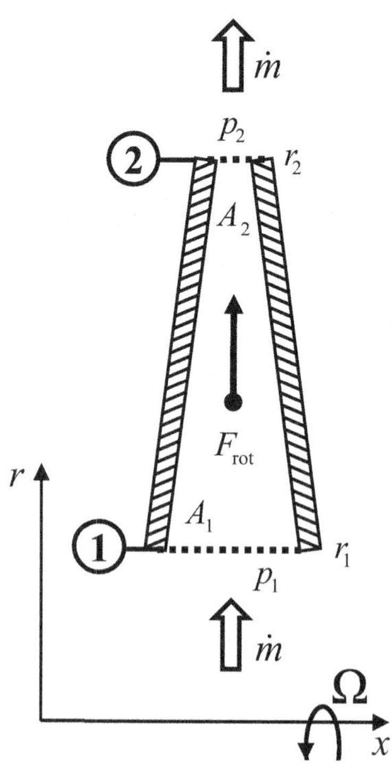

FIGURE 1.23 Radially outward flow through a rotating duct with variable area (Example 1.16).

MOMENTUM EQUATION

The linear momentum balance on the nonrotating duct control volume yields:

$$p_1 + \frac{1}{2}\rho V_1^2 = p_2 + \frac{1}{2}\rho V_2^2$$

$$\left(p_2 - p_1\right)_{\text{non-rotating}} = \frac{1}{2}\rho V_1^2 - \frac{1}{2}\rho V_2^2$$

When the duct rotates, we calculate the change in static pressure between section 1 and section 2 due to the rotational body force F_{rot} as:

$$\left(p_2 - p_1\right)_{\text{rotation}} = \int_{r_1}^{r_2} \rho\Omega^2 r dr = \frac{1}{2}\rho\Omega^2\left(r_2^2 - r_1^2\right)$$

Thus, we can write:

$$p_2 - p_1 = \left(p_2 - p_1\right)_{\text{non-rotating}} + \left(p_2 - p_1\right)_{\text{rotation}}$$

From the results obtained earlier, substituting the terms on the right-hand side of this equation yields:

$$p_2 - p_1 = \frac{1}{2}\rho V_1^2 - \frac{1}{2}\rho V_2^2 + \frac{1}{2}\rho\Omega^2\left(r_2^2 - r_1^2\right)$$

Rearranging the terms in this equation yields:

$$\left(p_2 + \frac{1}{2}\rho V_2^2\right) - \left(p_1 + \frac{1}{2}\rho V_1^2\right) = \frac{1}{2}\rho\Omega^2\left(r_2^2 - r_1^2\right)$$

$$p_{02} - p_{01} = \frac{1}{2}\rho\Omega^2\left(r_2^2 - r_1^2\right)$$

which shows that, between any two sections of a frictionless variable area rotating duct, the change in total pressure relative to the rotor equals the change in static pressure due to rotation only.

Example 1.17

Figure 1.24 shows an incompressible swirling flow in an abrupt pipe expansion with the upstream smaller pipe of diameter D_1 and the downstream larger pipe of D_2. The flow in the smaller pipe has a uniform swirl angle of 45°. The static pressure at the inlet pipe centerline is p_o. The flow exiting the larger pipe at section 2 is turbulent with uniform axial velocity with zero swirl and uniform static pressure p_2. Find the change in static pressure and total pressure between sections 1 and 2 of this flow system. Neglect any shear stress on the pipe walls.

Solution

Unlike Examples 1.11 and 1.14, this example has a strong swirl velocity in the inlet pipe. The main effect of the swirl component at the face of the sudden expansion is to yield nonuniform distributions of both axial velocity and static pressure. The swirl velocity in the incoming flow neither contributes to the mass flow rate nor contributes to the axial momentum flow rate through the control volume bounded by sections 1 and 2. If we assume the swirling flow to be a forced vortex in the smaller pipe, the tangential velocity varies linearly with radius: $V_\theta = 0$ at $r = 0$ and $V_\theta = V_{\theta\max}$

at $r = R_1$. As the swirl angle ($\tan^{-1}(V_\theta/V_1)$) is constant at 45°, the axial velocity in the smaller pipe also varies linearly with radius: $V_1 = 0$ at $r = 0$ and $V_1 = V_{1\max}$ at $r = R_1$. In equation form, we can express the profiles of tangential and axial velocities in the smaller pipe as:

$$V_\theta = V_{\theta\max}\frac{r}{R_1} \text{ and } V_1 = V_{1\max}\frac{r}{R_1}$$

where $R_1 = D_1/2$. At section 1, shown in Figure 1.24, we assume that the swirling flow behaves like a free vortex in the annulus region between the larger and smaller pipes. We can thus write the radial variation of the tangential velocity in this region as:

$$V_\theta = \frac{V_{\theta\max}R_1}{r} \text{ for } R_1 \le r \le R_2 \text{ where } R_2 = D_2/2.$$

CONTINUITY EQUATION

The mass flow rate entering the control volume through the inlet pipe is $\dot{m}_1 = A_1\rho\bar{V}_1$, where $\bar{V}_1$ is the average axial velocity through the pipe having a flow area A_1. Let us now evaluate the mass flow rate in the inlet pipe for the linear axial velocity profile as follows:

$$\dot{m}_1 = \int_0^{R_1} 2\pi\rho V_1 r\,dr = \int_0^{R_1} 2\pi\rho V_{1\max}\left(\frac{r}{R_1}\right) r\,dr$$

$$\dot{m}_1 = \frac{2}{3}\pi R_1^2\rho V_{1\max} = \frac{2}{3}A_1\rho V_{1\max} = A_1\rho\bar{V}_1$$

which yields $\bar{V}_1 = \frac{2}{3}V_{1\max}$.

The mass flow rate exiting the control volume through section 2 is $\dot{m}_2 = A_2\rho V_2$, where $A_2 = \pi D_2^2/4$ is the flow area of the larger pipe.

Equating $\dot{m}_1$ and $\dot{m}_2$ for a steady flow through the control volume yields:

$$\frac{A_2}{A_1} = \frac{\bar{V}_1}{V_2}$$

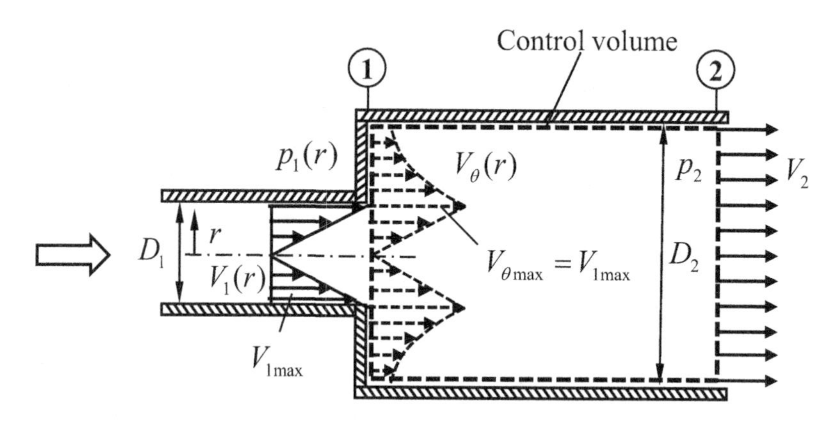

FIGURE 1.24 Swirling flow in an abrupt pipe expansion (Example 1.17).

MOMENTUM EQUATION

For applying the momentum equation to the control volume shown in Figure 1.24, let us first evaluate the axial momentum flow rate in the smaller pipe for the linear axial velocity profile as follows:

$$\dot{M}_1 = \int_0^{R_1} 2\pi\rho V_1^2 r\,\mathrm{d}r = \int_0^{R_1} 2\pi\rho V_{1\max}^2 \left(\frac{r}{R_1}\right)^2 r\,\mathrm{d}r$$

$$\dot{M}_1 = \frac{2\pi\rho V_{1\max}^2}{R_1^2}\int_0^{R_1} r^3\,\mathrm{d}r = \frac{1}{2}\rho A_1 V_{1\max}^2$$

Substituting $V_{1\max}$ in terms of $\overline{V}_1$ in this equation yields:

$$\dot{M}_1 = \frac{9}{8}\rho A_1 \overline{V}_1^2 = \beta\rho A_1 \overline{V}_1^2$$

which yields the momentum correction factor for the linear axial velocity profile in the inlet pipe as $\beta = 9/8$.

The rate of axial momentum outflow through section 2 is $\dot{M}_2 = \dot{m}_2 V_2$. Due to a forced vortex over the inlet pipe and a free vortex over the annulus area, the static pressure distribution in section 1 of the control volume is nonuniform. Let us first evaluate the total pressure force over this section.

We can use the radial equilibrium equation to compute the variation of static pressure in a vortex. Substituting the linear profile for the tangential velocity in the equation:

$$\frac{\mathrm{d}p_1}{\mathrm{d}r} = \frac{\rho V_\theta^2}{r}$$

for the forced vortex, we obtain:

$$\frac{\mathrm{d}p_1}{\mathrm{d}r} = \frac{\rho V_{\theta\max}^2}{r}\left(\frac{r}{R_1}\right)^2 = \frac{\rho V_{\theta\max}^2 r}{R_1^2}$$

whose integration over the inlet pipe yields:

$$p_1 = \frac{\rho V_{\theta\max}^2 r^2}{2R_1^2} + p_o$$

where p_o is the static pressure at the center of the inlet pipe. From this equation, we calculate the static pressure at $r = R_1$ as:

$$p_{R1} = p_o + \frac{\rho V_{\theta\max}^2}{2}$$

For the free vortex region, substituting the radial variation of the tangential velocity in the radial equilibrium equation, we obtain:

$$\frac{\mathrm{d}p_1}{\mathrm{d}r} = \frac{\rho V_\theta^2}{r} = \frac{\rho R_1^2 V_{\theta\max}^2}{r^3}$$

Integrating this equation yields static pressure variation from $r = R_1$ to $r = R_2$:

$$p_1 = -\frac{\rho V_{\theta\max}^2 R_1^2}{2r^2} + C$$

Using $p_1 = p_{R1} = p_o + \frac{\rho V_{\theta\max}^2}{2}$ at $r = R_1$ in this equation yields:

$$p_1 = p_o + \rho V_{\theta\max}^2 - \frac{\rho V_{\theta\max}^2 R_1^2}{2r^2} = p_o + \frac{\rho V_{\theta\max}^2}{2}\left(2 - \frac{R_1^2}{r^2}\right)$$

Using the aforementioned static pressure distributions, let us find the total pressure force at section 1 (inlet to the control volume):

$$F_{p1} = \int_0^{R_1}\left(\frac{\rho V_{\theta\max}^2 r^2}{2R_1^2} + p_o\right)2\pi r\,\mathrm{d}r + \int_{R_1}^{R_2}\left[p_o + \frac{\rho V_{\theta\max}^2}{2}\left(2 - \frac{R_1^2}{r^2}\right)\right]2\pi r\,\mathrm{d}r$$

$$= \pi R_2^2 p_o + 2\pi\int_0^{R_1}\frac{\rho V_{\theta\max}^2 r^3}{2R_1^2}\mathrm{d}r + 2\pi\int_{R_1}^{R_2}\frac{\rho V_{\theta\max}^2}{2}\left(2r - \frac{R_1^2}{r}\right)\mathrm{d}r$$

$$= \pi R_2^2 p_o + \frac{\pi}{4}\rho V_{\theta\max}^2 R_1^2 + 2\pi\int_{R_1}^{R_2}\frac{\rho V_{\theta\max}^2}{2}\left(2r - \frac{R_1^2}{r}\right)\mathrm{d}r$$

$$= \pi R_2^2 p_o + \frac{\pi}{4}\rho V_{\theta\max}^2 R_1^2 + \pi\rho V_{\theta_{\max}}^2 R_2^2 - \pi\rho V_{\theta\max}^2 R_1^2\left(1 + \ln\left(\frac{R_2}{R_1}\right)\right)$$

$$= \pi R_2^2 p_o + \pi\rho V_{\theta\max}^2 R_2^2 - \frac{3}{4}\pi\rho V_{\theta\max}^2 R_1^2 - \pi\rho V_{\theta\max}^2 R_1^2 \ln\left(\frac{R_2}{R_1}\right)$$

which yields the average static pressure at section 1 as:

$$\bar{p}_1 = \frac{\pi R_2^2 p_o + \pi\rho V_{\theta\max}^2 R_2^2 - \frac{3}{4}\pi\rho V_{\theta\max}^2 R_1^2 - \pi\rho V_{\theta\max}^2 R_1^2 \ln\left(\frac{R_2}{R_1}\right)}{\pi R_2^2}$$

$$\bar{p}_1 = p_o + \rho V_{\theta\max}^2 - \frac{3}{4}\rho V_{\theta\max}^2\left(\frac{R_1}{R_2}\right)^2 - \rho V_{\theta\max}^2\left(\frac{R_1}{R_2}\right)^2 \ln\left(\frac{R_2}{R_1}\right)$$

$$\bar{p}_1 = p_o + \rho V_{\theta\max}^2\left[1 - \left(\frac{R_1}{R_2}\right)^2\left\{\frac{3}{4} + \ln\left(\frac{R_2}{R_1}\right)\right\}\right]$$

The force–momentum balance on the control volume yields:

$$(\bar{p}_1 - p_2)A_2 = \dot{M}_2 - \dot{M}_1 = \dot{m}_2\left(V_2 - \frac{9}{8}\bar{V}_1\right) = \rho A_1\bar{V}_1\left(V_2 - \frac{9}{8}\bar{V}_1\right)$$

$$(p_2 - \bar{p}_1) = \rho\bar{V}_1^2\left(\frac{A_1}{A_2}\right)\left(\frac{9}{8} - \frac{A_1}{A_2}\right)$$

$$C_p = \frac{(p_2 - \bar{p}_1)}{\frac{1}{2}\rho\bar{V}_1^2} = 2\left(\frac{A_1}{A_2}\right)\left(\frac{9}{8} - \frac{A_1}{A_2}\right)$$

where the pressure rise (recovery) coefficient C_p is defined in terms of the dynamic pressure based on the average inlet velocity $\bar{V}_1$. Note that C_p reaches a maximum value of 81/128 when the area ratio (A_1/A_2) is 9/16.

We compute the change in total pressure between sections 1 and 2 as follows:

$$\text{Total pressure at section 1: } \bar{p}_{01} = \bar{p}_1 + \frac{1}{2}\alpha\rho\bar{V}_1^2 + \frac{\int_0^{R_1}\left(\frac{1}{2}\rho V_\theta^2\right)\left(2\pi\rho V_1 r\,dr\right)}{\pi R_1^2 \rho \bar{V}_1}$$

where α is the kinetic energy correction factor for the linear axial velocity profile and the third term on the right-hand side is the mass-averaged dynamic pressure associated with the tangential velocity in the pipe.

We calculate the kinetic energy correction factor α as follows ($\dot{K}_1 \equiv$ the flow rate of kinetic energy through section 1):

$$\dot{K}_1 = \int_0^{R_1} 2\pi\rho V_1\left(\frac{1}{2}V_1^2\right)r\,dr = \int_0^{R_1} \pi\rho V_{1\max}^3\left(\frac{r}{R_1}\right)^3 r\,dr$$

$$= \frac{\pi\rho V_{1\max}^3}{R_1^3}\int_0^{R_1} r^4\,dr = \frac{\pi R_1^2\rho V_{1\max}^3}{5} = \frac{1}{5}A_1\rho\left(\frac{3}{2}\right)^3\bar{V}_1^3$$

$$= \frac{27}{40}\left(A_1\rho\bar{V}_1\right)\bar{V}_1^2 = \frac{27}{20}\dot{m}\left(\frac{1}{2}\bar{V}_1^2\right) = \alpha\,\dot{m}\left(\frac{1}{2}\bar{V}_1^2\right)$$

which gives the kinetic energy correction factor for the linear axial velocity profile in the inlet pipe as $\alpha = 27/20$. As both axial and tangential velocities are equal over the inlet pipe cross section, their corresponding mass-averaged values of the dynamic pressure must be identical, as can be verified by evaluating the integral:

$$\frac{\int_0^{R_1}\left(\frac{1}{2}\rho V_\theta^2\right)\left(2\pi\rho V_1 r\,dr\right)}{\pi R_1^2\rho\bar{V}_1} = \frac{27}{40}\rho\bar{V}_1^2$$

giving:

$$\bar{p}_{01} = \bar{p}_1 + \frac{27}{40}\rho\bar{V}_1^2 + \frac{27}{40}\rho\bar{V}_1^2 = \bar{p}_1 + \frac{27}{20}\rho\bar{V}_1^2$$

We compute the total pressure at section 2 as:

$$p_{02} = p_2 + \frac{1}{2}\rho V_2^2$$

$$\bar{p}_{01} - p_{02} = \bar{p}_1 + \frac{27}{20}\rho\bar{V}_1^2 - p_2 - \frac{1}{2}\rho V_2^2 = \frac{27}{20}\rho\bar{V}_1^2 - \frac{1}{2}\rho V_2^2 - \left(p_2 - \bar{p}_1\right)$$

Substituting $\left(p_2 - \bar{p}_1\right)$ in this equation yields:

$$\bar{p}_{01} - p_{02} = \frac{27}{20}\rho\bar{V}_1^2 - \frac{1}{2}\rho V_2^2 - \rho\bar{V}_1^2\left(\frac{A_1}{A_2}\right)\left(\frac{9}{8} - \frac{A_1}{A_2}\right)$$

With the substitution of $V_2 / \bar{V}_1$ for A_1 / A_2, this equation becomes:

$$\bar{p}_{01} - p_{02} = \frac{27}{20}\rho\bar{V}_1^2 - \frac{1}{2}\rho V_2^2 - \rho\bar{V}_1^2\left(\frac{V_2}{\bar{V}_1}\right)\left(\frac{9}{8} - \frac{V_2}{\bar{V}_1}\right)$$

$$= \frac{27}{20}\rho\bar{V}_1^2 - \frac{1}{2}\rho V_2^2 - \frac{9}{8}\rho\bar{V}_1 V_2 + \rho V_2^2$$

$$\bar{p}_{01} - p_{02} = \frac{27}{20}\rho\bar{V}_1^2 - \frac{9}{8}\rho\bar{V}_1 V_2 + \frac{1}{2}\rho V_2^2$$

Expressing this result in terms of the loss coefficient K, we obtain:

$$K = \frac{(\bar{p}_{01} - p_{02})}{\frac{1}{2}\rho\bar{V}_1^2} = \frac{27}{10} - \frac{9}{4}\left(\frac{A_1}{A_2}\right) + \left(\frac{A_1}{A_2}\right)^2$$

When the downstream pipe diameter is much larger than the diameter of the upstream pipe, the loss coefficient equals 27/10. Note that the value of K obtained here is greater than the value of $K=2$ obtained in Example 1.14 with the parabolic inlet axial velocity profile because of the additional loss of dynamic pressure associated with the incoming swirl velocity, both velocities varying linearly over the inlet pipe cross section.

Example 1.18

For the system, in Example 1.12, assume that no braking force acts on the cart and that it is free to move due to the incoming water jet having velocity V_{jet}. The curved plate mounted on the cart deflects the jet, which leaves the plate at an angle θ with the x axis. The cross-sectional area A_{jet} of the jet remains constant over the curved plate. If the cart of total mass m_{cart} is initially at rest, find its acceleration and velocity as a function of time. Neglect the mass of the water jet within the noninertial control volume encompassing the cart and the curved plate. The density of water is ρ.

Solution

The equal mass flow rate entering and exiting the control volume satisfies the continuity equation, and the total mass of the cart control volume remains constant at m_{cart}. As the cart accelerates from rest, the coordinate axes attached to the cart become noninertial. The uniform static ambient pressure around the cart control volume applies no pressure force on it.

We compute the mass flow rate as:

$$\dot{m}_{\text{in}} = \dot{m}_{\text{out}} = A_{\text{jet}}\rho\left(V_{\text{jet}} - U_{\text{cart}}\right)$$

Assuming that the entire cart control volume accelerates uniformly and that there is negligible change in the momentum of the fluid within the control volume relative to the cart, we write the momentum equation in this case as:

$$-m_{\text{cart}}a_{\text{cart}} = \dot{m}_{\text{out}}V_{x\_\text{out}} - \dot{m}_{\text{in}}V_{x\_\text{in}}$$

Note that we can express the momentum velocities in the x direction in this equation as:

$$V_{x\_\text{in}} = V_{\text{jet}} - U_{\text{cart}}$$

$$V_{x\_\text{out}} = (V_{\text{jet}} - U_{\text{cart}})\cos\theta$$

giving:

$$m_{\text{cart}}a_{\text{cart}} = A\rho(V_{\text{jet}} - U_{\text{cart}})(V_{\text{jet}} - U_{\text{cart}}) - A\rho(V_{\text{jet}} - U_{\text{cart}})(V_{\text{jet}} - U_{\text{cart}})\cos\theta$$

$$a_{\text{cart}} = \frac{A\rho(V_{\text{jet}} - U_{\text{cart}})^2(1-\cos\theta)}{m_{\text{cart}}}$$

As the cart velocity U_{cart} varies with time, this equation for the cart acceleration is also time-dependent.

We obtain the cart velocity by integrating its acceleration as follows:

$$a_{\text{cart}} = \dot{U}_{\text{cart}} = \frac{A\rho(V_{\text{jet}} - U_{\text{cart}})^2(1-\cos\theta)}{m_{\text{cart}}}$$

For a constant V_{jet}, we can write this equation as:

$$-\frac{\mathrm{d}}{\mathrm{d}t}\left(V_{\text{jet}} - U_{\text{cart}}\right) = \frac{A\rho(V_{\text{jet}} - U_{\text{cart}})^2(1-\cos\theta)}{m_{\text{cart}}}$$

Separating variables in this equation, we obtain:

$$-\int_{V_{\text{jet}}}^{V_{\text{jet}}-U_{\text{cart}}} \frac{\mathrm{d}(V_{\text{jet}} - U_{\text{cart}})}{(V_{\text{jet}} - U_{\text{cart}})^2} = \int_0^t \frac{A\rho(1-\cos\theta)}{m_{\text{cart}}}\mathrm{d}t$$

$$\frac{1}{(V_{\text{jet}} - U_{\text{cart}})} - \frac{1}{V_{\text{jet}}} = \frac{A\rho(1-\cos\theta)}{m_{\text{cart}}}t$$

$$\frac{U_{\text{cart}}}{V_{\text{jet}}} = \frac{\tau}{1+\tau}$$

where the dimensionless time τ is given by:

$$\tau = \frac{A\rho V_{\text{jet}}(1-\cos\theta)}{m_{\text{cart}}}t$$

which, during a finite time from rest, reveals that the cart velocity U_{cart} always remains less than the water jet velocity V_{jet}. When $\tau \to \infty$, the cart velocity asymptotically attains the jet velocity at which point the cart acceleration becomes zero.

Example 1.19

As shown in Figure 1.25, a rocket is fired into outer space (vacuum) with negligible friction and gravitational body forces. The initial mass of the rocket at launch is m_o, and the constant mass flow rate of exhaust gases is $\dot{m}_e$. The flow area and the static pressure at the exit of the rocket nozzle are A_e and p_e, respectively. Find the rocket velocity U_R in the inertial reference frame XYZ. The noninertial coordinate axes xyz are attached to the rocket control volume.

Solution

Because the mass of the material within the rocket control volume in this example constantly decreases with time, it is sometimes called a variable mass problem. Note that the external motion of the rocket does not influence its interior ballistics, resulting in constant static pressure, velocity, and density of the exhaust gases at the rocket nozzle exit.

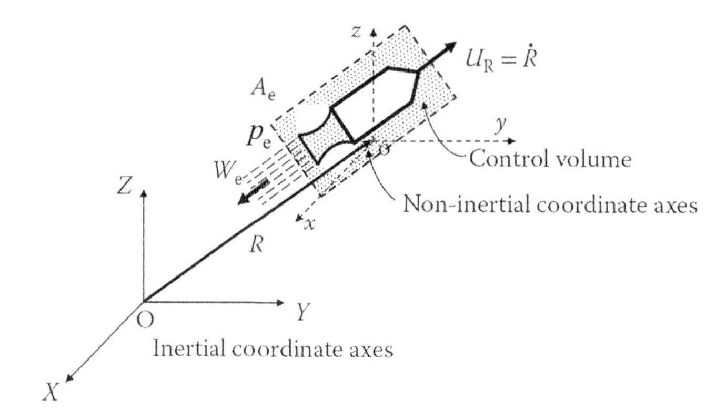

FIGURE 1.25 An accelerating rocket in outer space (Example 1.19).

CONTINUITY EQUATION

Applying the unsteady form of the continuity equation to the rocket control volume shown in Figure 1.25, we obtain:

$$\frac{\partial}{\partial t}\iiint_{\mathrm{cv}} \rho \, \mathrm{d}\forall = -\dot{m}_e$$

$$\frac{\mathrm{d}m}{\mathrm{d}t} = -\dot{m}_e$$

With $m = m_o$ at $t = 0$, the integration of this equation yields the time-dependent mass within the rocket control volume as:

$$m = m_o - \dot{m}_e t$$

MOMENTUM EQUATION IN THE NONINERTIAL REFERENCE FRAME

The only surface force acting on the control volume shown in Figure 1.25 is due to the static pressure p_e at the rocket nozzle exit area A_e. The left-hand side of the applicable momentum equation (Equation 3.34) becomes $p_e A_e - m\dot{U}_R$. The surface integral, which represents the net rate of momentum outflow from the noninertial control volume, reduces to $-\dot{m}_e W_e$ with the momentum velocity $-W_e$.

A careful examination of the unsteady term on the right-hand side of Equation 3.34 of Sultanian (2015) reveals that the linear momentum of the unburned fuel in the rocket relative to the noninertial coordinate system attached to the rocket is zero. In addition, the nonzero linear momentum associated with some of the exhaust gases within the control volume remains constant with time. The unsteady term does not contribute to the noninertial momentum equation from both considerations. Thus, the governing momentum equation in this problem becomes:

$$p_e A_e - m\dot{U}_R = -\dot{m}_e W_e$$

$$m\dot{U}_R = p_e A_e + \dot{m}_e W_e = S_{T_e}$$

where S_{T_e} is the stream thrust at the rocket nozzle exit, and it equals the rocket thrust measured by a load cell when the rocket is tested under vacuum (zero ambient pressure) conditions.

Substituting the instantaneous rocket mass m (obtained from the continuity equation) finally yields:

$$(m_o - \dot{m}_e t)\frac{dU_R}{dt} = S_{T_e}$$

Separating variables in this equation, we obtain:

$$\frac{dU_R}{S_{T_e}} = \frac{dt}{(m_o - \dot{m}_e t)}$$

Using the initial condition $U_R = 0$ at $t = 0$, we obtain the time-dependent rocket velocity measured in the inertial reference frame as:

$$U_R = \frac{S_{T_e}}{\dot{m}_e}\ln\left(\frac{m_o}{(m_o - \dot{m}_e t)}\right)$$

MOMENTUM EQUATION IN THE INERTIAL REFERENCE FRAME

In the preceding section, we used the noninertial control volume analysis to obtain the solution for the time-dependent rocket velocity measured in the inertial reference frame. We now demonstrate that we can obtain the same result by performing the rocket control volume analysis entirely in the inertial reference frame by applying Equation 1.20. In this analysis, we use velocities in the inertial reference frame. Again, in this case, the only surface force acting on the control volume is due to the static pressure p_e at the rocket nozzle exit area A_e. The left-hand side of Equation 1.20 becomes $p_e A_e$. The surface integral, representing the net outflow of momentum from the control volume, reduces to only one outflow $\dot{m}_e V_e$ with the momentum velocity V_e. We can write the rocket exhaust velocity V_e measured in the inertial reference frame as:

$$V_e = -W_e + U_R$$

which corresponds to the momentum velocity at the rocket nozzle exit. The net rate of momentum outflow from the control volume thus becomes $\dot{m}_e(U_R - W_e)$.

The unsteady integral term on the right-hand side of Equation 1.20 needs careful examination and evaluation. The total instantaneous mass within the rocket control volume consists of burned and unburned fuel and structural mass. Let us assume that the mass of the burned fuel within the rocket control volume remains constant at m_b with its velocity $-W_e$ relative to the rocket. We can thus write the instantaneous momentum of the entire mass of the rocket control volume in the inertial reference frame as:

$$(m - m_b)U_R + m_b(U_R - W_e)$$

$$mU_R - m_b W_e$$

We can write the unsteady term in Equation 1.20 as:

$$\frac{\partial}{\partial t_{XYZ}}\iiint_{cv} V_{XYZ}(\rho\, d\forall) = \frac{d}{dt}(mU_R - m_b W_e) = \frac{d}{dt}(mU_R) = U_R\frac{dm}{dt} + m\frac{dU_R}{dt}$$

where we have used the fact that the term $m_b W_e$ remains constant. Substituting m from the continuity equation, the unsteady term finally becomes:

$$\frac{\partial}{\partial t_{XYZ}}\iiint_{cv} V_{XYZ}(\rho\, d\forall) = -\dot{m}_e U_R + (m_o - \dot{m}_e t)\frac{dU_R}{dt}$$

Substituting this in the momentum equation in the inertial reference frame yields:

$$p_e A_e = \dot{m}_e (U_R - W_e) - \dot{m}_e U_R + (m_o - \dot{m}_e t)\frac{dU_R}{dt}$$

$$(m_o - \dot{m}_e t)\frac{dU_R}{dt} = p_e A_e + \dot{m}_e W_e = S_{T_e}$$

where S_{T_e} is the stream thrust at the rocket nozzle exit. This equation is identical to the one we obtained in the preceding section using the noninertial control volume analysis, yielding the following solution for the time-dependent rocket velocity:

$$U_R = \frac{S_{T_e}}{\dot{m}_e} \ln\left(\frac{m_o}{(m_o - \dot{m}_e t)}\right)$$

Example 1.20

As shown in Figure 1.26, a water jet of constant velocity V_{jet} and cross-sectional area A_{jet} is entering a water tanker that is free to move with negligible friction. Assuming the tanker to be initially at rest with mass m_o, find its acceleration $\dot{U}_{tanker}$, velocity U_{tanker}, and mass m as a function of time under the influence of the incoming water jet.

Solution

In Example 1.19, the accelerating rocket loses mass at a constant rate. In this example, the water tanker gains mass at a decreasing rate until it asymptotically reaches the jet velocity.

CONTINUITY EQUATION

Applying the unsteady form of the continuity equation to the tanker control volume shown in Figure 1.26, we obtain:

$$\frac{\partial}{\partial t}\iiint_{cv} \rho \, d\forall = \rho W_{jet} A_{jet}$$

$$\frac{dm}{dt} = \rho W_{jet} A_{jet}$$

where W_{jet} is the jet velocity relative to the noninertial coordinate axes attached to the tanker and can be written as:

$$W_{jet} = V_{jet} - U_{tanker}$$

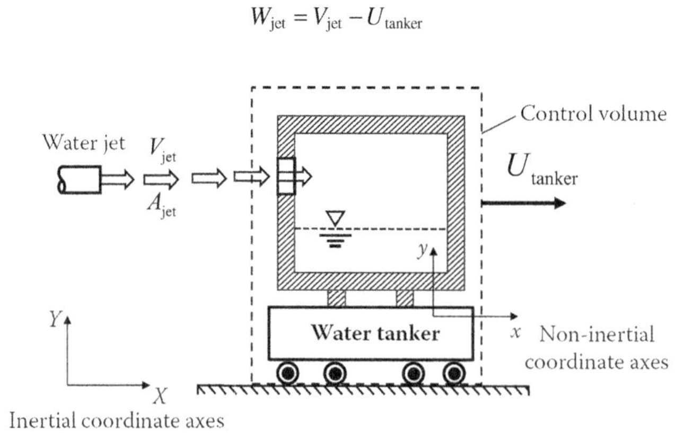

FIGURE 1.26 An accelerating water tanker with an incoming water jet (Example 1.20).

which for constant V_{jet} gives:

$$\dot{W}_{jet} = -\dot{U}_{tanker}$$

MOMENTUM EQUATION IN THE NONINERTIAL REFERENCE FRAME

No surface and body forces are acting on the tanker control volume, as shown in Figure 1.26. The left-hand side of the applicable momentum equation becomes $-(m + m_{jet})\dot{U}_{tanker}$. The surface integral, representing the net rate of linear momentum outflow from the noninertial control volume, reduces to only one inflow with the linear momentum velocity W_{jet}. The resulting rate of linear momentum outflow equals $-\dot{m}W_{jet}$.

The unsteady term on the right-hand side of Equation 3.35 of Sultanian (2015) needs careful evaluation. Let us assume that the water jet entering the tanker has zero velocity relative to the tanker, instantly taking the tanker's velocity U_{tanker}. The total instantaneous mass within the tanker control volume consists of the tanker with water with total mass m and a part of the water jet with constant mass m_{jet} and velocity V_{jet} before it enters the tanker. We can write the instantaneous momentum of the entire mass of the tanker control volume in the noninertial reference frame as $m_{jet}\left(V_{jet} - U_{tanker}\right)$.

The unsteady term in Equation 3.35 of Sultanian (2015) becomes:

$$\frac{\partial}{\partial t_{xyz}} \iiint_{cv} V_{xyz}\left(\rho\, d\mathcal{V}\right) = \frac{d}{dt}\left(m_{jet}V_{jet} - m_{jet}U_{tanker}\right) = -\frac{d}{dt}\left(m_{jet}U_{tanker}\right) = -m_{jet}\dot{U}_{tanker}$$

where we have used the fact that both m_{jet} and V_{jet} remain constant. The governing momentum equation becomes:

$$-(m + m_{jet})\dot{U}_{tanker} = -\dot{m}W_{jet} - m_{jet}\dot{U}_{tanker}$$

$$-m\dot{U}_{tanker} = -\dot{m}W_{jet}$$

Substituting $\dot{W}_{jet} = -\dot{U}_{tanker}$ in this equation from the continuity equation yields:

$$m\dot{W}_{jet} = -\dot{m}W_{jet}$$

$$\frac{\dot{m}}{m} = -\frac{\dot{W}_{jet}}{W_{jet}}$$

Let's now derive this equation using the control volume analysis in the inertial reference frame.

MOMENTUM EQUATION IN THE INERTIAL REFERENCE FRAME

With no surface and body forces acting on the tanker control volume, the left-hand side of Equation 1.20 becomes zero. We express the water jet velocity V_{jet} in the inertial reference frame as:

$$V_{jet} = W_{jet} + U_{tanker}$$

which corresponds to the momentum velocity at the inlet to the water tanker. The net momentum outflow rate from the control volume thus becomes $-\dot{m}V_{jet}$.

The unsteady term on the right-hand side of Equation 1.20 needs a careful evaluation. The total instantaneous mass within the tanker control volume consists of the tanker with water with total mass m and a part of the water jet with constant m_{jet} and velocity V_{jet} before it enters the tanker. Let us assume that the water jet entering the tanker instantly takes the tanker velocity U_{tanker}. We can write the instantaneous momentum of the entire mass of the tanker control volume in the inertial reference frame as $mU_{tanker} + m_{jet}V_{jet}$. The unsteady term in Equation 1.20 becomes:

$$\frac{\partial}{\partial t_{XYZ}} \iiint_{cv} V_{XYZ}(\rho\, d\forall) = \frac{d}{dt}(mU_{tanker} + m_{jet}V_{jet}) = \frac{d}{dt}(mU_{tanker}) = U_{tanker}\frac{dm}{dt} + m\frac{dU_{tanker}}{dt}$$

where $m_{jet}V_{jet}$ is constant. Equation 1.20 reduces to:

$$0 = -\dot{m}V_{jet} + U_{tanker}\dot{m} + m\dot{U}_{tanker}$$

$$\dot{m}(V_{jet} - U_{tanker}) = m\dot{U}_{tanker}$$

$$\dot{m}W_{jet} = -m\dot{W}_{jet}$$

$$\frac{\dot{m}}{m} = -\frac{\dot{W}_{jet}}{W_{jet}}$$

which is identical to the equation we obtained earlier using the noninertial control volume analysis. When the water tanker is at rest, we have $m = m_o$ and $W_{jet} = V_{jet}$. The integration of this equation yields:

$$\int_{m_o}^{m} \frac{dm}{m} = -\int_{V_{jet}}^{W_{jet}} \frac{dW_{jet}}{W_{jet}}$$

$$\ln\left(\frac{m}{m_o}\right) = -\ln\left(\frac{W_{jet}}{V_{jet}}\right) = \ln\left(\frac{V_{jet}}{W_{jet}}\right)$$

In this equation, equating expressions under the natural log on both sides and rearranging terms, we obtain:

$$W_{jet} = \frac{m_o V_{jet}}{m}$$

Substituting this in the continuity equation ($dm/dt = \rho W_{jet} A_{jet}$) yields:

$$\frac{dm}{dt} = \rho A_{jet} \frac{m_o V_{jet}}{m}$$

$$m dm = (\rho A_{jet} V_{jet}) m_o dt = \dot{m}_{jet} m_o dt$$

where $\dot{m}_{jet} = \rho A_{jet} V_{jet}$ is the constant water jet mass flow rate exiting the nozzle. Integration of this equation gives:

$$\int_{m_o}^{m} m\, dm = \int_{0}^{t} \dot{m}_{jet} m_o\, dt$$

$$\frac{m^2 - m_o^2}{2} = \dot{m}_{jet} m_o t$$

$$m^2 = m_o^2 + 2\dot{m}_{jet} m_o t$$

$$m = \left(m_o^2 + 2\dot{m}_{jet} m_o t\right)^{1/2}$$

$$m = m_o\left(1 + 2\zeta t\right)^{1/2}$$

This equation gives the variation of tanker mass with time, where

$$\zeta = \frac{\dot{m}_{jet}}{m_o} = \frac{\rho A_{jet} V_{jet}}{m_o}$$

Substituting the time-dependent variation of the tanker mass m into the earlier obtained equation for the jet velocity W_{jet} relative to the tanker yields:

$$W_{jet} = \frac{m_o V_{jet}}{m} = V_{jet}\left(1 + 2\zeta t\right)^{-1/2}$$

We can now write the time-varying tanker velocity as:

$$U_{tanker} = V_{jet} - W_{jet} = V_{jet} - V_{jet}\left(1 + 2\zeta t\right)^{-1/2}$$

$$U_{tanker} = V_{jet}\left\{1 - \left(1 + 2\zeta t\right)^{-1/2}\right\}$$

Differentiating this equation with time yields the following expression for the tanker acceleration:

$$\dot{U}_{tanker} = V_{jet}\zeta\left(1 + 2\zeta t\right)^{-3/2}$$

Example 1.21

As shown in Figure 1.27, a high-pressure rotary arm is used for impingement air cooling of a cylindrical surface. The total pressure and the total temperature of the air inside the rotary arm are 4.7 bar and 27 °C, respectively. The static pressure outside the rotary arm is 1 bar and that at each nozzle exit is 248292 Pa. Note that each nozzle is choked at the exit with a jet velocity of 316.938 m/s and a mass flow rate of 0.0861 kg/s. At the maximum RPM, the rotary arm needs to overcome a frictional torque of 1.4 Nm. For the given geometric data, calculate the maximum RPM of the rotary arm. Assume that the given total pressure and total temperature are relative to the arm and independent of arm rotation. Assume $\gamma = 1.4$ and $R = 287$ J/(kg K) for air. The following are the given geometric data: jet diameter $(d_j) = 10$ mm, $R_1 = 40$ cm, $R_2 = 50$ cm, and $a = 20$ cm.

Solution

In this example, the rotary arm control volume is rotating at a constant speed around a single axis of rotation. We can thus use Equation 1.23 to carry out the torque and angular momentum balance. We can write the rate of angular momentum outflow at location 1 (in the counterclockwise direction) as:

$$\dot{H}_{j1} = \dot{m}a(a\Omega - W_j) + \dot{m}R_1(R_1\Omega - 0)$$

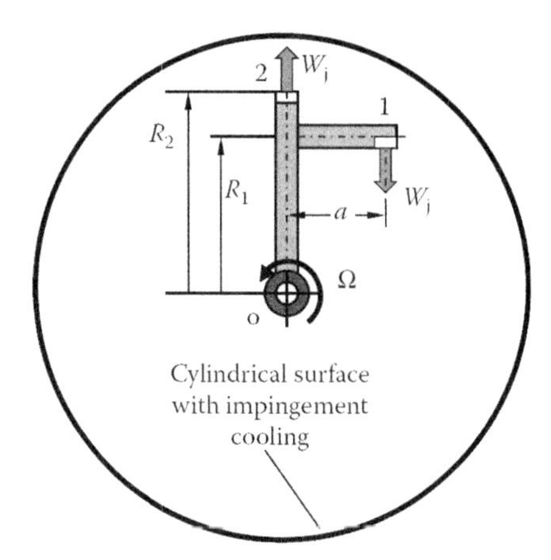

FIGURE 1.27 Impingement air cooling of a cylindrical surface with a rotary arm (Example 1.21).

and at location 2 (in the counterclockwise direction) as:

$$\dot{H}_{j2} = \dot{m}R_2(R_2\Omega - 0)$$

In this equation, we have used the absolute jet velocity $R_2\Omega$, which is in the tangential direction, whereas the relative velocity W_j is in the radial direction, contributing zero moment about the origin—an error often made in solving this problem.

We write the torque due to pressure force at location 1 (in the counterclockwise direction) as: $A_j a(p^* - p_{amb})$, where $p^* = 248292\ \text{Pa}$.

Because the arm is rotating in the counterclockwise direction, the frictional torque acts on the control volume in the clockwise direction. Note that the pressure force acting radially at location 2 does not produce any torque on the control volume, and the inflow of angular momentum at the origin is zero. Thus, using Equation 1.23, we obtain:

$$-\Gamma_f + A_j a(p^* - p_{amb}) = \dot{H}_{j1} + \dot{H}_{j2}$$

$$-\Gamma_f + A_j a(p^* - p_{amb}) = \dot{m}R_1(R_1\Omega - 0) + \dot{m}R_2(R_2\Omega - 0) + \dot{m}a(a\Omega - W_j)$$

$$-\Gamma_f + A_j a(p^* - p_{amb}) = \dot{m}(a^2\Omega - aW_j + R_1^2\Omega + R_2^2\Omega)$$

$$-\Gamma_f + A_j a(p^* - p_{amb}) = \dot{m}\Omega(a^2 + R_1^2 + R_2^2) - \dot{m}aW_j$$

$$\Omega = \frac{\dot{m}aW_j - \Gamma_f + A_j a(p^* - p_{amb})}{\dot{m}(a^2 + R_1^2 + R_2^2)}$$

$$\Omega = \frac{0.0861 \times 0.2 \times 316.938 - 1.4 + 2.329}{0.0861(0.2^2 + 0.40^2 + 0.50^2)} = 164.848\ \text{rad/s}$$

Thus, the rotary arm rotates at the maximum of 1,574.184 rpm.

In problems like this with multiple inflows and outflows, the calculation of angular momentum flows is often error-prone if done by inspection. A more formal approach using vector quantities, outlined here for this example, will help prevent such errors.

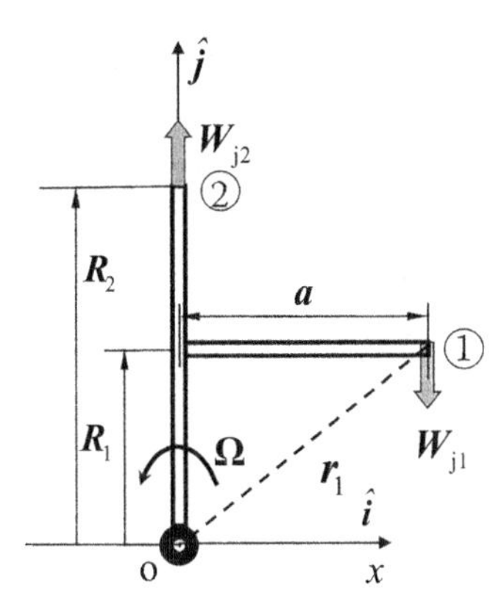

FIGURE 1.28 Vector analysis to calculate angular momentum outflows in Example 1.21.

Figure 1.28 depicts various quantities of Example 1.21 in vector form. Using unit vectors along coordinate axes, where $\hat{\boldsymbol{k}}$ is the unit vector along the z axis not shown in the figure, we can express these vectors as:

$$\boldsymbol{\Omega} = \Omega\,\hat{\boldsymbol{k}} \qquad \boldsymbol{R}_1 = R_1\hat{\boldsymbol{j}} \qquad \boldsymbol{R}_2 = R_2\hat{\boldsymbol{j}}$$

$$\boldsymbol{a} = a\hat{\boldsymbol{i}} \qquad \boldsymbol{W}_{j1} = -W_j\hat{\boldsymbol{j}} \qquad \boldsymbol{W}_{j2} = W_j\hat{\boldsymbol{j}}$$

$$\boldsymbol{r} = a\hat{\boldsymbol{i}} + R_1\hat{\boldsymbol{j}}$$

We evaluate various quantities needed to compute the flow of angular momentum through each outlet as follows:

RATE OF JET ABSOLUTE ANGULAR MOMENTUM OUTFLOW AT OUTLET 1

Rotary arm rotation vector: $\boldsymbol{\Omega} = \Omega\hat{\boldsymbol{k}}$

Jet velocity relative to the rotary arm: $\boldsymbol{W}_{j1} = -W_j\hat{\boldsymbol{j}}$

Radius vector for the jet exit: $\boldsymbol{r}_1 = a\hat{\boldsymbol{i}} + R_1\hat{\boldsymbol{j}}$

Rotary arm tangential velocity at jet exit: $\boldsymbol{U}_1 = \boldsymbol{\Omega} \times \boldsymbol{r}_1$

$$= \Omega\hat{\boldsymbol{k}} \times (a\hat{\boldsymbol{i}} + R_1\hat{\boldsymbol{j}})$$

$$= \Omega\, a(\hat{\boldsymbol{k}} \times \hat{\boldsymbol{i}}) + \Omega R_1(\hat{\boldsymbol{k}} \times \hat{\boldsymbol{j}})$$

$$= \Omega\, a\hat{\boldsymbol{j}} - \Omega R_1\hat{\boldsymbol{i}}$$

Absolute jet velocity: $\boldsymbol{V}_{j1} = \boldsymbol{U}_1 + \boldsymbol{W}_{j1} = (\Omega a\hat{\boldsymbol{j}} - \Omega R_1\hat{\boldsymbol{i}}) - W_j\hat{\boldsymbol{j}} = -\Omega R_1\hat{\boldsymbol{i}} + (\Omega\, a - W_j)\hat{\boldsymbol{j}}$

Jet absolute angular momentum outflow rate: $\dot{\boldsymbol{H}}_{j1} = \dot{m}\boldsymbol{r}_1 \times \boldsymbol{V}_{j1}$

$$= \dot{m}(a\hat{\boldsymbol{i}} + R_1\hat{\boldsymbol{j}}) \times \left\{-\Omega R_1\hat{\boldsymbol{i}} + (\Omega a - W_j)\hat{\boldsymbol{j}}\right\}$$

$$= \dot{m}(a^2\Omega - aW_j + R_1^2\Omega)\hat{\boldsymbol{k}}$$

RATE OF JET ABSOLUTE ANGULAR MOMENTUM OUTFLOW AT OUTLET 2

Rotary arm rotation vector: $\boldsymbol{\Omega} = \Omega\hat{\boldsymbol{k}}$

Jet velocity relative to the rotary arm: $W_{j2} - W_j\hat{\boldsymbol{j}}$

Radius vector for the jet exit: $\boldsymbol{R}_2 = R_2\hat{\boldsymbol{j}}$

Rotary arm tangential velocity at jet exit: $U_2 = \Omega \times \boldsymbol{R}_2 = \Omega\hat{\boldsymbol{k}} \times R_2\hat{\boldsymbol{j}} = -R_2\Omega\hat{\boldsymbol{i}}$

Absolute jet velocity: $\boldsymbol{V}_{j2} = \boldsymbol{U}_2 + \boldsymbol{W}_{j2} = -\Omega R_2\hat{\boldsymbol{i}} + W_j\hat{\boldsymbol{j}}$

The rate of jet absolute angular momentum outflow: $\dot{\boldsymbol{H}}_{j2} = \dot{m}\boldsymbol{R}_2 \times \boldsymbol{V}_{j2}$

$$= \dot{m}R_2\hat{\boldsymbol{j}} \times (-\Omega R_2\hat{\boldsymbol{i}} + W_j\hat{\boldsymbol{j}})$$

$$= \dot{m}R_2^2\Omega\hat{\boldsymbol{k}}$$

The net rate of angular momentum outflow becomes:

$$\boldsymbol{H}_{j\_out} = \dot{\boldsymbol{H}}_{j1} + \dot{\boldsymbol{H}}_{j2} = \dot{m}(a^2\Omega - aW_j + R_1^2\Omega)\hat{\boldsymbol{k}} + \dot{m}R_2^2\Omega\hat{\boldsymbol{k}}$$

$$\dot{H}_{j\_out} = \dot{m}\,\Omega(a^2 + R_1^2 + R_2^2) - \dot{m}aW_j$$

which is identical to the one we used earlier for the net rate of angular momentum outflow in the solution of this example without using the vector analysis.

Example 1.22

Figure 1.29 schematically shows a centrifugal pump. The working fluid enters the pump axially and exits its blades radially. Assuming that the fluid tangential velocities are equal to the blade tangential velocities at both the inlet and the outlet, find an expression to compute the specific electric power needed to run the pump at a constant angular velocity Ω if its mechanical efficiency is η. The fluid total pressure increases from p_{01} to p_{02} across the pump.

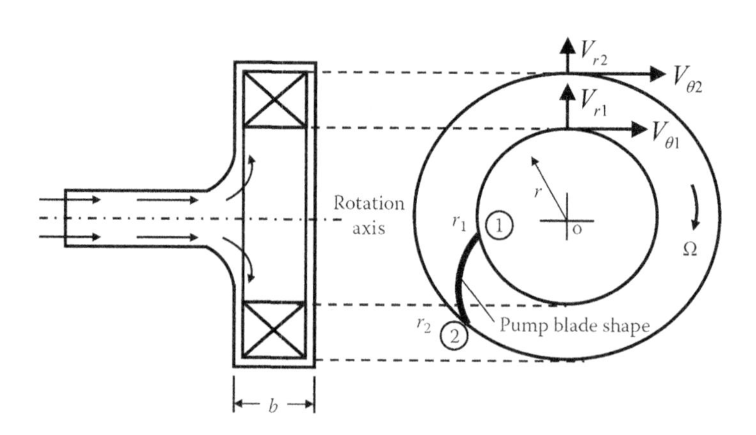

FIGURE 1.29 A schematic diagram of a centrifugal pump (Example 1.22).

Solution

In this example, the pressure forces at the pump inlet and the outlet do not contribute to any torque around the pump axis of rotation. As the flow tangential velocities are equal to the blade tangential velocities at both the inlet and the outlet, we can compute them as $V_{\theta 1} = r_1\Omega$ and $V_{\theta 2} = r_2\Omega$.

We express the pump torque per unit mass flow rate as:

$$\Gamma_{\text{pump}} = r_2 V_{\theta_2} - r_1 V_{\theta_1} = \Omega\left(r_2^2 - r_1^2\right)$$

The specific fluid power output of the pump equals the product of the torque and its angular speed, and we calculate it as:

$$\dot{W}_{\text{pump}} = \Omega\Gamma_{\text{pump}} = \Omega^2\left(r_2^2 - r_1^2\right)$$

giving the specific electric power needed to run the pump as:

$$\dot{E}_{\text{pump}} = \frac{\dot{W}_{\text{pump}}}{\eta} = \frac{\Omega^2\left(r_2^2 - r_1^2\right)}{\eta}$$

Example 1.23

Figure 1.30 schematically shows the rotor–stator cavity of a gas turbine engine. The coolant air at 673 K (absolute total) and swirling at 60% of the rotor rpm enters the cavity at the inner radius. It exits the cavity at the outer radius with a swirl of 40% of the rotor rpm. The mass flow rate of the coolant air is 20 kg/s. If the total frictional torque from the stator surface acting on the cavity air is 3,008 Nm, find the rotor torque and the exit total temperature of the coolant air. The rotor–stator surfaces are adiabatic (zero heat transfer). All quantities are given in the inertial (stator) reference frame. Assume $c_p = 1{,}004\ \text{J}/(\text{kg K})$ for air.

Solution

In this example, the coolant air temperature increases not by heat transfer (adiabatic flow) but by work transfer from the rotor. This form of work transfer in a rotor–stator cavity or a rotor–rotor cavity in turbomachinery is often called windage. Note that there is no rise in the total air temperature due to windage in a stator cavity because of its nonrotating walls.

We can write the angular momentum equation for the control volume between sections 1 and 2 as:

$$\Gamma_{\text{rotor}} - \Gamma_{\text{stator}} = \dot{m}(r_2 V_{\theta 2} - r_1 V_{\theta 1})$$

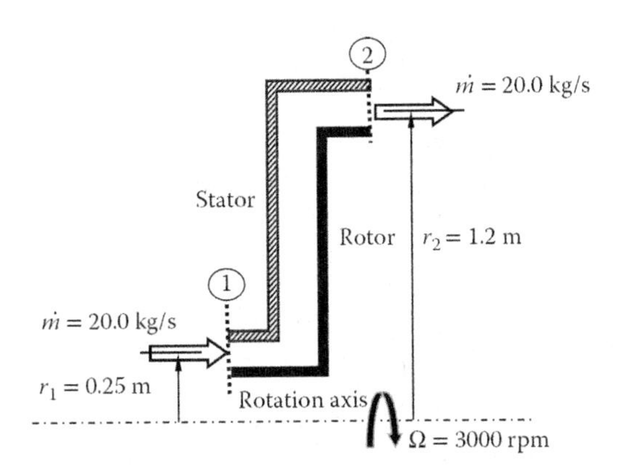

FIGURE 1.30 Windage temperature rise in a rotor–stator cavity (Example 1.23).

We can write the rate of work transfer (windage) to the coolant air in the cavity as:

$$\dot{W}_{\text{windage}} = \Gamma_{\text{rotor}}\Omega$$

Substituting rotor torque Γ_{rotor} in this equation in terms of stator torque Γ_{stator} from the torque and angular momentum balance equation yields the following equation:

$$\dot{W}_{\text{windage}} = \Gamma_{\text{stator}}\Omega + \dot{m}\,\Omega(r_2 V_{\theta_2} - r_1 V_{\theta_1})$$

Because the stator cannot do any work on the cavity air, we should never interpret the first term on the right-hand side of this equation as the stator torque doing work toward windage generation in the cavity. The rotor torque, which is solely responsible for work transfer into the cavity air, consists of two parts: one to balance the stator torque and the other to change the angular momentum of the coolant air. To be physically consistent, we should always use the rotor torque to compute work transfer in a rotor–stator or rotor–rotor cavity. Thus, we calculate the rise in air total temperature using the equation:

$$\Delta T_{0\_\text{windage}} = \frac{\Gamma_{\text{rotor}}\Omega}{\dot{m}_{\text{air}} c_p}$$

The numerical results obtained using the aforementioned equations and the data specified for this example are as follows:

Angular velocity of the rotor: $\Omega = \dfrac{3{,}000 \times 2\pi}{60} = 314.159 \text{ rad/s}$

Air tangential velocity at the inlet (section 1): $V_{\theta 1} = 0.6 \times 314.159 \times 0.25 = 47.124 \text{ m/s}$

Air tangential velocity at the outlet (section 2): $V_{\theta 2} = 0.4 \times 314.159 \times 1.2 = 150.796 \text{ m/s}$

Rotor torque: $\Gamma_{\text{rotor}} = \Gamma_{\text{stator}} + \dot{m}(r_2 V_{\theta 2} - r_1 V_{\theta 1})$

$$= 3{,}008 + 20(1.2 \times 150.796 - 0.25 \times 47.124) = 6{,}391.495 \text{ Nm}$$

Windage temperature rise: $\Delta T_{0\_\text{windage}} = \dfrac{\Gamma_{\text{rotor}}\Omega}{\dot{m}_{\text{air}} c_p} = \dfrac{6{,}391.495 \times 314.159}{20 \times 1{,}004} = 100 \text{ K}$

(Note: temperature difference in kelvin)

Exit total temperature of coolant air: $T_{02} = T_{01} + \Delta T_{0\_\text{windage}} = 673 + 100 = 773 \text{K}$

Example 1.24

Figure 1.31 shows a four-sided radial duct representing a part of an internal cooling passage of a steam-cooled gas turbine blade with rotational velocity Ω. Each duct wall has a different surface area A_w, temperature T_w, and heat transfer coefficient h_w. For a given coolant (steam) mass flow

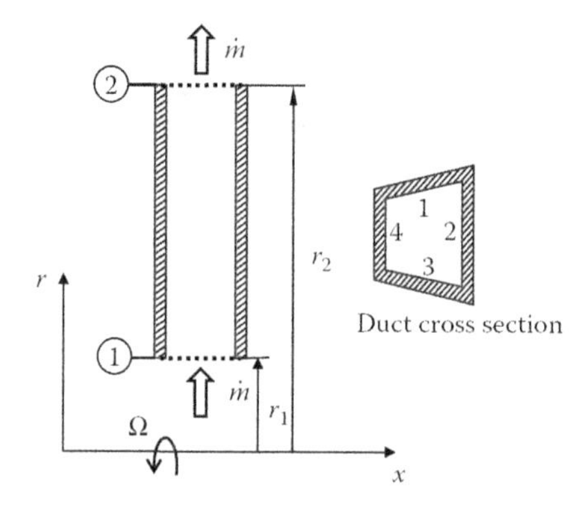

FIGURE 1.31 Heat transfer in a rotating duct of an arbitrary cross section (Example 1.24).

rate $\dot{m}$ and constant specific heat (at constant pressure) c_p, find expressions to evaluate the rise in coolant total temperature due to heat transfer and blade rotation. Because the coolant flow Mach number is less than 0.3, the flow may be assumed to be incompressible. Neglect the coupling between the coolant's total temperature changes due to rotation and convective heat transfer with the duct wall.

Solution

In this example, the coolant steam temperature in the gas turbine blade increases by both convective heat transfer and rotational work transfer. The coolant temperature change due to rotation will influence convective heat transfer from the duct walls to the coolant. In the present solution, we calculate separate changes in coolant temperature due to heat transfer and rotation, neglecting their coupling (see Appendix B of Sultanian, 2015, for deriving a coupled solution).

COOLANT TOTAL TEMPERATURE RISE DUE TO HEAT TRANSFER

The solution of Example 1.7 shows that the difference between the constant wall temperature and the fluid total temperature decays exponentially from duct inlet to exit, that is:

$$(T_w - T_{0\_outlet}) = (T_w - T_{0\_inlet})\, e^{-\eta}$$

where the exponent η is given by:

$$\delta = \frac{h_w A_w}{\dot{m} c_p}$$

In the present case, unlike Example 1.7, which involved a pipe with one wall, we have a four-sided duct, each side with different heat transfer properties. Using newly defined average quantities, we can easily extend the solution of Example 1.7 for the present case as follows:

$$(\overline{T}_w - T_{0\_outlet}) = (\overline{T}_w - T_{0\_inlet})\, e^{-\overline{\eta}}$$

where

$$\overline{T}_w = \frac{\sum_{i=1}^{i=4} h_{w_i} A_{w_i} T_{w_i}}{\sum_{i=1}^{i=4} h_{wi} A_{wi}} \quad \text{and} \quad \overline{\eta} = \frac{\sum_{i=1}^{i=4} h_{w_i} A_{w_i}}{\dot{m} c_p}$$

Thus, we write the change in coolant total temperature due to heat transfer as:

$$\Delta T_{0\_heat\ transfer} = \left(T_{0\_outlet} - T_{0\_inlet}\right)_{heat\ transfer} = \left(\overline{T}_w - T_{0\_inlet}\right)\left(1 - e^{-\overline{\eta}}\right)$$

Note that the coolant total temperatures in the aforementioned expressions are in the rotor reference frame.

COOLANT TOTAL TEMPERATURE RISE DUE TO ROTATION

The fluid rothalpy between any two sections of a rotating duct remains constant under adiabatic conditions. Accordingly, expressing rothalpy in terms of coolant total temperature in the rotor reference frame, we obtain:

$$c_p T_{0\_outlet} - \frac{r_2^2 \Omega^2}{2} = c_p T_{0\_inlet} - \frac{r_1^2 \Omega^2}{2}$$

$$\Delta T_{0\_\text{rotation}} = \left(T_{0\_\text{outlet}} - T_{0\_\text{inlet}}\right)_{\text{rotation}} = \frac{\Omega^2}{2c_p}\left(r_2^2 - r_1^2\right)$$

which shows that the total temperature change due to rotation depends on the flow direction, with equal magnitude. The work done on the fluid due to rotation is positive for a radially outward flow, and for a radially inward flow, it is negative.

Thus, we write the total change in coolant total temperature due to both heat transfer and rotation as:

$$\Delta T_0 = \Delta T_{0\_\text{heat transfer}} + \Delta T_{0\_\text{rotation}}$$

$$\Delta T_0 = \left(\bar{T}_w - T_{0\_\text{inlet}}\right)\left(1 - e^{-\bar{\eta}}\right) + \frac{\Omega^2}{2c_p}\left(r_2^2 - r_1^2\right)$$

Suppose we had considered the inherent coupling between heat transfer and rotation on coolant temperature change. In that case, for a radially outward flow, the actual coolant temperature rise will be lower than that obtained by this equation based on the linear superposition of individual temperature changes. The actual coolant temperature will be higher than that obtained by this equation for a radially inward flow.

Example 1.25

As shown in Figure 1.32, the steady flow of a fluid of specific heat c_p in a circular pipe is heated uniformly by an electrical wire running along the pipe axis and by convection from the pipe wall maintained at a uniform wall temperature T_w with a constant heat transfer coefficient h between the wall and the fluid. Using a differential (small) control volume analysis within the pipe flow, show that the dimensionless fluid temperature variation from the pipe inlet to the outlet is given by:

$$\theta = \left(1 + \frac{\varepsilon}{\eta}\right)e^{-\eta\xi} - \frac{\varepsilon}{\eta}$$

where the dimensionless quantities are defined as follows:

$$\xi = \frac{x}{L}, \theta = \frac{T_w - T_0}{T_w - T_{0\_\text{inlet}}}, \eta = \frac{h\pi DL}{\dot{m}c_p}, \quad \text{and } \varepsilon = \frac{\dot{E}L}{\dot{m}c_p\left(T_w - T_{0\_\text{inlet}}\right)}$$

where $\dot{E}$ is the rate of uniform electrical heating per unit pipe length. Find and plot the relationship between η and ε such that the fluid outlet temperature equals the pipe wall temperature and show that for $\eta = 0$, $\varepsilon = 1$, and for $\eta \to \infty$, $\varepsilon = 0$. Finally, find the total convective heat transfer over the entire pipe.

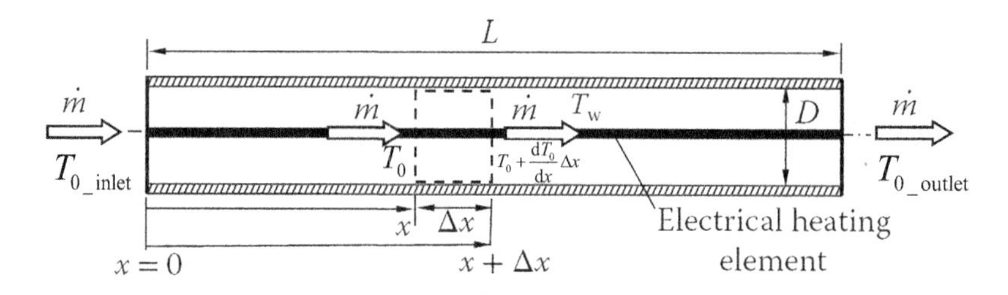

FIGURE 1.32 Steady circular pipe flow heated by convection from the isothermal pipe wall and by uniform electrical heat generation at the pipe axis (Example 1.25).

Solution

Using the steady flow energy equation, the energy balance over the small control volume between x and $x+\Delta x$, shown in Figure 1.32, yields:

$$\dot{m}c_p\left(T_0+\frac{\mathrm{d}T_0}{\mathrm{d}x}\Delta x\right)-\dot{m}c_pT_0=h\pi D\Delta x(T_\mathrm{w}-T_0)+\dot{E}\Delta x$$

where

$h \equiv$ constant heat transfer coefficient
$T_\mathrm{w} \equiv$ uniform pipe wall temperature

$$\dot{m}c_p\frac{\mathrm{d}T_0}{\mathrm{d}x}\Delta x=h\pi D\Delta x(T_\mathrm{w}-T_0)+\dot{E}\Delta x$$

Dividing this equation by Δx yields:

$$\dot{m}c_p\frac{\mathrm{d}T_0}{\mathrm{d}x}=h\pi D(T_\mathrm{w}-T_0)+\dot{E}$$

We write this equation in terms of the given dimensionless variables as:

$$\frac{\mathrm{d}\theta}{\mathrm{d}\xi}+\eta\theta=-\varepsilon$$

which is a first-order linear non-homogeneous differential equation, having the solution of its homogeneous part is given by:

$$\theta_\mathrm{h}=\mathrm{e}^{-\eta\xi}$$

and a particular solution is given by:

$$\theta_\mathrm{p}=-\frac{\varepsilon}{\eta}$$

We can write the resulting solution of the differential equation as:

$$\theta=C\theta_\mathrm{h}+\theta_\mathrm{p}=C\mathrm{e}^{-\eta\xi}-\frac{\varepsilon}{\eta}$$

We determine the constant coefficient C in this solution using the pipe inlet boundary condition $\theta=1$ at $\xi=0$, giving:

$$C=1+\frac{\varepsilon}{\eta}$$

and

$$\theta=\left(1+\frac{\varepsilon}{\eta}\right)\mathrm{e}^{-\eta\xi}-\frac{\varepsilon}{\eta}$$

The condition of the fluid total temperature reaching the wall temperature at the pipe outlet corresponds to $\theta=0$ at $\xi=1$. For this boundary condition, the aforementioned solution yields the following relation between ε and η:

$$0=\left(1+\frac{\varepsilon}{\eta}\right)\mathrm{e}^{-\eta}-\frac{\varepsilon}{\eta}$$

$$\varepsilon = \frac{\eta e^{-\eta}}{\left(1 - e^{-\eta}\right)}$$

which, we plot in Figure 1.33, shows that, for $\eta = 0$, we obtain $\varepsilon = 1$, indicating that the fluid total temperature increases from the pipe inlet to the outlet entirely due to electrical heating. For a large value of η, the fluid total temperature asymptotically reaches the wall temperature at the pipe outlet with no need for additional electrical heating.

We can compute the total convective heat transfer over the entire pipe by using the overall energy balance for the pipe, giving:

$$\dot{Q} = \dot{Q}_{\text{convection}} + \dot{Q}_{\text{electrical}} = \dot{m}c_p\left(T_{0\_\text{outlet}} - T_{0\_\text{inlet}}\right)$$

$$\dot{Q} = \dot{m}c_p\left(T_{\text{w}} - T_{0\_\text{inlet}}\right)\left(\theta_{\text{inlet}} - \theta_{\text{outlet}}\right)$$

With $\theta_{\text{inlet}} = 1$ and $\theta_{\text{outlet}} = 1 + \frac{\varepsilon}{\eta}e^{-\eta} - \frac{\varepsilon}{\eta}$, we obtain:

$$\dot{Q} = \dot{m}c_p\left(T_{\text{w}} - T_{0\_\text{inlet}}\right)\left\{1 - \left[\left(1 + \frac{\varepsilon}{\eta}\right)e^{-\eta} - \frac{\varepsilon}{\eta}\right]\right\}$$

$$\dot{Q} = \dot{m}c_p\left(T_{\text{w}} - T_{0\_\text{inlet}}\right)\left(1 + \frac{\varepsilon}{\eta}\right)\left(1 - e^{-\eta}\right)$$

With $\dot{Q}_{\text{electrical}} = \dot{E}L = \dot{m}c_p\left(T_w - T_{0\_\text{inlet}}\right)\varepsilon$, this equation yields:

$$\dot{Q}_{\text{convection}} = \dot{m}c_p\left(T_{\text{w}} - T_{0\_\text{inlet}}\right)\left(1 + \frac{\varepsilon}{\eta}\right)\left(1 - e^{-\eta}\right) - \dot{m}c_p\left(T_{\text{w}} - T_{0\_\text{inlet}}\right)\varepsilon$$

$$\dot{Q}_{\text{convection}} = \dot{m}c_p\left(T_{\text{w}} - T_{0\_\text{inlet}}\right)\left\{\left(1 + \frac{\varepsilon}{\eta}\right)\left(1 - e^{-\eta}\right) - \varepsilon\right\}$$

$$\dot{Q}_{\text{convection}} = \dot{m}c_p\left(T_{\text{w}} - T_{0\_\text{inlet}}\right)\left\{\left(1 - e^{-\eta}\right) - \left(1 - \frac{1}{\eta} + \frac{e^{-\eta}}{\eta}\right)\varepsilon\right\}$$

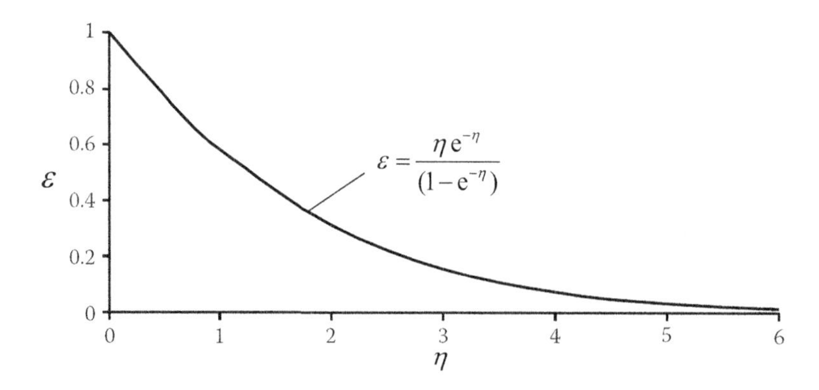

FIGURE 1.33 Relationship between ε and η yielding fluid outlet total temperature equal to the pipe wall temperature (Example 1.25).

which shows that without electrical heating the total convective heat transfer becomes:

$$\dot{\tilde{Q}}_{\text{convection}} = \dot{m}c_p\left(T_{\text{w}} - T_{0\_\text{inlet}}\right)\left(1 - e^{-\eta}\right)$$

which in the presence of electrical heating reduces by an amount given by:

$$\dot{\tilde{Q}}_{\text{convection}} - \dot{Q}_{\text{convection}} = \dot{m}c_p\left(T_w - T_{0\_\text{inlet}}\right)\left(1 - \frac{1}{\eta} + \frac{e^{-\eta}}{\eta}\right)\varepsilon$$

With electrical heating, the fluid total temperature increases along the pipe, reducing the difference between the pipe wall temperature and the fluid total temperature and reducing the total heat transfer due to convection compared to the situation with no electrical heating. The aforementioned results support this physical behavior.

Example 1.26

As shown in Figure 1.34, water is falling freely under gravity through a vertical pipe of flow area A_1 at a constant mass flow rate $\dot{m}$. Using various quantities shown in the figure and neglecting any drag force acting on the water jet, find the area ratio A_2 / A_1.

Solution

In this example, gravity is the only force acting on the falling water. The ambient pressure cancels on both sides of the Bernoulli equation applied between sections 1 and 2. For the steady water flow, the continuity equation yields:

$$\dot{m} = \rho A_1 V_1 = \rho A_2 V_2$$

$$V_1 = \frac{\dot{m}}{\rho A_1}$$

$$V_2 = \frac{\dot{m}}{\rho A_2}$$

Because the sum of the potential energy and kinetic energy of each particle in the water flow remains constant, we obtain:

$$\frac{V_1^2}{2} + gh_1 = \frac{V_2^2}{2} + gh_2$$

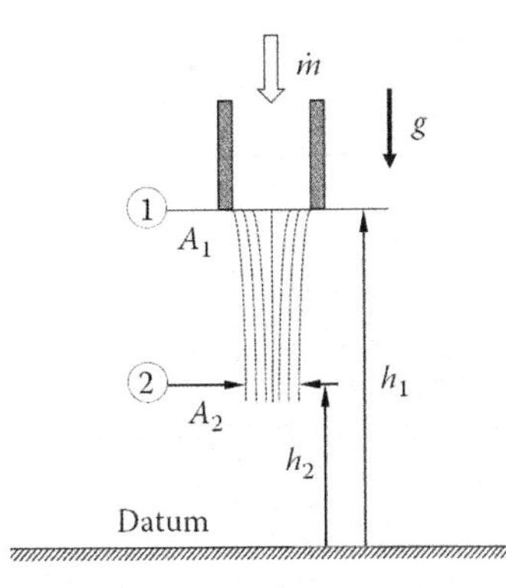

FIGURE 1.34 Free fall of water flow from a pipe (Example 1.26).

Substituting V_1 and V_2 in this equations from the continuity equations at both sections yields:

$$\frac{\dot{m}^2}{2\rho^2 A_1^2} + gh_1 = \frac{\dot{m}^2}{2\rho^2 A_2^2} + gh_2$$

$$\frac{1}{A_2^2} = \frac{1}{A_1^2} + \frac{2\rho^2 g(h_1 - h_2)}{\dot{m}^2}$$

$$\frac{A_1^2}{A_2^2} = 1 + \frac{2\rho^2 A_1^2 g(h_1 - h_2)}{\dot{m}^2} = 1 + \frac{2g(h_1 - h_2)}{V_1^2}$$

$$\frac{A_2}{A_1} = \frac{1}{\sqrt{1 + \frac{2g(h_1 - h_2)}{V_1^2}}}$$

which shows that the area ratio is less than 1.0, implying $A_2 < A_1$. Thus, the reduction in water jet flow area depends upon the initial flow velocity V_1, which in turn depends on the mass flow rate $\dot{m}$ and the initial flow area A_1, and the total fall $(h_1 - h_2)$ from the initial location.

Example 1.27

Figure 1.35 shows an open-channel flow of water. The cross section of the channel parallel to the flow remains constant from top to bottom. As shown in Figure 1.35a, over the length L in the middle, both side walls of the open channel converge, reducing its width from b_1 to b_2. For a volumetric flow rate Q, Figure 1.35b shows that the water depth in the open channel varies from z_1 to z_2. Assuming no energy loss in this flow system, find an equation to express the volumetric flow rate Q in terms of other quantities shown in the figure. Does the water depth vary linearly in the transition section?

Solution

In the solution of this example, we assume that the flow velocity is uniform over each cross section of the open channel. The volumetric flow rate Q remains constant through the channel. At the beginning of the transition, the flow area is $b_1 z_1$ and the velocity is V_1. At the end of the transition, the flow area is $b_2 z_2$ and the velocity is V_2. For a steady flow, the continuity equation over the transition yields:

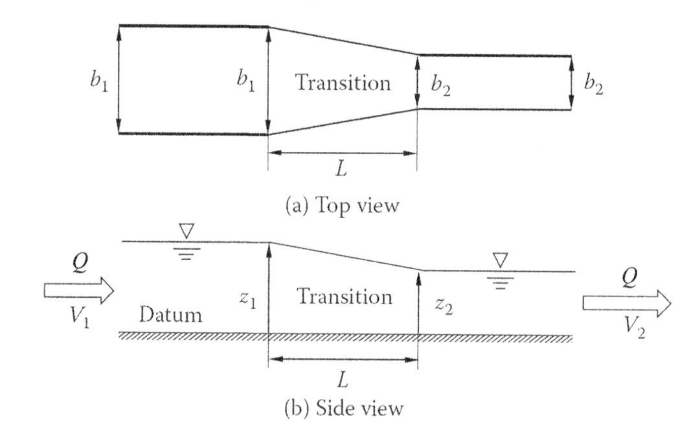

FIGURE 1.35 An open-channel water flow as a flow measuring device (Example 1.27).

$$V_1 = \frac{Q}{b_1 z_1}$$

$$V_2 = \frac{Q}{b_2 z_2}$$

We can write from the conservation of the sum of kinetic and potential energies for a streamline along the open surface of the channel flow:

$$\frac{V_1^2}{2} + g z_1 = \frac{V_2^2}{2} + g z_2$$

In this equation, substituting V_1 and V_2 from the continuity equation yields:

$$\frac{Q^2}{2b_1^2 z_1^2} + g z_1 = \frac{Q^2}{2b_2^2 z_2^2} + g z_2$$

$$Q^2\left(\frac{1}{2b_2^2 z_2^2} - \frac{1}{2b_1^2 z_1^2}\right) = g(z_1 - z_2)$$

$$Q = \sqrt{\frac{g(z_1 - z_2)}{\left(\frac{1}{2b_2^2 z_2^2} - \frac{1}{2b_1^2 z_1^2}\right)}}$$

which is the equation to calculate the volumetric flow rate Q from the measurements of z_1 and z_2 in a steady flow in an open channel with known b_1 and b_2. This equation also shows that the decrease in water depth during the transition is not linear.

The open-channel device analyzed in this example is often used to measure flow rate in the cold-flow calibration of various equipment.

Example 1.28

In January 1738, Daniel Bernoulli asked a young fluids engineer to perform some calculations on water flow in a variable area duct shown in Figure 1.36. He asked the engineer to neglect any frictional force in the duct. Reviewing the calculation results shown in the figure, Bernoulli commented that the young engineer's calculated results are physically impossible. Why do you agree or disagree with Bernoulli's conclusion?

Solution

The solution provided by the young engineer clearly satisfies the continuity equation as the mass flow rate entering and leaving the duct is 100 kg/s. Let us examine whether the solution also obeys the momentum equation.

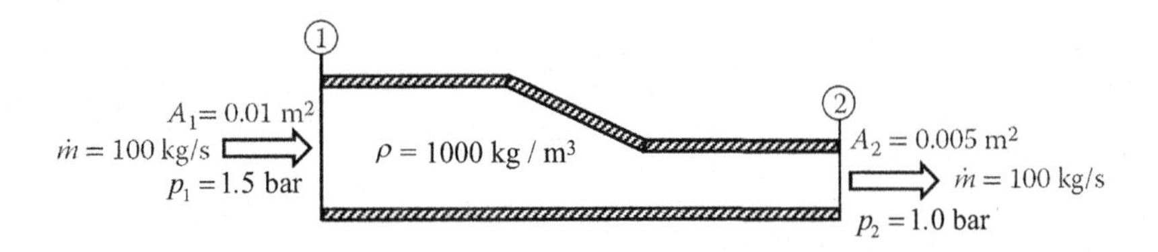

FIGURE 1.36 Water flow in a frictionless duct (Example 1.28).

SECTION 1 (INLET)

Velocity:

$$V_1 = \frac{\dot{m}}{A_1 \rho} = \frac{100}{0.01 \times 1000} = 10 \text{ m/s}$$

Stream thrust:

$$S_{T1} = p_1 A_1 + \dot{m} V_1 = 1.5 \times 10^5 \times 0.01 + 100 \times 10 = 2500 \text{ N}$$

SECTION 2 (OUTLET)

Velocity:

$$V_2 = \frac{\dot{m}}{A_2 \rho} = \frac{100}{0.005 \times 1000} = 20 \text{ m/s}$$

Stream thrust:

$$S_{T2} = p_2 A_2 + \dot{m} V_2 = 1.0 \times 10^5 \times 0.005 + 100 \times 20 = 2500 \text{ N}$$

The results show no change in the stream thrust between the duct inlet and the outlet. Although we neglect wall friction in this example, the convergent upper wall in the mid-section of the duct will create an opposing pressure force on the fluid control volume. As a result, the stream thrust at the duct exit would be lower than that at its inlet.

From energy conservation with no loss in the system, the sum of the flow work and kinetic energy remains constant along the duct. Accordingly, let us now examine how the calculated total pressure changes between the duct inlet and the outlet.

Total pressure at the duct inlet:

$$p_{01} = 1.5 \times 10^5 + \frac{1000 \times 10 \times 10}{2} = 2.0 \text{ bar}$$

Total pressure at the duct outlet:

$$p_{02} = 1.0 \times 10^5 + \frac{1000 \times 20 \times 20}{2} = 3.0 \text{ bar}$$

The calculation of the total pressure shows that it increases from the duct inlet to the outlet, violating the flow physics. For a duct flow with friction, the total pressure decreases downstream; for a frictionless flow, it remains constant. The present analysis fully supports the conclusion reached by Bernoulli.

Example 1.29

Figure 1.37 shows the operation of a siphon. The fluid is water with density 1000 kg/m^3. The ambient pressure is 1 bar. Calculate the discharge velocity at D and the static pressures inside the constant area tube at B and C, the highest point on the siphon centerline. Neglect any frictional loss in the siphon system.

Solution

The operation of an ideal siphon shown in Figure 1.37 is governed by the Bernoulli equation, which yields a constant total head at points A, B, C, and D.

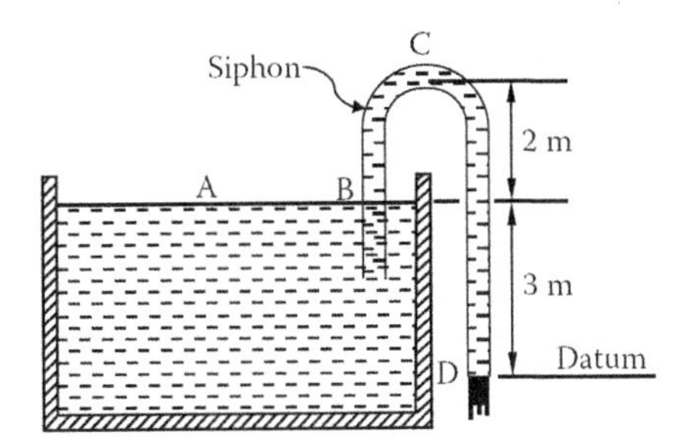

FIGURE 1.37 Operation of a siphon (Example 1.29).

DISCHARGE VELOCITY AT D

Assigning the datum at D for measuring the geodetic head, we write the Bernoulli equation between points A and D as:

$$\frac{p_{amb}}{\rho g}+\frac{V_A^2}{2g}+3.0=\frac{p_{amb}}{\rho g}+\frac{V_D^2}{2g}$$

Neglecting V_A in comparison with V_D in this equation, we obtain:

$$\frac{V_D^2}{2g}=3.0$$

which yields:

$$V_D=\sqrt{2\times 9.81\times 3.0}=7.672\ \text{m/s}$$

STATIC PRESSURE AT B INSIDE THE SIPHON TUBE

Writing the Bernoulli equation between points B and D yields:

$$\frac{p_B}{\rho g}+\frac{V_B^2}{2g}+3.0=\frac{p_{amb}}{\rho g}+\frac{V_D^2}{2g}$$

From the continuity equation between points B and D inside the tube with constant flow area, we obtain $V_B=V_D$, reducing this equation to:

$$\frac{p_B}{\rho g}=\frac{p_{amb}}{\rho g}-3$$

$$p_B=p_{amb}-3\rho g=1.0\times 10^5-3\times 1000\times 9.81$$

$$p_B=70570\,\text{Pa}$$

Thus, it gives a sub-ambient static pressure at point B.

STATIC PRESSURE AT C INSIDE THE SIPHON TUBE

Writing the Bernoulli equation between points C and D yields:

$$\frac{p_C}{\rho g}+\frac{V_C^2}{2g}+5.0=\frac{p_{amb}}{\rho g}+\frac{V_D^2}{2g}$$

The continuity equation between points C and D inside the tube with constant flow area yields $V_C = V_D$, reducing this equation to:

$$\frac{p_C}{\rho g} = \frac{p_{amb}}{\rho g} - 5$$

$$p_C = p_{amb} - 5\rho g = 1.0\times10^5 - 5\times1000\times9.81$$

$$p_C = 50950\,\text{Pa}$$

Thus, the static pressure at point C is sub-ambient and lower than at point B. The calculated static pressures in the siphon tube show that they decrease from point B to point C, reaching their lowest value at the highest point in the siphon tube. The static pressure would, however, increase from point C to point D, where it equals the ambient pressure. As we increase the height of point C, the static pressure in the tube becomes more and more sub-ambient until it reaches the vapor pressure of the liquid used in the siphon. Beyond this point, the formation of liquid vapor will break the liquid flow, and the siphon will cease to operate.

Example 1.30

Figure 1.38 shows a flow system that uses a centrifugal pump to pump oil of density $850\,\text{kg/m}^3$ at the volumetric flow rate $Q = 0.015\,\text{m}^3/\text{s}$. The gauge pressure measured at point A is –30 kPa and that at point B is 300 kPa. The inner diameter of the pipe upstream of the pump is 7.62 cm and that of the downstream pipe is 5.08 cm. At the given flow rate, the minor head loss of the check valve is 0.75 m, and the major head loss due to friction in pipes is 1.25 m. Calculate the power delivered by the pump in this flow system.

Solution

From the given inner diameters of pipes upstream and downstream of the pump, we calculate the corresponding flow areas as $A_A = 4.560\times10^{-3}\,\text{m}^2$ and $A_B = 2.027\times10^{-3}\,\text{m}^2$.

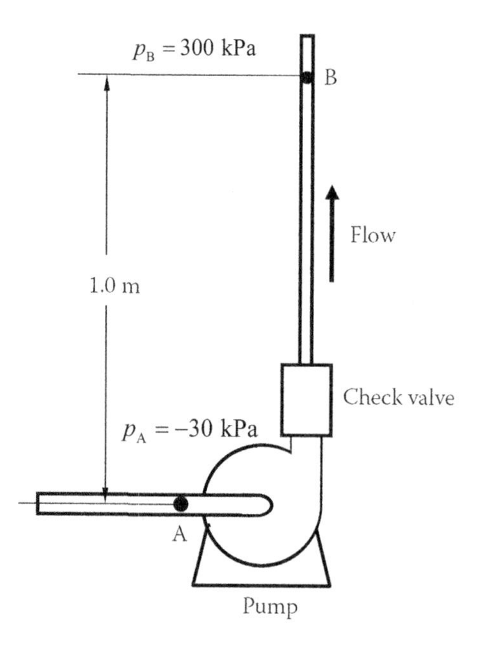

FIGURE 1.38 A pump flow system (Example 1.30).

Velocity at A:

$$V_A = \frac{Q}{A_A} = \frac{0.015}{4.560 \times 10^{-3}} = 3.289 \text{ m/s}$$

Velocity at B:

$$V_B = \frac{Q}{A_B} = \frac{0.015}{2.027 \times 10^{-3}} = 7.401 \text{ m/s}$$

We can write the extended Bernoulli equation between points A and B as:

$$\frac{p_A}{\rho g} + \frac{V_A^2}{2g} + z_A + h_{\text{pump}} - h_{\text{loss}} = \frac{p_B}{\rho g} + \frac{V_B^2}{2g} + z_B$$

$$h_{\text{pump}} = \frac{p_B - p_A}{\rho g} + \frac{V_B^2 - V_A^2}{2g} + (z_B - z_A) + h_{\text{loss}}$$

We calculate each term on the right-hand side of this equation as follows:

$$\frac{p_B - p_A}{\rho g} = \frac{300 \times 10^3 - (-30 \times 10^3)}{850 \times 9.81} = 39.575 \text{ m}$$

$$\frac{V_B^2 - V_A^2}{2g} = \frac{(7.401)^2 - (3.289)^2}{2 \times 9.81} = 2.240 \text{ m}$$

$$(z_B - z_A) = 1.0 - 0.0 = 1.0 \text{ m}$$

$$h_{\text{loss}} = 0.75 + 1.25 = 2.0 \text{ m}$$

Hence, the head imparted by the pump becomes:

$$h_{\text{pump}} = 39.575 + 2.240 + 1.0 + 2.0 = 44.815 \text{ m}$$

Pumping power:

$$\dot{W}_{\text{pump}} = h_{\text{pump}} \dot{m} g = (h_{\text{pump}} \rho g) Q = \Delta p_{0\_\text{pump}} Q$$

$$\dot{W}_{\text{pump}} = 44.815 \times 850 \times 9.81 \times 0.015 = 5605 \text{ W} = 5.605 \text{ kW}$$

Problems

1.1 The velocity distribution in a three-dimensional steady incompressible flow is given by $V_x = 5xe^{-2z}$, $V_y = -3ye^{-2z}$, and $V_z = e^{-2z}$. Find the equation of the streamline that passes through point A (1, 1, 1).

1.2 For the flow field given in Problem 1.1, density varies as $\rho = e^{-z}$, which alters the velocity component in the z direction only. Find the new distribution of V and the equation of the new streamline that passes through point A (1, 1, 1).

1.3 The u velocity profile in a laminar boundary layer over an impermeable flat plate is given by:

$$\frac{u}{U} = 2\frac{y}{\delta} - \left(\frac{y}{\delta}\right)^2$$

where U is the free-stream velocity and δ is the local boundary layer thickness, given by $\delta = K\sqrt{x}$. Find an expression for the velocity component v that is perpendicular to u.

1.4 For a steady one-dimensional incompressible flow in a conical diffuser shown in Figure 1.39, find the convective acceleration at a distance x from the inlet for the volumetric flow rate of Q.

1.5 The following velocity vector represents a three-dimensional unsteady incompressible flow:

$$V = (x^2y + 5xt + yz^2)\hat{\boldsymbol{i}} + (xy^2 + 4t)\hat{\boldsymbol{j}} - (4xyz - xy + 5zt)\hat{\boldsymbol{k}}$$

a. Verify that the given velocity field satisfies the continuity equation.
b. Determine the acceleration vector at point A (1, 1, 1) at $t=1.0$.
c. Determine the rotation vector at point A (1, 1, 1) at $t=1.0$.

1.6 Figure 1.40 shows a few streamlines (circles) of a forced vortex. Calculate the circulation around squares A and B, each having side a. Square A is centered at the origin, and square B is centered at point P ($2a$, a).

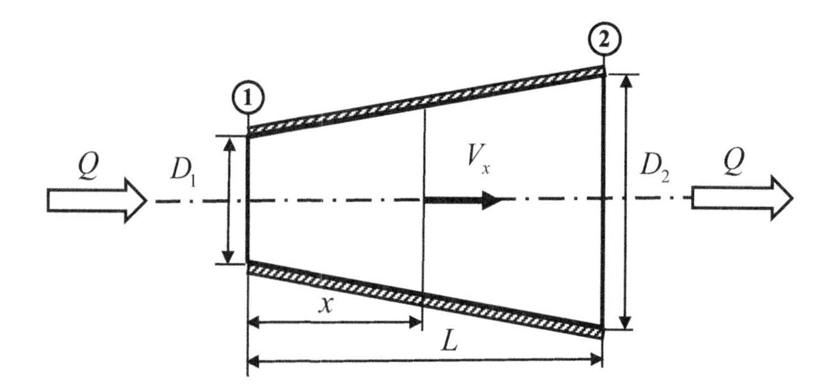

FIGURE 1.39 Steady one-dimensional incompressible flow in a conical diffuser (Problem 1.4).

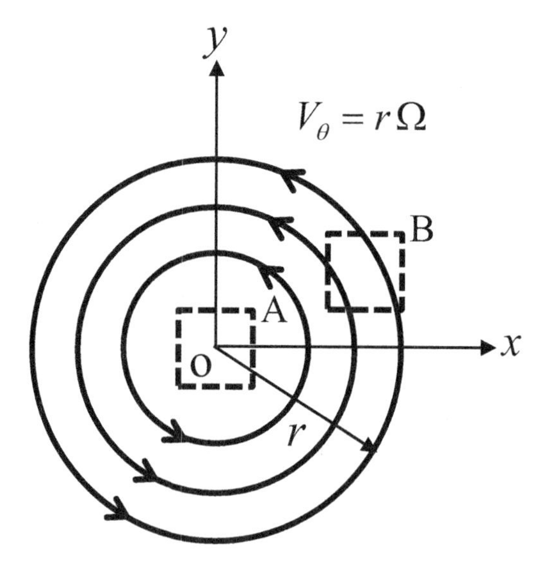

FIGURE 1.40 Circulation in a forced vortex (Problem 1.6).

1.7 Figure 1.41 shows a few streamlines (circles) of a free vortex. Calculate the circulation around squares A and B, each having side a. Square A is centered at the origin, and square B is centered at point P ($2a$, a).

1.8 Figure 1.42 shows incompressible flows through two ducts: (a) constant area circular pipe and (b) pipe of (a) appended with a short conical diffuser. Identical inlet total pressure p_0 and exit static pressure p are used in both flows with negligible friction and heat transfer. Will the mass flow rate be equal in each duct? If you expect the mass flow rates to be different, which duct will flow higher and why?

1.9 Figure 1.43 shows an incompressible flow through an adiabatic diffuser with a bypass duct connecting sections A and B. Based on how the static and total pressures change in a diffuser, determine the direction of flow in the bypass duct, whether it is from A to B or from B to A.

1.10 As shown in Figure 1.44, an incompressible flow enters a rotating duct at location o and exits it at location 1. Using the quantities shown in the figure, determine the increase in static pressure from the duct inlet to the exit for a fluid of density ρ.

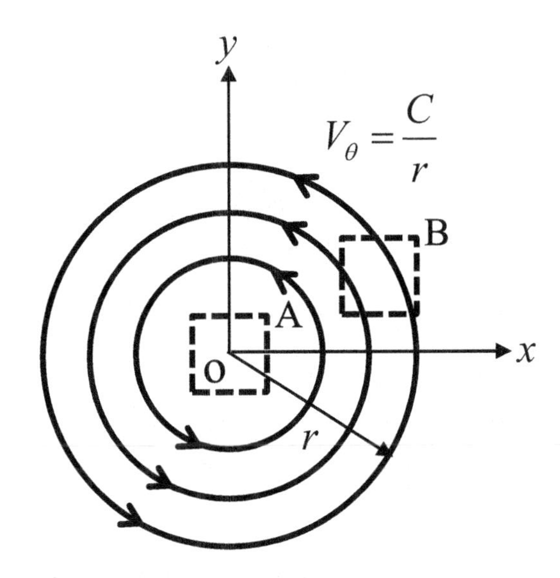

FIGURE 1.41 Circulation in a free vortex (Problem 1.7).

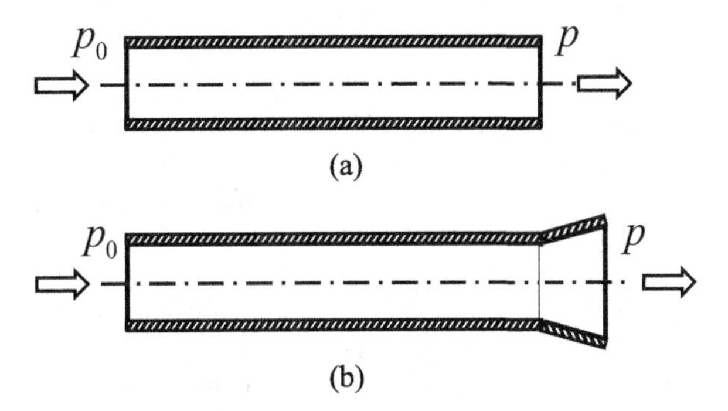

FIGURE 1.42 Incompressible flows through frictionless adiabatic ducts: (a) constant area circular pipe and (b) pipe in (a) appended with a short conical diffuser (Problem 1.8).

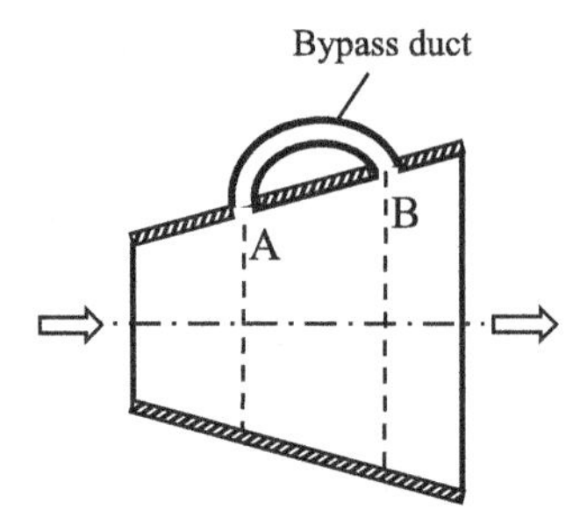

FIGURE 1.43 Incompressible flow in an adiabatic diffuser with a bypass duct (Problem 1.9).

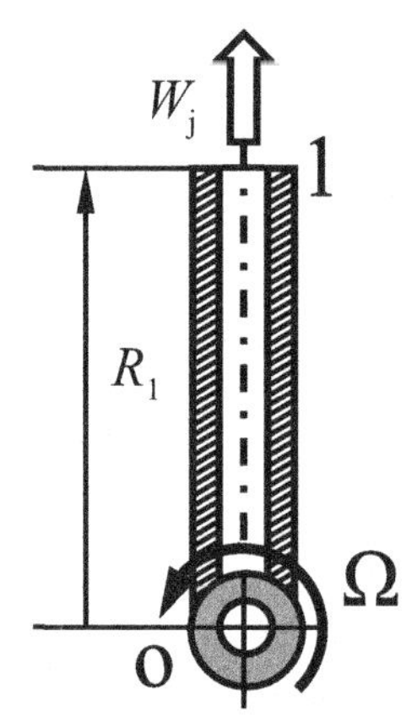

FIGURE 1.44 Incompressible flow through a rotating duct (Problem 1.10).

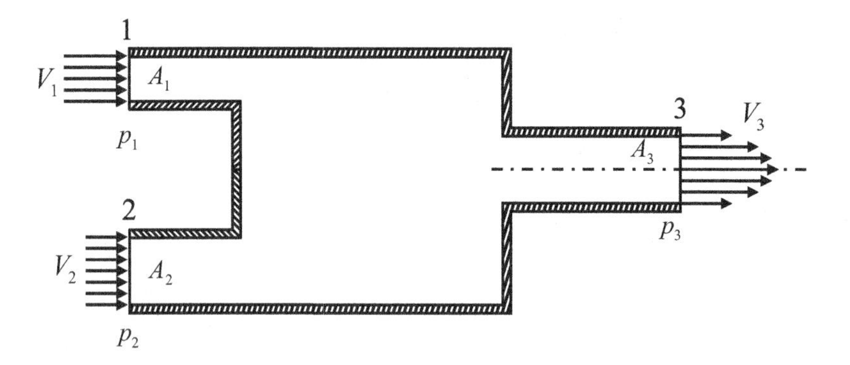

FIGURE 1.45 Incompressible flow through a manifold with two inlets and one outlet (Problem 1.11).

1.11 Figure 1.45 shows an incompressible flow of a fluid of density ρ through a manifold with two inlets and one outlet. The flow velocity at each inlet is uniform. The velocity exiting the outlet pipe has a parabolic profile with its average value denoted by V_3. Determine the force needed to hold the manifold in position under the given flow field.

1.12 In a qualifying examination, the professor asked a student a simple question about fluid mechanics. A vertical jet of fluid density ρ hits a horizontal plate with velocity V. Over stagnation area A, where the jet hits the plate, what force is exerted on the plate? Assume zero ambient pressure (vacuum). The student calculated the stagnation pressure on the plate as $\rho V^2/2$ and, knowing that force equals pressure times area, quickly responded with the answer $A\rho V^2/2$. The professor immediately commented that the student's answer was only 50% right. Do you agree with the professor or the student, and why?

1.13 Using the operating and geometric quantities for a fire hose nozzle shown in Figure 1.46, find an expression to compute the force F that a firefighter will have to resist in handling the fire hose. Will the firefighter experience a pull force or a push force?

1.14 Figure 1.47 shows a wooden log with a square cross section in a two-dimensional crossflow with a uniform velocity in the x direction. The figure also shows the x direction linear velocity profile leaving the control volume over the log and no flow across CD. The fluid has a constant density ρ. Calculate the net force per unit depth acting on the control volume in the x direction.

1.15 Figure 1.48 shows an ejector in which two incompressible turbulent flows of the same fluid mix between sections 1 and 2. At section 1, the central jet enters with uniform velocity V_j and the induced flow through the annulus enters with uniform velocity $V_j / 4$. Both flows at section 1 occupy equal cross-sectional areas. At downstream section 2, the two flows have fully mixed with the resulting uniform velocity V_2. Neglecting wall shear force, calculate the increase in static pressure and the decrease in average total pressure between sections 1 and 2. The duct cross section remains constant between these sections.

1.16 A young boy helping his dad with the yard work and playing with the garden hose asked, "Hey, Dad, when I half cover the hose opening, the water jets out faster and goes further." The dad replied, "Obviously, son. If the area becomes half, the jet velocity doubles for the same flow." After a short while, the boy asked again, "Daddy, but now it takes longer to fill the bucket." The dad replied, "Hmm ..., I need to think about that." With your understanding of fluid flow, how would you resolve the dad's dilemma?

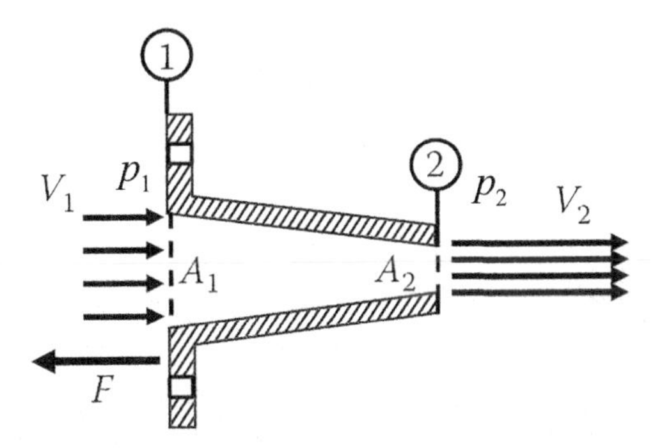

FIGURE 1.46 Thrust produced by a fire hose nozzle (Problem 1.13).

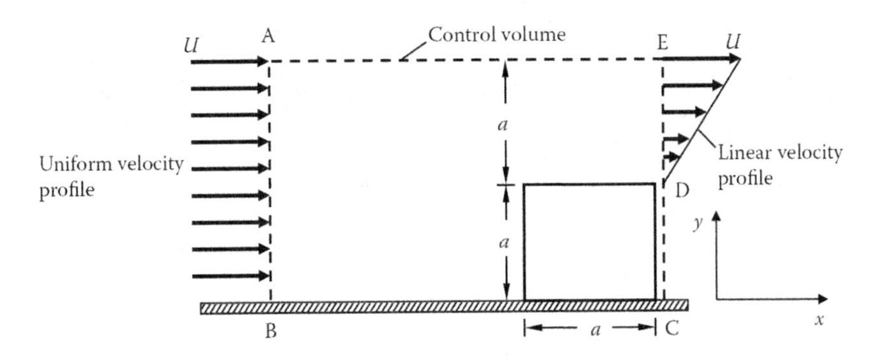

FIGURE 1.47 A wooden log of square cross section in a two-dimensional crossflow with uniform velocity (Problem 1.14).

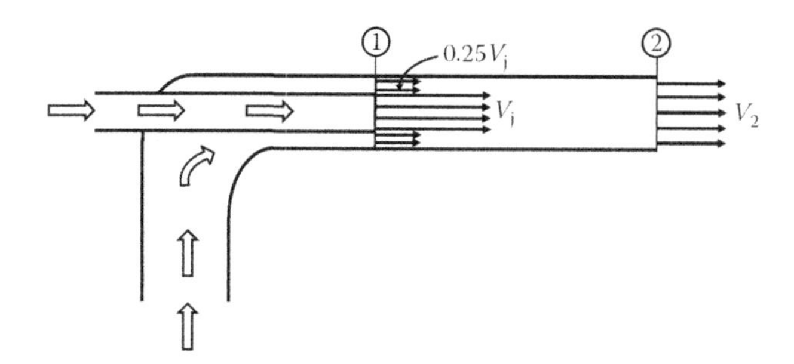

FIGURE 1.48 Mixing of two incompressible turbulent flows in an ejector (Problem 1.15).

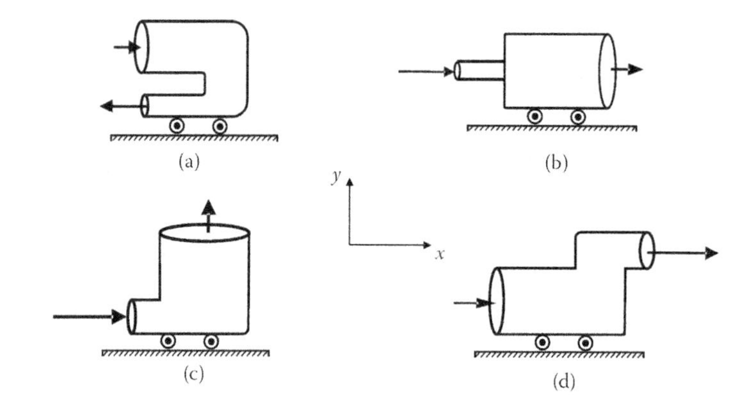

FIGURE 1.49 Four flow devices constrained to move along the x direction when released from rest (Problem 1.17).

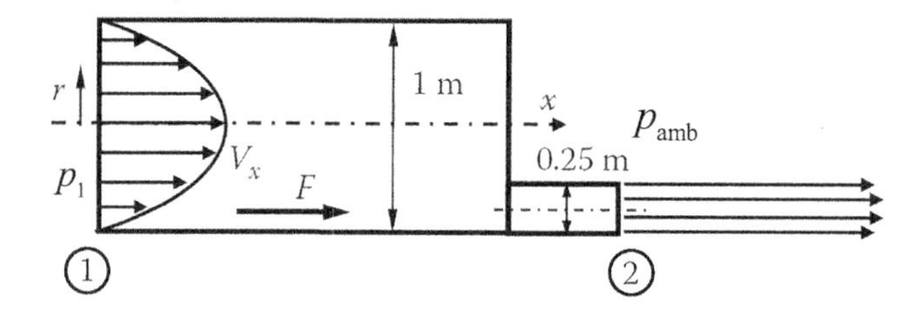

FIGURE 1.50 Force needed to hold a water tank in position under the given inflow and outflow conditions (Problem 1.18).

1.17 Figure 1.49 shows four flow devices resting on frictionless wheels. These devices are restricted to moving in the x direction only and are initially held stationary. The pressure at the inlet and outlet of each device is atmospheric, and the flows entering and leaving them are incompressible. When released, which device will move to the right and the left?

1.18 Figure 1.50 shows a cylindrical water tank of diameter 1 m with an exit pipe of diameter 0.25 m. The velocity profile at the tank inlet is given by:

$$V_x = 1 - \left(\frac{r}{0.5}\right)^2$$

and that at the pipe exit corresponds to a fully developed turbulent pipe flow with the one-seventh power-law profile. What is the net force needed to hold the tank in position? Use a momentum correction factor of 1.333 at the inlet to the large tank and 1.02 in the exit pipe. Assume that the water density is 1000 kg/m^3 and the static pressure at the tank inlet is 1.4 bar. The ambient pressure equals 1 bar.

1.19 Consider the fully developed incompressible turbulent flow of a fluid of density ρ having an average velocity V through an annulus of inner diameter D_1, outer diameter D_2, and length L. The surface roughness of the inner wall of the annulus is higher than that of its outer wall. For the prevailing flow Reynolds number and the relative roughness parameters, test data indicate that the Moody friction factor for the inner wall is f_1 and that for the outer wall is f_2. The static pressure drop across the annulus is given by:

$$p_1 - p_2 = \frac{f^* L}{D_h}\left(\frac{1}{2}\rho V^2\right)$$

Find an expression to compute f^* in terms of the given quantities.

1.20 Figure 1.51 shows an accelerating cart containing a fluid of constant density ρ. The slow motion of a heavy plate generates a constant mass flow rate $\dot{m}$ through the nozzle with an exit velocity W_0 relative to the cart. The total initial mass of the cart system is m_0. Assuming the air drag and contact friction to be negligible, calculate the time-dependent velocity U_c of the cart if it starts from rest.

1.21 A horizontal water jet of constant velocity V_j and area A_j enters a vane-on-cart system of mass m_c and leaves it at an angle θ, causing the vane-on-cart system to accelerate (Figure 1.52). The mass of water with velocity V_j within the cart control volume remains constant at m_w. Find the angle at which the jet leaves the vane if the velocity of the cart is known to be U_{c1} at time t_1. Consider the area and absolute velocity of the jet to remain constant as it enters and exits the vane. Neglect any frictional force resisting the cart motion.

1.22 Find the time-dependent acceleration of the cart shown in Figure 1.53. The initial total mass of the cart containing a fluid of density ρ is m_0. Other quantities needed to find the solution are shown in the figure. Neglect any frictional force opposing the cart movement.

1.23 Figure 1.54 shows a lawn sprinkler with two unequal arms. For the given geometric quantities and the jet velocity at the exit of the nozzle attached to each sprinkler arm, find an

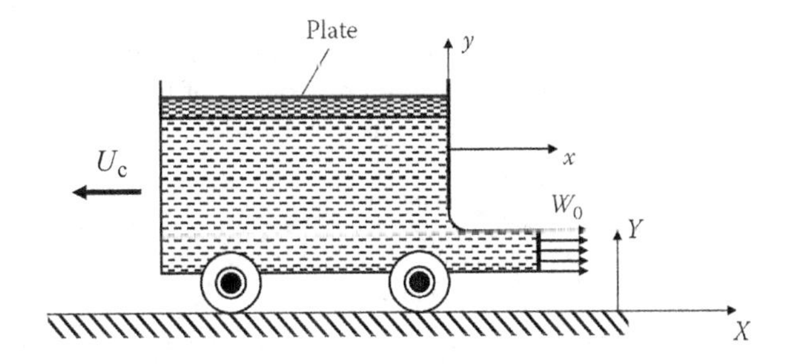

FIGURE 1.51 An accelerating cart with constant outflow (Problem 1.20).

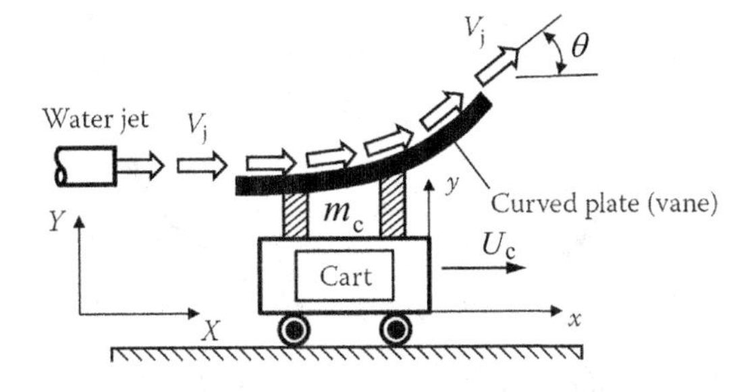

FIGURE 1.52 An accelerating cart under a deflected water jet (Problem 1.21).

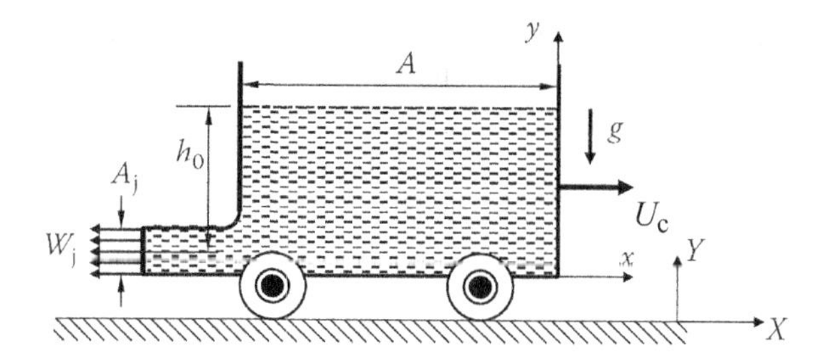

FIGURE 1.53 An accelerating cart with variable outflow (Problem 1.22).

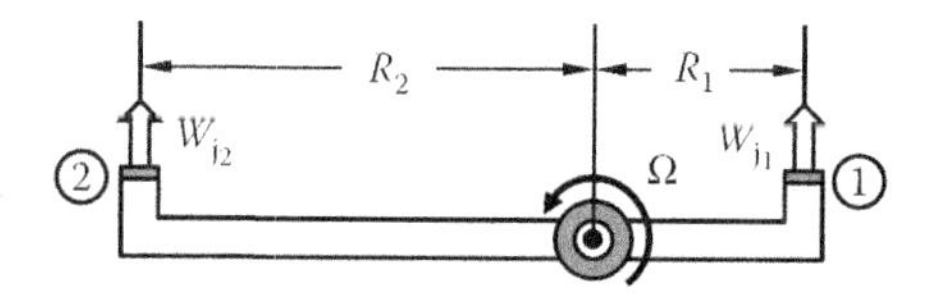

FIGURE 1.54 A lawn sprinkler with two unequal arms (Problem 1.23).

expression to calculate the speed of rotation of the sprinkler with negligible frictional torque. Assume that A_{j1} is the jet area at outlet 1 and A_{j2} is the jet area at outlet 2.

1.24 As shown in Figure 1.55, a high-pressure rotary arm is used for air impingement cooling of a cylindrical surface. The total pressure and the total temperature inside the rotary arm are 3 bar and 134.5°C, respectively. The static pressure outside the rotary arm is 1 bar. At maximum rpm, the rotary arm needs to overcome a frictional torque of 12.5 Nm. For the given geometric data, calculate the maximum rpm of the rotary arm. For air, assume $\gamma = 1.4$ and $R = 287\,\text{J/(kg K)}$.

Geometric data: jet diameter (d_j) = 7 mm, R_1 = 50 cm, R_2 = 100 cm, and R_3 = 140 cm. (Note: each air nozzle operates under the choked flow condition with identical jet velocity relative to the rotary arm.)

1.25 Figure 1.55 shows a frictionless rotary arm used for air impingement cooling of a cylindrical surface. The total pressure and the total temperature of the air inside the rotary arm feeding each isentropic convergent–divergent nozzle with an exit-to-throat area ratio of

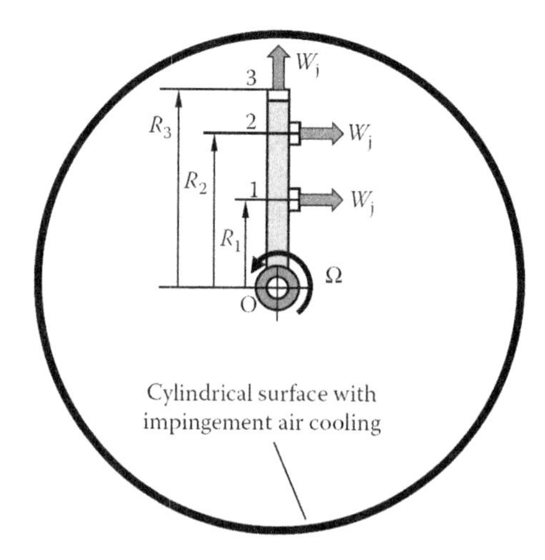

FIGURE 1.55 Impingement air cooling of a cylindrical surface with a rotary arm with three jets (Problems 1.24 through 1.26).

2.0 are 10.646 bar and 78.6°C, respectively. The static pressure in the cylinder is 1 bar. For the given geometric data and air properties, calculate the following:

a. Maximum rpm attained by the rotary arm.
b. Time it takes for the rpm to become half its maximum value computed in part (a) when the jets at locations 1 and 2 are simultaneously turned off. The mass moment of inertia of the rotary arm about the axis of rotation is 5.0 kgm^2.

Geometric data: nozzle throat diameter (d_j) = 10 mm, R_1 = 50 cm, R_2 = 100 cm, and R_3 = 125 cm.
Air properties: $\gamma = 1.4$ and $R = 287\ \text{J/(kg K)}$.

1.26 Figure 1.55 shows an adiabatic, frictionless rotary arm used for air impingement cooling of a cylindrical surface. The total pressure and the total temperature of the cooling air at the rotary arm inlet (at the axis of rotation) are 10^6 Pa and 373 K, respectively. The static pressure in the cylinder is 10^5 Pa. Only the jet at location 3 is turned on. The jets at locations 1 and 2 are turned off. Note that under all operating conditions the flow through the convergent impingement nozzle remains choked (the Mach number = 1.0) at its throat. Calculate the following:

a. The mass flow rate through the rotary arm when it is stationary.
b. The mass flow rate through the rotary arm when it is rotated at 3000 rpm by an electric motor.
c. The power consumption of the electric motor.

Assume the following data for your calculations:
Geometric data: nozzle throat diameter (d_j) = 0.01 m, R_1 = 0.40 m, R_2 = 0.75 m, and R_3 = 1.0 m.
Air properties: $\gamma = 1.4$ and $R = 287\ \text{J/(kg K)}$.

1.27 Figure 1.56 shows the operation of a siphon. The fluid density and kinematic viscosity are $750\ \text{kg/m}^3$ and $1.0 \times 10^{-4}\ \text{m}^2/\text{s}$, respectively. The ambient pressure is 1.0 bar. Calculate the static pressure at the highest point C inside the siphon tube under three assumptions: (a) the flow velocity is uniform throughout the siphon tube, and there are no head losses in the system from A to D; (b) the flow within the siphon tube from B to D is laminar with a parabolic velocity profile, and there are no head losses in the system; and (c) same as in (b) with frictional loss within the siphon tube with smooth wall. The inner diameter of the siphon tube is 0.01 m.

1.28 Venturi flow meter shown in Figure 1.57 is used to measure pipe flow rate. This simple device consists of a converging nozzle with a throat followed by a gradually diverging section (diffuser). The drop in static pressure between the upstream section and the throat is measured by a pair of piezometer tubes or a U-tube differential gauge, as shown in the figure. If the loss in the head between 1–1 and 2–2 is given by:

$$h_{\text{loss}} = K\frac{V_2^2}{2g}$$

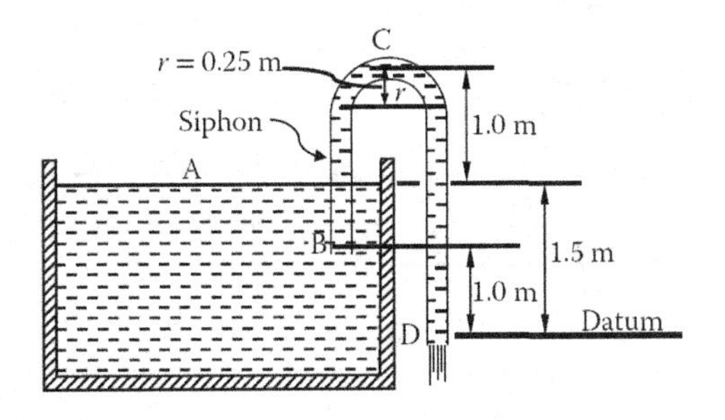

FIGURE 1.56 Minimum pressure in a siphon tube (Problem 1.27).

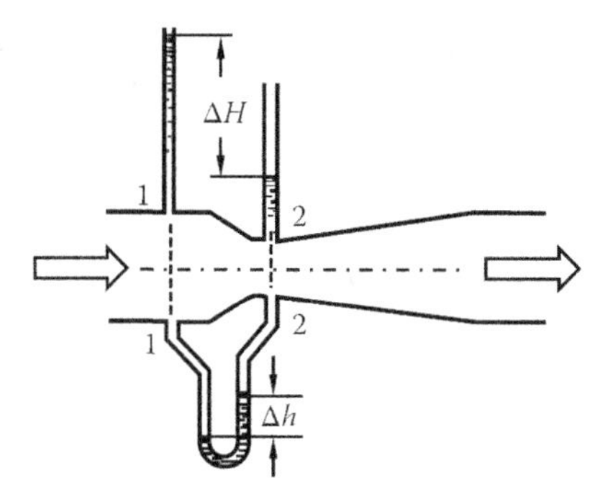

FIGURE 1.57 Venturi flow meter (Problem 1.28).

find an expression to compute the volumetric flow rate measured by the venturi flow meter.

1.29 An electrical engineer, Tom, loves his old scooter that uses a carburetor. He often fine-tunes the scooter carburetor to obtain an optimal air–fuel mixture ratio, saving gas. He, however, needs to understand fully how the carburetor works. Figure 1.58 shows the schematic diagram of a carburetor where the stream of air passes through a venturi tube housing a fuel jet. As the velocity of air increases in the venturi throat, the static pressure drops according to the Bernoulli equation. This reduction in static pressure causes the fuel to flow as a fuel jet into the air stream creating a homogeneous air–fuel mixture that undergoes combustion downstream. Assuming the minor loss coefficients of K_a and K_g , respectively, in the air and gasoline flow passages and neglecting all related significant losses, use the extended Bernoulli equation to find an expression to compute the air–fuel mixture ratio for this carburetor. Use your result to explain to Tom how the air–fuel mixture ratio changes with the change in venturi throat diameter D.

1.30 Figure 1.59 shows an ejector pump used in many engineering devices. Describe the working principle of this pump. Using the Bernoulli equation, determine the ratio Q_2 / Q_1. Make any simplifying assumptions needed for your analysis.

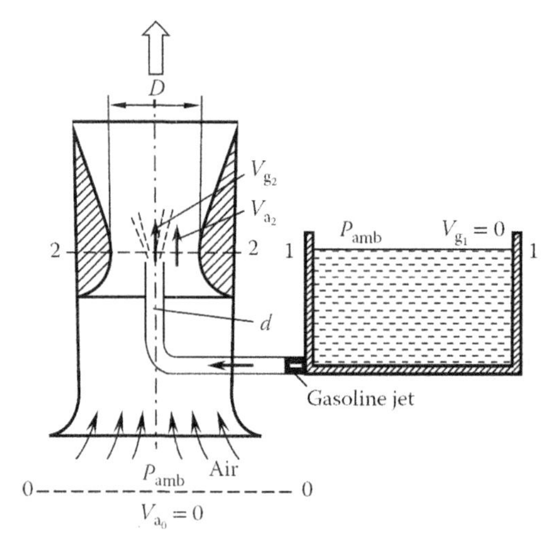

FIGURE 1.58 Air–fuel mixture ratio in a carburetor (Problem 1.29).

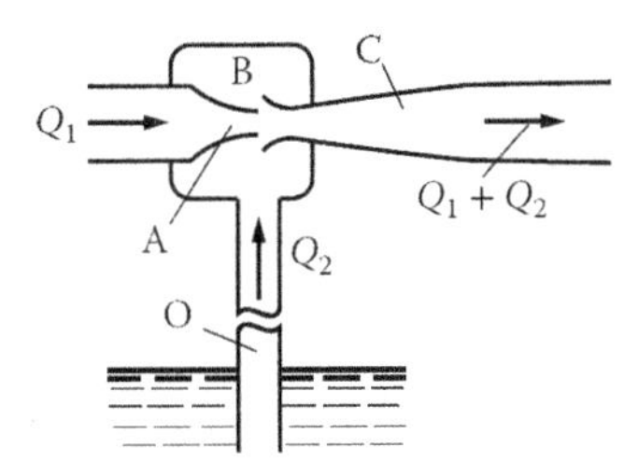

FIGURE 1.59 An ejector pump (Problem 1.30).

1.31 As shown in Figure 1.60, water of density $750\ \mathrm{kg/m^3}$ flows through a pipe whose inner diameter varies from 10 to 20 cm over a length of 1 m. If the Darcy friction factor for the entire pipe is constant at 0.005, find the major head loss due to friction within the pipe for a volumetric flow rate of $1.0\ \mathrm{m^3/s}$.

1.32 Figure 1.61 shows a pump delivering water from the tank at a lower elevation to the tank at a higher elevation. The relationship between the head and discharge of the pump is given by:

$$H_{\text{pump}} = 100 - 8000\,Q^2$$

Elevations of the free surface A and free surface G, measured from a fixed datum, are 20 and 50 m, respectively. Using the following data for the piping on the suction and discharge sides of the pump, calculate the discharge and power delivered by the pump.

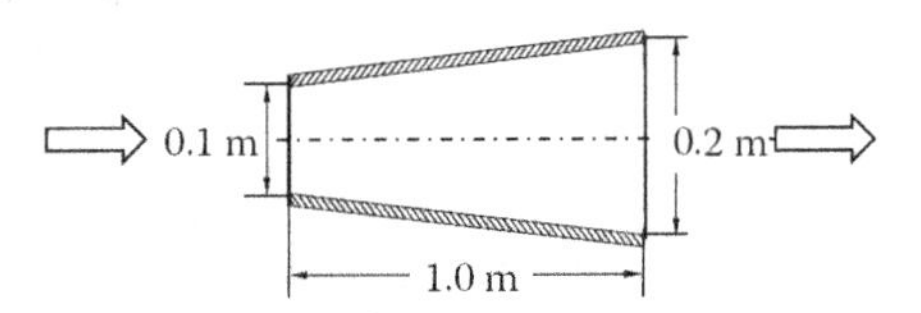

FIGURE 1.60 Head loss in a diverging pipe flow (Problem 1.31).

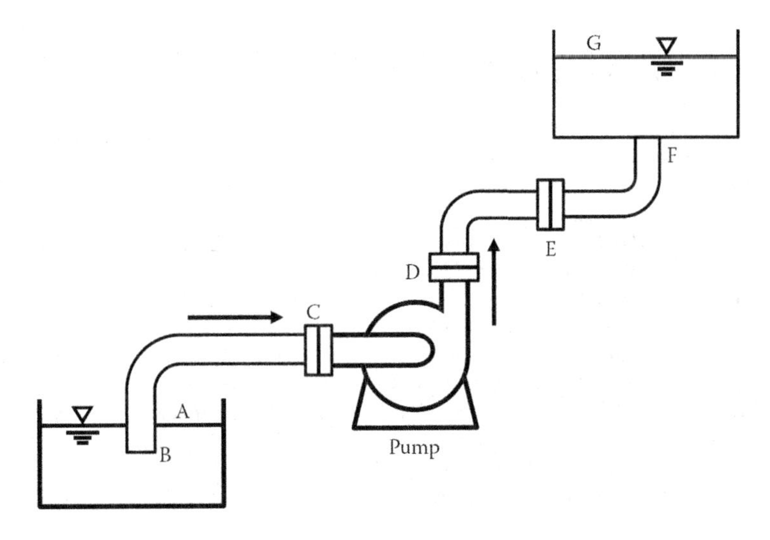

FIGURE 1.61 Discharge and power delivered by a pump in a flow system (Problem 1.32).

Suction side:

$$D_1 = 0.2 \text{ m}; L_{BC} = 5.0 \text{ m}; f_1 = 0.02; K_1 = 0.1$$

Discharge side:

$$D_2 = 0.1 \text{ m}; L_{DF} = 10.0 \text{ m}; f_2 = 0.025; K_2 = 0.2$$

REFERENCES

Bird, R.B., W.E. Stewart, and E.N. Lightfoot. 2007. *Transport Phenomena,* Revised 2nd edition. New York: John Wiley.

Hansen, A.G. 1967. *Fluid Mechanics.* New York: Wiley.

Potter, M.C. and J.F. Foss. 1975. *Fluid Mechanics.* Okemos: Great Lakes Press.

Reynolds, W.C. and P. Colonna. 2018. *Thermodynamics: Fundamentals and Engineering Applications.* Cambridge: Cambridge University Press.

Sultanian, B.K. 2015. *Fluid Mechanics: An Intermediate Approach.* Boca Raton, FL: Taylor & Francis.

Sultanian, B.K. 2021. *Fluid Mechanics and Turbomachinery: Problems and Solutions.* Boca Raton, FL: Taylor & Francis.

Sultanian, B.K. 2022. *Thermal-Fluids Engineering: Problems with Solutions.* Independently published by Kindle Direct Publishing (Amazon.com).

BIBLIOGRAPHY

Durst, F. 2022. *Fluid* Mechanics*: An Introduction to the Theory of Fluid Flows,* 2nd edition. Heidelberg: Springer Verlag GmbH.

Falkovich, G. 2018. *Fluid Mechanics*, 2nd edition. Cambridge: Cambridge University Press.

Fox, W., P. Prichard, A. McDonald. 2010. *Introduction to Fluid Mechanics*, 7th edition. New York: John Wiley & Sons.

Greitzer, E.M., C.S. Tan, and M.B. Graf. 2004. *Internal Flow Concepts and Applications.* Cambridge: Cambridge University Press.

Kundu, P.K., I.M. Cohen, D.R. Dowling, J. Capecilatro. 2024. *Fluid Mechanics*, 7th edition. Burlington: Academic Press.

Lugt, H.J. 1995. *Vortex Flow in Nature and Technology.* Malabar: Krieger Publishing Company.

Marusic, I. and S. Broomhall. 2021. Leonardo da Vinci and fluid mechanics. *Annual Review of Fluid Mechanics.* 53: 1–25.

Panton, R.L. 2024. *Incompressible Flow,* 5th edition. New York: Wiley.

Powers, J.M. 2023. *Mechanics of Fluids*, 1st edition. Cambridge: Cambridge University Press.

Samimy, M., K.S. Breuer, L.G. Leal, et al. 2003. *A Gallery of Fluid Motion.* Cambridge: Cambridge University Press.

Sultanian, B.K. 2018. *Gas Turbines: Internal Flow Systems Modeling* (Cambridge Aerospace Series). Cambridge: Cambridge University Press.

Sultanian, B.K. 2019. *Logan's Turbomachinery: Flowpath Design and Performance Fundamental,* 3rd edition. Boca Raton: Taylor & Francis.

Sultanian, B.K. 2022a. *FLUID MECHANICS An Intermediate Approach*: Errata for *the First Edition Published in 2015.* Independently published on KDP (Amazon.com).

Sultanian, B.K. 2022b. *GAS TURBINES Internal Flow Systems Modeling: Errata for the First Edition Published in 2018.* Independently published on KDP (Amazon.com).

Sultanian, B.K. 2022c. *Power Generation Gas Turbines: High-Performance Exhaust Diffuser Design.* Independently published on KDP (Amazon.com).

Turns, S.R. and L.L. Pauley. 2020. *Thermodynamics: Concepts and Applications*, 2nd edition. Cambridge: Cambridge University Press.

Van Dyke, M. 1982. *An Album of Fluid Motion.* Stanford: The Parabolic Press.

Visconti, G. and P. Ruggieri. 2020. *Fluid Dynamics: Fundamental and Applications.* Cham, Switzerland: Spinger.

White, F.M and J. Majdalani. 2022. *Viscous Fluid Flow*, 4th edition. New York: McGraw-Hill.

NOMENCLATURE

Symbol	Definition
A	Flow area (magnitude)
$\mathbf{A}$	Flow area (vector)
c_p	Specific heat of gas at constant pressure
c_v	Specific heat of gas at constant volume
C	Speed of sound
D	Pipe diameter
F	Force (magnitude)
$\mathbf{F}$	Force (vector)
e	Specific total energy of a system
E	Total energy of a system
g	Acceleration due to gravity
h	Specific enthalpy
H	Angular momentum
$\dot{H}$	Angular momentum flow rate
$\tilde{I}$	Summation over inlets or outlets
k	Thermal conductivity
m	Mass
$\dot{m}$	Mass flow rate
M	Mach number; linear momentum (magnitude)
$\mathbf{M}$	Linear momentum (vector)
MEE	Mechanical energy equation
$\mathbf{n}$	Surface normal unit vector
N	Number of inlets or outlets
p	Pressure
P	Power
q	Amount of heat transfer (per unit mass)
$\dot{q}$	Heat flux
$\dot{\mathbf{q}}$	Heat flux vector
$\dot{Q}$	Heat transfer rate
Q	Volumetric flow rate
r	Cylindrical polar coordinate r
$\mathbf{r}$	Radial vector
R	Gas constant
s	Specific entropy
t	Time
T	Temperature
u	Specific internal energy
V	Total absolute velocity (magnitude)
$\mathbf{V}$	Total absolute velocity (vector)
$\forall$	Volume
W	Total relative velocity (magnitude)
$\mathbf{W}$	Total relative velocity (vector)
$\dot{W}$	Rate of work transfer
x	Cartesian coordinate x
y	Cartesian coordinate y
z	Cartesian coordinate z, distance measured from a datum

SUBSCRIPTS AND SUPERSCRIPTS

0	Stagnation (total)
1	Section 1
2	Section 2
amb	Ambient
b	Body
bx	Body force in the x direction
cv	Control volume
cs	Control surface
f	Friction
hyd	Hydraulic
i	Index for inlets
inlets	Inlets
irrev	Irreversible
j	Index for outlets
m	Meridional
ms	Material surface
mv	Material volume
n	Normal to flow area
outlets	Outlets
p	Pressure
r	Component in the radial direction
R	Rotor reference frame
rel	Relative
rev	Reversible
s	Surface; streamline
sx	Surface force in the x direction
sys	System
v	Volume
x	Component in the x direction; axial direction
xr	Axial–radial (meridional) plane
y	Component in the y direction
z	Component in the z direction
θ	Component in θ direction
*	Value at the sonic condition ($M = 1$)
‾	Average value

GREEK SYMBOLS

α	Kinetic energy correction factor
Γ	Torque
γ	Ratio of specific heats ($\gamma = c_p/c_v$)
π	Change agent per unit volume
Π	Change agent
ρ	Density
Φ	Extensive general property
ϕ	Intensive general property
Ω	Rotational velocity around the axial direction

2 Potential Flow

2.1 INTRODUCTION

All natural fluids have nonzero viscosity, whose effect is mainly confined to wall boundary layers or to shear layers in complex shear flows away from the wall, for example, in the recirculation regions of a sudden expansion pipe flow. Outside the shear layers, the fluid may be considered ideal with zero viscosity, and the flow field is modeled as a potential or irrotational flow, rendering considerable simplifications. The term "potential flow" originates from obtaining the three-dimensional velocity vector field of such a flow from the gradient of a three-dimensional scalar potential function. Irrotational flow, on the other hand, connotes that the curl of the three-dimensional velocity vector field associated with the flow is zero. Since, from vector calculus, the curl of the gradient of a scalar function is identically zero, a potential flow must always be irrotational or vice versa. Potential flows are also known as ideal flows. They are isentropic with constant total pressure everywhere. Most fluid flows, external flows in particular, are dominated by potential flows, and as discussed in Chapter 4, we impose the static pressure distribution in a wall boundary layer by the potential flow in the free stream away from the wall.

This chapter is of significant historical importance, introducing to readers many founding concepts of early fluid mechanics dominated by analytical solutions. Although this chapter primarily deals with two-dimensional potential flows, which are incompressible and inviscid, most of the basic concepts presented here equally apply to three-dimensional potential flows.

This chapter features six worked examples and 13 chapter-end problems. The original version of this chapter appears as Chapter 6 in the first edition of this book by Sultanian (2015).

2.2 BASIC CONCEPTS

2.2.1 Velocity Potential

We can express the velocity field of a potential or irrotational flow as:

$$V = \nabla \Phi \tag{2.1}$$

where $\Phi(x, y, z)$ is, in general, a three-dimensional scalar function. This equation identically satisfies the condition of irrotationality ($\nabla \times V = 0$). The vector field V obtained from this equation will represent a flow field only if it satisfies the continuity equation:

$$\nabla \bullet V = 0 \tag{2.2}$$

whose substitution in Equation 2.1 yields the following Laplace equation:

$$\nabla^2 \Phi = 0 \tag{2.3}$$

This equation represents many other analogous physical phenomena, for example, steady heat conduction in a solid with temperature-independent thermal conductivity.

DOI: 10.1201/9781003325192-2

2.2.2 Stream Function

For a two-dimensional incompressible flow, we define a scalar stream function Ψ, giving the velocity components in the Cartesian coordinate system as:

$$V_x = \frac{\partial \Psi}{\partial y} \tag{2.4}$$

$$V_y = -\frac{\partial \Psi}{\partial x} \tag{2.5}$$

which automatically satisfy the continuity equation at every point in the flow field:

$$\frac{\partial V_x}{\partial x} + \frac{\partial V_y}{\partial y} = \frac{\partial^2 \Psi}{\partial x \partial y} - \frac{\partial^2 \Psi}{\partial x \partial y} = 0$$

The condition of irrotationality for a two-dimensional potential flow in the z plane (the plane normal to the z-coordinate direction), or the x–y plane, results in:

$$\frac{\partial V_y}{\partial x} - \frac{\partial V_x}{\partial y} = 0$$

Substituting V_x from Equation 2.4 and V_y from Equation 2.5 in this equation yields:

$$\frac{\partial^2 \Psi}{\partial x^2} + \frac{\partial^2 \Psi}{\partial y^2} = \nabla^2 \Psi = 0 \tag{2.6}$$

Thus, the stream function defined for a two-dimensional potential flow must also satisfy the Laplace equation (Equation 2.6).

Let us consider a line of constant stream function in a two-dimensional flow. For a differential change in Ψ along this flow line, we write:

$$\mathrm{d}\Psi = \frac{\partial \Psi}{\partial x}\mathrm{d}x + \frac{\partial \Psi}{\partial y}\mathrm{d}y = 0$$

Substituting $\partial \Psi / \partial x$ from Equation 2.4 and $\partial \Psi / \partial y$ from Equation 2.5 in this equation yields:

$$-V_y \mathrm{d}x + V_x \mathrm{d}y = 0$$

$$\left(\frac{\mathrm{d}y}{\mathrm{d}x}\right)_\Psi = \frac{V_y}{V_x} \tag{2.7}$$

which shows that the local velocity vector is tangent to the flow line of constant Ψ. As the velocity vector is also tangent to a streamline, we conclude that Ψ remains constant along a streamline. Therefore, each constant value of Ψ gives a different streamline.

Figure 2.1 shows a two-dimensional duct of unit depth formed by two streamlines. The volumetric flow rate Q per unit depth through an arbitrary section AB can be expressed as:

$$\begin{aligned} Q &= \int_A^B V_x \mathrm{d}y - \int_A^B V_y \mathrm{d}x = \int_A^B (V_x\,\mathrm{d}y - V_y\,\mathrm{d}x) \\ Q &= \int_A^B \left(\frac{\partial \Psi}{\partial y}\mathrm{d}y + \frac{\partial \Psi}{\partial x}\mathrm{d}x\right) = \int_A^B \mathrm{d}\Psi = \Psi_2 - \Psi_1 \end{aligned} \tag{2.8}$$

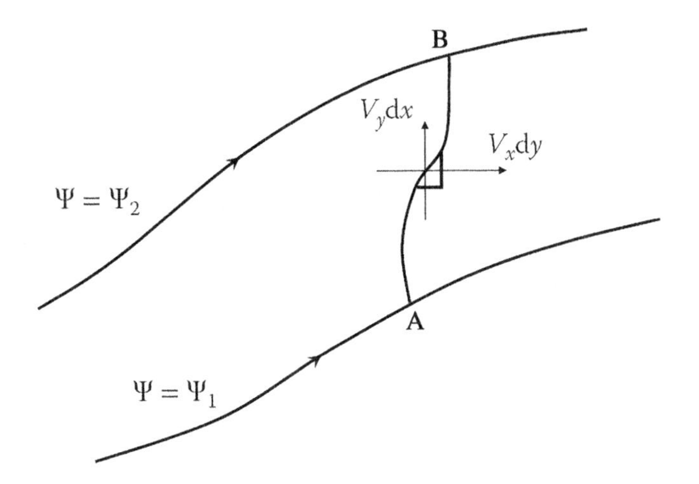

FIGURE 2.1 Volumetric flow rate through a two-dimensional duct formed by two streamlines.

which shows that the volumetric flow rate per unit depth across an arbitrary section in a two-dimensional flow is the difference in stream function values of the bounding streamlines.

Steam function Ψ is constant along a streamline in a two-dimensional flow, and Φ is constant along an equipotential line. Let us find the angle at which these two lines intersect each other. As $\mathrm{d}\Phi = 0$ along an equipotential line, we write:

$$\mathrm{d}\Phi = \frac{\partial \Phi}{\partial x}\mathrm{d}x + \frac{\partial \Phi}{\partial y}\mathrm{d}y = 0$$

$$V_x \mathrm{d}x + V_y \mathrm{d}y = 0 \tag{2.9}$$

$$\left(\frac{\mathrm{d}y}{\mathrm{d}x}\right)_{\Phi} = -\frac{V_x}{V_y}$$

From Equations 2.7 and 2.9, we obtain the following relation between the slope of a streamline and that of an equipotential line at their intersection point:

$$\left(\frac{\mathrm{d}y}{\mathrm{d}x}\right)_{\Psi}\left(\frac{\mathrm{d}y}{\mathrm{d}x}\right)_{\Phi} = -1 \tag{2.10}$$

Equation 2.10 proves that a streamline and an equipotential line are orthogonal at their intersection point, as shown in Figure 2.2a. Figure 2.2b depicts a net formed by the intersections of streamlines and equipotential lines in a two-dimensional duct flow—a technique often used for generating an orthogonal computational fluid dynamics (CFD) mesh in a two-dimensional planar or three-dimensional axisymmetric duct.

2.2.3 Complex Potential, Complex Velocity, and Complex Circulation

In a two-dimensional potential flow, we obtain the velocity components V_x and V_y from the potential function Φ and stream function Ψ in the Cartesian coordinates using the following equations:

$$V_x = \frac{\partial \Phi}{\partial x} = \frac{\partial \Psi}{\partial y} \text{ and } V_y = \frac{\partial \Phi}{\partial y} = -\frac{\partial \Psi}{\partial x} \tag{2.11}$$

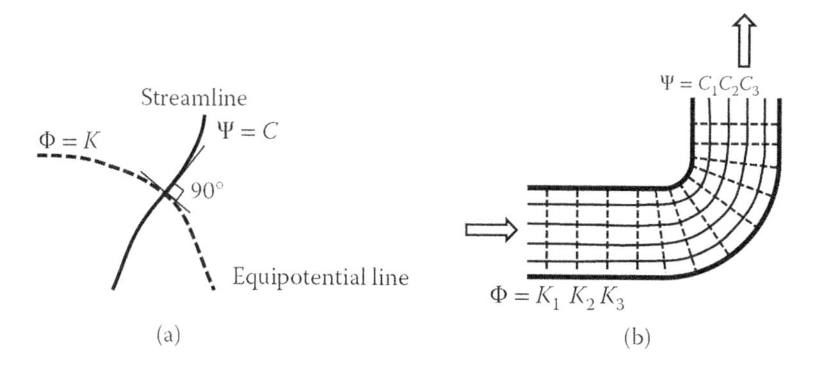

FIGURE 2.2 (a) Streamline and equipotential line intersecting at 90° and (b) flow net formed by intersecting streamlines and equipotential lines in a two-dimensional duct.

Similarly, we obtain V_r and V_θ from Φ and Ψ in polar coordinates using the equations:

$$V_r = \frac{\partial \Phi}{\partial r} = \frac{\partial \Psi}{r\,\partial \theta} \text{ and } V_\theta = \frac{\partial \Phi}{r\,\partial \theta} = -\frac{\partial \Psi}{\partial r} \tag{2.12}$$

Equations 2.11 and 2.12 represent the Cauchy–Riemann equations for the functions Φ and Ψ, making them conjugate.

2.2.3.1 Complex Potential

For a two-dimensional potential flow, let us define a complex potential as:

$$\mathrm{F}(z) = \Phi(x, y) + \mathrm{i}\Psi(x, y) \tag{2.13}$$

where the complex coordinate $z = x + \mathrm{i}y$. Being conjugate functions, $\Phi(x, y)$ and $\Psi(x, y)$ satisfy the necessary and sufficient conditions for the complex potential $F(z)$ to be analytic.

2.2.3.2 Complex Velocity

At an arbitrary point $z = x + \mathrm{i}y$ in a two-dimensional potential flow, let us evaluate the derivative of $F(z)$ with respect to z as:

$$\frac{\mathrm{d}F(z)}{\mathrm{d}z} = \lim_{\delta z \to 0} \frac{F(z + \delta z) - F(z)}{\delta z}$$

As $F(z)$ is an analytic complex-valued function, the value of the derivative is independent of the direction we choose to carry out the limit of going from $z + \delta z$ to z. Let us first take the limit in the x direction (y = constant) with $\delta z = \delta x$, giving:

$$\frac{\mathrm{d}F(z)}{\mathrm{d}z} = \lim_{\delta x \to 0} \frac{\Phi(x + \delta x, y) + \mathrm{i}\Psi(x + \delta x, y) - \Phi(x, y) - \mathrm{i}\Psi(x, y)}{\delta x}$$
$$\frac{\mathrm{d}F(z)}{\mathrm{d}z} = \frac{\delta \Phi}{\delta x} + \mathrm{i}\frac{\delta \Psi}{\delta x} = V_x - \mathrm{i}V_y = \overline{W} \tag{2.14}$$

Let us now take the limit along the y direction—the imaginary axis (x = constant)—with $\delta z = \mathrm{i}\delta y$, giving:

$$\frac{dF(z)}{dz} = \lim_{\delta y \to 0} \frac{\Phi(x, y+\delta y) + i\Psi(x, y+\delta y) - \Phi(x, y) - i\Psi(x, y)}{i\delta y}$$

$$\frac{dF(z)}{dz} = -i\frac{\delta\Phi}{\delta y} + \frac{\delta\Psi}{\delta y} = V_x - iV_y$$

which is identical to that given by Equation 2.14. Thus, we obtain the complex conjugate velocity $\bar{W}(z) = V_x - iV_y$ as the ordinary derivative of the complex potential $F(z)$ with respect to z.

To obtain the scalar product of the velocity vector, we multiply the complex velocity by its complex conjugate:

$$W\bar{W} = (V_x + iV_y)(V_x - iV_y) = V_x^2 + V_y^2 \tag{2.15}$$

Figure 2.3 shows the components of complex velocity in the Cartesian and polar coordinates, related as follows:

$$V_x = V_r\cos\theta - V_\theta\sin\theta$$

$$V_y = V_r\sin\theta + V_\theta\cos\theta$$

We can express the complex velocity in polar coordinates as:

$$W = V_x + iV_y = (V_r\cos\theta - V_\theta\sin\theta) + i(V_r\sin\theta + V_\theta\cos\theta)$$

$$W = V_r(\cos\theta + i\sin\theta) + iV_\theta(\cos\theta + i\sin\theta) \tag{2.16}$$

$$W = (V_r + iV_\theta)e^{i\theta}$$

Similarly, we can express the complex conjugate velocity in polar coordinates as:

$$\bar{W} = (V_r - iV_\theta)e^{-i\theta} \tag{2.17}$$

2.2.3.3 Complex Circulation

For a simply connected closed contour in a two-dimensional potential flow, we define the complex circulation $C(z)$ as:

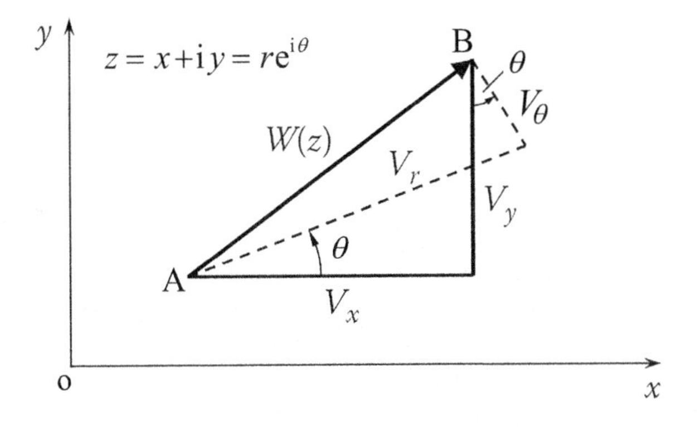

FIGURE 2.3 Components of complex velocity in the Cartesian and polar coordinates.

$$C(z) = \oint_C \overline{W}\,\mathrm{d}z = \oint_C (V_x - \mathrm{i}V_y)(\mathrm{d}x + \mathrm{i}\,\mathrm{d}y)$$

$$C(z) = \oint_C (V_x\,\mathrm{d}x + V_y\,\mathrm{d}y) + \mathrm{i}\oint_C (V_x\,\mathrm{d}y - V_y\,\mathrm{d}x) \tag{2.18}$$

$$C(z) = \oint_C \boldsymbol{V} \bullet \hat{\boldsymbol{m}}\,\mathrm{d}\ell + \mathrm{i}\oint_C \boldsymbol{V} \bullet \hat{\boldsymbol{n}}\,\mathrm{d}\ell = \Gamma + \mathrm{i}Q$$

where $\hat{\boldsymbol{m}}$ is the unit vector along the line segment $\mathrm{d}\ell$, and $\hat{\boldsymbol{n}}$ is the unit vector normal to this line segment. The real part of $C(z)$ is the circulation Γ of the fluid, and the imaginary part gives the volumetric flow rate Q (per unit length perpendicular to the plane of the flow) that results from the sources (singularities, discussed in Section 2.2.4) enclosed within the contour of integration. For a singularity-free, simply connected flow region, we obtain $C(z) = 0$.

2.2.4 Two Basic Types of Singularities

All potential flows must satisfy two kinematic conditions: first, $\nabla \bullet \boldsymbol{V} = 0$ (zero divergence) to satisfy the continuity equation with constant density, and second, $\boldsymbol{\omega} = \nabla \times V = 0$ (zero curl), being irrotational. A source or sink at a point in a potential flow violates the first condition, while a vortex (counterclockwise or clockwise) at a point violates the second condition. In Section 2.4, we present various practical potential flows using combinations of sources/sinks and vortices along with a uniform flow far away from these singularities.

2.3 ELEMENTARY PLANE POTENTIAL FLOWS

2.3.1 Uniform Flows

2.3.1.1 Uniform Flow Parallel to the *x* axis

For the complex potential:

$$F(z) = Uz \tag{2.19}$$

we obtain the complex conjugate velocity:

$$\overline{W} = V_x - \mathrm{i}V_y = \frac{\mathrm{d}F}{\mathrm{d}z} = U \tag{2.20}$$

where $V_x = U$ and $V_y = 0$. Figure 2.4a shows the streamlines in this flow.

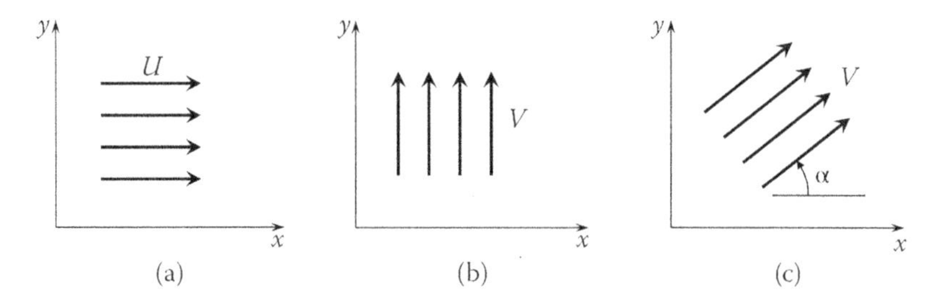

FIGURE 2.4 (a) Uniform flow parallel to the x axis, (b) uniform flow parallel to the y axis, and (c) uniform flow inclined at an angle α to the x axis.

2.3.1.2 Uniform Flow Parallel to the *y* axis

For the complex potential:

$$F(z) = -\mathrm{i}Vz \tag{2.21}$$

we obtain the complex conjugate velocity:

$$\overline{W} = V_x - \mathrm{i}V_y = \frac{\mathrm{d}F}{\mathrm{d}z} = -\mathrm{i}V \tag{2.22}$$

where $V_x = 0$ and $V_y = V$. Figure 2.4b depicts the streamlines in this flow.

Uniform Flow Inclined at an Angle α to the x axis

For the complex potential function:

$$F(z) = V\mathrm{e}^{-\mathrm{i}\alpha}z \tag{2.23}$$

we obtain the complex conjugate velocity:

$$\overline{W} = V_x - \mathrm{i}V_y = \frac{\mathrm{d}F}{\mathrm{d}z} = V\mathrm{e}^{-\mathrm{i}\alpha} = V\cos\alpha - \mathrm{i}V\sin\alpha \tag{2.24}$$

where $V_x = V\cos\alpha$ and $V_y = V\sin\alpha$. Figure 2.4c shows that the resulting velocity field corresponds to a uniform flow inclined at an angle α to the x axis.

Equations 2.19 and 2.23 indicate that when we multiply the complex potential $\mathrm{e}^{-\mathrm{i}\alpha}$, we rotate the entire flow field counterclockwise by an angle α. Accordingly, we can obtain the complex potential for the uniform flow velocity parallel to the y axis by multiplying $F(z) = Vz$, which is the potential function for the uniform flow with velocity V parallel to the x axis, by $\mathrm{e}^{-\mathrm{i}\pi/2} = -\mathrm{i}$. The resulting potential function $F(z) = -\mathrm{i}Vz$ is identical to Equation 2.21. In other words, because the equipotential lines are orthogonal to streamlines, we can switch them by multiplying the corresponding complex potential function by (–i). As a result, the new stream function becomes the negative of the old potential function and the old stream function becomes the new potential function.

2.3.2 Source and Sink

For the complex potential function:

$$F(z) = c\ln z = c\ln(r\,\mathrm{e}^{\mathrm{i}\theta}) = c\ln r + \mathrm{i}c\,\theta \tag{2.25}$$

which yields $\Phi = c\ln r$ and $\Psi = c\theta$. In the flow field generated by this potential function, the lines of constant r are the equipotential lines and the lines of constant θ are the streamlines. From this potential function, we obtain the complex conjugate velocity as:

$$\overline{W} = \frac{\mathrm{d}F(z)}{\mathrm{d}z} = \frac{c}{z} = \frac{c}{r}\mathrm{e}^{-\mathrm{i}\theta} = \left(\frac{c}{r} - \mathrm{i}0\right)\mathrm{e}^{-\mathrm{i}\theta} = (V_r - \mathrm{i}V_\theta)\mathrm{e}^{-\mathrm{i}\theta} \tag{2.26}$$

which yields $V_r = c/r$ and $V_\theta = 0$. Therefore, the complex potential $F(z) = c\ln z$ represents a source when c is positive and a sink when c is negative, each located at the origin. If the source or the sink is located at $z_0 = x_0 + \mathrm{i}y_0$, we express its complex potential by $F(z) = c\ln(z - z_0)$. Figure 2.5a shows the streamlines and equipotential lines generated by this complex potential for a source, and Figure 2.5b shows them for a sink. As shown in these figures, the velocity in a source or sink flow is purely radial and its magnitude decreases as we move away from the singularity point, characterized by an infinite radial velocity.

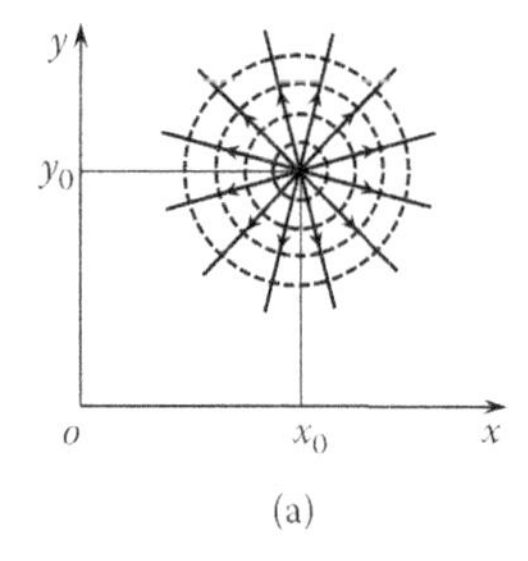

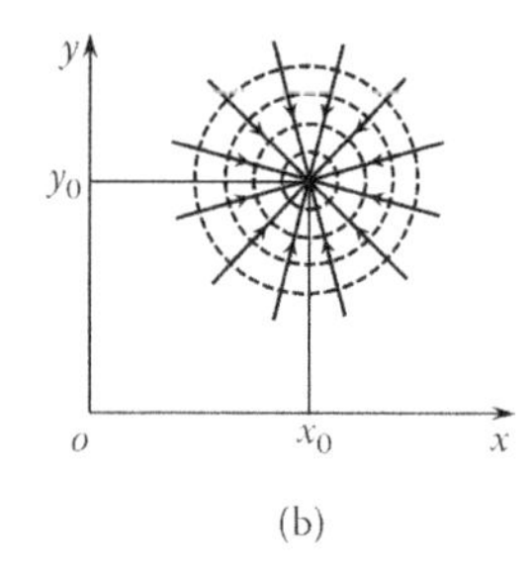

FIGURE 2.5 (a) Streamlines (solid lines) and equipotential lines (dash lines) in a flow due to a source located at z_0 and (b) streamlines (solid lines) and equipotential lines (dash lines) in a flow due to a sink located at z_0.

From the continuity equation, the total volumetric flow rate m (per unit depth) crossing each equipotential line, which is a circle shown in Figure 2.5, must remain constant. Accordingly, we obtain:

$$m = \int_0^{2\pi} \frac{c}{r} r\mathrm{d}\theta = 2\pi c$$
$$c = \frac{m}{2\pi} \tag{2.27}$$

We can write the complex potential function for a source located at z_0 as:

$$F(z) = \frac{m}{2\pi} \ln(z - z_0) \tag{2.28}$$

where m is called the source strength. For a sink, which is a negative source, m is replaced by $(-m)$ in Equation 2.28.

2.3.3 Vortex

By multiplying the complex potential of a source, given by Equation 2.25, by $-\mathrm{i}$, we locally rotate the entire flow field counterclockwise by 90° and obtain the complex potential of a vortex as:

$$F(z) = -\mathrm{i}c \ln z = -\mathrm{i}c \ln(r\,\mathrm{e}^{\mathrm{i}\theta}) = c\theta - \mathrm{i}c \ln r \tag{2.29}$$

which yields $\Phi = c\theta$ and $\Psi = -c\ln r$. In this flow field, the lines of constant radius r are the streamlines and the lines of constant θ are the equipotential lines. We obtain the complex conjugate velocity as:

$$\overline{W} = \frac{\mathrm{d}F(z)}{\mathrm{d}z} = \frac{-\mathrm{i}c}{z} = \frac{-\mathrm{i}c}{r}\mathrm{e}^{-\mathrm{i}\theta} = \left(0 - \mathrm{i}\frac{c}{r}\right)\mathrm{e}^{-\mathrm{i}\theta} = (V_r - \mathrm{i}V_\theta)\mathrm{e}^{-\mathrm{i}\theta} \tag{2.30}$$

which yields $V_r = 0$ and $V_\theta = c / r$. Thus, the complex potential $F(z) = -\mathrm{i}c \ln z$ represents a vortex located at the origin. A positive value of c yields a counterclockwise rotating vortex, considered positive. A negative value of c yields a clockwise rotating vortex, considered negative. If the vortex is located at $z_0 = x_0 + \mathrm{i}y_0$, its complex potential becomes $F(z) = -\mathrm{i}c\ln(z - z_0)$. Figure 2.6a for a positive vortex and Figure 2.6b for a negative vortex show the streamlines and equipotential lines generated by this complex potential. As shown in the figures, the velocity in a vortex flow is purely

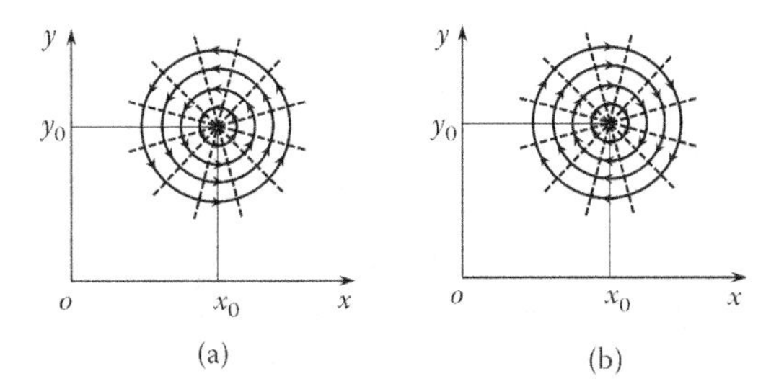

FIGURE 2.6 (a) Streamlines (solid lines) and equipotential lines (broken lines) in a flow due to a counterclockwise vortex located at z_0 and (b) streamlines (solid lines) and equipotential lines (broken lines) in a flow due to a clockwise vortex located at z_0.

tangential, and its magnitude decreases as we move away from the singularity point, characterized by an infinite swirl velocity.

The circulation Γ of a vortex is a measure of its strength given by:

$$\Gamma = \oint \boldsymbol{V} \bullet \hat{\boldsymbol{m}}\, \mathrm{d}\ell$$

$$\Gamma = \int_0^{2\pi} V_\theta r\, \mathrm{d}\theta = \int_0^{2\pi} \frac{c}{r} r\, \mathrm{d}\theta = 2\pi c \tag{2.31}$$

$$c = \frac{\Gamma}{2\pi}$$

We can write the complex potential function for a vortex located at z_0 as:

$$F(z) = -\mathrm{i}\frac{\Gamma}{2\pi}\ln(z - z_0) \tag{2.32}$$

where Γ is the vortex strength. Note that Γ is positive for a counterclockwise vortex and negative for a clockwise vortex.

2.3.4 Corner Flows

We can generate a class of irrotational flows, called corner flows, using the complex potential:

$$F(z) = Az^n \tag{2.33}$$

where A and n are constant. For $n > 1$, the flow occurs over a concave corner, and for $n < 1$, it occurs over a convex corner. We can express this equation in cylindrical polar coordinates as:

$$F(z) = \Phi + \mathrm{i}\Psi = Ar^n \mathrm{e}^{\mathrm{i}n\theta}$$

$$\Phi + \mathrm{i}\Psi = Ar^n \cos(n\theta) + \mathrm{i}Ar^n \sin(n\theta)$$

which yields the stream function for a corner flow as:

$$\Psi = Ar^n \sin(n\theta) \tag{2.34}$$

Let us now consider some corner flows for various values of n and plot each flow using a few streamlines.

Case: $n = 3$

In this case, the stream function becomes:

$$\Psi = Ar^3 \sin(3\theta) \tag{2.35}$$

Figure 2.7 shows a few streamlines for this concave corner flow. In this figure, the thick lines with $\theta = 0$ and $\theta = \pi / 3$ correspond to streamlines with $\Psi = 0$. As no flow crosses a streamline, each streamline in a potential flow could represent an impermeable wall.

Case: $n = 2$

In this case, the stream function becomes:

$$\Psi = Ar^2 \sin(2\theta) = 2Axy \tag{2.36}$$

representing a two-dimensional stagnation concave corner flow for which Figure 2.8 shows a few streamlines in the first quadrant. In this figure, the thick lines along $x = 0$ and $y = 0$ correspond to $\Psi = 0$. The included angle formed by these lines is $\pi / 2$.

Case: $n = 3/2$

In this case, we obtain the stream function:

$$\Psi = Ar^{3/2} \sin(3\theta/2) \tag{2.37}$$

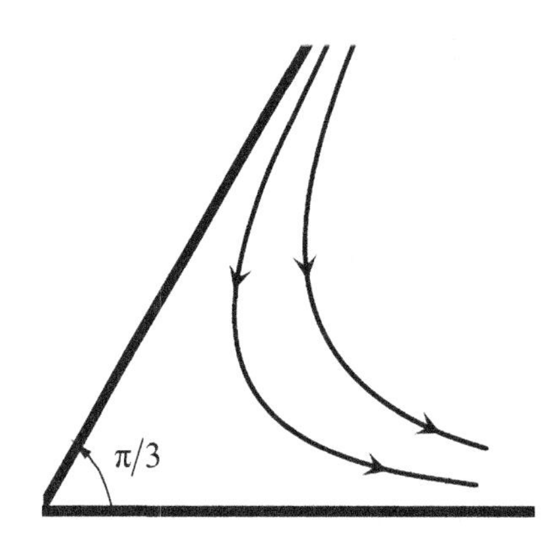

FIGURE 2.7 Streamlines in a concave corner flow with $\Psi = Ar^3 \sin(3\theta)$.

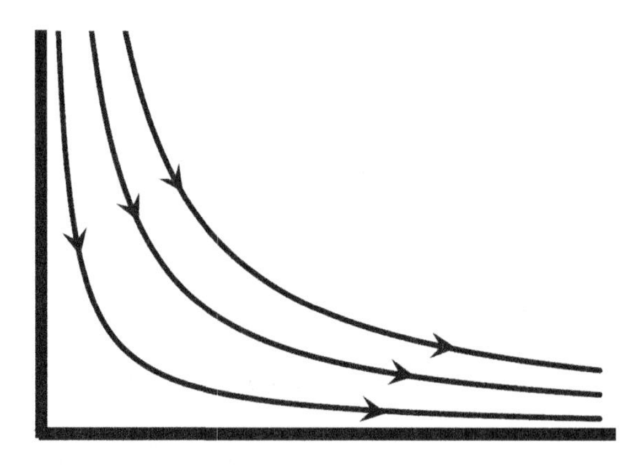

FIGURE 2.8 Streamlines in a concave corner flow with $\Psi = 2Axy$.

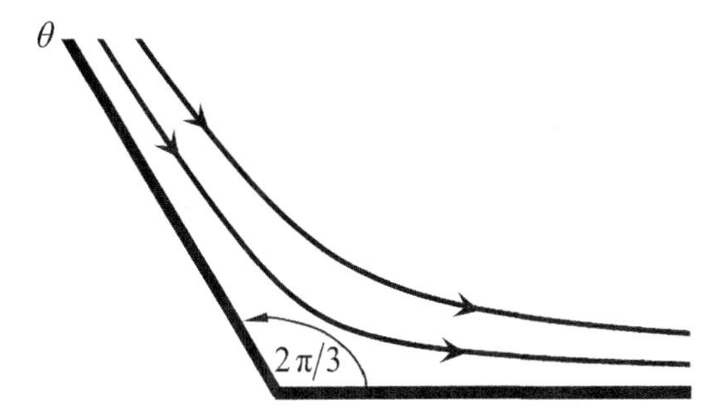

FIGURE 2.9 Streamlines in a concave corner flow with $\Psi = Ar^{3/2}\sin(3\theta/2)$.

Figure 2.9 shows the resulting concave corner of included angle $2\pi/3$ with a few streamlines. In this figure, the thick lines correspond to $\Psi = 0$.

Case: $n = 1$

In this case, the resulting stream function:

$$\Psi = Ar\sin\theta = Ay \tag{2.38}$$

represents a uniform flow over a flat plate shown in Figure 2.10.

Case: $n = 2/3$

In this case, the resulting stream function:

$$\Psi = Ar^{2/3}\sin(2\theta/3) \tag{2.39}$$

represents the flow over a convex corner of angle $3\pi/2$ shown in Figure 2.11, where the thick lines correspond to $\Psi = 0$.

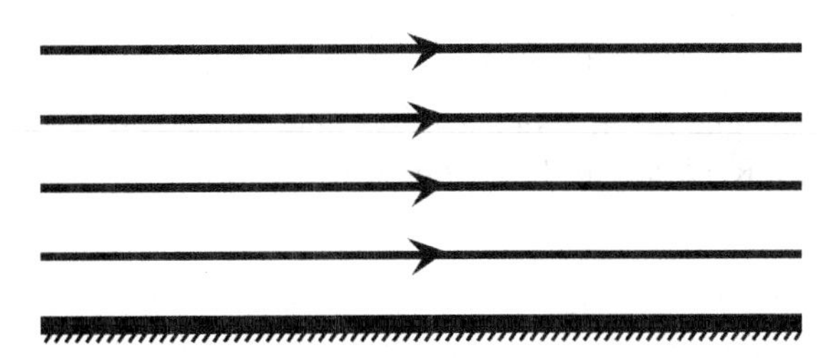

FIGURE 2.10 Uniform flow over a flat plate with $\Psi = Ay$.

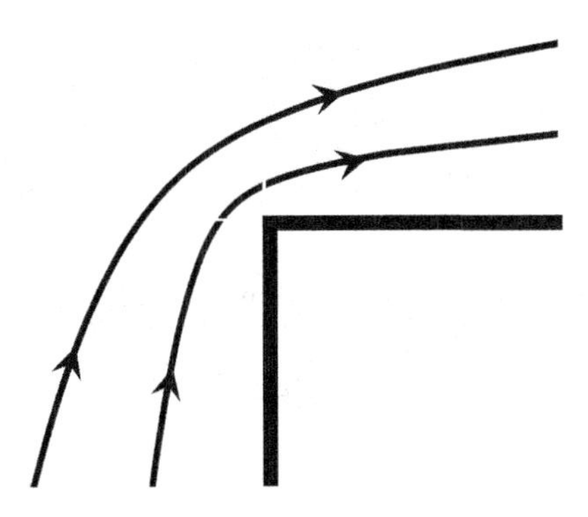

FIGURE 2.11 Streamlines in a convex corner flow with $\Psi = Ar^{2/3}\sin(2\theta/3)$.

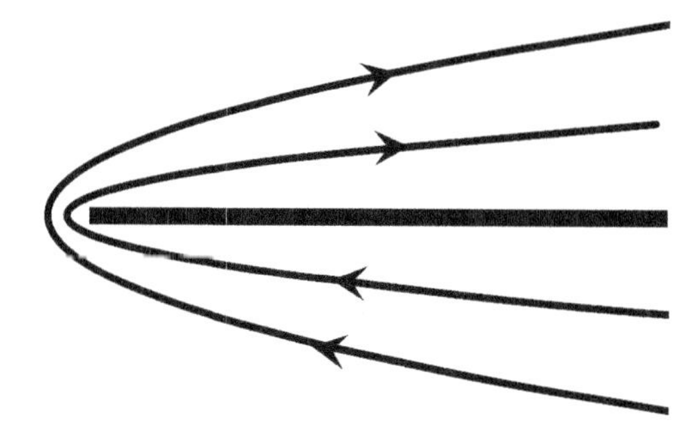

FIGURE 2.12 Flow turning around a flat plate $\Psi = Ar^{1/2}\sin(\theta/2)$.

Case: $n = 1/2$
In this case, the stream function:

$$\Psi = Ar^{1/2}\sin(\theta/2) \tag{2.40}$$

describes the flow turning around a flat plate shown in Figure 2.12, where the thick line represents a flat plate and corresponds to $\Psi = 0$.

2.4 SUPERPOSITION OF TWO OR MORE PLANE POTENTIAL FLOWS

In Sections 2.3.1 through 2.3.4, we discussed simple potential flows with uniform velocity and those generated by singularities such as sources, sinks, positive vortices, and negative vortices. We also discussed a class of flows called corner flows. We can generate more complex potential flows with significant practical applications by linear superposition of these simple flows.

2.4.1 Superposition of Uniform Flow and Source: Flow around a Rankine Half-Body

When we combine the potential function of a uniform flow with that of a source located at the origin, a potential flow over a Rankine half-body results as follows:

$$\begin{aligned}
F(z) &= Uz + \frac{m}{2\pi}\ln z \\
F(z) &= Ure^{i\theta} + \frac{m}{2\pi}\ln(re^{i\theta}) \\
F(z) &= Ur(\cos\theta + i\sin\theta) + \frac{m}{2\pi}\ln r + i\frac{m}{2\pi}\theta \\
F(z) &= \left(Ur\cos\theta + \frac{m}{2\pi}\ln r\right) + i\left(Ur\sin\theta + \frac{m}{2\pi}\theta\right)
\end{aligned} \tag{2.41}$$

which yields the stream function for the potential flow over a Rankine half-body as:

$$\Psi = Ur\sin\theta + \frac{m}{2\pi}\theta \tag{2.42}$$

in which the source of strength m is at the origin. Figure 2.13 shows a few streamlines for this flow, where S is the stagnation point at which all velocity components are zero. To verify this, let us compute V_r and V_θ for this flow:

$$V_r = \frac{\partial \Psi}{r\partial \theta} = U\cos\theta + \frac{m}{2\pi r} \tag{2.43}$$

$$V_\theta = -\frac{\partial \Psi}{\partial r} = -U\sin\theta \tag{2.44}$$

For $V_r = V_\theta = 0$ at the stagnation point S, we obtain $\theta_S = \pi$ and $r_S = m/(2\pi U)$ from Equations 2.43 and 2.44. Substituting these values in Equation 2.42 yields $\Psi_s = m/2$, which is the value of the stream function for the streamline that corresponds to the Rankine half-body shown in the figure. We can write the equation of this streamline as:

$$r_b = \frac{m}{2U\sin\theta}\left(1 - \frac{\theta}{\pi}\right) \tag{2.45}$$

We can calculate the width of the Rankine half-body by the equation:

$$2y_b = 2r_b \sin\theta = \frac{m}{U}\left(1 - \frac{\theta}{\pi}\right) \tag{2.46}$$

Equation 2.46 and Figure 2.13 reveal that the maximum width of m/U for the Rankine half-body occurs as θ asymptotically approaches zero.

2.4.2 Superposition of Source and Sink: Dipole

Figure 2.14a shows the geometry of a dipole in which a source and a sink, each of strength m, are located at $(\varepsilon, 0)$ and $(-\varepsilon, 0)$, respectively. We can write the complex potential for this dipole as:

$$F(z) = \frac{m}{2\pi}\ln(z - \varepsilon) - \frac{m}{2\pi}\ln(z + \varepsilon) \tag{2.47}$$

At point P, shown in Figure 2.14a, we can write the stream function resulting from the superposition of the stream functions of the source and sink in a dipole as:

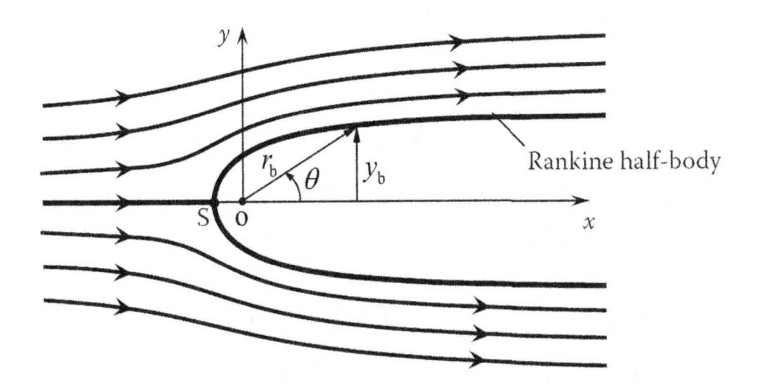

FIGURE 2.13 Superposition of a uniform flow and a source: Rankine half-body.

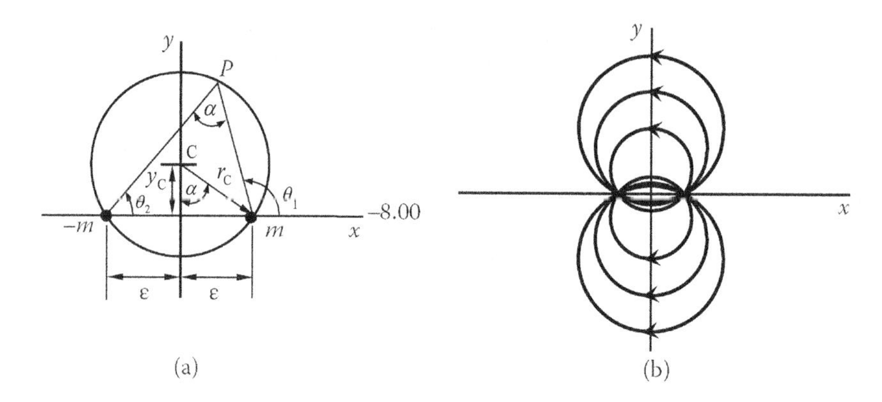

FIGURE 2.14 (a) Dipole geometry and (b) streamlines in a dipole.

$$\Psi = \Psi_{\text{source}} + \Psi_{\text{sink}} = \frac{m}{2\pi}\theta_1 - \frac{m}{2\pi}\theta_2 = \frac{m}{2\pi}(\theta_1 - \theta_2) = \frac{m}{2\pi}\alpha \tag{2.48}$$

which shows that, for each value of Ψ, α is constant and the corresponding streamline is a circle whose center is located on the y axis. From the geometry shown in Figure 2.14a, we can compute the radius of the circle and the y coordinate of its center for each value of Ψ as follows:

$$r_C = \frac{\varepsilon}{\sin\alpha} \tag{2.49}$$

$$y_C = \frac{\varepsilon}{\tan\alpha} \tag{2.50}$$

where $\alpha = 2\pi\Psi / m$. Figure 2.14b shows a few streamlines of a dipole, including the symmetric streamlines with center at $(0, -y_C)$. All the streamlines are circles, passing through the source and the sink on the x axis. The flow emerges from the source and converges at the sink. The center of each circle lies on the y axis. Note that the equipotential lines of a dipole, not shown in the figure, are circles with their centers on the x axis.

2.4.3 Superposition of Source and Sink: Doublet

A doublet is the limiting case of a dipole where the distance between the source and the sink is vanishingly tiny while keeping the product of the equal strength of the source and sink and the distance between them constant. We can write the complex potential for the dipole as:

$$F(z) = -\frac{m}{2\pi}\ln\left(\frac{z+\varepsilon}{z-\varepsilon}\right) = -\frac{m}{2\pi}\ln\left(\frac{1+\dfrac{\varepsilon}{z}}{1-\dfrac{\varepsilon}{z}}\right)$$

For $-1 < \varepsilon / z < 1$, the power series expansion of the logarithmic term in this equation yields:

$$F(z) = -\frac{m}{2\pi}\left[2\frac{\varepsilon}{z} + \frac{2}{3}\left(\frac{\varepsilon^3}{z^3}\right) + \frac{2}{5}\left(\frac{\varepsilon^5}{z^5}\right) + \ldots\right]$$

Letting $\varepsilon \to 0$ and $m \to \infty$ in such a way as to yield $m\varepsilon = \pi\mu$, where μ is a constant, yields:

$$F(z) = -\frac{\mu}{z} \tag{2.51}$$

which represents the complex potential of a doublet, where the flow direction continues to be from the source to the sink of the dipole used to derive the doublet. Thus, a doublet results from the superposition of a very strong source and an equally strong sink nearby.

From the complex potential of a doublet given by Equation 2.51, we obtain the potential function and stream function as follows:

$$F(z) = -\frac{\mu}{z} = -\frac{\mu}{re^{i\theta}} = -\frac{\mu}{r}e^{-i\theta} = -\frac{\mu}{r}\cos\theta + i\frac{\mu}{r}\sin\theta$$

which yields

$$\Phi = -\frac{\mu}{r}\cos\theta = -\frac{\mu x}{x^2 + y^2} \tag{2.52}$$

and

$$\Psi = \frac{\mu}{r}\sin\theta = \frac{\mu y}{x^2 + y^2} \tag{2.53}$$

We can rearrange Equation 2.53 as:

$$x^2 + \left(y - \frac{\mu}{2\Psi}\right)^2 = \left(\frac{\mu}{2\Psi}\right)^2 \tag{2.54}$$

For a constant value of Ψ, this equation represents the equation of a circle of radius $\mu/(2\Psi)$ with its center at $\{0, \mu/(2\Psi)\}$. Figure 2.15 presents a few streamlines of a doublet derived from the dipole shown in Figure 2.14.

We can write the complex potential of a doublet located at $z = z_0$ as:

$$F(z) = -\frac{\mu}{(z - z_0)} \tag{2.55}$$

where μ is the strength of the doublet. A doublet is useful in generating more complex and practical potential flows by linear superposition with other simple flows.

2.4.4 Superposition of Uniform Flow and Doublet: Flow around a Cylinder

We can generate the potential flow around a cylinder by the superposition of a uniform flow and a doublet. The resulting complex potential becomes:

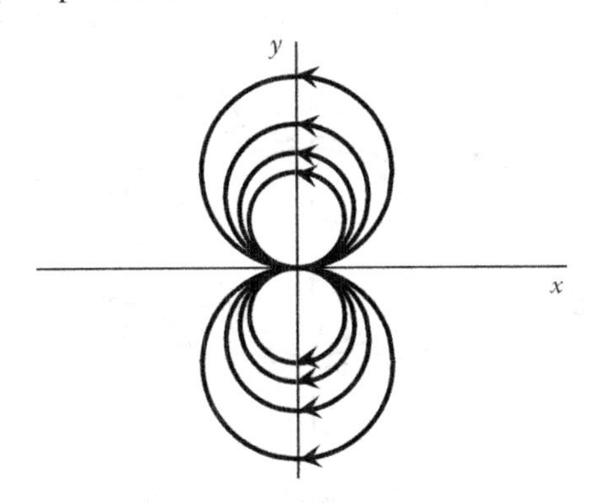

FIGURE 2.15 Streamlines in a doublet.

$$F(z) = Uz + \frac{\mu}{z} = Ur\,\mathrm{e}^{\mathrm{i}\theta} + \frac{\mu}{r}\mathrm{e}^{-\mathrm{i}\theta} = \left(Ur + \frac{\mu}{r}\right)\cos\theta + \mathrm{i}\left(Ur - \frac{\mu}{r}\right)\sin\theta \tag{2.56}$$

which yields the stream function:

$$\Psi = \left(Ur - \frac{\mu}{r}\right)\sin\theta \tag{2.57}$$

giving its value on a circle of radius $r = a$ as:

$$\Psi_a = \left(Ua - \frac{\mu}{a}\right)\sin\theta$$

For $\Psi_a = 0$, this equation yields $\mu = Ua^2$, which means superimposing a uniform flow of velocity U on a doublet of strength Ua^2 results in a potential flow over a cylinder of radius $r = a$ whose surface corresponds to zero stream function. Within the cylinder, the adjacent sink absorbs the total mass flow rate generated by the source of the doublet. The resulting complex potential function and the stream function for this flow are as follows:

$$F(z) = U\left(z + \frac{a^2}{z}\right) \tag{2.58}$$

$$\Psi = U\left(r - \frac{a^2}{r}\right)\sin\theta \tag{2.59}$$

Figure 2.16 shows a few streamlines corresponding to Equation 2.59. The flow field exhibits symmetry about the x and y axes. Therefore, the resulting pressure distribution on the cylinder will integrate to zero force in the x direction (drag force) and zero force in the y direction (lift force). However, the drag force on the cylinder is not zero for an actual fluid flow with nonzero viscosity, while it experiences zero lift. We discuss it further in this chapter under D'Alembert's paradox (Section 2.3.3).

2.4.4.1 Stagnation Points

At a stagnation point in a flow, all velocity components become zero. From Equation 2.58, we obtain the complex conjugate velocity as:

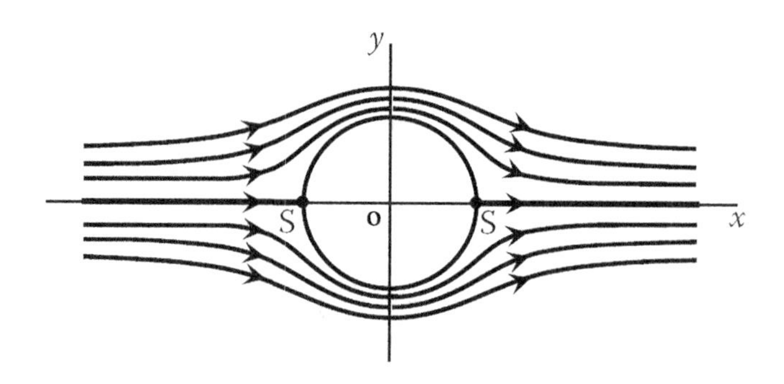

FIGURE 2.16 Potential flow around a cylinder with no circulation.

$$\bar{W}(z) = \frac{\mathrm{d}F}{\mathrm{d}z} = U\left(1 - \frac{a^2}{z^2}\right)$$
$$\bar{W}(z) = \left[U\left(1 - \frac{a^2}{r^2}\right)\cos\theta + \mathrm{i}U\left(1 + \frac{a^2}{r^2}\right)\sin\theta\right]\mathrm{e}^{-\mathrm{i}\theta} \tag{2.60}$$

Comparing Equations 2.17 and 2.60, we obtain:

$$V_r = U\left(1 - \frac{a^2}{r^2}\right)\cos\theta \tag{2.61}$$

$$V_\theta = -U\left(1 + \frac{a^2}{r^2}\right)\sin\theta \tag{2.62}$$

According to Equation 2.61, the radial velocity is zero everywhere on the cylinder surface ($r = a$), where the tangential velocity becomes zero at $\theta = 0$ and $\theta = \pi$, resulting in two stagnation points, one at the front end and the other at the back end of the cylinder surface, as shown in Figure 2.20.

2.4.5 Superposition of Uniform Flow, Doublet, and Vortex: Flow around a Cylinder with Circulation

Adding a clockwise rotating vortex to the flow around a cylinder results in the complex potential:

$$F(z) = U\left(z + \frac{a^2}{z}\right) + \mathrm{i}\frac{\Gamma}{2\pi}\ln z + c_1 \tag{2.63}$$

where we have added the constant c_1 to render $\Psi = 0$ on the cylinder surface ($r = a$) with no effect on the velocity and pressure distributions in the flow field. We express this equation in polar coordinates as:

$$F(z) = \left[U\left(r + \frac{a^2}{r}\right)\cos\theta - \frac{\Gamma\theta}{2\pi}\right] + \mathrm{i}\left[U\left(r - \frac{a^2}{r}\right)\sin\theta + \frac{\Gamma}{2\pi}\ln r\right] + c_1 \tag{2.64}$$

where the substitution $c_1 = -\mathrm{i}(\Gamma/2\pi)\ln a$ renders $\Psi = 0$ at $r = a$, giving:

$$F(z) = U\left(z + \frac{a^2}{z}\right) + \mathrm{i}\frac{\Gamma}{2\pi}\ln\frac{z}{a} \tag{2.65}$$

which yields the stream function for a flow around a cylinder with nonzero circulation as:

$$\Psi = U\left(r - \frac{a^2}{r}\right)\sin\theta + \frac{\Gamma}{2\pi}\ln r - \frac{\Gamma}{2\pi}\ln a \tag{2.66}$$

2.4.5.1 Stagnation Points

Equation 2.65 yields the complex conjugate velocity:

$$\bar{W}(z) = \frac{dF}{dz} = U\left(1 - \frac{a^2}{z^2}\right) + i\frac{\Gamma}{2\pi z}$$

$$\bar{W}(z) = \left[U\left(1 - \frac{a^2}{r^2}\right)\cos\theta + i\left\{U\left(1 + \frac{a^2}{r^2}\right)\sin\theta + \frac{\Gamma}{2\pi r}\right\}\right]e^{-i\theta} \tag{2.67}$$

By comparing Equations 2.17 and 2.67, we obtain:

$$V_r = U\left(1 - \frac{a^2}{r^2}\right)\cos\theta \tag{2.68}$$

$$V_\theta = -U\left(1 + \frac{a^2}{r^2}\right)\sin\theta - \frac{\Gamma}{2\pi r} \tag{2.69}$$

Equation 2.68 is identical to Equation 2.61, which we obtained for the potential flow around a cylinder with zero circulation, giving zero radial velocity everywhere on the cylinder surface ($r = a$). Comparing Equation 2.69 with Equation 2.62, we see that including a vortex at the origin alters the tangential velocity distribution in the flow around the cylinder. For $V_\theta = 0$ on the cylinder surface, Equation 2.69 yields:

$$\sin\theta = -\frac{\Gamma}{4\pi Ua} \tag{2.70}$$

which reveals that, for the stagnation points to be on the cylinder surface, we must have $\Gamma \leq 4\pi Ua$, and the corresponding θ values will lie in the third and fourth quadrants. Figure 2.17 shows a few streamlines computed from the stream function given by Equation 2.66 for the potential flow with the vortex strength $\Gamma < 4\pi Ua$ around a cylinder of radius $r = a$. The flow field is symmetric about the y axis. As expected, the leading stagnation point is in the third quadrant, and the rear one is in the fourth quadrant, and the streamline passing through them correspond to $\Psi = 0$. Note that the streamlines below the streamline with $\Psi = 0$ are initially concave near $\theta = 3\pi / 2$, gradually becoming straight far away from the origin.

As the vortex strength Γ increases, the stagnation points on the cylinder surface move toward each other. For $\Gamma = 4\pi Ua$, Figure 2.18 shows that the stagnation points coincide at $\theta = 3\pi / 2$ near which the lower streamlines are initially convex, gradually becoming straight far from the origin.

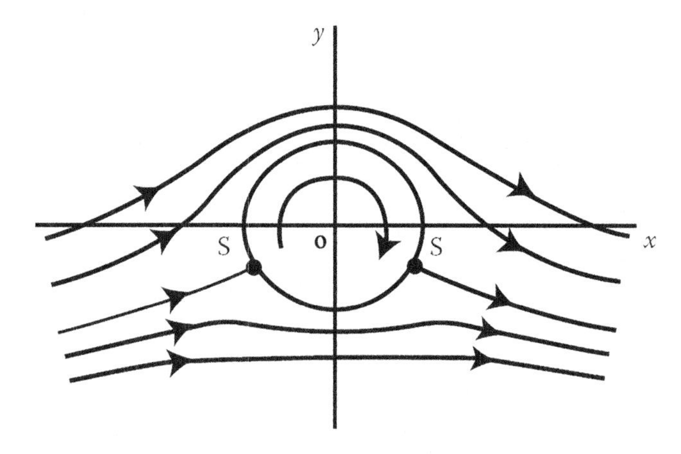

FIGURE 2.17 Potential flow around a cylinder with circulation $\Gamma < 4\pi Ua$.

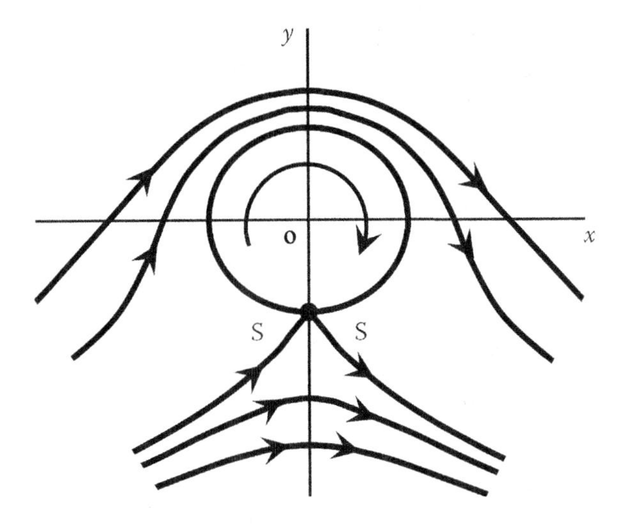

FIGURE 2.18 Potential flow around a cylinder with circulation $\Gamma = 4\pi Ua$.

For $\Gamma > 4\pi Ua$, Figure 2.19 shows that the coincident stagnation points leave the cylinder surface and appear along $\theta = 3\pi/2$ within the flow field where, according to Equation 2.68, $V_r = 0$. For $V_\theta = 0$ and $r_s > a$, Equation 2.69 yields:

$$r_S = \frac{1}{4\pi U}\left(\Gamma + \sqrt{\Gamma^2 - (4\pi Ua)^2}\right) \tag{2.71}$$

Figure 2.19 also shows that the fluid trapped between the cylinder and the streamline with a nonzero stream function passing through the stagnation points is in perpetual rotation around the cylinder. In this case, also, the streamlines below the stagnation points are initially convex, gradually becoming straight far from the origin.

2.4.6 Milne-Thomson Circle Theorem

Milne-Thomson (1968) devised the circle theorem that provides a straightforward method to modify the complex potential of a two-dimensional flow in the z plane when we place a cylinder, with its

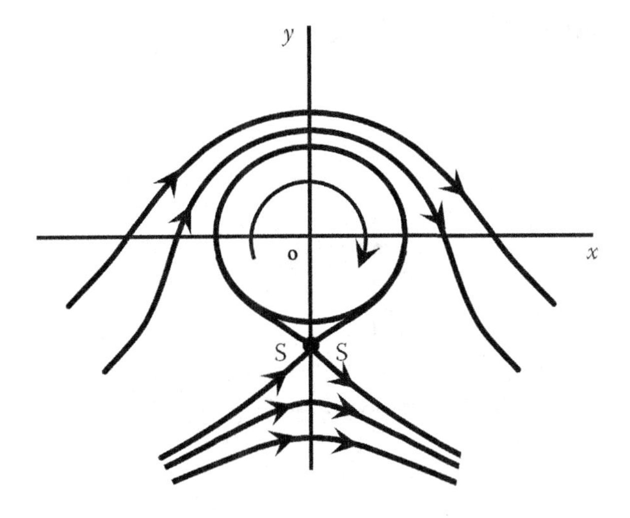

FIGURE 2.19 Potential flow around a cylinder with circulation $\Gamma > 4\pi Ua$.

axis parallel to the z axis (appears as a circle on the plane), in the flow in a singularity-free region. If the complex potential of the original flow is given by $F_1(z)$, using the circle theorem, we can write the complex potential of the resulting flow, with the circle of radius a and its center at the origin, as:

$$F(z) = F_1(z) + \bar{F}_1(a^2/z) \tag{2.72}$$

where $\bar{F}_1(a^2/z)$ is obtained by replacing z by a^2/z and i by −i in $F_1(z)$. Note that all the singularities of $F(z)$ must lie outside the circle.

2.5 FORCE AND MOMENT ON A BODY IN POTENTIAL PLANE FLOWS

2.5.1 Blasius Integral Theorems

Blasius proposed two integral theorems: one for calculating the force and the other for calculating the moment on a two-dimensional body in an incompressible potential flow. To understand these theorems, consider the body of arbitrary cross section shown in Figure 2.20. The direct method to calculate the force acting on a body in a flow is to integrate the static pressure distribution on its surface. As the pressure force on the surface is normal and compressive (opposite to the outward pointing area vector), based on the pressure forces acting in the x and y directions on the differential contour element dz shown in the figure, we can evaluate dF_x and dF_y as

$$\mathrm{d}F_x = -p\mathrm{d}y \text{ and } \mathrm{d}F_y = p\mathrm{d}x$$

Performing the integration in the counterclockwise direction on the closed contour C and combining F_x and F_y into a complex conjugate force, we obtain:

$$F_x - \mathrm{i}F_y = -\oint_C p(\mathrm{d}y + \mathrm{i}\mathrm{d}x) = -\mathrm{i}\oint_C p\,\mathrm{d}\bar{z} \tag{2.73}$$

where $\bar{z} = x - \mathrm{i}y$ is the complex conjugate of $z = x + \mathrm{i}y$. As the total pressure in a potential flow remains constant, the Bernoulli equation without the gravitational term (geodetic pressure) yields:

$$p = p_0 - \frac{1}{2}\rho V^2 = p_0 - \frac{1}{2}\rho W(z)\bar{W}(z)$$

Using this equation to replace p in Equation 2.73, we obtain:

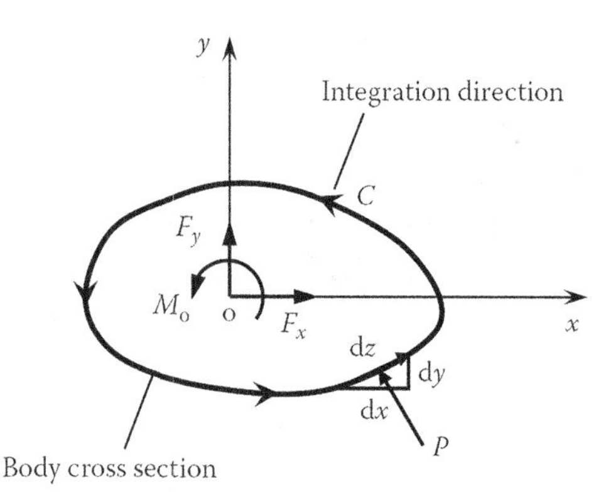

FIGURE 2.20 Force and moment acting on a body in plane flow.

$$F_x - iF_y = -i\oint_C \left(p_0 - \frac{1}{2}\rho W(z)\bar{W}(z) \right) d\bar{z}$$

$$F_x - iF_y = -i\oint_C p_0 \, d\bar{z} + \frac{i\rho}{2}\oint_C \frac{1}{2}\rho W(z)\bar{W}(z)\, d\bar{z}$$

For constant p_0, the first integral on the right-hand side of this equation becomes zero, giving:

$$F_x - iF_y = \frac{i\rho}{2}\oint_C W(z)\bar{W}(z)\, d\bar{z} \tag{2.74}$$

As the closed contour around the body corresponds to a streamline, the velocity vector $W(z) = V_x + iV_y = \left(\sqrt{V_x^2 + V_y^2}\right) e^{i\theta}$ and $dz = |dz| e^{i\theta}$ are collinear. As a result, we can write:

$$W(z)d\bar{z} = \bar{W}(z)dz$$

whose substitution in Equation 2.74 yields the Blasius theorem to compute forces on a two-dimensional body in a potential flow:

$$F_x - iF_y = \frac{i\rho}{2}\oint_C \bar{W}(z)^2 dz = \frac{i\rho}{2}\oint_C \left(\frac{dF(z)}{dz} \right)^2 dz \tag{2.75}$$

The forces dF_x and dF_y acting at a point (x, y) on the closed contour of the body cross section shown in Figure 2.20 generate a counterclockwise moment about the origin, given by:

$$dM_o = x dF_y - y dF_x$$

$$dM_o = \text{Re}\left\{ iz(dF_x - i dF_y) \right\}$$

Using Equation 2.73, we rewrite the preceding equation as:

$$dM_o = \text{Re}\left\{ zp d\bar{z} \right\}$$

which by substituting for p from the Bernoulli equation yields:

$$dM_o = \text{Re}\left\{ z\left(p_0 - \frac{1}{2}\rho W(z)\bar{W}(z) \right) d\bar{z} \right\}$$

whose integration around the close contour C further yields:

$$M_o = \text{Re}\oint_C zp_0 \, d\bar{z} - \frac{1}{2}\rho \,\text{Re}\oint_C zW(z)\bar{W}(z)\, d\bar{z} \tag{2.76}$$

Let us now evaluate the first integral on the right-hand side of Equation 2.76:

$$\oint_C zp_0 \, d\bar{z} = p_0 \oint_C z \, d\bar{z} = p_0 \oint_C re^{i\theta}(dr)e^{-i\theta} = p_0 \oint_C r \, dr = p_0 \oint_C d\left(\frac{r^2}{2} \right)$$

As the integrand of this integral over a closed contour is an exact differential, the value of the integral becomes zero, giving:

$$M_o = -\frac{1}{2}\rho \text{Re}\left\{\oint_C zW(z)\bar{W}(z)\,d\bar{z}\right\}$$

which with $W(z)d\bar{z} = \bar{W}(z)dz$ becomes:

$$M_o = -\frac{1}{2}\rho \text{Re}\left[\oint_C z\{\bar{W}(z)\}^2\,dz\right] = -\frac{1}{2}\rho \text{Re}\left[\oint_C \left\{\frac{dF(z)}{dz}\right\}^2 z\,dz\right] \tag{2.77}$$

which represents the second Blasius theorem for computing the moment on a two-dimensional body in a potential flow. In this equation, *Re* stands for the real part of the expression that follows it.

Note that only the terms that have different values at the start and end of the closed contour, when θ varies from 0 to 2π, will contribute nonzero values in the Blasius integrals of Equations 2.75 and 2.77. All other terms will integrate to zero. In general, going around a loop enclosing the point z_0, the term $\ln(z - z_0)$ increases by $2\pi i$. As, only the terms involving $(z - z_0)^{-1}$ in the integrand will integrate to $\ln(z - z_0)$, the evaluation of Blasius integrals simplifies considerably. A powerful method to evaluate these integrals is to use the residue theorem discussed, for example, by Churchill (1986). Accordingly, we rewrite Equations 2.75 and 2.77 as:

$$F_x - iF_y = \frac{i\rho}{2}\oint_C \{\bar{W}(z)\}^2 dz = \frac{i\rho}{2}\left[2\pi i \sum \left(\text{residues of } \bar{W}^2 \text{ inside C}\right)\right] \tag{2.78}$$

$$M_o = -\frac{1}{2}\rho \text{Re}\left[\oint_C \{\bar{W}(z)\}^2 z\,dz\right] = -\frac{1}{2}\rho \text{Re}\left[2\pi i \sum \left(\text{residues of } z\bar{W}^2 \text{ inside C}\right)\right] \tag{2.79}$$

2.5.2 Force and Moment on a Cylinder with Circulation

Equation 2.65 gives the complex potential of a uniform flow with circulation around a cylinder of radius $r = a$. We obtain the corresponding complex conjugate velocity as:

$$\bar{W}(z) = \frac{dF(z)}{dz} = U\left(1 - \frac{a^2}{z^2}\right) + i\frac{\Gamma}{2\pi z} \tag{2.80}$$

giving:

$$\{\bar{W}(z)\}^2 = U^2 - \frac{2U^2a^2}{z^2} + \frac{U^2a^4}{z^4} + \frac{iU\Gamma}{\pi z} - \frac{iU\Gamma a^2}{\pi z^3} - \frac{\Gamma^2}{4\pi^2 z^2} \tag{2.81}$$

which yields residue $iU\Gamma/\pi$. Using Equation 2.78, we obtain the complex conjugate force acting on the cylinder as:

$$F_x - iF_y = i\frac{\rho}{2}(2\pi i)\left(\frac{iU\Gamma}{\pi}\right) = -i\rho U\Gamma \tag{2.82}$$

Equating real and imaginary parts on both sides of Equation 2.82 yields the forces acting on the cylinder as:

Drag force:

$$F_x = 0 \tag{2.83}$$

Lift force:

$$F_y = \rho U \Gamma \tag{2.84}$$

2.5.3 D'Alembert's Paradox

Equation 2.83 indicates that, based on potential flow with symmetry about the y axis, the cylinder's drag force per unit length is zero. This result forms the basis of D'Alembert's paradox. Consistent with Equation 2.83, D'Alembert (1717–1783), in his experiments, expected the drag force on a sphere placed in an actual fluid flow to be zero as the fluid viscosity approached zero. Instead, he found the drag force to converge to a nonzero value.

In an actual flow, the drag force on an object consists of two parts: the skin friction drag and form drag due to different pressure distributions in the front and back ends of the object. In a potential flow with the slip boundary condition on the body (zero skin friction drag), the nonzero value of the drag for a fluid with negligible viscosity must be due to the form drag. In 1904, Prandtl introduced the concept of boundary layer and suggested that a natural fluid, no matter how small its viscosity may be, must satisfy the no-slip boundary condition at the surface of a body. Accordingly, he conjectured that the rear stagnation point on the cylinder does not exist in an actual fluid flow as the thin boundary layer peels off the body under an adverse pressure gradient. This flow behavior destroys the symmetric pressure distribution predicted by the potential flow. Hoffman and Johnson (2010) have questioned the resolution of D'Alembert's paradox using Prandtl's boundary layer concept and have instead proposed an alternate mechanism of three-dimensional slip separation caused by flow unsteadiness.

Equation 2.84, also known as the Kutta–Joukowski law for calculating lift on a body in a potential flow, offers a simple method to evaluate the lift force per unit length of the cylinder. Unlike the drag force, the lift force on the cylinder calculated using this equation compares well with the empirical data. Furthermore, in Equation 2.84, it is interesting to note that the lift force does not depend on the cylinder's radius. Zdravkovich (1997) provides comprehensive details of the flow phenomena, experiments, applications, mathematical models, and simulations for the flow around a circular cylinder.

2.5.4 The Magnus Effect

The side force acting on a cylinder in cross-flow with circulation is also known as the Magnus effect. To demonstrate this effect, let us conduct a simple experiment. Take a cardboard spool of a spent paper towel roll. Wind a long thread (dental floss) around the spool. As shown in Figure 2.21, hold the free end of the thread at a height and release the spool. As the spool unwinds, it spins in air. Interestingly, the spool does not fall straight vertically. Instead, a side force (the Magnus effect) deflects its path. Due to air viscosity, the spinning spool generates circulation in the surrounding air in the same direction as its rotation. The air flows in a direction opposite to the direction of the translational velocity of the spool. The figure further shows that the instantaneous direction of the side force $\boldsymbol{L}$ is in the direction of the cross product $\boldsymbol{U} \times \Omega \hat{\boldsymbol{k}}$. Physically, we can determine the direction of $\boldsymbol{L}$ as follows. Due to the spool's rotation, the air velocity near the spool where the thread is attached is higher than its velocity on the opposite side. From the Bernoulli principle of "high velocity, low pressure," the pressure force acting on the spool on the thread side is lower than that on the opposite side, creating a net side force on the spool in the direction shown in the figure.

Let us now use Equation 2.79 to calculate the moment (about the origin) of the surface forces acting on the cylinder. Multiplying Equation 2.81 by z yields:

$$z\bar{W}(z)^2 = zU^2 - \frac{2U^2a^2}{z} + \frac{U^2a^4}{z^3} + \frac{\mathrm{i}U\Gamma}{\pi} - \frac{\mathrm{i}U\Gamma a^2}{\pi z^2} - \frac{\Gamma^2}{4\pi^2 z} \tag{2.85}$$

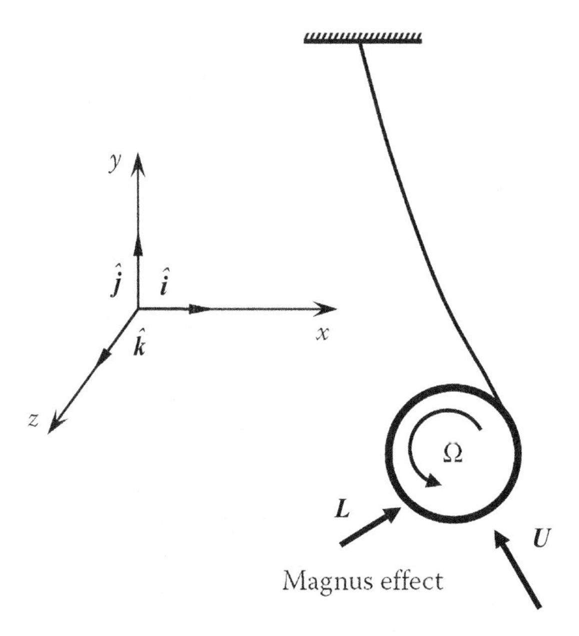

FIGURE 2.21 The Magnus effect on a spinning cylinder falling in air.

From this equation, we obtain:

$$\sum\left(\text{residues of } z\bar{W}^2 \text{ inside C}\right) = -2U^2a^2 - \frac{\Gamma^2}{4\pi^2}$$

whose substitution in Equation 2.79 yields:

$$M_o = -\frac{1}{2}\rho\,\mathrm{Re}\left[2\pi \mathrm{i}\left(-2U^2a^2 - \frac{\Gamma^2}{4\pi^2}\right)\right] = 0 \tag{2.86}$$

which shows that the potential flow imparts zero hydrodynamic moment on the cylinder.

2.6 CONFORMAL TRANSFORMATION

Conformal mapping is a highly developed branch of the complex variable theory presented at length in books on advanced mathematics and classical fluid mechanics dealing with potential flows. In this section, we limit our discussion to two-dimensional potential flows where we can use conformal mapping to transform them into a flow around a circular cylinder (a circle in the z plane), which we have discussed extensively in Section 2.4.

Under a conformal transformation, the angle between any two intersecting curves in the physical plane equals the angle between the corresponding curves in the transformed plane. A mapping by an analytical function $f(z)$ is conformal except at points where $f'(z)$ is zero. As shown in Figure 2.22, the primary motivation for conformal transformation is to map streamlines and equipotential lines of a potential flow around a body in the z plane (physical plane) into corresponding streamlines and equipotential lines of another potential flow around a simple body (e.g., a circular cylinder) in the ζ plane.

Let us now examine as to how the complex conjugate velocity $\bar{W}(z)$ in the physical plane transforms into the corresponding complex conjugate velocity $\bar{W}(\zeta)$ in the transformed plane:

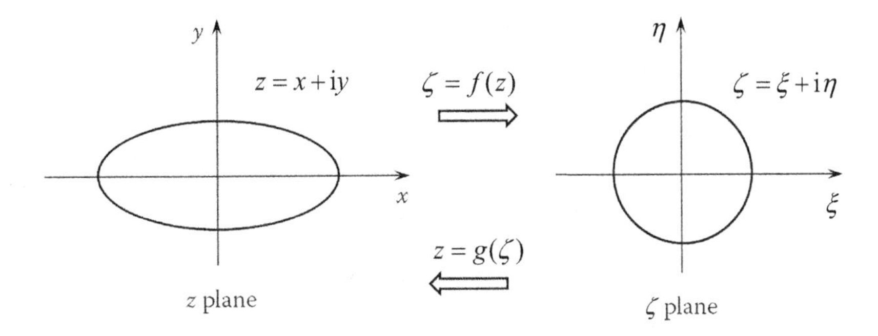

FIGURE 2.22 Conformal mapping between z plane and ζ plane.

$$\bar{W}(z) = \frac{dF(z)}{dz} = \frac{dF(\zeta)}{d\zeta}\frac{d\zeta}{dz}$$

$$\bar{W}(z) = \frac{d\zeta}{dz}\bar{W}(\zeta) = f'(z)\bar{W}(\zeta) \tag{2.87}$$

$$\bar{W}(z) = \frac{d\zeta}{dz}\bar{W}(\zeta) = f'(z)\bar{W}(\zeta)$$

which indicates that, under conformal mapping, complex velocities do not map one-to-one; instead, the derivative of the mapping analytical function $f'(z)$ multiplies them as the scaling factor.

In the z plane, for a closed path C_z containing a source of strength m_z and a vortex of strength Γ_z, we can use Equation 2.18 to write the complex circulation as:

$$C(z) = \Gamma_z + im_z = \oint_{C_z} \bar{W}(z)\,dz$$

Substituting $\bar{W}(z)$ from Equation 2.87 in this equation yields:

$$\Gamma_z + im_z = \oint_{C_z} \frac{d\zeta}{dz}\bar{W}(\zeta)\,dz = \oint_{C_\zeta} \bar{W}(\zeta)\,d\zeta$$

where C_ζ is the closed path in the ζ plane mapped from C_z in the z plane. We finally obtain

$$\Gamma_z + im_z = \Gamma_\zeta + im_\zeta$$

Equating real and imaginary parts in this equation yields $\Gamma_z = \Gamma_\zeta$ and $m_z = m_\zeta$. Thus, under conformal mapping between two plane potential flows, the strengths of sources, sinks, and vortices remain unchanged.

2.6.1 Joukowski Transformation

The Joukowski transformation is a historically famous conformal transformation in which we can transform a flow over a cylinder placed in a uniform cross-flow with circulation into various practical flows, including flows over symmetric and asymmetric airfoils. The Joukowski transformation function is given by:

$$z = \zeta + \frac{c^2}{\zeta} \tag{2.88}$$

which yields the inverse transformation function:

$$\zeta = \frac{z}{2} \pm \frac{\sqrt{z^2 - 4c^2}}{2} \tag{2.89}$$

in which the negative root corresponds to points inside the circle in the ζ plane. Note that the Joukowski transformation function maps all the points on the ξ axis into points on the x axis, and the points on the η axis into the points on the y axis.

The idea behind the Joukowski transformation is very simple. First, as discussed in Section 2.4, we can quickly generate a potential flow around a circle with arbitrary magnitudes of uniform free stream velocity U and circulation strength Γ. Then, using Equation 2.88, we can quickly transform this flow into another potential flow in the physical plane, retaining the strengths of various singularities of the originating flow. Finally, we can iteratively adjust the flow in the originating plane to obtain the desired geometric shape and the corresponding potential flow in the physical plane. A few examples presented in the following demonstrate the ease and power of the Joukowski transformation.

2.6.1.1 Flow around a Flat Plate

Let us consider a circle of radius $r = a$ in the ζ plane. As shown in Figure 2.23, the circle has its center at the origin. For $c = a$ in Equation 2.88, the Joukowski transformation function becomes:

$$z = \zeta + \frac{c^2}{\zeta} = re^{i\theta} + \frac{c^2}{re^{i\theta}} \tag{2.90}$$

which shows that the points on the circle ($r = a = c$) will map in the z plane as:

$$z = x + iy = ce^{i\theta} + ce^{-i\theta} = 2c\cos\theta$$

giving $x = 2c\cos\theta$ and $y = 0$. The figure also shows that the top and bottom halves of the circle map onto the line segment (flat plate of zero thickness), spanning from $x = -2c$ to $x = 2c$. Using this equation, we can similarly map the entire flow around the circle into the flow around the flat plate in the physical plane.

2.6.1.2 Flow around an Ellipse

As shown in Figure 2.24, when the radius of the circle is greater than the constant c in Equation 2.88, the Joukowski transformation function maps the circle into an ellipse rather than a flat plate. According to Equation 2.90, the points on the circle ($r = a$) will map as:

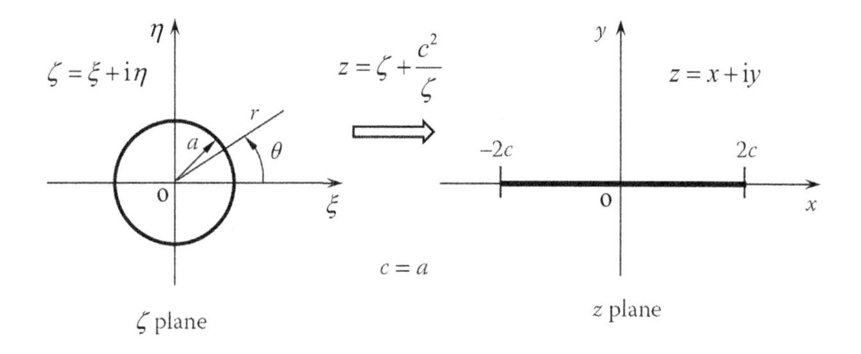

FIGURE 2.23 Joukowski transformation: potential flow around a flat plate.

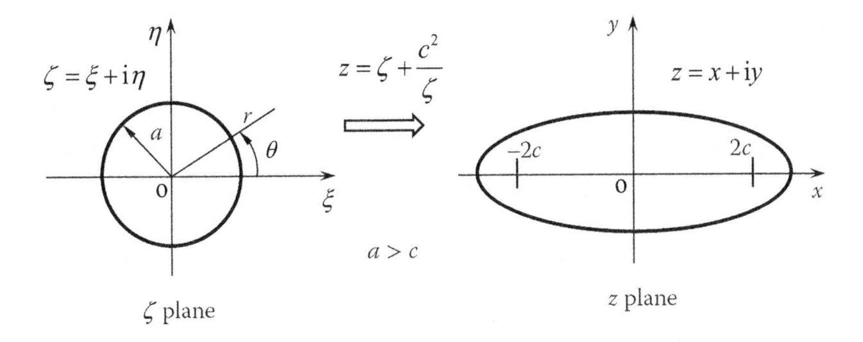

FIGURE 2.24 Joukowski transformation of a circle into an ellipse.

$$z = x + \mathrm{i}y = ae^{\mathrm{i}\theta} + \frac{c^2}{a}\mathrm{e}^{-\mathrm{i}\theta} = \left(a + \frac{c^2}{a}\right)\cos\theta + \mathrm{i}\left(a - \frac{c^2}{a}\right)\sin\theta$$

giving the following equation for the ellipse with its foci located at $(2c, 0)$ and $(-2c, 0)$:

$$\frac{x^2}{\left(a + \dfrac{c^2}{a}\right)^2} + \frac{y^2}{\left(a - \dfrac{c^2}{a}\right)^2} = 1 \tag{2.91}$$

In this case, if the potential flow around the circle had been transformed, it would have created the corresponding potential flow around the ellipse.

2.6.1.3 Flow around a Symmetric Airfoil

When we displace the circle along the ξ axis in the ζ plane, the Joukowski transformation into the z plane yields a symmetric airfoil, as shown in Figure 2.25 for $c / a = 0.75$ and $\varepsilon_\xi / a = 0.2$.

We can express the Joukowski mapping function in the Cartesian coordinates as:

$$z = x + \mathrm{i}y = (\xi + \mathrm{i}\eta) + \frac{c^2}{(\xi + \mathrm{i}\eta)} = (\xi + \mathrm{i}\eta) + \frac{c^2(\xi - \mathrm{i}\eta)}{\xi^2 + \eta^2}$$

$$x + \mathrm{i}y - \xi\left(1 + \frac{c^2}{\xi^2 + \eta^2}\right) + \mathrm{i}\eta\left(1 - \frac{c^2}{\xi^2 + \eta^2}\right)$$

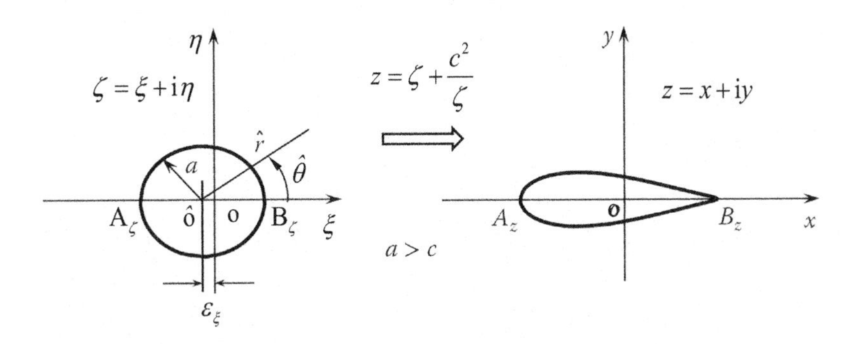

FIGURE 2.25 Joukowski transformation of a circle into a symmetric airfoil.

which yields:

$$x = \xi\left(1 + \frac{c^2}{\xi^2 + \eta^2}\right) \tag{2.92}$$

$$y = \eta\left(1 - \frac{c^2}{\xi^2 + \eta^2}\right) \tag{2.93}$$

We can write the coordinates of a point on the circle shown in Figure 2.25 as $\xi = a\cos\hat{\theta} - \varepsilon_\xi$ and $\eta = a\sin\hat{\theta}$—varying $\hat{\theta}$ from 0 to 2π, we traverse the entire circle in the ζ plane. Substituting these coordinates in Equations 2.92 and 2.93 yields the x and y coordinates of the points mapped into the z plane. For example, the point $A_\zeta\left\{-(a+\varepsilon_\xi),0\right\}$, which corresponds to $\hat{\theta} = \pi$ on the circle, will map into $A_z\left[-\left\{(a+\varepsilon_\xi)^2 + c^2\right\}/(a+\varepsilon_\xi),0\right]$ on the symmetric airfoil. Similarly, the point $B_\zeta\left\{(a-\varepsilon_\xi),0\right\}$, which corresponds to $\hat{\theta} = 0$ on the circle, will map into $B_z\left[\left\{(a-\varepsilon_\xi)^2 + c^2\right\}/(a-\varepsilon_\xi),0\right]$ in the z plane.

In this case, if we had transformed the potential flow around the circle in Figure 2.25, just as we transformed the points on the circle, we would have obtained the corresponding potential flow around the symmetric airfoil in the z plane.

2.6.1.4 Flow around a Circular Arc Airfoil

When we displace the circle along the η axis in the ζ plane, the Joukowski transformation into the z plane produces a circular arc airfoil, as shown in Figure 2.26. For the displacement ε_η, the geometry shown in the figure yields $a^2 = c^2 + \varepsilon_\eta^2$ and $\sin\beta = \varepsilon_\eta / a$. As derived by Glauert (1986), we can write the equation of the circle of which the arc is a part in the z plane as

$$x^2 + (y + 2c\cot 2\beta)^2 = (2c\,\mathrm{cosec}\,2\beta)^2 \tag{2.94}$$

which yields the coordinates of the center of the circle as $O'(0,-2c\cot 2\beta)$ and the radius $R = 2c\,\mathrm{cosec}\,2\beta$. We can express the trigonometric functions in this equation as:

$$2c\cot 2\beta = c\left(\frac{c}{\varepsilon_\eta} - \frac{\varepsilon_\eta}{c}\right) \tag{2.95}$$

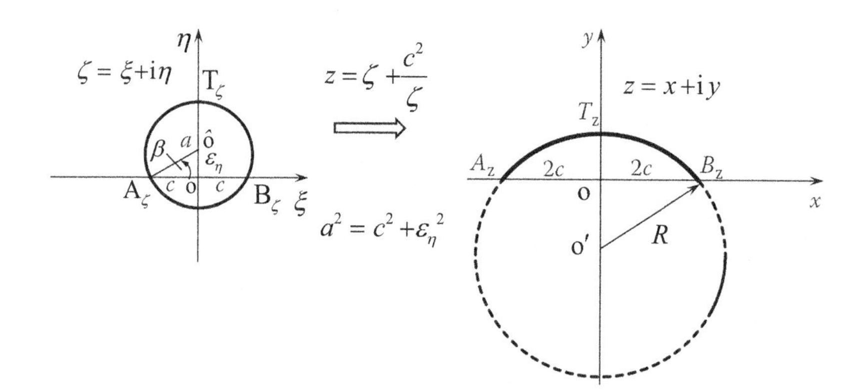

FIGURE 2.26 Joukowski transformation of a circle into a circular arc airfoil.

$$2c\,\text{cosec}\,2\beta = c\left(\frac{c}{\varepsilon_\eta} + \frac{\varepsilon_\eta}{c}\right) \tag{2.96}$$

Equations 2.94–2.96 yield $\text{OT}_z = 2\varepsilon_\eta$ and $A_zB_z = 4c$.

2.6.1.5 Flow around an Airfoil with a Camber

When we displace the circle to locate its center at $(-\varepsilon_\xi, \varepsilon_\eta)$ in the ζ plane, the Joukowski transformation into the z plane yields an airfoil with a camber, as shown in Figure 2.27 for $c / a = 0.85$, $\varepsilon_\xi / a = 0.1$, and $\varepsilon_\eta / a = 0.1$. We can write the coordinates of a point on the circle shown in this figure as $\xi = a\cos\hat{\theta} - \varepsilon_\xi$ and $\eta = a\sin\hat{\theta} + \varepsilon_\eta$. On the circle in the ζ plane, $\hat{\theta}$ varies from 0 to 2π. By substituting the coordinates ξ and η of all the points on the circle in Equations 2.92 and 2.93, we obtain the x and y coordinates of the corresponding points on the airfoil with a camber. Based on the geometry shown in this figure, the point A_ζ on the circle corresponds to $\hat{\theta} = \pi + \sin^{-1}(\varepsilon_\eta / a)$ and the point B_ζ to $\hat{\theta} = 2\pi - \sin^{-1}(\varepsilon_\eta / a)$.

Following the aforementioned method of transforming the circle into an airfoil with camber, we can map the entire potential flow around the circle into the corresponding flow around the airfoil, as shown in Figure 2.28. Note that the front and rear stagnation point locations do not generally coincide with the airfoil leading and trailing edges. Their locations on the airfoil depend upon the strength of the superimposed vortex in the originating potential flow around the cylinder (circle in the plane) and also on the angle of attack, the angle by which the entire flow field is rotated around the circle relative to the ξ axis.

2.6.2 Kutta Condition

As shown in Figure 2.28, generating a potential flow around an airfoil using the Joukowski transformation of the potential flow over a cylinder, the second stagnation point on the airfoil does not, in general, coincide with the airfoil trailing edge. Mathematically, the flow must turn around the trailing edge with a very high velocity to map the second stagnation point either on the pressure surface (as shown in the figure) or the suction surface. This mathematical situation does not represent the real-fluid flow behavior at the trailing edge. To remedy this situation, we impose an additional condition called the Kutta condition, which requires the second stagnation point to coincide with the airfoil trailing edge.

As shown in Figure 2.29, the airfoil trailing edge corresponds to the second point of intersection between the circle and the ξ axis. We determine this point from the geometry alone, locating it on the circle at an angle, $-\beta$, with the ξ' axis, which is parallel to the ξ axis and passes through the center of the circle. In this figure, α represents the angle of attack on the airfoil. We simulate this angle of attack by rotating the entire flow around the circle by α.

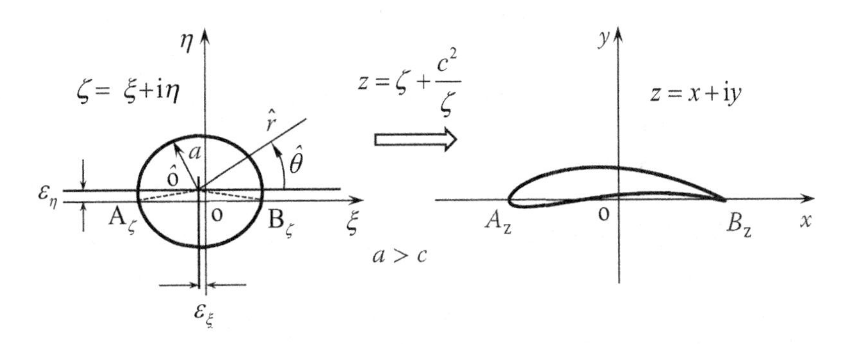

FIGURE 2.27 Joukowski transformation of a circle into an airfoil with camber.

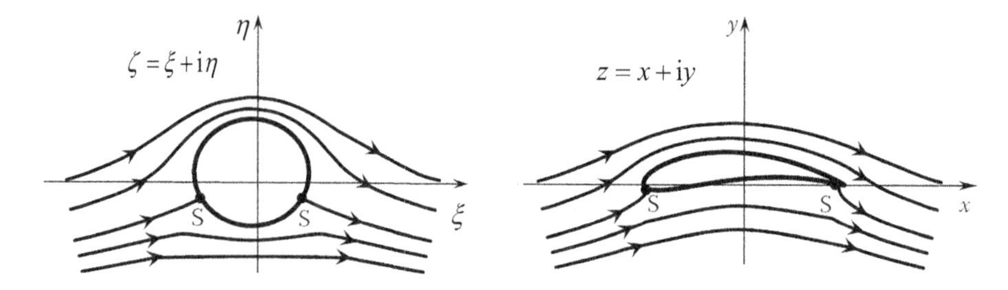

FIGURE 2.28 Potential flow over a Joukowski airfoil.

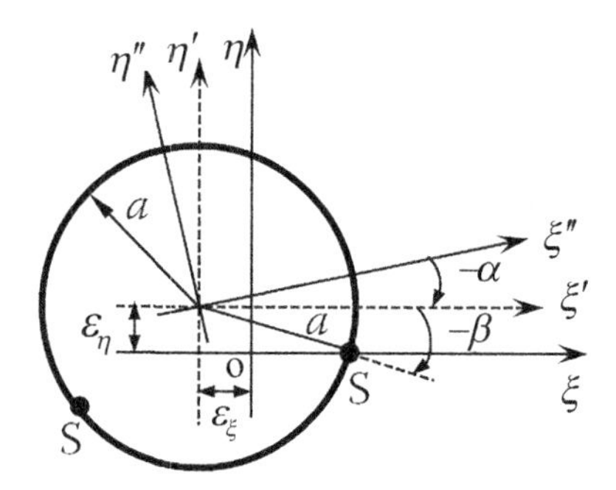

FIGURE 2.29 Kutta condition for the trailing edge stagnation point.

The stagnation point on the circle corresponds to θ_S (measured from the ξ'' axis) satisfying Equation 2.70:

$$\sin\theta_S = -\frac{\Gamma}{4\pi Ua}$$

which, to satisfy the Kutta condition [$\theta_S = -(\alpha + \beta)$], yields:

$$\Gamma = 4\pi Ua \sin(\alpha + \beta) \tag{2.97}$$

Knowing the vortex strength from this equation, we can easily calculate the lift force acting on the airfoil as:

$$L = \rho U\Gamma = 4\pi\rho U^2 a \sin(\alpha + \beta) \tag{2.98}$$

using which, we can express the lift coefficient of a thin airfoil with $c \approx a$, $\varepsilon_\xi << c$, $\varepsilon_\eta << c$, and the chord length $4c$ as:

$$C_L = \frac{L}{\frac{1}{2}\rho U^2 (4c)} = \frac{4\pi\rho U^2 a \sin(\alpha + \beta)}{\frac{1}{2}\rho U^2 (4c)} = 2\pi \sin(\alpha + \beta) \tag{2.99}$$

2.7 CONCLUDING REMARKS

This chapter presents essential concepts of classical fluid mechanics using complex variables. The steady two-dimensional potential flows presented here include uniform rectilinear flows, source/sink flows, vortex flows, and some valuable flows generated by the linear superposition of other potential flows, including dipole, doublet, Rankine half-body, and flow around a cylinder with and without circulation; the one with circulation generates a lift force on the cylinder. This lift force is the Magnus effect computed by the Kutta–Joukowski law. As a matter of historical importance and because of the influential roles they have played in the development of potential flow theory, we have also discussed here the Milne-Thomson circle theorem, Blasius integral theorems, D'Alembert's paradox, and the Joukowski conformal transformation, used to generate potential flows around a variety of airfoils from the potential flows around a cylinder represented by a circle in a plane. Although the potential flow theory does not accurately predict the drag force acting on a body, its significant contribution is in providing good insight into the mechanism of generating the lift force with its reasonably accurate predictions in most cases.

While a scalar potential function exists for two- and three-dimensional potential flows, the stream function, whose value remains constant along each streamline, exists only for two-dimensional flows. However, both potential and stream functions satisfy the Laplace equation, which arises in many other analogous physical situations, for example, temperature distribution due to steady heat conduction in a solid of constant thermal conductivity. All the potential flow solutions presented in this chapter are closed-form analytical solutions. We can use today's computing power for more complex geometries to obtain semi-numerical and numerical solutions. For example, we can use the boundary collocation method involving harmonic functions used by Sultanian and Sastri (1980) to solve the Laplace equation governing conduction heat transfer in an arbitrary multiply connected domain to solve complex two-dimensional potential flows. With many commercial CFD codes available, CFD (see Chapter 6 for an overview) offers a powerful method to generate numerical solutions to three-dimensional unsteady potential flows in complex internal and external geometries. In these CFD solutions, the assumptions of zero fluid viscosity (impossible in an experiment), no turbulence model, and slip boundary condition at the wall (no need to resolve the wall boundary layer) considerably simplify the task. However, the available closed-form potential flow solutions are helpful for the initial validation of a CFD code.

To develop their skills in solving various fluid flow problems, readers may want to review numerous problems with solutions presented in Sultanian (2021, 2022).

WORKED EXAMPLES

Example 2.1

Verify that $\Phi = 2xy + x^2 - y^2$ is a potential function for a two-dimensional incompressible flow. Find the corresponding stream function. Plot a few streamlines and indicate the flow direction along each streamline.

Solution

A potential function to represent an incompressible flow field must satisfy the Laplace equation $\nabla^2\Phi = 0$. For the given potential function, we obtain:

$$\nabla^2\Phi = \nabla^2(2xy + x^2 - y^2) = \frac{\partial^2}{\partial x^2}(2xy + x^2 - y^2) + \frac{\partial^2}{\partial y^2}(2xy + x^2 - y^2) = 2 - 2 = 0$$

which verifies that the given potential function corresponds to a two-dimensional incompressible potential flow with the velocity components given by:

$$V_x = \frac{\partial}{\partial x}(2xy + x^2 - y^2) = 2y + 2x$$

and

$$V_y = \frac{\partial}{\partial y}(2xy + x^2 - y^2) = 2x - 2y$$

Using Equation 2.4, we obtain:

$$\frac{\partial \Psi}{\partial y} = V_x = 2y + 2x$$

whose integration yields:

$$\Psi = y^2 + 2xy + f(x)$$

which using Equation 2.5 yields:

$$-\frac{\partial \Psi}{\partial x} = -2y - f'(x) = V_y = 2x - 2y$$

$$-f'(x) = 2x$$

$$f(x) = -x^2$$

where, without any loss of generality, we have assumed the constant of integration to be zero.
Thus, we obtain the stream function for the given velocity potential as:

$$\Psi = y^2 + 2xy - x^2$$

To plot the streamlines in the *x-y* plane, let us rearrange the preceding quadratic equation for y as:

$$y^2 + 2xy - (x^2 + \Psi) = 0$$

whose roots are $y = -x + \sqrt{2x^2 + \Psi}$ and $y = -x - \sqrt{2x^2 + \Psi}$.

Figure 2.30 shows the streamlines with the flow direction. The solid streamlines in the upper half correspond to the first root, and those by the dashed line in the lower half correspond to the second root.

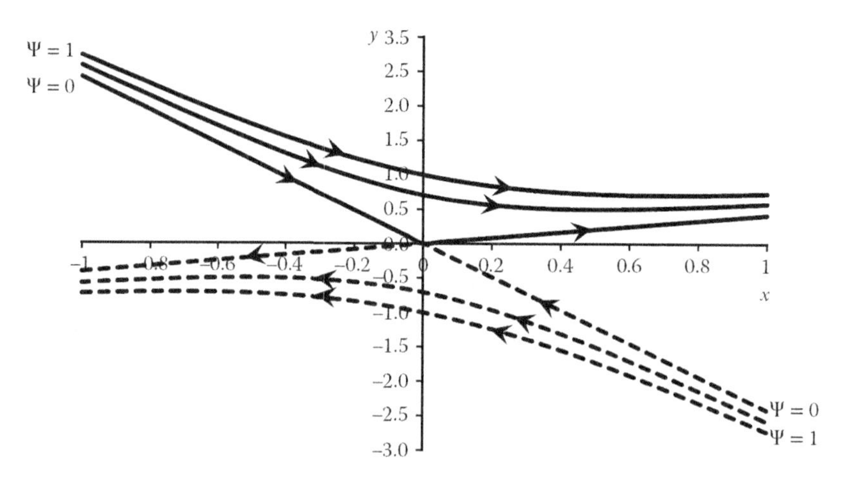

FIGURE 2.30 Streamlines for the potential function of Example 2.1.

Example 2.2

The free stream static pressure far away from the Rankine half-body shown in Figure 2.13 is p_∞ where the velocity is U. Find an expression to calculate the static pressure distribution along the surface of the Rankine half-body with a source of strength m.

Solution

From the given free stream static pressure and corresponding flow velocity, we obtain the total pressure for an incompressible flow as:

$$p_0 = p_\infty + \frac{1}{2}\rho U^2$$

As the total pressure remains constant in a potential flow, we can obtain the static pressure at any point in this flow using the following equation (the Bernoulli equation):

$$p = p_\infty + \frac{1}{2}\rho U^2 - \frac{1}{2}\rho(V_r^2 + V_\theta^2)$$

To obtain the static pressure at a point (r_b, θ) on the Rankine half-body, we substitute for V_r and V_θ from Equation 6.43 and Equation 6.44, respectively, with $r = r_b$, yielding:

$$p = p_\infty + \frac{1}{2}\rho U^2 - \frac{1}{2}\rho\left(U\cos\theta + \frac{m}{2\pi r_b}\right)^2 - \frac{1}{2}\rho(U\sin\theta)^2$$

$$\frac{p - p_\infty}{\frac{1}{2}\rho U^2} = -\left(\frac{m\cos\theta}{\pi r_b U} + \frac{m^2}{4\pi^2 r_b^2 U^2}\right)$$

Substituting for r_b from Equation 6.45 in this equation and simplifying the resulting expression, we finally obtain:

$$\frac{p - p_\infty}{\frac{1}{2}\rho U^2} = -\frac{\sin 2\theta}{(\pi - \theta)} - \frac{\sin^2\theta}{(\pi - \theta)^2}$$

which is independent of the source strength m, and it is plotted in Figure 2.31. At the stagnation point, which corresponds to $\theta = 180^o$, the static pressure equals the stagnation pressure. With

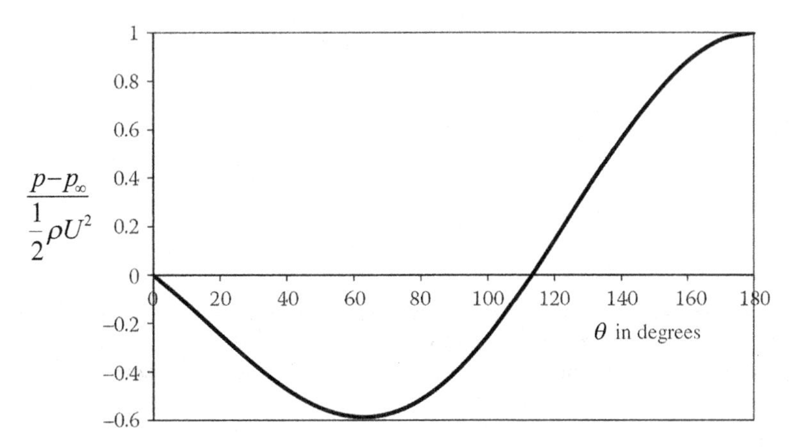

FIGURE 2.31 Static pressure distribution over a Rankine half-body (Example 2.2).

decreasing θ, the flow accelerates along the Rankine half-body with decrease in static pressure. At $\theta = 113.2^o$, the velocity equals the incoming uniform velocity U, and the static pressure equals the free stream value p_∞. At $\theta \cong 63^o$, the velocity on the surface reaches its maximum value with the minimum static pressure. As θ approaches zero, the static pressure and velocity attain their free stream value.

Example 2.3

The stream function $\Psi_1 = Axy$, where A is a constant, represents a two-dimensional potential flow with its stagnation point at the origin. When we add a source of strength m to this flow at the origin, the stagnation point moves up by a distance h along the y axis. Find the value of h in terms of A and m. For $A = 4$ and $m = 1$, plot a few streamlines in the first quadrant of the x-y plane, including the streamline that contains the stagnation point.

Solution

We can write the stream function $\Psi_1 = Axy$ in polar coordinates as:

$$\Psi_1 = Axy = A(r\cos\theta)(r\sin\theta) = \frac{Ar^2}{2}\sin 2\theta$$

We can write the stream function corresponding to a point source of strength m located at the origin as $\Psi_2 = m\theta/2\pi$. The combined stream function becomes:

$$\Psi = \Psi_1 + \Psi_2 = \frac{Ar^2}{2}\sin 2\theta + \frac{m}{2\pi}\theta$$

from which we obtain the radial and tangential velocity components as follows:

$$V_r = \frac{\partial \Psi}{r\partial\theta} = Ar\cos 2\theta + \frac{m}{2\pi r}$$

$$V_\theta = -\frac{\partial \Psi}{\partial r} = -Ar\sin 2\theta$$

both of which must vanish at the stagnation point, giving:

$$V_\theta = Ar\sin 2\theta = 0$$

$$\theta = \pi/2$$

and

$$V_r = Ar\cos 2\theta + \frac{m}{2\pi r} = 0$$

With the substitution $\theta = \pi/2$ and $r = h$ in this equation, we finally obtain $h = \sqrt{m/2\pi A}$ and $\Psi = m/4$.

For $A = 4$ and $m = 1$, $h = 0.199$ and $\Psi = 0.25$. Figure 2.32 shows the streamlines that correspond to the stream function $\Psi = 2r^2\sin 2\theta + \theta/2\pi$, indicating that the stagnation point lies on the streamline with $\Psi = 0.25$ at a distance $h = 0.199$ from the origin.

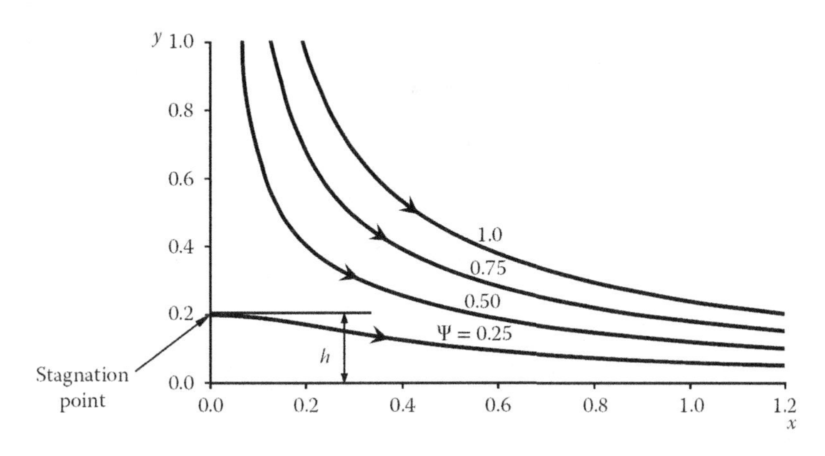

FIGURE 2.32 Streamlines for the stream function in Example 2.3.

Example 2.4

The velocity potential for a spiral vortex flow is given by:

$$\Phi = \Phi_{\text{vortex}} + \Phi_{\text{sink}} = \left(\frac{\Gamma}{2\pi}\right)\theta - \left(\frac{m}{2\pi}\right)\ln r$$

where Γ and m are constants representing the vortex strength and sink strength, respectively. Find the corresponding stream function and plot a few streamlines. Also, show that the angle between the velocity vector and the radial direction is constant for a spiral vortex throughout the flow field.

Solution

We write the complex potential for a vortex as:

$$F_{\text{vortex}} = \Phi_{\text{vortex}} + i\Psi_{\text{vortex}} = \left(\frac{\Gamma}{2\pi}\right)\theta - i\left(\frac{\Gamma}{2\pi}\right)\ln r$$

and for a sink as:

$$F_{\text{sink}} = \Phi_{\text{sink}} + i\Psi_{\text{sink}} = -\left(\frac{m}{2\pi}\right)\ln r - i\left(\frac{m}{2\pi}\right)\theta$$

The linear superposition of F_{vortex} and F_{sink} yields:

$$F = F_{\text{vortex}} + F_{\text{sink}} = \Phi + i\Psi = \left[\left(\frac{\Gamma}{2\pi}\right)\theta - \left(\frac{m}{2\pi}\right)\ln r\right] - i\left[\left(\frac{\Gamma}{2\pi}\right)\ln r + \left(\frac{m}{2\pi}\right)\theta\right]$$

giving the stream function for a spiral vortex as:

$$\Psi = -\left(\frac{\Gamma}{2\pi}\right)\ln r - \left(\frac{m}{2\pi}\right)\theta$$

For $m = 1$ and $\Gamma = 2.5$, Figure 2.33 shows three streamlines for $\Psi = 0.50, 0.75,$ and 1.0, all spiraling down to the sink located at the origin.

From the given velocity potential, we obtain the following radial and tangential velocity components:

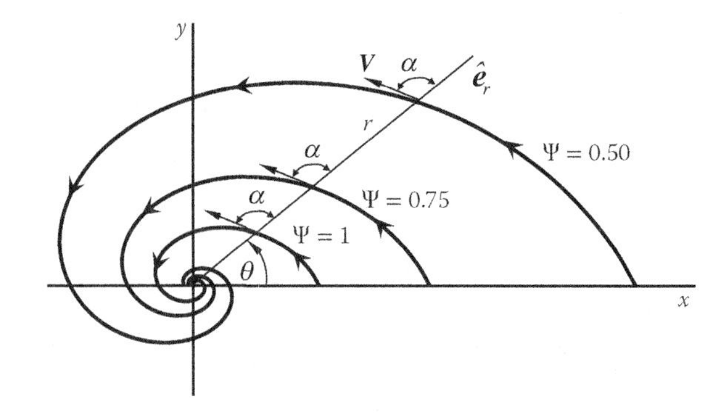

FIGURE 2.33 Streamlines in the spiral vortex of Example 2.4.

$$V_r = \frac{\partial \Phi}{\partial r} = -\left(\frac{m}{2\pi}\right)\frac{1}{r} = -\frac{m}{2\pi r}$$

$$V_\theta = \frac{1}{r}\frac{\partial \Phi}{\partial \theta} = \frac{1}{r}\left(\frac{\Gamma}{2\pi}\right) = \frac{\Gamma}{2\pi r}$$

which we write in vector notation as:

$$\boldsymbol{V} = V_r \hat{e}_r + V_\theta \hat{e}_\theta = -\frac{m}{2\pi r}\hat{e}_r + \frac{\Gamma}{2\pi r}\hat{e}_\theta$$

We obtain the angle between the velocity vector and the radial direction by taking the dot-product between the two vectors as follows:

$$\boldsymbol{V} \bullet \hat{e}_r = |\boldsymbol{V}| \cos \alpha$$

$$\cos \alpha = \frac{\boldsymbol{V} \bullet \hat{e}_r}{|\mathrm{V}|} = \frac{-m}{\sqrt{m^2 + \Gamma^2}} = \frac{-1}{\sqrt{1 + \left(\frac{\Gamma}{m}\right)^2}}$$

Thus, for the given values of m and Γ, the angle α is constant throughout the flow field. We obtain $\alpha = 111.8^\circ$ for the spiral vortex shown in Figure 2.33.

Example 2.5

Consider a uniform two-dimensional potential flow with rectilinear velocity V at an angle α to the x axis. Use the Milne-Thomson circle theorem to find the complex potential of the flow augmented by the placement of a circular cylinder of radius a with its axis along the z axis. Show that the circle $z = a\,\mathrm{e}^{i\theta}$, which represents the cylinder on the z plane, is a streamline.

Solution

We can express the complex potential of a uniform flow with velocity V at an angle α to the x as:

$$F_1(z) = Vz\mathrm{e}^{-i\alpha}$$

Using the Milne-Thomson circle theorem, we obtain the complex potential for the flow with the placement of the cylinder at the origin as:

$$F(z) = Vze^{-i\alpha} + V\frac{a^2}{z}e^{i\alpha}$$

$$F(z) = Vze^{-i\alpha} + V\frac{a^2}{ze^{-i\alpha}} = V\left(ze^{-i\alpha} + \frac{a^2}{ze^{-i\alpha}}\right)$$

$$F(z') = V\left(z' + \frac{a^2}{z'}\right)$$

where $z' = ze^{-i\alpha}$. The preceding equation is similar to Equation 2.58, representing the complex potential of a potential flow around a cylinder in a uniform cross-flow where the rectilinear velocity is parallel to the x axis.

The complex potential on the circle $z = ae^{-i\theta}$ becomes:

$$F_{\text{circle}} = V\left(ae^{i\theta}e^{-i\alpha} + \frac{a^2}{ae^{i\theta}e^{-i\alpha}}\right)$$

$$F_{\text{circle}} = V\left(ae^{i(\theta-\alpha)} + ae^{-i(\theta-\alpha)}\right)$$

$$F_{\text{circle}} = 2Va\cos(\theta - \alpha)$$

which being real, represents a streamline with $\Psi = 0$.

Example 2.6

Consider in the z plane a Joukowski airfoil, generated by displacing a circle of radius 1.1 m by $d\xi = -0.05$ m and $d\eta = 0.05$ m in the ζ plane. For $\alpha = 0$ and $\alpha = 11°$, compute the vortex strength Γ needed to impose the Kutta condition for the potential flow over the airfoil with free stream velocity $U = 11$ m/s. Also evaluate the lift coefficient C_{lift} in each case.

Solution

With reference to Figure 2.33, this example specifies $a = 1.1$ m and $\varepsilon_\eta = d\eta = 0.05$ m. From the geometry, we calculate:

$$\beta = \sin^{-1}(0.05/1.1) = 0.0546 \text{ radians} = 3.127°$$

For $\alpha = 0$, Equations 2.97 and 2.99 yield:

$$C_{\text{lift}\_\alpha=0} = 2\pi\sin(\alpha+\beta) = 2\times\pi\times\sin(3.127°) = 0.343$$

$$\Gamma_{\alpha=0} = 4\pi Ua\sin(\alpha+\beta) = 4\times\pi\times11\times1.1\times\sin(3.127°) = 8.294 \text{ m}^2/\text{s}$$

For $\alpha = 11°$, we obtain:

$$\Gamma_{\alpha=11°} = 4\pi Ua\sin(\alpha+\beta) = 4\times\pi\times11\times1.1\times\sin(11°+3.127°) = 37.111 \text{ m}^2/\text{s}$$

$$C_{\text{lift}\_\alpha=11°} = 2\pi\sin(\alpha+\beta) = 2\times\pi\times\sin(11°+3.127°) = 1.534$$

As expected, both the vortex strength and the lift coefficient for $\alpha = 11°$ are significantly higher than their values for zero angle of attack.

Problems

2.1 In a two-dimensional flow on a flat plate, $V_y = 0$ and $V_x = Ky$. Find the stream function for this flow. Is this flow irrotational?

2.2 In a two-dimensional potential flow, radial and tangential velocities vary as $V_r = V_\theta = 2/r$. Find the stream function, potential function, and static pressure distribution for this flow field for a fluid of density ρ.

2.3 Consider an incompressible flow with velocity potential given by $\Phi = y^2 - x^2$. Does this function satisfy the Laplace equation? Find the stream function for the flow and sketch the streamline and flow direction for $\Psi = 3$.

2.4 In a two-dimensional flow of fluid of constant density ρ, the radial and tangential components of velocity are given by $V_r = A/r - B\cos\theta$ and $V_\theta = B\sin\theta$, where A and B are positive real constants. Does the flow satisfy continuity? Is this flow irrotational? Find the potential and stream functions. Find an expression in terms of A, B, r, and ρ to determine the static pressure variation along the line $\theta = 0$. The total pressure remains constant at p_0.

2.5 A uniform potential flow of rectilinear velocity $V = 20\,\text{m/s}$, inclined at an angle of 18° to the x axis, is superimposed on a source of strength $m = 15\,\text{m}^2/\text{s}$ situated at the origin. Determine the velocity components for the resulting flow and the location of the stagnation points.

2.6 In a two-dimensional (z plane) airflow with uniform velocity $U = 20\,\text{m/s}$ parallel to the x axis past a cylinder of radius $a = 0.1\,\text{m}$ with its axis along the z axis, the stagnation points are found to be located at 45° above the x axis. What is the circulation strength associated with this flow? Find the magnitude and direction of the lift force per unit length of the cylinder. Assume $\rho = 1.184\,\text{kg/m}^3$ for air.

2.7 The complex potential for a cylinder (radius $= a$) in a cross-flow with uniform velocity U (parallel to the x axis) superimposed with a clockwise vortex of strength Γ is given by $F(z) = U\left(z + \frac{a^2}{z}\right) + \frac{i\Gamma}{2\pi}\ln\frac{z}{a}$. Determine the static pressure distribution on the surface of the cylinder. Integrate the pressure distribution to calculate the lift force acting on the cylinder, and hence verify the validity of the Kutta–Joukowski law for this particular flow.

2.8 The following complex potential function represents the flow over a sinusoidal wavy wall:

$$F(z) = V_0[z + y_0 e^{2i\pi z/\lambda}]$$

where $z = x + iy$, y_0 is the amplitude, and λ is the wavelength. Find the stream function Ψ and the variation of y for the streamline $\Psi = 0$.

2.9 Use the Milne-Thomson circle theorem to find the complex potential of the flow generated by a source of strength m located at $z = z_0$ with the inclusion of a circular cylinder of radius $a < |z_0|$ with its axis along the z axis. Show that the circle $z = ae^{i\theta}$, which represents the cylinder on the z plane, is a streamline.

2.10 Use the Milne-Thomson circle theorem to find the complex potential of the flow generated by a clockwise (negative) vortex of strength Γ located at $z = z_0$ with the inclusion of a circular cylinder of radius $a < |z_0|$ with its axis along the z axis. Show that the circle $z = ae^{i\theta}$, which represents the cylinder on the z plane, is a streamline.

2.11 Consider a flow with the following velocity components:

$$U = A\left[1 - \left(\frac{y}{a}\right)^2\right] \text{ and } V = 0$$

where A and a are positive real constants and $y \leq a$. Find the stream function and the vorticity for this flow. Is continuity satisfied? Can a velocity potential be found for this flow?

2.12 In an incompressible flow, the radial and circumferential components of the velocity are given by:

$$V_r = \frac{A}{r} - B\cos\theta \text{ and } V_\theta = ae^{i\theta}$$

where A and B are positive real constants. Does a velocity potential exist for this flow? Does the flow satisfy continuity? Find the stream function and give the pressure variation along the line $\theta = 0$ in terms of A, B, and r. The pressure at infinity is constant.

2.13 Consider the vortex-sink combination located at origin.
   a. Write the complex velocity potential.
   b. Find the magnitude and direction of the velocity vector at any point in the flow.
   c. Obtain an expression for the difference in pressure between any two points in the flow.

REFERENCES

Churchill, R.V. 1960. *Complex Variables and Applications.* New York: McGraw-Hill.

Glauert, H. 1982. *The Elements of Aerofoil and Airscrew Theory*, 2nd Edition. London, UK: Cambridge University Press.

Hoffman, J. and C. Johnson. 2010. Resolution of D'Alembert's paradox. *Journal of Mathematical Fluid Mechanics.* 12(3): 321–334.

Milne-Thomson, L.M. 1968. *Theoretical Hydrodynamics.* New York: Dover Publications.

Sultanian, B.K. 2015. *Fluid Mechanics: An Intermediate Approach*, 1st Edition. Boca Raton, FL: Taylor & Francis.

Sultanian, B.K. 2021. *Fluid Mechanics and Turbomachinery: Problems and Solutions.* Boca Raton, FL: Taylor & Francis.

Sultanian, B.K. 2022. *Thermal-Fluids Engineering: Problems with Solutions.* USA: Independently published by Kindle Direct Publishing (Amazon.com).

Sultanian, B.K. and V.M.K. Sastri. 1980. Effect of geometry on heat conduction in coolant channels of a liquid rocket engine. *Wärme- und Stoffübertragung.* 14: 245–251.

Zdravkovich, M.M. 1997. *Flow around Circular Cylinders Volume 1: Fundamentals.* Oxford, UK: Oxford University Press.

BIBLIOGRAPHY

Batchelor, G.K. 1967. *An Introduction to Fluid Dynamics.* London, UK: Cambridge University Press.

Currie, I.G. 2013. *Fundamental Mechanics of Fluids*, 4th Edition. Boca Raton, FL: Taylor & Francis.

Durst, F. 2022. *Fluid Mechanics: An Introduction to the Theory of Fluid Flows*, 2nd Edition. Heidelberg: Springer Verlag GmbH.

Jeffrey, A. 1992. *Complex Analysis and Applications.* Boca Raton, FL: Taylor & Francis.

Kirchhoff, R.H. 1985. *Potential Flows: Computer Graphic Solutions*, 1st Edition. New York: Marcel Dekker.

Kundu, P.K., I.M. Cohen, D.R. Dowling, J. Capecilatro. 2024. *Fluid Mechanics*, 7th Edition. Burlington, VT: Academic Press.

Nunn, R.H. 1989. *Intermediate Fluid Mechanics.* Boca Raton, FL: Taylor & Francis.

Panton, R.L. 2005. *Incompressible Flow*, 3rd Edition. New York: John Wiley.

Paterson, A.R. 1984. *A First Course in Fluid Dynamics*, 1st Edition. London, UK: Cambridge University Press.

Powers, J.M. 2023. *Mechanics of Fluids*, 1st Edition. Cambridge: Cambridge University Press.

Sultanian, B.K. 2022. *Fluid Mechanics: An Intermediate Approach*: Errata for *the First Edition Published in 2015.* USA: Independently published on KDP (Amazon.com).

NOMENCLATURE

a	Cylinder radius
c	Constant in Joukowski transformation function
C	Complex circulation; closed contour

$\hat{\boldsymbol{e}}_r$	Unit vector in the radial direction
$\hat{\boldsymbol{e}}_\theta$	Unit vector in the θ direction
f	Conformal transformation function
F	Complex potential function; force
i	Imaginary number ($\mathrm{i} = \sqrt{-1}$)
$\hat{k}$	Unit vector along z axis
d$\boldsymbol{l}$	Differential length vector along a closed contour
L	Force due to the Magnus effect
m	Source strength
n	Exponent in the complex potential for corner flows
$\hat{\boldsymbol{n}}$	Unit vector normal to dl
O	Origin
p	Pressure; static pressure
Q	Volumetric flow rate
r	Coordinate r of cylindrical polar coordinate system
$\hat{r}$	Coordinate r in the ζ plane
R	Radius of the circular arc mapped using Joukowski transformation
Re	Real part of a complex quantity
S	Stagnation point
U	Uniform velocity in x coordinate direction
V	Uniform velocity in y coordinate direction
$\boldsymbol{V}$	Velocity vector
W	Complex velocity
$\overline{W}$	Complex conjugate velocity
x	Coordinate x of the Cartesian coordinate system
y	Coordinate y of the Cartesian coordinate system
z	Complex coordinate ($z = x + \mathrm{i}y$)
$\overline{z}$	Complex conjugate coordinate ($\overline{z} = x - \mathrm{i}y$)

SUBSCRIPTS AND SUPERSCRIPTS

0	Total (stagnation)
b	Point on the Rankine half-body
c	Point on a circular streamline
r	Component in r coordinate direction
S	Stagnation point
x	Component in x coordinate direction
y	Component in y coordinate direction
z	Belonging to z plane (physical plane)
θ	Component in θ coordinate direction
∞	Infinity (very far away)

GREEK SYMBOLS

α	Angle between the velocity vector and the x axis; angle of attack
Γ	Circulation; vortex strength
ε	Distance of source or sink from the origin
ε_η	Displacement along the η axis in the ζ plane
ε_ξ	Displacement along the ξ axis in the ζ plane
ζ	Complex variable of the transformed plane ($\zeta = \xi + \mathrm{i}\eta$)

η	Imaginary coordinate of the transformed plane
θ	Coordinate θ of the cylindrical polar coordinate system
$\hat{\theta}$	Coordinate θ in the ζ plane
ξ	Real coordinate of the transformed plane
ρ	Density
Φ	Velocity potential function
Ψ	Stream function
$\boldsymbol{\omega}$	Rotation vector
Ω	Angular speed

3 The Navier–Stokes Equations
Exact Solutions

3.1 INTRODUCTION

We have discussed potential flows in Chapter 2. They are ideal fluid flows with constant density and zero viscosity. Although a fluid flow with constant density is physically possible, one with zero viscosity is not. Nevertheless, one of the significant contributions of the potential flow theory, which forms most classical fluid mechanics, is that it has provided good insight into the origin of a lift force on a body. Euler's equations governing potential flows offer a system of partial differential equations whose numerical solutions (using a CFD method) yield three-dimensional results in complex geometries. Without the wall boundary layers in actual flows, these solutions provide a good insight into these flows with their reasonably accurate static pressure distributions. A logical next step in understanding actual fluid flows is introducing constant fluid viscosity in mathematical modeling. The resulting governing partial differential equations are the Navier–Stokes equations, which are discussed at some length in this chapter, along with their few exact solutions under special situations of two-dimensional, fully developed laminar flows.

We may model a fluid flow as a laminar flow in several practical applications, such as electronics cooling, bio-fluid mechanics, and lubrication. In the beginning, fluids engineers analyzed turbulent flows like a laminar flow with the fluid viscosity modified by the so-called turbulent viscosity. Chapter 6 overviews CFD methods used in many engineering design applications. The material presented in this chapter provides a good framework and building blocks for developing a better intuitive understanding of internal shear flows.

This chapter features three worked examples and eight chapter-end problems. The original version of this chapter appears as Chapter 7 in the first edition of this book by Sultanian (2015).

3.2 FORCES ON A FLUID ELEMENT

3.2.1 Surface Forces due to Stresses

At a point on a differential fluid element, Figure 3.1 shows all nine components of the 3×3 stress tensor. As shown in the figure, the first subscript of τ_{ij} denotes the direction normal to the face on

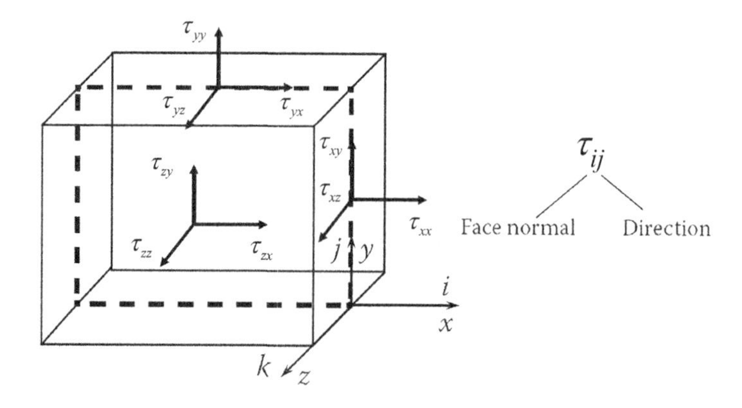

FIGURE 3.1 Surface forces acting on a fluid element.

DOI: 10.1201/9781003325192-3

which the stress is acting, and the second subscript denotes the direction of the force associated with the stress. For example, τ_{xy} denotes a shear stress acting in the y direction on the fluid element face normal to the x axis. For $i = j$, τ_{ii} denotes the normal stresses, and for $i \neq j$, it denotes the shear stresses. The stress tensor being symmetric ($\tau_{ij} = \tau_{ji}$), we only have six different stresses—three normal and three shear stresses—acting at every point in a flow. For positive τ_{ij}, both i and j should be either positive or negative.

Figure 3.2 shows a small control volume measuring δx, δy, and δz along the Cartesian coordinate directions. The figure also shows how the stresses τ_{yx}, τ_{yy}, and τ_{yz} vary along the y direction from the point (x, y, z) located at the center of the control volume. Let us evaluate the contributions of these stresses (on the positive and negative y faces) to the total surface force acting on the control volume along each coordinate direction. As an example, for τ_{yx}, we obtain:

$$\delta F_{s\_yx} = \left(\tau_{yx} + \frac{\partial \tau_{yx}}{\partial y}\frac{\delta y}{2}\right)\delta x \delta z - \left(\tau_{yx} - \frac{\partial \tau_{yx}}{\partial y}\frac{\delta y}{2}\right)\delta x \delta z$$

$$\frac{\delta F_{s\_yx}}{\delta x \delta y \delta z} = \frac{\partial \tau_{yx}}{\partial y}$$

which shows that $\partial \tau_{yx} / \partial y$ gives the force per unit volume acting on the control volume in the x direction. Similarly, $\partial \tau_{yy} / \partial y$ gives the force per unit volume in the y direction, and $\partial \tau_{yz} / \partial y$ gives the force per unit volume in the z direction. Considering similar variations of stresses in the x and y directions, we can write the net surface force per unit volume acting on the control volume in each coordinate direction as:

$$\delta f_{sx} = \frac{\partial \tau_{xx}}{\partial x} + \frac{\partial \tau_{yx}}{\partial y} + \frac{\partial \tau_{zx}}{\partial z} \tag{3.1}$$

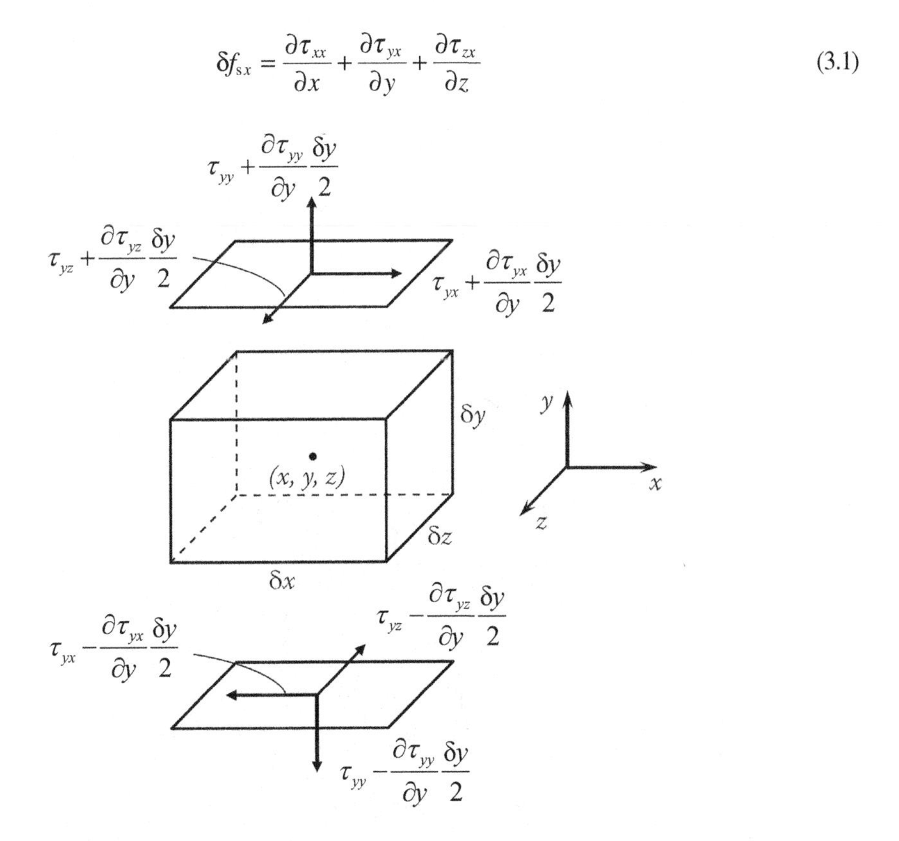

FIGURE 3.2 Gradients of stresses at a point in fluid flow.

$$\delta f_{sy} = \frac{\partial \tau_{xy}}{\partial x} + \frac{\partial \tau_{yy}}{\partial y} + \frac{\partial \tau_{zy}}{\partial z} \tag{3.2}$$

$$\delta f_{sz} = \frac{\partial \tau_{xz}}{\partial x} + \frac{\partial \tau_{yz}}{\partial y} + \frac{\partial \tau_{zz}}{\partial z} \tag{3.3}$$

We can write Equations 3.1–3.3 in compact tensor notation (see Appendix A) as:

$$\delta f_{sj} = \frac{\partial \tau_{ij}}{\partial x_i}, \qquad j = 1,2,3 \tag{3.4}$$

The foregoing development shows that, at any point in a fluid flow, not the stresses but their gradients act on the differential control volume as surface forces per unit volume.

3.2.2 Body Force due to Gravity

In the momentum equation, in addition to the surface forces discussed in the previous section, we will include the body force due to gravity, which acts vertically. As shown in Figure 3.3, regardless of the orientation of the Cartesian coordinate axes, a height h measured from a fixed datum is associated with each point in the flow. We can express the gravitational body force acting on the control volume in the x direction as:

$$\begin{aligned} \delta F_{bx} &= -\rho g (\delta x \delta y \delta z) \sin\theta \\ \frac{\delta F_{bx}}{\delta x \delta y \delta z} &= -\rho g \sin\theta \end{aligned} \tag{3.5}$$

As:

$$\sin\theta = \frac{\left(h + \dfrac{\partial h}{\partial x}\delta x\right) - h}{\delta x} = \frac{\partial h}{\partial x}$$

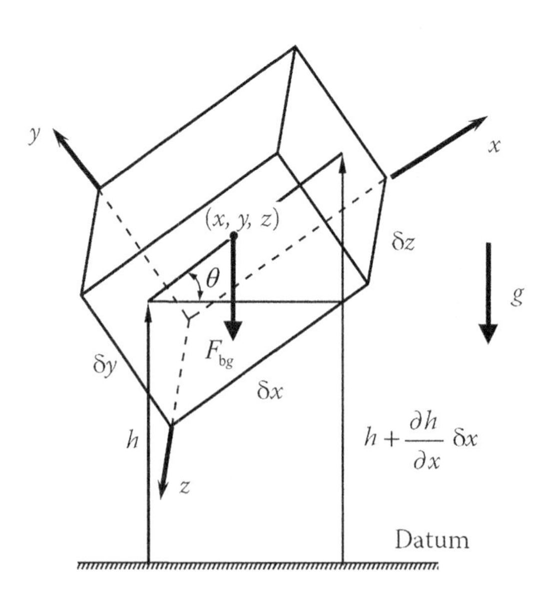

FIGURE 3.3 Body force due to gravity.

Equation 3.5 becomes:

$$\frac{\delta F_{\mathrm{b}x}}{\delta x \delta y \delta z} = \delta f_{\mathrm{b}x} = -\rho g \frac{\partial h}{\partial x} \tag{3.6}$$

where $\delta f_{\mathrm{b}x}$ is the body force per unit volume. We can generalize Equation 3.6 to obtain body forces due to gravity in the y and z directions. Thus, in the compact tensor notation, we can write gravitational body force per unit volume acting at every point in a fluid flow as:

$$\delta f_{\mathrm{bi}} = -\rho g \frac{\partial h}{\partial x_i} \tag{3.7}$$

3.3 DEFORMATION RATE TENSOR

We can write the deformation rate tensor for the velocity components u_1, u_2, and u_3 along the Cartesian coordinate directions x_1, x_2, and x_3, respectively, as:

$$S_{ij} = \begin{bmatrix} S_{11} & S_{12} & S_{13} \\ S_{21} & S_{22} & S_{23} \\ S_{31} & S_{32} & S_{33} \end{bmatrix} = \begin{bmatrix} \frac{\partial u_1}{\partial x_1} & \frac{\partial u_1}{\partial x_2} & \frac{\partial u_1}{\partial x_3} \\ \frac{\partial u_2}{\partial x_1} & \frac{\partial u_2}{\partial x_2} & \frac{\partial u_2}{\partial x_3} \\ \frac{\partial u_3}{\partial x_1} & \frac{\partial u_3}{\partial x_2} & \frac{\partial u_3}{\partial x_3} \end{bmatrix} \tag{3.8}$$

which in tensor notation becomes:

$$S_{ij} = \frac{\partial u_i}{\partial x_j} = u_{i,j} \tag{3.9}$$

Let us express the deformation rate tensor in terms of a symmetric tensor, which stays the same on the interchange of suffixes, and an antisymmetric tensor, which changes sign on the interchange of suffixes, as follows:

$$S_{ij} = \frac{1}{2}\left(S_{ij} + S_{ji}\right) + \frac{1}{2}\left(S_{ij} - S_{ji}\right) = e_{ij} + \omega_{ij} \tag{3.10}$$

From Equations 3.9 and 3.10, we obtain the symmetric strain rate tensor e_{ij} as:

$$e_{ij} = e_{ji} = \frac{1}{2}\left(u_{i,j} + u_{j,i}\right)$$

$$e_{ij} = \begin{bmatrix} \frac{\partial u_1}{\partial x_1} & \frac{1}{2}\left(\frac{\partial u_1}{\partial x_2} + \frac{\partial u_2}{\partial x_1}\right) & \frac{1}{2}\left(\frac{\partial u_1}{\partial x_3} + \frac{\partial u_3}{\partial x_1}\right) \\ \frac{1}{2}\left(\frac{\partial u_1}{\partial x_2} + \frac{\partial u_2}{\partial x_1}\right) & \frac{\partial u_2}{\partial x_2} & \frac{1}{2}\left(\frac{\partial u_2}{\partial x_3} + \frac{\partial u_3}{\partial x_2}\right) \\ \frac{1}{2}\left(\frac{\partial u_1}{\partial x_3} + \frac{\partial u_3}{\partial x_1}\right) & \frac{1}{2}\left(\frac{\partial u_2}{\partial x_3} + \frac{\partial u_3}{\partial x_2}\right) & \frac{\partial u_3}{\partial x_3} \end{bmatrix} \tag{3.11}$$

and the antisymmetric rotation tensor ω_{ij} as:

$$\omega_{ij} = -\omega_{ji} = \frac{1}{2}\left(u_{i,j} - u_{j,i}\right)$$

$$\omega_{ij} = \begin{bmatrix} 0 & \frac{1}{2}\left(\frac{\partial u_1}{\partial x_2} - \frac{\partial u_2}{\partial x_1}\right) & \frac{1}{2}\left(\frac{\partial u_1}{\partial x_3} - \frac{\partial u_3}{\partial x_1}\right) \\ -\frac{1}{2}\left(\frac{\partial u_1}{\partial x_2} - \frac{\partial u_2}{\partial x_1}\right) & 0 & \frac{1}{2}\left(\frac{\partial u_2}{\partial x_3} - \frac{\partial u_3}{\partial x_2}\right) \\ -\frac{1}{2}\left(\frac{\partial u_1}{\partial x_3} - \frac{\partial u_3}{\partial x_1}\right) & -\frac{1}{2}\left(\frac{\partial u_2}{\partial x_3} - \frac{\partial u_3}{\partial x_2}\right) & 0 \end{bmatrix} \tag{3.12}$$

Note that the rotation vector ω_i associated with an infinitesimal fluid element is related to the rotation tensor ω_{ij} by $\omega_i = \omega_{kj}$.

We define the fluid dilatation by the trace (sum of the diagonal terms) of the symmetric strain rate tensor e_{ij}. This trace remains invariant under changes of axes:

$$e_{ii} = e_{11} + e_{22} + e_{33} = \frac{\partial u_1}{\partial x_1} + \frac{\partial u_2}{\partial x_2} + \frac{\partial u_3}{\partial x_3} = u_{i,i} = \nabla \cdot \boldsymbol{V} \tag{3.13}$$

which represents the local rate of fluid volume change and must be zero for an incompressible fluid flow—recall from Chapter 1 the steady continuity equation $\nabla \cdot \boldsymbol{V} = 0$ for an incompressible flow. A close examination of the strain rate tensor given by Equation 3.11 reveals that its diagonal terms represent the actual normal strain rate while each off-diagonal term equals one-half the actual rate of shear strain.

3.4 DIFFERENTIAL FORMS OF EQUATIONS OF MOTION

3.4.1 The Continuity Equation

In Chapter 1, we derived the equation governing the conservation of mass (the continuity equation), which we summarize here in various forms as follows:

Tensor notation:

$$\frac{\partial \rho}{\partial t} + \frac{\partial}{\partial x_i}(\rho u_i) = 0 \tag{3.14}$$

Vector notation:

$$\frac{\partial \rho}{\partial t} + \nabla \cdot (\rho \boldsymbol{V}) = 0 \tag{3.15}$$

In terms of the substantial derivative of density, we can rewrite Equations 3.14 and 3.15 as:

Tensor notation:

$$\frac{\mathrm{D}\rho}{\mathrm{D}t} + \rho \frac{\partial u_i}{\partial x_i} = 0 \tag{3.16}$$

Vector notation:

$$\frac{\mathrm{D}\rho}{\mathrm{D}t} + \rho \nabla \cdot \boldsymbol{V} = 0 \tag{3.17}$$

which for a steady incompressible flow—the primary focus of this chapter—reduces to:

$$\frac{\mathrm{D}\rho}{\mathrm{D}t} = \frac{\partial u_i}{\partial x_i} = \nabla \bullet V = 0 \tag{3.18}$$

3.4.2 The Linear Momentum Equation

From a force–momentum balance on a differential control volume at any point in a fluid flow, discussed at length in Chapter 1, we can write the momentum equations in all three Cartesian coordinate directions in tensor notation as:

$$\frac{\partial}{\partial t}(\rho u_i) + \frac{\partial}{\partial x_j}(\rho u_i u_j) = \frac{\partial \tau_{ji}}{\partial x_j} - \rho g \frac{\partial h}{\partial x_i} \tag{3.19}$$

in which each term for $i = 1, 2, \textit{and}\ 3$ has the following interpretation:

$\frac{\partial}{\partial t}(\rho u_i) \equiv$ time rate of change of linear momentum per unit volume

$\frac{\partial}{\partial x_j}(\rho u_i u_j) \equiv$ net efflux of linear momentum per unit volume

$\frac{\partial \tau_{ji}}{\partial x_j} \equiv$ surface forces per unit volume

$-\rho g \frac{\partial h}{\partial x_i} \equiv$ gravitational body force per unit volume

Expanding the left-hand side of Equation 3.19 yields:

$$\frac{\partial}{\partial t}(\rho u_i) + \frac{\partial}{\partial x_j}(\rho u_i u_j) = \rho \frac{\partial u_i}{\partial t} + u_i \left[\frac{\partial \rho}{\partial t} + \frac{\partial}{\partial x_j}(\rho u_j) \right] + \rho u_j \frac{\partial u_i}{\partial x_j}$$

Using Equation 3.14, the expression within the brackets on the right-hand side of the preceding becomes zero, giving:

$$\frac{\partial}{\partial t}(\rho u_i) + \frac{\partial}{\partial x_j}(\rho u_i u_j) = \rho \frac{\partial u_i}{\partial t} + \rho u_j \frac{\partial u_i}{\partial x_j} = \rho \frac{D u_i}{D t}$$

which, when substituted in Equation 3.19, finally yields the commonly used form of the momentum equation involving the substantial derivative of u_i:

$$\rho \frac{\mathrm{D}u_i}{\mathrm{D}t} = \frac{\partial \tau_{ji}}{\partial x_j} - \rho g \frac{\partial h}{\partial x_i} \tag{3.20}$$

3.5 THE NAVIER–STOKES EQUATIONS

3.5.1 Constitutive Relationship between Stress and Rate of Strain

For a Newtonian fluid, the following general relationship exists between stress and rate of strain; see Bird, Stewart, and Lightfoot (1960):

$$\tau_{ij} = -\left(p + \frac{2}{3}\mu \nabla \bullet V \right)\delta_{ij} + 2\mu e_{ij} \tag{3.21}$$

which yields the following expressions for normal and shear stresses in the Cartesian coordinates:

Normal stresses:

$$\tau_{xx} = \sigma_{xx} = -p - \frac{2}{3}\mu\nabla \bullet \boldsymbol{V} + 2\mu\frac{\partial u}{\partial x} \tag{3.22}$$

$$\tau_{yy} = \sigma_{yy} = -p - \frac{2}{3}\mu\nabla \bullet \boldsymbol{V} + 2\mu\frac{\partial v}{\partial y} \tag{3.23}$$

$$\tau_{zz} = \sigma_{zz} = -p - \frac{2}{3}\mu\nabla \bullet \boldsymbol{V} + 2\mu\frac{\partial w}{\partial z} \tag{3.24}$$

Note that Equations 3.22–3.24 yield at a point the following useful relationship between the static pressure and three normal stresses:

$$p = -\frac{1}{3}(\sigma_{xx} + \sigma_{yy} + \sigma_{zz}) \tag{3.25}$$

Shear stresses:

$$\tau_{xy} = \tau_{yx} = \mu\left(\frac{\partial v}{\partial x} + \frac{\partial u}{\partial y}\right) \tag{3.26}$$

$$\tau_{yz} = \tau_{zy} = \mu\left(\frac{\partial w}{\partial y} + \frac{\partial v}{\partial z}\right) \tag{3.27}$$

$$\tau_{zx} = \tau_{xz} = \mu\left(\frac{\partial u}{\partial z} + \frac{\partial w}{\partial x}\right) \tag{3.28}$$

3.5.2 The Navier–Stokes Equations and Their Simplifications

Substituting Equation 3.21 into Equation 3.20, we obtain the Navier–Stokes equations in tensor notation as:

$$\rho\frac{\mathrm{D}u_i}{\mathrm{D}t} = -\rho g\frac{\partial h}{\partial x_i} - \frac{\partial p}{\partial x_i} + \frac{\partial}{\partial x_i}\left[\mu\left(\frac{\partial u_j}{\partial x_i} + \frac{\partial u_i}{\partial x_j}\right)\right] - \frac{2}{3}\frac{\partial}{\partial x_i}\left(\mu\frac{\partial u_m}{\partial x_m}\right) \tag{3.29}$$

which is the general form of the time-dependent Navier–Stokes equations, which are widely used for the analysis of a Newtonian fluid flow, be it compressible or incompressible, laminar or turbulent.

As an example, for $i = 1$, Equation 3.29 yields the Navier–Stokes equation in the x direction as:

$$\begin{aligned}\rho\frac{\mathrm{D}u}{\mathrm{D}t} = -\rho g\frac{\partial h}{\partial x} - \frac{\partial p}{\partial x} + \frac{\partial}{\partial x}\left[\mu\left(\frac{\partial u}{\partial x} + \frac{\partial u}{\partial x}\right)\right] + \frac{\partial}{\partial y}\left[\mu\left(\frac{\partial v}{\partial x} + \frac{\partial u}{\partial y}\right)\right]\\ + \frac{\partial}{\partial z}\left[\mu\left(\frac{\partial w}{\partial x} + \frac{\partial u}{\partial z}\right)\right] - \frac{2}{3}\frac{\partial}{\partial x}\left[\mu\left(\frac{\partial u}{\partial x} + \frac{\partial v}{\partial y} + \frac{\partial w}{\partial z}\right)\right]\end{aligned} \tag{3.30}$$

We can similarly obtain the Navier–Stokes equations in the y and z coordinate directions.

The Navier–Stokes equations (Equation 3.29) are too complicated to yield a closed-form analytical solution. The primary source of difficulty lies in the nonlinear convection terms in the substantial derivative of the velocity vector. Let us now make some simplifying assumptions to obtain various forms of the Navier–Stokes equations.

3.5.2.1 Constant Viscosity

When the flows are nearly isothermal, or the fluid viscosity does not vary significantly in the range of temperature variation, we appropriately assume constant viscosity. Under this assumption, Equation 3.29 reduces to:

$$\rho \frac{Du_i}{Dt} = -\rho g \frac{\partial h}{\partial x_i} - \frac{\partial p}{\partial x_i} + \mu \nabla^2 u_i + \frac{2\mu}{3} \frac{\partial}{\partial x_i}\left(\frac{\partial u_m}{\partial x_m}\right) \quad (3.31)$$

3.5.2.2 Constant Viscosity and Constant Density

In addition to assuming constant viscosity, if the fluid is incompressible or the gas flows at a Mach number less than 0.3, we can use constant density in Equation 3.31. As $\partial u_m / \partial x_m = 0$ from Equation 3.18 for an incompressible flow, Equation 3.31 further reduces to:

$$\rho \frac{Du_i}{Dt} = -\rho g \frac{\partial h}{\partial x_i} - \frac{\partial p}{\partial x_i} + \mu \nabla^2 u_i \quad (3.32)$$

As we will be using this equation along with the continuity equation (Equation 3.18) extensively in this chapter to obtain the closed-form analytical solutions to various incompressible laminar flows with constant viscosity in both the Cartesian and cylindrical polar coordinates, let us express these equations explicitly in these coordinate systems.

3.5.3 The Continuity and Navier–Stokes Equations in the Cartesian Coordinates

3.5.3.1 The Continuity Equation

$$\frac{\partial u}{\partial x} + \frac{\partial v}{\partial y} + \frac{\partial w}{\partial z} = 0 \quad (3.33)$$

3.5.3.2 The Navier–Stokes Equations

$$\rho \frac{Du}{Dt} = -\rho g \frac{\partial h}{\partial x} - \frac{\partial p}{\partial x} + \mu \nabla^2 u \quad (3.34)$$

$$\rho \frac{Dv}{Dt} = -\rho g \frac{\partial h}{\partial y} - \frac{\partial p}{\partial y} + \mu \nabla^2 v \quad (3.35)$$

$$\rho \frac{Dw}{Dt} = -\rho g \frac{\partial h}{\partial z} - \frac{\partial p}{\partial z} + \mu \nabla^2 w \quad (3.36)$$

where

$$\frac{D}{Dt} = \frac{\partial}{\partial t} + u\frac{\partial}{\partial x} + v\frac{\partial}{\partial y} + w\frac{\partial}{\partial z}$$

and

$$\nabla^2 = \frac{\partial^2}{\partial x^2} + \frac{\partial^2}{\partial y^2} + \frac{\partial^2}{\partial z^2}$$

3.5.4 The Continuity and Navier–Stokes Equations in Cylindrical Polar Coordinates

3.5.4.1 The Continuity Equation

$$\frac{1}{r}\frac{\partial}{\partial r}(rv_r) + \frac{1}{r}\frac{\partial v_\theta}{\partial \theta} + \frac{\partial v_z}{\partial z} = 0 \quad (3.37)$$

3.5.4.2 The Navier–Stokes Equations

$$\rho\left(\frac{\mathrm{D}v_r}{\mathrm{D}t} - \frac{v_\theta^2}{r}\right) = -\rho g\frac{\partial h}{\partial r} - \frac{\partial p}{\partial r} + \mu\left(\nabla^2 v_r - \frac{v_r}{r^2} - \frac{2}{r^2}\frac{\partial v_\theta}{\partial \theta}\right) \quad (3.38)$$

$$\rho\left(\frac{\mathrm{D}v_\theta}{\mathrm{D}t} + \frac{v_r v_\theta}{r}\right) = -\frac{\rho g}{r}\frac{\partial h}{\partial \theta} - \frac{1}{r}\frac{\partial p}{\partial \theta} + \mu\left(\nabla^2 v_\theta - \frac{v_\theta}{r^2} + \frac{2}{r^2}\frac{\partial v_r}{\partial \theta}\right) \quad (3.39)$$

$$\rho\frac{\mathrm{D}v_z}{\mathrm{D}t} = -\rho g\frac{\partial h}{\partial z} - \frac{\partial p}{\partial z} + \mu\nabla^2 v_z \quad (3.40)$$

where

$$\frac{\mathrm{D}}{\mathrm{D}t} = \frac{\partial}{\partial t} + v_r\frac{\partial}{\partial r} + \frac{v_\theta}{r}\frac{\partial}{\partial \theta} + v_z\frac{\partial}{\partial z}$$

and

$$\nabla^2 = \frac{\partial^2}{\partial r^2} + \frac{1}{r}\frac{\partial}{\partial r} + \frac{1}{r^2}\frac{\partial^2}{\partial \theta^2} + \frac{\partial^2}{\partial z^2}$$

3.6 EXACT SOLUTIONS

This section presents a few classical solutions to the Navier–Stokes equations governing the fully developed incompressible laminar flows. In all these cases, while the governing equations are the same, the solutions differ primarily due to the boundary conditions that adequately model the applicable flow physics. One common boundary condition in all cases involving a real fluid with nonzero viscosity is the no-slip condition, which requires the fluid to have the same velocity as the wall in contact. Riley and Drazin (2006) present a variety of possible exact solutions to the Navier–Stokes equations.

3.6.1 The Couette Flow

3.6.1.1 Flow Physics and Geometry

The Couette flow, shown in Figure 3.4, is a two-dimensional, fully developed incompressible laminar flow between two parallel plates separated by H. Both plates are assumed to be infinite in

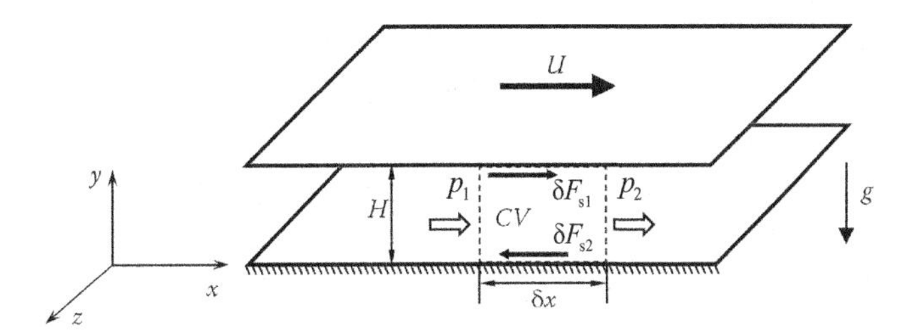

FIGURE 3.4 The Couette flow physics and geometry.

dimensions with no variation of flow properties in the z direction. Relative to the bottom plate, which is assumed to be stationary, the top plate is moving at a constant velocity U along the x direction. As shown in the figure, the Cartesian coordinate system is better suited to handle the Couette flow geometry.

For a fully developed flow between the plates, the axial momentum flow rate across any section remains constant. The gravitational body force, acting in the negative y direction, has no contribution to the forces acting on the control volume (CV) in the x direction. Therefore, all the surface forces acting on the CV must be balanced. If there is no pressure gradient in the flow direction ($p_1 = p_2$), the shear force generated by the moving top plate must overcome the equal and opposite shear force at the bottom plate. In the presence of imposed favorable or adverse pressure gradients and/or a contribution from the gravitational body force along the flow direction, when the parallel plates are not normal to the direction of g, the shear forces on the top and bottom plates will not be equal and opposite. The gravitational body force and the surface forces due to static pressure and wall shear stresses acting on the control volume in a general fully developed flow between the two plates will balance to zero. The preceding statement is true even for a fully developed turbulent flow between the plates.

3.6.1.2 Assumptions

For the Couette flow physics and geometry shown in Figure 3.4, we make the following assumptions:

1. Incompressible with constant viscosity ($\rho = \text{constant}$ and $\mu = \text{constant}$)
2. Laminar
3. Steady ($\partial/\partial t = 0$)
4. Fully developed along the flow direction ($\partial/\partial x = 0$)
5. Two-dimensional ($\partial/\partial z = 0$ and $w = 0$)

3.6.1.3 The Continuity Equation

Based on the assumptions above, the continuity equation simplifies this problem. For an incompressible flow, Equation 3.33 in the Cartesian coordinates reads:

$$\frac{\partial u}{\partial x} + \frac{\partial v}{\partial y} + \frac{\partial w}{\partial z} = 0$$

which in a two-dimensional (x–y plane), fully developed flow reduces to:

$$\frac{\partial v}{\partial y} = 0$$

whose integration yields:

$$v = C$$

where C is the integration constant. As $v = 0$ at the wall, it must be zero everywhere in the flow, giving $C = 0$. Note that, for permeable plates, v must be equal at both plates to satisfy the continuity equation for the fully developed flow in the x direction.

3.6.1.4 The Reduced Navier–Stokes Equations

Under the assumptions made in this case, the z-momentum equation (Equation 3.36) is identically satisfied. As $v = 0$, the y-momentum equation yields:

$$\rho g \frac{\partial h}{\partial y} + \frac{\partial p}{\partial y} = 0$$

With a new static pressure:

$$p' = p + h\rho g \tag{3.41}$$

the y-momentum equation becomes:

$$\frac{\partial p'}{\partial y} = 0 \tag{3.42}$$

which shows that the static pressure variation in the y direction results from the hydrostatic pressure distribution under gravitational body force.

The convective acceleration term on the left-hand side of the x-momentum equation (Equation 3.34) becomes:

$$\rho \frac{\mathrm{D}u}{\mathrm{D}t} = \rho\left(\frac{\partial u}{\partial t} + u\frac{\partial u}{\partial x} + v\frac{\partial u}{\partial y} + w\frac{\partial u}{\partial z}\right) = 0 \qquad \text{(Why?)}$$

while the viscous diffusion term on the right-hand side of Equation 3.34 reduces to:

$$\mu \nabla^2 u = \mu\left(\frac{\partial^2 u}{\partial x^2} + \frac{\partial^2 u}{\partial y^2} + \frac{\partial^2 u}{\partial z^2}\right) = \mu \frac{\partial^2 u}{\partial y^2} = \mu \frac{\mathrm{d}^2 u}{\mathrm{d}y^2}$$

With these simplifications, the x-momentum equation (Equation 3.34) finally reduces to:

$$\mu \frac{\mathrm{d}^2 u}{\mathrm{d}y^2} = \frac{\partial p'}{\partial x} = \frac{\mathrm{d}p'}{\mathrm{d}x} \qquad \text{(Why?)}$$

As the left-hand side of this equation is a function of y only, and the right-hand side is a function of x only, each must be equal to a constant, giving the following reduced form of the Navier–Stokes equations for a Couette flow:

$$\mu \frac{\mathrm{d}^2 u}{\mathrm{d}y^2} = \frac{\mathrm{d}p'}{\mathrm{d}x} = \text{constant} \tag{3.43}$$

3.6.1.5 Boundary Conditions

To enforce the no-slip condition at each plate, we obtain the following boundary conditions:

$$u = 0 @ y = 0 \tag{3.44}$$

$$u = U @ y = H \tag{3.45}$$

3.6.1.6 Solution and Discussion

Based on the preceding development, we find that the flow physics of a Couette flow leads to the mathematical problem of solving the second-order ordinary differential equation (Equation 3.43) subject to the boundary conditions given by Equations 3.44 and 3.45. Integrating Equation 3.43 twice yields the general solution:

$$u = \frac{1}{2\mu}\frac{dp'}{dx}y^2 + C_1 y + C_2 \tag{3.46}$$

Applying the boundary conditions yields the integration constants in Equation 3.46 as:

$$C_2 = 0 \text{ and } C_1 = \frac{U}{H} - \frac{1}{2\mu}\frac{dp'}{dx}H$$

giving the Couette flow solution for the x velocity distribution as:

$$u = \frac{y}{H}U - \frac{H^2}{2\mu}\frac{dp'}{dx}\frac{y}{H}\left(1 - \frac{y}{H}\right) \tag{3.47}$$

In dimensionless form, we can write Equation 3.47 as:

$$u^* = \eta - \beta\eta(1-\eta) \tag{3.48}$$

where

$$u^* = \frac{u}{U},\ \eta = \frac{y}{H},\ \text{and } \beta = \frac{H^2}{2\mu U}\frac{dp'}{dx}$$

For various β values (positive for adverse pressure gradient and negative for favorable pressure gradient), Figure 3.5 shows the dimensionless velocity profiles given by Equation 3.48. Without an imposed pressure gradient ($\beta = 0$), the figure shows a linear velocity profile ($u^* = \eta$), resulting in constant shear stress throughout the flow.

The shear stress at any point in the flow is proportional to $du^*/d\eta$, which we evaluate from Equation 3.48 as:

$$\frac{du^*}{d\eta} = 1 - \beta(1 - 2\eta) \tag{3.49}$$

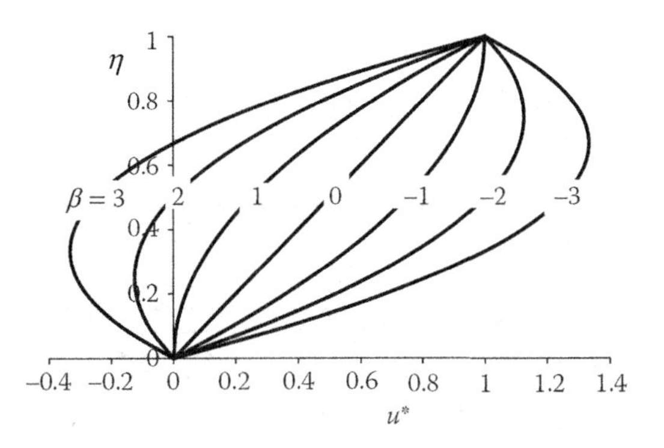

FIGURE 3.5 Velocity profiles in the Couette flows

which yields the following relationship between β and η where the shear stress is zero:

$$\eta = \frac{\beta - 1}{2\beta} \tag{3.50}$$

which shows that the zero shear stress occurs at the stationary plate ($\eta = 0$) for $\beta = 1$. For $\beta > 1$, Figure 3.5 shows a flow reversal near this plate. Similarly, at the moving plate ($\eta = 1$), the zero shear stress corresponds to $\beta = -1$. For $\beta < -1$, the flow near the top plate moves faster than the plate itself.

3.6.2 The Hagen–Poiseuille Flow

3.6.2.1 Flow Physics and Geometry

The Hagen–Poiseuille flow, shown in Figure 3.6, is a two-dimensional, fully developed incompressible laminar flow in a circular pipe of diameter D. This flow is a cylindrical analog to the plane flow between two parallel plates. The figure shows that the cylindrical polar coordinate system is the right choice to handle the Hagen–Poiseuille flow geometry.

When the flow enters the pipe with a uniform velocity, the no-slip condition at the pipe wall initiates an axisymmetric boundary layer, which grows in the axial direction. As a result, the flow momentum in the entrance region increases along the pipe for the fixed mass flow rate through the pipe. In this region, a part of the axial static pressure gradient is responsible for this increase in flow momentum, and the remaining part supports the loss of momentum to the pipe wall, generating wall shear stress. At the end of the entrance region, the boundary layer growing around the pipe merges at the pipe axis; at this point, the flow becomes fully developed with no axial variation in the velocity profile. In the fully developed flow (the Hagen–Poiseuille flow), the static pressure gradient along the pipe overcomes the shear force at the wall. As the streamlines are parallel in the Hagen–Poiseuille flow, there is no radial variation of static pressure at any pipe cross section.

3.6.2.2 Assumptions

For the Hagen–Poiseuille flow physics and geometry shown in Figure 3.10, we invoke the following assumptions:

1. Incompressible with constant viscosity (ρ = constant and μ = constant)
2. Laminar
3. Steady ($\partial/\partial t = 0$)
4. Fully developed along the flow direction ($\partial/\partial z = 0$)
5. Axisymmetric ($\partial/\partial\theta = 0$ and $V_\theta = 0$)

3.6.2.3 The Continuity Equation

The continuity equation (Equation 3.37) in cylindrical polar coordinates reads:

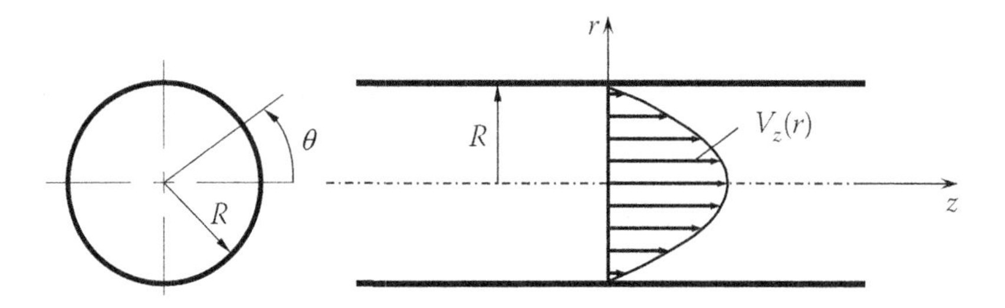

FIGURE 3.6 The Hagen–Poiseuille flow physics and geometry.

$$\frac{1}{r}\frac{\partial}{\partial r}(rv_r)+\frac{1}{r}\frac{\partial v_\theta}{\partial \theta}+\frac{\partial v_z}{\partial z}=0$$

Using $\frac{\partial v_z}{\partial z}=\frac{\partial v_\theta}{\partial \theta}=0$ from the assumptions, the above equation reduces to:

$$\frac{\partial (rv_r)}{\partial r}=0$$

whose integration yields $rv_r = C$ in the r direction, where C is the integration constant. As $v_r = 0$ at the wall ($r = R$), it must be zero everywhere in the flow, giving $C = 0$. We obtained this significant result from the continuity equation applied to the Hagen–Poiseuille flow.

3.6.2.4 The Reduced Navier–Stokes Equations

With $\partial/\partial\theta = v_\theta = 0$, the momentum equation (Equation 3.39) in the θ direction is identically satisfied. With $v_r = 0$, the r-momentum equation reduces to:

$$\rho g\frac{\partial h}{\partial r}+\frac{\partial p}{\partial r}=0$$

Defining a new static pressure:

$$p' = p + h\rho g \tag{3.51}$$

the r-momentum equation reduces to:

$$\frac{\partial p'}{\partial r}=0 \tag{3.52}$$

which implies that p' is a function of z (flow direction) only.

The convective acceleration term on the left-hand side of the z-momentum equation (Equation 3.40) becomes zero, that is:

$$\rho\frac{\mathrm{D}v_z}{\mathrm{D}t}=\rho\left(\frac{\partial v_z}{\partial t}+v_r\frac{\partial v_z}{\partial r}+\frac{v_\theta}{r}\frac{\partial v_z}{\partial \theta}+v_z\frac{\partial v_z}{\partial z}\right)=0 \tag{3.53}$$

and the viscous diffusion term on the right-hand side reduces to:

$$\mu\nabla^2 v_z=\mu\left(\frac{\partial^2 u}{\partial x^2}+\frac{\partial^2 u}{\partial y^2}+\frac{\partial^2 u}{\partial z^2}\right)=\mu\frac{\partial^2 u}{\partial y^2}=\mu\frac{\mathrm{d}^2 u}{\mathrm{d}y^2}$$

$$\mu\nabla^2 v_z=\mu\left(\frac{\partial^2 v_z}{\partial r^2}+\frac{1}{r}\frac{\partial v_z}{\partial r}+\frac{1}{r^2}\frac{\partial^2 v_z}{\partial \theta^2}+\frac{\partial^2 v_z}{\partial z^2}\right)=\mu\left(\frac{\partial^2 v_z}{\partial r^2}+\frac{1}{r}\frac{\partial v_z}{\partial r}\right) \tag{3.54}$$

With the substitution of Equations 3.53 and 3.54, the z-momentum equation (Equation 3.40) finally reduces to:

$$\mu\left(\frac{\partial^2 v_z}{\partial r^2}+\frac{1}{r}\frac{\partial v_z}{\partial r}\right)=\frac{\partial p'}{\partial z} \tag{3.55}$$

whose left-hand side is a function of r only, and the right-hand side is a function of z only, which implies that each must be equal to a constant, giving the following reduced form of the Navier–Stokes equation for the Hagen–Poiseuille flow:

$$\frac{1}{r}\frac{\mathrm{d}}{\mathrm{d}r}\left(r\frac{\mathrm{d}v_z}{\mathrm{d}r}\right)=\frac{1}{\mu}\frac{\mathrm{d}p'}{\mathrm{d}z}=\text{constant} \tag{3.56}$$

3.6.2.5 Boundary Conditions

The no-slip condition at the pipe wall yields:

$$v_z = 0 @ r = R \tag{3.57}$$

From the flow symmetry about the pipe axis, we obtain:

$$\frac{\mathrm{d}v_z}{\mathrm{d}r} = 0 @ r = 0 \tag{3.58}$$

3.6.2.6 Solution and Discussion

Integrating the governing second-order ordinary differential equation (Equation 3.56) once yields:

$$\frac{\mathrm{d}v_z}{\mathrm{d}r}=\left(\frac{1}{2\mu}\frac{\mathrm{d}p'}{\mathrm{d}z}\right)r + C_1$$

where C_1 is the integration constant. Applying the second boundary condition (Equation 3.58), this equation yields $C_1 = 0$, giving:

$$\frac{\mathrm{d}v_z}{\mathrm{d}r}=\left(\frac{1}{2\mu}\frac{\mathrm{d}p'}{\mathrm{d}z}\right)r \tag{3.59}$$

The integration of this equation yields:

$$v_z=\left(\frac{1}{4\mu}\frac{\mathrm{d}p'}{\mathrm{d}z}\right)r^2 + C_2 \tag{3.60}$$

where C_2 is the integration constant. Applying the first boundary condition (Equation 3.57) to this equation, we obtain:

$$C_2 = -\left(\frac{1}{4\mu}\frac{\mathrm{d}p'}{\mathrm{d}z}\right)R^2$$

Thus, Equation 3.60 becomes:

$$v_z=\left(-\frac{1}{4\mu}\frac{\mathrm{d}p'}{\mathrm{d}z}\right)(R^2 - r^2) \tag{3.61}$$

which is the required distribution of v_z in the Hagen–Poiseuille flow. This equation shows that the radial variation of the axial velocity is parabolic in a fully developed laminar pipe flow. We obtain the maximum velocity at the pipe axis, given by:

$$v_{\max} = -\frac{R^2}{4\mu}\frac{\mathrm{d}p'}{\mathrm{d}z} \tag{3.62}$$

which reveals that for v_{max} to be positive, the static pressure gradient dp'/dz must be negative, implying a decrease in static pressure in the flow direction. When the flow is horizontal, gravity plays no role, yielding:

$$\frac{dp'}{dz} = \frac{dp}{dz}$$

By combining Equations 3.61 and 3.62, we can express the velocity distribution as:

$$\frac{v_z}{v_{max}} = 1 - \left(\frac{r}{R}\right)^2 \tag{3.63}$$

As the flow is steady, the mass flow rate at each cross section in the pipe (including the entrance region not shown here) must remain constant. Let us now evaluate the mass flow rate for the parabolic velocity profile given by Equation 3.63:

$$\dot{m} = \int_A \rho v_z dA = \rho v_{max} \int_0^R \left(1 - \frac{r^2}{R^2}\right) 2\pi r dr$$

where A is the pipe cross-sectional area. This integration yields:

$$\dot{m} = \frac{\pi R^2 \rho v_{max}}{2} = \pi R^2 \rho \bar{v} - p \tag{3.64}$$

where $\bar{v}$ is the section-average velocity. The preceding result establishes a useful relationship between the average velocity and the centerline maximum velocity of the parabolic profile in a Hagen–Poiseuille flow, namely:

$$\bar{v} = \frac{v_{max}}{2} \tag{3.65}$$

Substituting v_{max} from Equation 3.62 into Equation 3.64, we obtain:

$$\dot{m} = \frac{\pi R^2 \rho}{2}\left(-\frac{R^2}{4\mu}\frac{dp'}{dz}\right)$$

$$\dot{m} = -\frac{\pi R^4 \rho}{8\mu}\left(\frac{dp'}{dz}\right) \tag{3.66}$$

which reveals that the mass flow rate is linearly dependent on the static pressure gradient in a Hagen–Poiseuille flow.

Using Equation 3.59, we can express the shear stress exerted by the fluid on the wall ($r = R$) as:

$$\tau_{wall} = -\mu \frac{dv_z}{dr} = -\frac{R}{2}\frac{dp'}{dz} \tag{3.67}$$

which, using Equations 3.62 and 3.65, we can write alternatively as:

$$\tau_{wall} = \frac{16}{Re}\left(\frac{1}{2}\rho \bar{v}^2\right) = C_f\left(\frac{1}{2}\rho \bar{v}^2\right) \tag{3.68}$$

where $Re = 2\rho \bar{v} R / \mu$ is the flow Reynolds number in the pipe and C_f is called the shear coefficient for which this equation yields:

$$C_{\mathrm{f}} = \frac{16}{Re} \tag{3.69}$$

For a pipe of length L and diameter D, Equation 3.67 yields the following relationship between the wall shear stress and the static pressure drop:

$$\Delta p'_{\mathrm{loss}} = 4\tau_{\mathrm{wall}}\left(\frac{L}{D}\right) \tag{3.70}$$

in which substituting τ_{wall} from Equation 3.68 yields:

$$\Delta p'_{\mathrm{loss}} = 4C_{\mathrm{f}}\left(\frac{L}{D}\right)\left(\frac{1}{2}\rho\bar{v}^2\right) \tag{3.71}$$

which, using the Darcy friction factor f for static pressure loss in a Hagen–Poiseuille flow, becomes:

$$\Delta p'_{\mathrm{loss}} = f\left(\frac{L}{D}\right)\left(\frac{1}{2}\rho\bar{v}^2\right) \tag{3.72}$$

Comparing Equations 3.71 and 3.72, we note that $f = 4C_{\mathrm{f}}$. Thus, we can determine f using the equation:

$$f = \frac{64}{Re} \tag{3.73}$$

Note here that the relationship $f = 4C_{\mathrm{f}}$ between the Darcy friction factor (f) and the shear coefficient (C_{f}) is also true for a fully developed turbulent pipe flow (why?).

From the computed velocity profile in a Hagen–Poiseuille flow, we can evaluate the nonzero component of the vorticity vector in the θ direction as follows. As $v_r = 0$, we obtain:

$$\zeta_\theta = \frac{\partial v_r}{\partial z} - \frac{\partial v_z}{\partial r} = -\frac{\partial v_z}{\partial r}$$

in which, substituting v_z from Equation 3.63 yields:

$$\zeta_\theta = -v_{\max}\frac{\partial}{\partial r}\left[1 - \left(\frac{r}{R}\right)^2\right] = \frac{2v_{\max}r}{R^2} = -\frac{1}{2\mu}\frac{\mathrm{d}p'}{\mathrm{d}z}r \tag{3.74}$$

which shows that the vorticity varies linearly from zero at the pipe axis to a maximum value at the wall, given by:

$$\zeta_{\theta\_\mathrm{wall}} = \frac{2v_{\max}R}{R^2} = -\frac{R}{2\mu}\frac{\mathrm{d}p'}{\mathrm{d}z} \tag{3.75}$$

From this equation and Equation 3.67, we conclude that the wall shear stress is proportional to the vorticity at the wall; that is, $\tau_{\mathrm{wall}} = \mu\zeta_{\theta\_\mathrm{wall}}$.

3.7 THE NAVIER–STOKES EQUATIONS IN TERMS OF VORTICITY AND STREAM FUNCTION

So far in this chapter, we have presented solutions to the Navier–Stokes equations governing steady incompressible laminar flows with constant viscosity in primitive variables (velocity components and static pressure). As the vorticity vector at a point in the flow is the curl of the velocity vector at that point ($\zeta = \nabla \times V$), knowing the velocity distribution, we can easily calculate the distribution of vorticity in the flow. Alternatively, we can directly solve the vorticity transport form of the Navier–Stokes equations. Similarly, for a two-dimensional flow, we can define a stream function at all points in a flow in terms of the local velocity gradients. Later in this chapter, we also derive a transport equation in terms of the stream function.

3.7.1 The Vorticity Transport Equation

In the x–y plane, let us consider an incompressible flow with constant viscosity, governed by the Navier–Stokes equations (Equations 3.34 and 3.35), using the primitive variables u, v, and p. We obtain the vorticity component for this flow as $\zeta_z = \partial v / \partial x - \partial u / \partial y$. Let us express the terms $\mathrm{D}u / \mathrm{D}t$ and $\nabla^2 u$ in Equation 3.34 in terms of vorticity as follows:

$\mathrm{D}u / \mathrm{D}t$ in Equation 3.34:

$$\frac{\mathrm{D}u}{\mathrm{D}t} = \frac{\partial u}{\partial t} + u\frac{\partial u}{\partial x} + v\frac{\partial u}{\partial y}$$

$$\frac{\mathrm{D}u}{\mathrm{D}t} = \frac{\partial u}{\partial t} + u\frac{\partial u}{\partial x} + v\left(\frac{\partial u}{\partial y} - \frac{\partial v}{\partial x}\right) + v\frac{\partial v}{\partial x} \tag{3.76}$$

$$\frac{\mathrm{D}u}{\mathrm{D}t} = \frac{\partial u}{\partial t} + \frac{1}{2}\frac{\partial}{\partial x}(u^2 + v^2) - v\zeta_z$$

$\nabla^2 u$ in Equation 3.34:

$$\nabla^2 u = \frac{\partial^2 u}{\partial x^2} + \frac{\partial^2 u}{\partial y^2}$$

$$\nabla^2 u = \frac{\partial}{\partial x}\left(-\frac{\partial v}{\partial y}\right) + \frac{\partial^2 u}{\partial y^2}$$

where we have used $\frac{\partial u}{\partial x} = -\frac{\partial v}{\partial y}$ from the continuity equation $\frac{\partial u}{\partial x} + \frac{\partial v}{\partial y} = 0$, giving:

$$\nabla^2 u = \frac{\partial}{\partial y}\left(-\frac{\partial v}{\partial x} + \frac{\partial u}{\partial y}\right)$$

$$\nabla^2 u = -\frac{\partial \zeta_z}{\partial y} \tag{3.77}$$

Using the approach used above, we can write Equation 3.35 as:

$$\frac{\mathrm{D}v}{\mathrm{D}t} = \frac{\partial v}{\partial t} + \frac{1}{2}\frac{\partial}{\partial y}(u^2 + v^2) + u\zeta_z \tag{3.78}$$

$$\nabla^2 v = \frac{\partial \zeta_z}{\partial x} \tag{3.79}$$

Substituting Equations 3.76 and 3.77 in Equation 3.34 yields:

$$\frac{\partial u}{\partial t} + \frac{1}{2}\frac{\partial}{\partial x}(u^2 + v^2) - v\zeta_z = -\frac{1}{\rho}\frac{\partial p'}{\partial x} - \nu\frac{\partial \zeta_z}{\partial y} \tag{3.80}$$

where we have used $p' = p + h\rho g$ and $\nu = \mu / \rho$. Similarly, substituting Equations 3.78 and 3.79 in Equation 3.35 yields:

$$\frac{\partial v}{\partial t} + \frac{1}{2}\frac{\partial}{\partial y}(u^2 + v^2) + u\zeta_z = -\frac{1}{\rho}\frac{\partial p'}{\partial y} + \nu\frac{\partial \zeta_z}{\partial x} \tag{3.81}$$

Taking the derivative of Equation 3.80 with respect to y and the derivative of Equation 3.81 with respect to x and combining (subtracting) the resulting equations, we obtain the following equation in terms of ζ_z:

$$\frac{\partial \zeta_z}{\partial t} + u\frac{\partial \zeta_z}{\partial x} + v\frac{\partial \zeta_z}{\partial y} = \frac{\mathrm{D}\zeta_z}{\mathrm{D}t} = \nu\nabla^2\zeta_z \tag{3.82}$$

which is the vorticity transport equation in the x–y plane. This equation equates the convection of vorticity (in the z direction) in the flow to its diffusion by viscosity. If we take the curl of the three-dimensional Navier–Stokes equation in the vector form and simplify it using the well-known vector identities, we obtain the following vorticity transport equation (see Problem 3.8):

$$\frac{\mathrm{D}\boldsymbol{\zeta}}{\mathrm{D}t} = (\boldsymbol{\zeta} \bullet \nabla)\boldsymbol{v} + \nu\nabla^2\boldsymbol{\zeta} \tag{3.83}$$

where $\boldsymbol{\zeta}$ and $\boldsymbol{v}$ are three-dimensional vorticity and velocity vectors, respectively. Note that the extra term $(\boldsymbol{\zeta} \bullet \nabla)v$ that appears on the right-hand side of Equation 3.83 is absent in the two-dimensional vorticity transport equation (Equation 3.82). Let us first show that the term $(\boldsymbol{\zeta} \bullet \nabla)\boldsymbol{v}$ in fact vanishes under the assumption of a two-dimensional flow. We can write $(\boldsymbol{\zeta} \bullet \nabla)\boldsymbol{v}$ as:

$$(\boldsymbol{\zeta} \bullet \nabla)\boldsymbol{v} = \left(\zeta_x\frac{\partial}{\partial x} + \zeta_y\frac{\partial}{\partial y} + \zeta_z\frac{\partial}{\partial z}\right)\boldsymbol{v} \tag{3.84}$$

For a two-dimensional flow in the x–y plane, where $\zeta_x = \zeta_y = \partial / \partial z = 0$, the right-hand side of Equation 3.84 becomes zero, giving $(\boldsymbol{\zeta} \bullet \nabla)\boldsymbol{v} = 0$, which is what we have in Equation 3.82.

The term given by Equation 3.84, which appears in the three-dimensional vorticity transport equation (Equation 3.83), has important ramifications in generating turbulent flows, which are inherently unsteady and three-dimensional. For example, if we extract the transport equation for ζ_z from Equation 3.83, without making the two-dimensional flow assumption, we obtain:

$$\frac{\mathrm{D}\zeta_z}{\mathrm{D}t} = \left(\zeta_x\frac{\partial w}{\partial x} + \zeta_y\frac{\partial w}{\partial y} + \zeta_z\frac{\partial w}{\partial z}\right) + \nu\nabla^2\zeta_z \tag{3.85}$$

On the right-hand side of this equation, the first two terms within the parentheses represent turning of ζ_x and ζ_y, respectively, while the last term within the parentheses represents stretching of ζ_z in the z direction. This turning and stretching of the vorticity in a shear flow is the primary mechanism of generating complex turbulent flows. Thus, throwing the term $(\boldsymbol{\zeta} \bullet \nabla)\boldsymbol{v}$ under the assumption of a two-dimensional flow is like throwing out the baby with the bath water, as it would happen if we extended Equation 3.82 to develop a three-dimensional vorticity transport equation.

3.7.2 The Stream Function Transport Equation

For a two-dimensional incompressible flow, we can define the stream function Ψ, which identically satisfies the continuity equation, as:

$$u = \frac{\partial \Psi}{\partial y} \tag{3.86}$$

and

$$v = -\frac{\partial \Psi}{\partial x} \tag{3.87}$$

Substituting Equations 3.86 and 3.87 in:

$$\zeta_z = \frac{\partial v}{\partial x} - \frac{\partial u}{\partial y}$$

we obtain:

$$\begin{aligned} \zeta_z &= \frac{\partial}{\partial x}\left(-\frac{\partial \Psi}{\partial x}\right) - \frac{\partial}{\partial y}\left(\frac{\partial \Psi}{\partial y}\right) \\ \zeta_z &= -\frac{\partial^2 \Psi}{\partial x^2} - \frac{\partial^2 \Psi}{\partial y^2} = -\nabla^2 \Psi \end{aligned} \tag{3.88}$$

Using Equations 3.86–3.88, the two-dimensional vorticity transport equation (Equation 3.82) becomes:

$$\frac{\partial}{\partial t}\left(\nabla^2 \Psi\right) - \frac{\partial \Psi}{\partial y}\frac{\partial}{\partial x}\left(\nabla^2 \Psi\right) + \frac{\partial \Psi}{\partial x}\frac{\partial}{\partial y}\left(\nabla^2 \Psi\right) = \nu \nabla^4 \Psi \tag{3.89}$$

which is a fourth-order partial differential equation in Ψ, representing the stream function form of the Navier–Stokes equation for a two-dimensional, incompressible flow with constant viscosity.

3.8 SLOW FLOW

A slow viscous flow is one where the viscous forces dominate over the inertia forces. As a result, such flows have low Reynolds numbers, as opposed to turbulent flows having high Reynolds numbers. The flow within the thin gaps between lubricated surfaces, discussed in the hydrodynamic lubrication theory, is an example of a slow flow.

In Equation 3.32, neglecting the inertia terms compared to viscous terms, we obtain the following Navier–Stokes equations for a steady slow flow:

$$\mu \nabla^2 u_i = \rho g \frac{\partial h}{\partial x_i} + \frac{\partial p}{\partial x_i} \tag{3.90}$$

Note that the elimination of the convective terms leading to Equation 3.90 renders the Navier–Stokes equation linear and easier to solve.

To get further insight into a slow flow, let us consider the three-dimensional vorticity transport equation (Equation 3.83) and make it nondimensional using the reference length L and the reference velocity V. Denoting nondimensional quantities with an asterisk "*," we obtain:

$$x^* = \frac{x}{L} \quad y^* = \frac{y}{L} \quad z^* = \frac{z}{L} \quad t^* = t\left(\frac{V}{L}\right)$$

$$u^* = \frac{u}{V} \quad v^* = \frac{v}{V} \quad w^* = \frac{w}{V}$$

$$\zeta = \nabla \times v = \left(\frac{V}{L}\right)\nabla \times v^* = \left(\frac{V}{L}\right)\zeta^* \text{ or } \zeta^* = \left(\frac{L}{V}\right)\zeta$$

Substituting these nondimensional quantities in Equation 3.83 yields:

$$\frac{\mathrm{D}\zeta^*}{\mathrm{D}t^*} = (\zeta^* \bullet \nabla)v^* + \frac{1}{Re}\nabla^2\zeta^*$$

$$\nabla^2\zeta^* = Re\left(\frac{\mathrm{D}\zeta^*}{\mathrm{D}t^*} - (\zeta^* \bullet \nabla)v^*\right) \tag{3.91}$$

where Reynolds number $Re = VL / \nu$. For very low Reynolds numbers in a slow flow, we may set the right-hand side of this equation to zero, giving:

$$\nabla^2\zeta^* = 0 \tag{3.92}$$

which shows that the Laplace equation governs the transport of dimensionless vorticity in a slow three-dimensional flow, much like it does for the heat condition in a solid with constant thermal conductivity. Schlichting (1968) presents additional details on very slow flow and its analysis in the parallel flow past a sphere, the hydrodynamic theory of lubrication, and the Hele-Shaw flow.

3.9 CONCLUDING REMARKS

The time-dependent Navier–Stokes equations in their various forms are the linear momentum equations, which, along with the continuity equation, are widely used for detailed analysis of many laminar and turbulent flows and are at the core of CFD technology. In this chapter, while introducing the Navier–Stokes equations, we have presented a few exact solutions for incompressible laminar flows of constant viscosity. Although most of these solutions seem of academic interest, they provide an insightful understanding of velocity distributions in these flows under the influence of a static pressure gradient, the gravitational body force. These velocity distributions allow us to compute shear stress distributions in the flow and at the bounding walls with no-slip boundary conditions. In some cases, we can readily verify the calculated results of wall shear stress using a simple control volume analysis presented in Chapter 1.

To develop their skills in solving various fluid flow problems, readers may want to review numerous problems with solutions presented in Sultanian (2021, 2022).

WORKED EXAMPLES

Example 3.1

As shown in Figure 3.7, a thin fluid film falls under gravity along a flat plate. The static pressure is constant throughout, and there is no wind shear at the edge of the film. Modeling the fluid film as a two-dimensional, incompressible, laminar flow, find the (a) w velocity distribution, (b) ratio of maximum and average velocities, (c) distribution of shear stress in the film and its magnitude and direction at the wall, and (d) wall shear stress using the force–momentum balance on the control volume shown in the figure.

Solution

In this case of fluid film flow along the vertical plate, the force due to gravity is acting on the flow. In the absence of wind shear at the edge of the film, the shear force at the wall acts in the opposite direction. The flow geometry favors the use of Cartesian coordinates.

Assumptions: (1) steady, (2) incompressible, (3) laminar, (4) fully developed ($\partial / \partial z = 0$), zero static pressure gradient ($\partial p / \partial z = 0$), and two-dimensional ($v = 0$).

As the flow is fully developed in the z direction, w can be a function of x only, that is, $w = w(x)$.

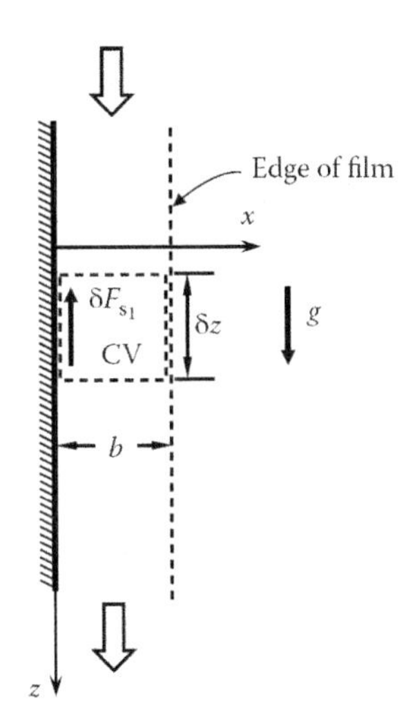

FIGURE 3.7 A fluid film falling under gravity along a flat plate (Example 3.1).

THE CONTINUITY EQUATION

Under the assumption of steady, incompressible flow, the continuity equation in the Cartesian coordinates reduces to:

$$\frac{\partial u}{\partial x}+\frac{\partial v}{\partial y}+\frac{\partial w}{\partial z}=0$$

which, for the present two-dimensional, fully developed flow, further reduces to:

$$\frac{\partial u}{\partial x}=\frac{du}{dx}=0$$

On integration, this equation yields:

$$u=\text{constant}$$

As $u=0$ at $x=0$, we conclude that $u=0$ everywhere.

PRESSURE GRADIENT UNDER GRAVITATIONAL BODY FORCE

$$p'=p+\rho g h$$

$$\frac{\mathrm{d}p'}{\mathrm{d}z}=\frac{\mathrm{d}}{\mathrm{d}z}(p+\rho g h)=\frac{\mathrm{d}p}{\mathrm{d}z}+\rho g\frac{\mathrm{d}h}{\mathrm{d}z}=\frac{\mathrm{d}p}{\mathrm{d}z}-\rho g$$

where we have used $\mathrm{d}h/\mathrm{d}z=-1$. With $\mathrm{d}p/\mathrm{d}z=0$, this equation yields:

$$\frac{\mathrm{d}p'}{\mathrm{d}z}=-\rho g$$

THE REDUCED NAVIER–STOKES EQUATION

With the foregoing assumptions, the Navier–Stokes equations reduce to the following governing equation for the velocity component in the z direction:

$$\mu \frac{d^2 w}{dx^2} = \frac{dp'}{dz} = -\rho g$$

$$\frac{d^2 w}{dx^2} = -\frac{g}{\nu}$$

where the kinematic viscosity $\nu = \mu / \rho$.

BOUNDARY CONDITIONS

For a unique solution, the foregoing second-order ordinary differential equation requires two boundary conditions. These boundary conditions for the present flow are as follows:

$$w = 0 \ @ \ x = 0 \text{ and } \frac{dw}{dx} = 0 \ @ \ x = b$$

SOLUTIONS AND DISCUSSION

a. Distribution of *w* velocity

Integrating the reduced governing Navier–Stokes equation and applying the boundary conditions yield the following solution for the velocity distribution:

$$w = -\frac{g}{2\nu} x^2 + \frac{bg}{\nu} x$$

which yields the maximum velocity $w_{max} = bg / 2\nu$ at $x = b$. Using $w^* = w / w_{max}$ and $\xi = x / b$, we express this velocity distribution in the dimensionless form as:

$$w^* = 2\xi - \xi^2$$

As shown in Figure 3.8, the above equation represents a semi-parabolic velocity profile for $0 \le \xi = \le 1$. Treating the edge of the film as the line of symmetry, the solution also represents the velocity profile of the flow falling under gravity between two vertical plates separated by $2b$ with zero imposed static pressure gradient. The dashed line in Figure 3.8 represents the second half of the parabola.

b. Ratio of Maximum-to-Average *w* Velocity

We calculate the average value from the computed velocity distribution as:

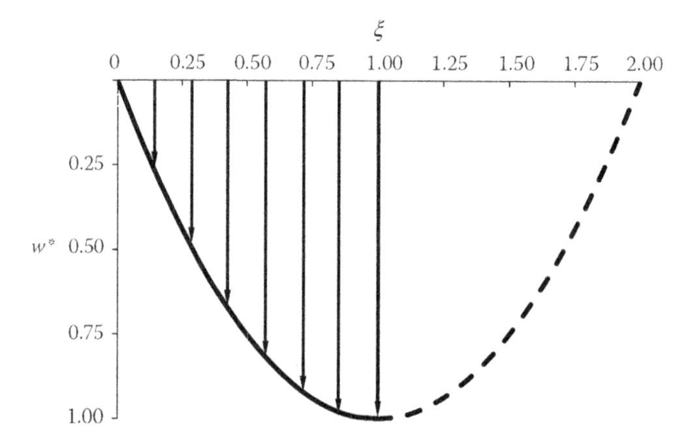

FIGURE 3.8 Velocity profile of the fluid film flow along the vertical plate (Example 3.1).

$$\bar{w} = \frac{\int_0^b \left(-\frac{g}{2\nu}x^2 + \frac{bg}{\nu}x \right) \mathrm{d}x}{b} = \frac{gb^2}{3\nu}$$

which yields:

$$\frac{w_{\max}}{\bar{w}} = \frac{3}{2}$$

c. **Shear Stress Distribution and Wall Shear Stress**
From the velocity distribution obtained above, we evaluate the shear stress at any point in the film flow as:

$$\tau_{xz} = \mu\frac{\mathrm{d}w}{\mathrm{d}x} = \mu\frac{\mathrm{d}}{\mathrm{d}x}\left(-\frac{g}{2\nu}x^2 + \frac{bg}{\nu}x \right) = -\rho gx + \rho gb$$

which shows that the shear stress becomes zero at $x = b$. At the wall ($x = 0$), we obtain $(\tau_{xz})_{x=0} = \rho gb$, which is positive. It is essential to recognize that the sign of the shear stress depends on the sign of the face normal and its direction. As shown in Figure 3.1, the face normal and shear stress direction should be either positive or negative for positive shear stress at a point relative to the chosen coordinate axes. As the fluid face normal at $x = 0$ is in the negative x direction, the positive value of $(\tau_{xz})_{x=0} = \rho gb$ implies that it is along the negative z direction (vertically up). As the shear force acting on the wall must be equal and opposite to the shear force acting on the fluid surface in contact, we obtain $\tau_{\text{wall}} = \rho gb$ in the positive z direction (vertically down).

d. **Wall Shear Stress from Control Volume Analysis**
For a fully developed flow, the net efflux of z-momentum from the control volume shown in Figure 3.7 is zero. The only forces acting on the control volume are the shear and gravitational body forces (weight). Thus, the force and z-momentum balance on the control volume reduces to:

$$\tau_{\text{wall}}\,\delta z\,\delta y = \rho gb\,\delta z\,\delta y$$

$$\tau_{\text{wall}} = \rho gb$$

which is identical to the value we obtained earlier using the product of fluid viscosity and the gradient of w velocity at the wall. This example shows the power of the control volume analysis if one is interested in knowing only the shear stress at the wall.

Example 3.2

Consider a steady, fully developed, incompressible, laminar flow over a porous plate of infinite length, as shown in Figure 3.9. The plate moves in the x direction at a constant velocity U_0. The fluid is removed through the porous plate at a constant velocity V_0. Clearly stating all relevant

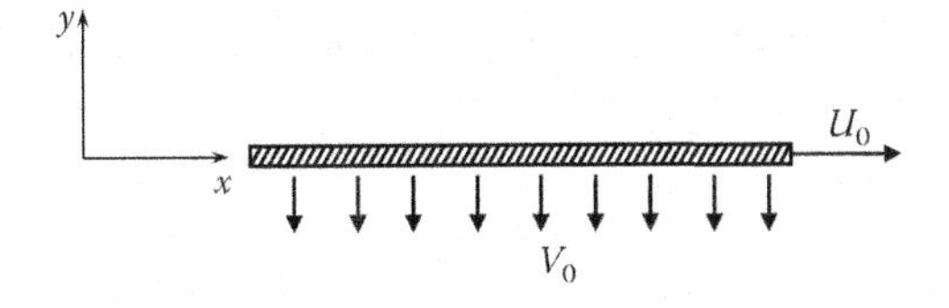

FIGURE 3.9 A porous plate moving at constant velocity through a stagnant fluid (Example 3.2)

assumptions, derive an expression for the velocity profile over the plate. Evaluate the shear stress at the wall and verify the result using a control volume analysis.

Solution

In this case, a porous plate of infinite dimension is pulled along the x direction at a constant velocity U_0. Due to the no-slip condition, the fluid in contact with the plate also moves with the same velocity. At a large distance from the plate in the y direction, the effect of viscosity becomes negligible, and the fluid retains zero velocity in the x direction. The fluid flows through the porous plate in the negative y direction at a uniform velocity V_0, which must become uniform in the entire flow field to satisfy the continuity equation. The flow geometry of this example favors the use of Cartesian coordinates.

Assumptions:

1. Steady
2. Incompressible
3. Laminar
4. Fully developed ($\partial / \partial x = 0$)
5. Zero static pressure gradient ($\partial p / \partial x = 0$)
6. Two-dimensional ($w = 0$)

As the flow is fully developed in the x direction, u can be a function of y only (i.e., $u = u(y)$).

THE CONTINUITY EQUATION

Under the assumptions of steady, incompressible flow, the continuity equation in the Cartesian coordinates reduces to:

$$\frac{\partial u}{\partial x} + \frac{\partial v}{\partial y} + \frac{\partial w}{\partial z} = 0$$

which for the present two-dimensional, fully developed flow further reduces to:

$$\frac{\partial v}{\partial y} = \frac{\mathrm{d}v}{\mathrm{d}y} = 0$$

which upon integration yields $v = \text{constant}$. As $v = -V_0$ at $y = 0$, we conclude that $v = -V_0$ everywhere.

THE REDUCED NAVIER–STOKES EQUATIONS

With the assumptions made in this case and with no contribution from the gravitational body force (horizontal flow), the Navier–Stokes equations reduce to the following governing equation for the velocity component in the x direction:

$$-\rho V_0 \frac{\mathrm{d}u}{\mathrm{d}y} = \mu \frac{\mathrm{d}^2 u}{\mathrm{d}y^2}$$

$$\frac{\mathrm{d}^2 u}{\mathrm{d}y^2} + \frac{V_0}{\nu} \frac{\mathrm{d}u}{\mathrm{d}y} = 0$$

where the kinematic viscosity $\nu = \mu / \rho$. The convection term, the second term on the LHS of this second-order ordinary differential equation, is linear.

BOUNDARY CONDITIONS

Two boundary conditions needed for a unique solution of the reduced Navier–Stokes equation in this case are $u = U_0$ @ $y = 0$ and $u = 0$ @ $y = \infty$.

SOLUTION FOR THE VELOCITY DISTRIBUTION

Integrating the reduced governing Navier–Stokes equation once yields:

$$\frac{du}{dy} + \frac{V_0 u}{\nu} = C_1$$

where C_1 is the integration constant. We can easily verify that the solution of the homogeneous part of this first-order ordinary differential equation is:

$$u_h = e^{-\frac{V_0 y}{\nu}}$$

A particular solution to this equation is:

$$u_p = \frac{\nu}{V_0} C_1$$

The resulting general solution, therefore, becomes:

$$u = C_2 u_h + u_p = C_2 e^{-\frac{V_0 y}{\nu}} + \frac{\nu}{V_0} C_1$$

Using the second boundary condition ($u = 0$ @ $y = \infty$) yields $C_1 = 0$. For the resulting solution, using the first boundary condition ($u = U_0$ @ $y = 0$) finally yields the u velocity distribution as:

$$u = U_0 e^{-\frac{V_0 y}{\nu}}$$

WALL SHEAR STRESS FROM VELOCITY DISTRIBUTION

The velocity distribution obtained above results in the following shear stress at the wall:

$$\tau_{wall} = \mu \left.\frac{du}{dy}\right|_{y=0} = \mu \frac{d}{dy}\left.\left(U_0 e^{-\frac{V_0 y}{\nu}} \right)\right|_{y=0} = -\rho U_0 V_0$$

As the plate is moving in the positive x direction, the shear stress generated on the plate is in the negative x direction.

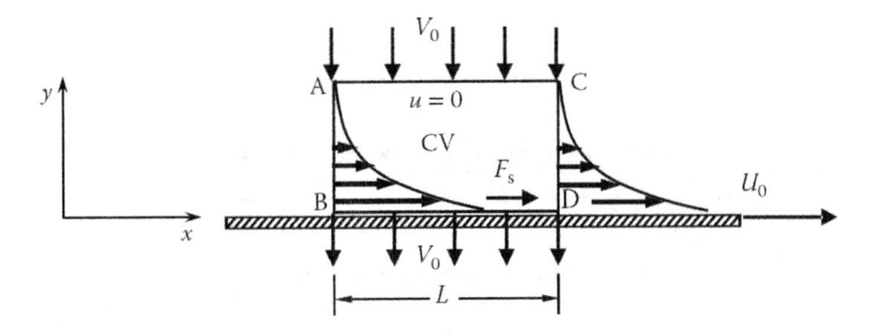

FIGURE 3.10 Control volume analysis of the flow over a moving porous plate (Example 3.2)

WALL SHEAR STRESS FROM CONTROL VOLUME ANALYSIS

As the flow is fully developed in the x direction, for a unit length perpendicular to the plane of the control volume shown in Figure 3.10, the x-momentum flux entering AB equals that exiting CD. Considering the remaining x-momentum fluxes through AC and BD, the x direction force–momentum balance on the control volume ABDC yields:

$$F_s = -\tau_{\text{wall}} L = \left(\rho V_0 U_0 L\right)_{\text{BD}} - (0)_{\text{AC}} = \rho V_0 U_0 L$$

$$\tau_{\text{wall}} = -\rho V_0 U_0$$

which is identical to the value we obtained earlier using the product of fluid viscosity and the gradient of x velocity at the wall. This example also shows that the control volume analysis is instrumental in determining the shear stress at the wall.

Example 3.3

Consider a fully developed, steady, incompressible, laminar flow induced when a wire of radius R_1 is drawn at a constant velocity W down the centerline of a pipe of radius R_2, as shown in Figure 3.11. Develop the appropriate form of the Navier–Stokes equations governing the flow with no applied pressure gradient. Find (a) an expression for the velocity distribution $V_z(r)$ in the annular region between the pipe and the wire and (b) an expression for the force needed to draw a wire of radius R_1 and length L in this flow system.

Solution

In this case, a wire of radius R_1 and infinite length is pulled along the z direction at a constant velocity w. Due to the no-slip condition, the fluid in contact with the wire also moves with the same velocity, setting the flow in the annular region between the pipe, where the fluid velocity is zero at the wall, and the wire, even in the absence of axial static pressure gradient along the pipe. The flow geometry of this case favors the use of cylindrical polar coordinates.

Assumptions:

1. Steady
2. Incompressible
3. Laminar
4. Fully developed ($\partial / \partial z = 0$)
5. Zero static pressure gradient ($\partial p / \partial z = 0$)
6. Axisymmetric ($v_\theta = \partial / \partial \theta = 0$)

As the flow is fully developed in the z direction, v_z can be a function of r only, that is, $v_z = v_z(r)$.

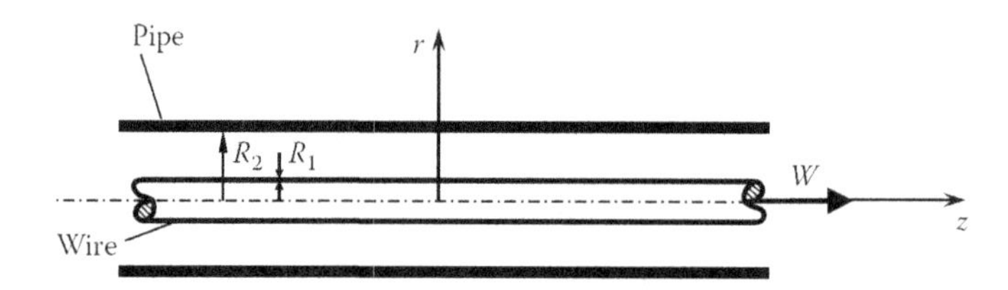

FIGURE 3.11 Annular flow between the pipe and the wire moving at constant velocity along the pipe axis (Example 3.3).

THE CONTINUITY EQUATION

For the steady incompressible flow, the continuity equation in cylindrical polar coordinates reads:

$$\frac{1}{r}\frac{\partial}{\partial r}(rv_r)+\frac{1}{r}\frac{\partial v_\theta}{\partial \theta}+\frac{\partial v_z}{\partial z}=0$$

which, using $\frac{\partial v_z}{\partial z}=\frac{\partial v_\theta}{\partial \theta}=0$ from the assumptions, reduces to:

$$\frac{\partial rv_r}{\partial r}=0$$

whose integration yields $rv_r=C_1$ in the r direction, where C_1 is the integration constant. As $v_r=0$ at the pipe wall ($r=R_2$) and the wire surface ($r=R_1$), it must be zero everywhere in the flow, giving $C_1=0$. This is an important result derived from the continuity equation in this flow problem.

THE REDUCED NAVIER–STOKES EQUATIONS

Following the development leading to Equation 3.56, in the absence of static pressure gradient in this case, the Navier–Stokes equations reduce to the following governing equation for the velocity component in the z direction:

$$\frac{1}{r}\frac{\mathrm{d}}{\mathrm{d}r}\left(r\frac{\mathrm{d}v_z}{\mathrm{d}r}\right)=0$$

BOUNDARY CONDITIONS

Two boundary conditions needed for a unique solution of the above reduced Navier–Stokes equation are $v_z=W\ @\, r=R_1$ and $v_z=0\ @\, r=R_2$.

SOLUTION FOR THE VELOCITY DISTRIBUTION

Integrating the reduced Navier–Stokes equation yields:

$$v_z=C_1\ln r+C_2$$

where C_1 and C_2 are the integration constants to be determined from the boundary conditions.

Applying the first boundary condition at the wire surface, we obtain:

$$W=C_1\ln R_1+C_2$$

Similarly, applying the second boundary condition at the pipe wall yields:

$$0=C_1\ln R_2+C_2$$

The preceding two equations yield $C_1=-\frac{W}{\ln(R_2/R_1)}$ and $C_2=\frac{W\ln R_2}{\ln(R_2/R_1)}$.

Thus, we finally obtain the radial distribution of the axial velocity in the annulus between the pipe and the wire as:

$$v_z=-\frac{W\ln(r/R_2)}{\ln(R_2/R_1)}$$

FORCE NECESSARY TO DRAW A WIRE OF RADIUS R_1 AND LENGTH L

We calculate the shear stress distribution from the velocity distribution as:

$$\tau_{rz} = \mu \frac{dv_z}{dr} = -\frac{1}{r}\left(\frac{\mu W}{\ln(R_2 / R_1)}\right)$$

which shows that the shear stress on the fluid surface in contact with the wire surface is negative. As the surface normal is in the negative r direction, the direction of the shear stress on the surface (from the convention we presented earlier in this chapter) must be in the positive z direction. Accordingly, the direction of the shear stress on the wire surface must be in the negative z direction, giving:

$$\tau_{\text{wire}} = -\frac{1}{R_1}\left(\frac{\mu W}{\ln(R_2 / R_1)}\right)$$

Thus, we calculate the required force in the positive z direction to draw a wire of length L as:

$$F_{\text{wire}} = -\tau_{\text{wire}}(2\pi R_1 L)$$

$$F_{\text{wire}} = \frac{2\pi W \mu L}{\ln(R_2 / R_1)}$$

The result here shows that the force needed to pull the wire is directly proportional to the wire velocity and the fluid viscosity. This result is invalid for $R_1 = 0$, in which case we will have no wire. The result is also invalid for $R_1 = R_2$, implying zero gap between the wire and the pipe with no fluid flow. For $0 < R_1 < R_2$, F_{wire} increases with the wire radius, initially at a low rate of increase for thin wires, becoming very large as the wire radius approaches the pipe radius (thin annulus).

Problems

3.1 Figure 3.12 shows a two-dimensional (x–y plane) fully developed incompressible laminar flow between two parallel plates separated by H. Both plates are infinite in dimensions with no variation of flow properties in the z direction. The flow between the plates is entirely driven by an imposed static pressure gradient with no contribution from gravitational body force. Find (a) the fully developed velocity $u(y)$ and the relationship between the average and maximum values in this profile, (b) the shear stress distribution between the plates, (c) the expression for the Darcy friction factor in terms of Reynolds number, and (d) the expression to compute the loss in static pressure over a length L in the flow direction.

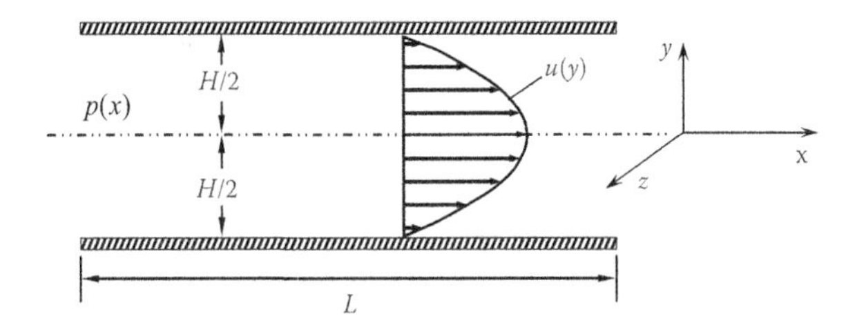

FIGURE 3.12 Flow between infinite parallel plates (Problem 3.1).

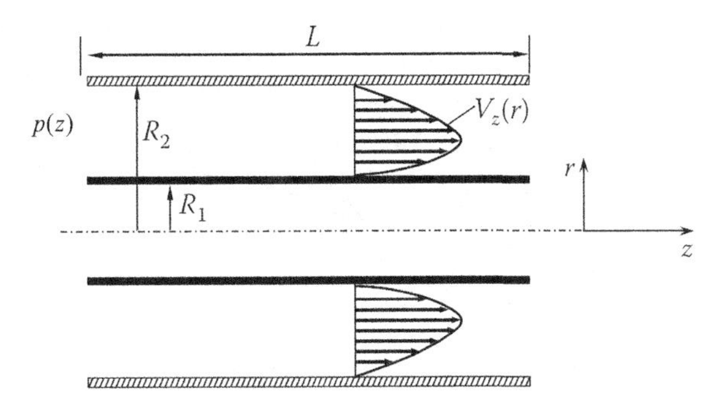

FIGURE 3.13 Flow in an annulus between concentric pipes (Problem 3.4).

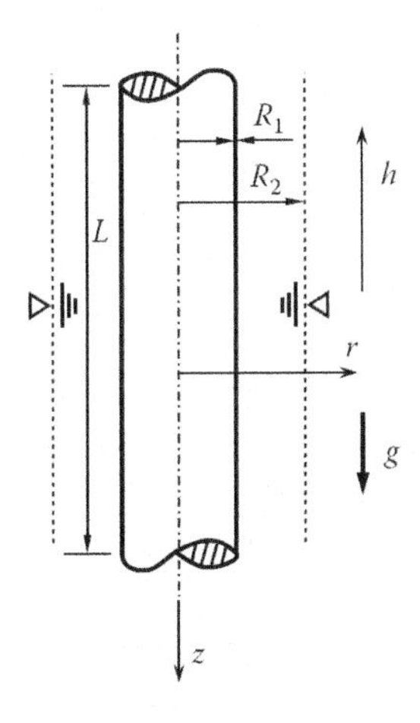

FIGURE 3.14 Annular flow falling vertically along a circular rod (Problem 3.5).

3.2 Two immiscible incompressible fluids of different viscosities and equal density are flowing between two parallel plates under the influence of a pressure gradient. Under the fully developed flow conditions, find the velocity distribution in each fluid.

3.3 Repeat Problem 3.2 for the fully developed flow in a circular pipe.

3.4 Figure 3.13 shows an axisymmetric fully developed incompressible laminar flow through an annulus between two concentric pipes of radii R_1 and R_2. The flow in the annulus is entirely driven by an imposed static pressure gradient with no contribution from gravitational body force. Find (a) the fully developed velocity $v_z(r)$, (b) the radial location of the maximum velocity and its relation to the average value, (c) the shear stress distribution within the annulus, (d) the expression for the Darcy friction factor in terms of Reynolds number, and (e) the expression to compute the loss in static pressure over a length L in the flow direction.

3.5 Figure 3.14 shows a fully developed annular film flow of an incompressible fluid falling vertically along a rod of radius R_1 under the influence of gravity. There is no imposed pressure gradient in this flow. The free surface of the fluid film corresponds to the radius R_2. Find the fully developed velocity $v_z(r)$. Show that the shear force exerted by the falling film on the surface of the rod of length L is equal to the weight of the falling fluid.

3.6 Consider a steady incompressible laminar flow between concentric cylinders of radii R_i and R_o, as shown in Figure 3.15. The inner cylinder is rotating at a constant angular velocity Ω. There is no flow in the axial direction, and the tangential velocity v_θ depends on r only. Show that $v_\theta(r)$ in the annulus is given by:

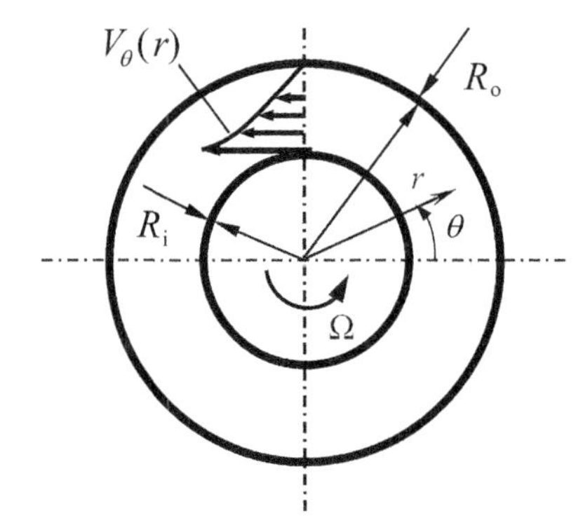

FIGURE 3.15 Flow between concentric cylinders with the inner cylinder rotating (Problem 3.6).

$$v_\theta(r) = \frac{\Omega R_i^2}{r}\left(\frac{R_o^2 - r^2}{R_o^2 - R_i^2}\right)$$

For cylinders of length L, find an expression for the torque necessary to turn the inner cylinder at an angular velocity ω.

3.7 Stokes' first problem: use the Navier–Stokes equations to develop the exact solution to the problem of a suddenly accelerated flat plate in a quiescent fluid.

3.8 Derive the three-dimensional vorticity transport equation for an incompressible laminar flow from the Navier–Stokes equations expressed in vector form.

REFERENCES

Bird, R.B., W.E. Stewart, and E.N. Lightfoot. 2007. *Transport Phenomena*, Revised 2nd edition. New York: John Wiley.

Riley, N. and P.G. Drazin. 2006. *The Navier–Stokes Equations: A Classification of Flows and Exact Solutions*. London: Cambridge University Press.

Schlichting, H. 1979. *Boundary Layer Theory*, 7th edition. New York: McGraw-Hill.

Sultanian, B.K. 2015. *Fluid Mechanics: An Intermediate Approach*, 1st edition. Boca Raton, FL: Taylor & Francis.

Sultanian, B.K. 2021. *Fluid Mechanics and Turbomachinery: Problems and Solutions*. Boca Raton, FL: Taylor & Francis.

Sultanian, B.K. 2022. *Thermal-Fluids Engineering: Problems with Solutions*. USA: Independently published by Kindle Direct Publishing (Amazon.com).

BIBLIOGRAPHY

Batchelor, G.K. 1963. *An Introduction to Fluid Dynamics*. London: Cambridge University Press.

Currie, I.G. 2013. *Fundamental Mechanics of Fluids*, 4th edition. Boca Raton, FL: Taylor & Francis.

Debler, W.R. 1990. *Fluid Mechanics Fundamentals*, 1st edition. Englewood Cliff: Prentice Hall.

Kundu, P.K., I.M. Cohen, D.R. Dowling, J. Capecilatro. 2024. *Fluid Mechanics*, 7th edition. Burlington: Academic Press.

Langlois, W.E. and M.O. Deville. 2014. *Slow Viscous Flow*, 2nd edition. Heidelberg: Springer.

Nunn, R.H. 1989. *Intermediate Fluid Mechanics*. Boca Raton, FL: Taylor & Francis.

Panton, R.L. 2024. *Incompressible Flow*, 5th edition. New York: Wiley.

Schlichting, H. and K. Gersten. 2017. *Boundary Layer Theory*. 9th edition. New York: McGraw-Hill.

Sultanian, B.K. 2022. *FLUID MECHANICS An Intermediate Approach*: Errata for *the First Edition Published in 2015*. USA: Independently published on KDP (Amazon.com).

Wang, C.Y. 1991. Exact solutions of the steady-state Navier–Stokes equations. *Annual Review of Fluid Mechanics*. 23: 159–177.

NOMENCLATURE

C_1, C_2	Integration constants
C_f	Shear coefficient
D	Pipe diameter
e_{ij}	Symmetric strain rate tensor
f	Darcy friction factor
δf_b	Body force per unit volume acting on a differential control volume
δf_s	Surface force per unit volume acting on a differential control volume
δF_b	Body force acting on a differential control volume
δF_{s_yx}	Surface force acting on a differential control volume due to $\partial \tau_{yx} / \partial y$
F	Force
g	Acceleration due to gravity
h	Height measured vertically up (for positive h) from a fixed datum
H	Distance between two parallel plates
L	Length
$\dot{m}$	Mass flow rate
p	Static pressure
p'	Modified static pressure that includes local hydrostatic pressure
r	Coordinate r of cylindrical polar coordinate system
R	Pipe radius
R_1	Inner radius; wire radius
R_2	Outer radius
Re	Reynolds number
S_{ij}	Deformation rate tensor
t	Time
u, v, w	Velocity components in the x, y, and z coordinate directions, respectively
u_i	Velocities in tensor notation
u^*	Dimensionless velocity in the Couette flow ($u^* = u / U$)
U	Top plate velocity in the Couette flow
U_0	Constant plate velocity
v_r, v_θ, v_z	Radial, tangential, axial (z direction) velocity components, respectively
$\overline{v}$	Section-average velocity
$\mathbf{V}$	Velocity vector
V_0	Constant velocity through a porous plate
W	Constant wire velocity along z direction (wire axis)
x	Coordinate x of the Cartesian coordinate system
x_i	Coordinates in tensor notation
y	Coordinate y of the Cartesian coordinate system
z	Coordinate z of the Cartesian coordinate system

SUBSCRIPTS AND SUPERSCRIPTS

0	Total (stagnation)
1	Location 1; Section 1
2	Location 2; Section 2
g	Due to gravity
h	Homogeneous solution
max	Maximum value

p	Particular solution
r	Component in r coordinate direction
x	Component in x coordinate direction
y	Component in y coordinate direction
z	Component in z coordinate direction
θ	Component in θ coordinate direction
$(^*)$	Nondimensional

GREEK SYMBOLS

β	Pressure gradient parameter in the Couette flow
δ_{ij}	Kronecker delta=1 for $i = j$ and=0 for $i \neq j$
ζ	Vorticity
$\boldsymbol{\zeta}$	Vorticity vector
η	Dimensionless distance in the Couette flow ($\eta = y / H$)
θ	Coordinate θ in cylindrical polar coordinate system
μ	Dynamic viscosity
ν	Kinematic viscosity ($\nu = \mu / \rho$)
ρ	Density
σ_{ii}	Normal stress in tensor notation
τ_{ij}	Stress acting on the i face of a differential control volume in the j direction ($\tau_{ij} = \tau_{ji}$)
Ψ	Stream function
ω_{ij}	Antisymmetric rotation tensor

4 Boundary Layer Flow

4.1 INTRODUCTION

Because of the nonzero viscosity of natural fluids, all flows feature the no-slip condition at a solid boundary, resulting in a boundary layer flow with high transverse velocity gradients and wall shear stress. Boundary layers predominantly confine the effect of viscosity. Outside the boundary layer, the effect of viscosity may be considered negligible, and the flow is modeled as a potential flow, as discussed in Chapter 2. Anderson (2005) captures the groundbreaking role of Ludwig Prandtl, who first presented the concept of the boundary layer in 1904 to the Mathematical Congress in Heidelberg, Germany, in an article titled "On the Fluid Motion with Very Little Viscosity." In his 10-minute presentation, Prandtl revolutionized the understanding and analysis of fluid flows of engineering interest. The idea of a thin boundary layer at the surface of a body in fluid flow explains the mechanism of separation under an adverse pressure gradient in the flow direction. If we had today's computing power to solve the complete set of Navier–Stokes equations numerically, we would have perhaps missed the development of the boundary layer theory and the simplifications it brings in solving many engineering problems involving fluid flow, including heat and mass transfer.

The drag force acting on a body consists of two parts: (a) frictional drag—due to the integrated effect of the shear stress distribution along the surface of the body—and (b) form drag—due to the integrated effect of the pressure distribution normal to the surface of the body. When the boundary layer separates, it creates a wake region of lower pressure behind the body, significantly increasing the form drag, which could be higher than the frictional drag on a bluff body.

All boundary layers feature a predominant flow direction where the downstream does not influence the upstream conditions. To a mathematician, this is characteristic of a parabolic flow, allowing us to use a step-by-step marching solution of the boundary layer equations. However, when the boundary layer flow separates from the wall, the reverse flow confined between the separating streamline and the wall is, mathematically speaking, elliptic, where the downstream conditions influence the upstream. In this region, the boundary layer assumptions and the parabolic governing partial differential equations are no longer applicable. One must, therefore, solve the full Navier–Stokes equations to compute various flow properties in the reverse flow region.

In this chapter, we present the essential concepts of laminar boundary layer flows of an incompressible fluid with constant viscosity. The classical book by Schlichting (1979) presents comprehensive analyses of all kinds of boundary layers, including turbulent boundary layers.

This chapter features two worked examples and six chapter-end problems. The original version of this chapter appears as Chapter 8 in the first edition of this book by Sultanian (2015).

4.2 DESCRIPTION OF A BOUNDARY LAYER

4.2.1 External Flow

Figure 4.1 shows a flat plate in an oncoming flow with a uniform velocity u_∞. The flow hitting the plate at its leading edge forms a stagnation point S. Because of the no-slip condition, the fluid has zero velocity at the solid boundary. As the flow velocity goes from zero to its free-stream value u_∞ over a minimal thickness in the stagnation region, the plate experiences the highest shear stress in this region. Starting from the stagnation point, as shown in the figure, a laminar boundary layer grows on the plate surface in the flow direction. The effect of viscosity remains confined within the laminar boundary with distributions of high shear stress and vorticity (maximum at the wall and zero at the edge of the boundary layer). Note here that even for a free-stream turbulent flow, the

DOI: 10.1201/[illegible]

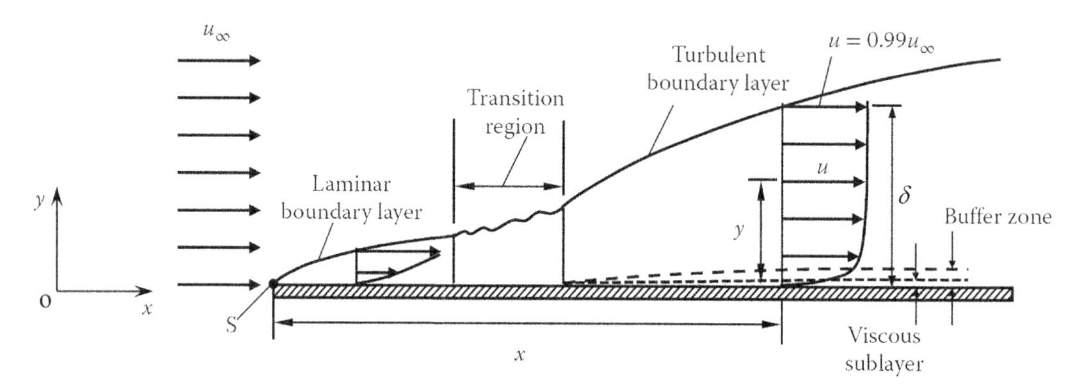

FIGURE 4.1 External boundary layer flow near a solid boundary.

boundary layer that originates from the stagnation point is laminar, for example, observed on the suction surface of an airfoil.

Beyond a certain length, the laminar boundary layer transitions into a turbulent boundary layer consisting of three regions. The innermost region is known as the viscous sublayer, called the "laminar sublayer" in the old literature on fluid mechanics. Recent measurements in the viscous sublayer show that it does not resemble a laminar flow. Next to the viscous sublayer in a turbulent boundary layer is the buffer zone with mixed laminar and turbulent behavior. The region between the buffer zone and the edge of the boundary layer features a fully turbulent behavior. In this region, the velocity distribution along the plate is nearly uniform, and its gradients remain primarily confined to the buffer zone. The shear stress and vorticity in a turbulent boundary layer are typically much higher than in a laminar boundary layer.

4.2.2 Internal Flow

Figure 4.2 shows the development of an axisymmetric laminar boundary layer in a circular pipe with a uniform velocity profile at its inlet. The no-slip condition at the wall initiates the growth of the boundary layer all around the pipe surface in contact with the fluid. Outside the boundary layer, the flow is unaffected by the fluid viscosity. Therefore, we may treat this flow as a potential flow with a uniform velocity that increases along the pipe axis. At the end of the entrance length, shown in the figure, the boundary layer reaches the pipe axis and stops growing further. Beyond this point, the velocity profile becomes parabolic and remains unchanged for the rest of the pipe. The resulting flow is commonly known as the fully developed pipe flow. Unlike in an external flow, where

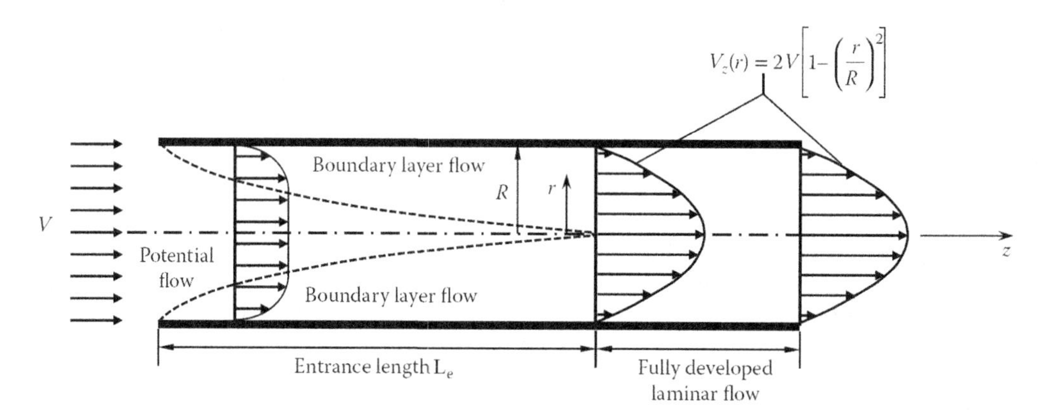

FIGURE 4.2 Boundary layer flow in the pipe entrance region.

the boundary layer continues to grow, maintaining two flow regimes—the boundary layer flow and the free-stream potential flow—in an internal flow, the entire flow in the fully developed region is affected by viscosity. In this chapter, we will only consider external laminar boundary layer flows.

4.3 DIFFERENTIAL BOUNDARY LAYER EQUATIONS

In this section, we derive the boundary layer equations from the continuity and Navier–Stokes equations (Equations 3.33–3.35). Writing these equations for the two-dimensional flow shown in Figure 4.3, we obtain:

4.3.1 The Continuity Equation

$$\frac{\partial u}{\partial x}+\frac{\partial v}{\partial y}=0 \tag{4.1}$$

4.3.2 The Momentum Equation in x Direction

$$\frac{\partial u}{\partial t}+u\frac{\partial u}{\partial x}+v\frac{\partial u}{\partial y}=-\frac{1}{\rho}\frac{\partial p}{\partial x}+\nu\left(\frac{\partial^2 u}{\partial x^2}+\frac{\partial^2 u}{\partial y^2}\right) \tag{4.2}$$

4.3.3 The Momentum Equation in y Direction

$$\frac{\partial v}{\partial t}+u\frac{\partial v}{\partial x}+v\frac{\partial v}{\partial y}=-\frac{1}{\rho}\frac{\partial p}{\partial y}+\nu\left(\frac{\partial^2 v}{\partial x^2}+\frac{\partial^2 v}{\partial y^2}\right) \tag{4.3}$$

In Equations 4.2 and 4.3, we have neglected the gravitational body force, which we can easily include by modifying the static pressure p, as discussed in Chapter 3.

To develop greater insight into the boundary layer and the associated boundary layer assumptions, let us perform an order-of-magnitude analysis of various terms in the continuity and momentum equations. For this, we use the free-stream velocity u_∞ and the characteristic length L as the appropriate velocity and length scales, respectively, giving the time scale L / u_∞. For the boundary layer flow to exist at a solid surface, we require that $L >> \delta$, where δ is the boundary layer thickness. Both shear stress and vorticity remain confined to the boundary layer, where the kinematic viscosity

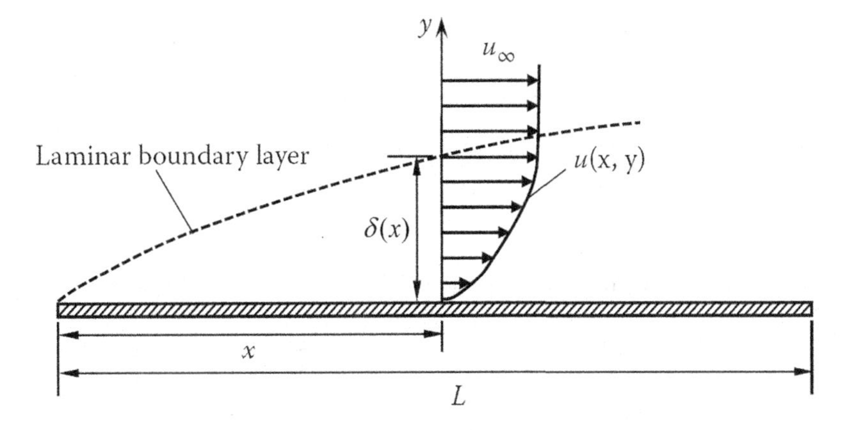

FIGURE 4.3 Laminar boundary layer flow over a solid surface.

ν governs the transverse diffusion. During the time of convection of vorticity across L, its viscous diffusion in the boundary layer equals $\delta \sim (\nu L / u_\infty)^{1/2}$. Thus, we can write:

$$L >> (\nu L / u_\infty)^{1/2}$$

$$Re = \left(\frac{L u_\infty}{\nu} \right) >> 1$$

which indicates that, for a boundary layer to exist, we must have a Reynolds number, which is the ratio of inertia force to viscous force, much larger than unity.

In terms of our selected scales of length, velocity, and time, we make each variable in the continuity and momentum equations nondimensional as follows:

Independent variables: $x^* = \frac{x}{L} \quad y^* = \frac{y}{L} \quad t^* = \frac{t u_\infty}{L}$

Dependent variables: $u^* = \frac{u}{u_\infty} \quad v^* = \frac{v}{u_\infty} \quad p^* = \frac{p}{\rho u_\infty^2}$

For a boundary layer flow, we have $\delta << L$ or $\delta^* = \delta / L << 1$. In view of their expected variations in the boundary layer, we can directly assign the following order-of-magnitude estimates to some of these variables:

$$x^* = \mathrm{O}(1) \quad y^* = \mathrm{O}(\delta^*) \quad t^* = \mathrm{O}(1) \quad u^* = \mathrm{O}(1)$$

where O(1) stands for "order-of-magnitude 1." Substituting these estimates in the nondimensional form of the continuity equation (Equation 4.1), we obtain:

$$\frac{\partial u^*}{\partial x^*} + \frac{\partial v^*}{\partial y^*} = 0$$

$$\mathrm{O}(1) \quad \mathrm{O}(1)$$

As $y^* = \mathrm{O}(\delta^*)$, the above equation yields $v^* = \mathrm{O}(\delta^*)$, which is consistent with our expectations in a boundary layer to yield $v^* << u^*$.

We can write the nondimensional form of the x-momentum equation (Equation 4.2) as:

$$\frac{\partial u^*}{\partial t^*} + u^* \frac{\partial u^*}{\partial x^*} + v^* \frac{\partial u^*}{\partial y^*} = -\frac{\partial p^*}{\partial x^*} + \frac{1}{\mathrm{Re}} \left(\frac{\partial^2 u^*}{\partial x^{*2}} + \frac{\partial^2 u^*}{\partial y^{*2}} \right) \tag{4.4}$$

We can express an order-of-magnitude estimate of each term in this equation as follows:

$$\frac{\partial u^*}{\partial t^*} = \mathrm{O}(1)$$

$$u^* \frac{\partial u^*}{\partial x^*} = \mathrm{O}(1)\,\mathrm{O}(1) = \mathrm{O}(1)$$

$$v^* \frac{\partial u^*}{\partial y^*} = \mathrm{O}(\delta^*) \frac{1}{\mathrm{O}(\delta^*)} = \mathrm{O}(1)$$

$$\frac{1}{Re} \left(\frac{\partial^2 u^*}{\partial x^{*2}} + \frac{\partial^2 u^*}{\partial y^{*2}} \right) = \frac{1}{Re} \left[\mathrm{O}(1) + \frac{1}{\mathrm{O}(\delta^{*2})} \right] = \frac{1}{Re} \left[\frac{1}{\mathrm{O}(\delta^{*2})} \right]$$

In the middle expression above, within the brackets, we have neglected O(1) in comparison with $1/\mathrm{O}(\delta^{*2})$. From these estimates, we see that each of the inertia terms on the left-hand side of Equation 4.4 is of the order-of-magnitude unity. For the viscous term on the right-hand side of Equation 4.4 to be of the order-of-magnitude unity, we must have $Re = 1/\mathrm{O}(\delta^{*2})$ as we conjectured in the beginning of this section. In that case, the inertia terms will balance the viscous terms in a boundary layer flow. Note that, for the pressure gradient term $\partial p^* / \partial x^*$ to influence the boundary layer flow in the x direction, it should be of the order-of-magnitude unity. We will have additional insight into $\partial p / \partial x$ when we perform the order-of-magnitude analysis of the y-momentum equation. Thus, pertinent to a boundary layer flow, the order-of-magnitude analysis of the x-momentum equation leads to two significant results. First, $\partial^2 u^* / \partial x^{*2}$ is negligible, implying negligible viscous diffusion in the x direction compared to that in the y direction, and second, $Re = 1/\mathrm{O}(\delta^{*2})$, which requires the Reynolds number to be high for the existence of a boundary layer flow.

Next, we write the nondimensional form of the y-momentum equation (Equation 4.3) as:

$$\frac{\partial v^*}{\partial t^*} + u^* \frac{\partial v^*}{\partial x^*} + v^* \frac{\partial v^*}{\partial y^*} = -\frac{\partial p^*}{\partial y^*} + \frac{1}{Re}\left(\frac{\partial^2 v^*}{\partial x^{*2}} + \frac{\partial^2 v^*}{\partial y^{*2}}\right) \tag{4.5}$$

An order-of-magnitude estimate of each term in this equation yields:

$$\frac{\partial v^*}{\partial t^*} = \mathrm{O}(\delta^*)$$

$$u^* \frac{\partial v^*}{\partial x^*} = \mathrm{O}(1)\,\mathrm{O}(\delta^*) = \mathrm{O}(\delta^*)$$

$$v^* \frac{\partial v^*}{\partial y^*} = \mathrm{O}(\delta^*)\;\mathrm{O}(1) = \mathrm{O}(\delta^*)$$

$$\frac{1}{Re}\left(\frac{\partial^2 v^*}{\partial x^{*2}} + \frac{\partial^2 v^*}{\partial y^{*2}}\right) = \frac{1}{Re}\left[\mathrm{O}(\delta^*) + \frac{1}{\mathrm{O}(\delta^*)}\right] = \frac{1}{Re}\left[\frac{1}{\mathrm{O}(\delta^*)}\right] = \mathrm{O}(\delta^*)$$

in which we have used $Re = 1/\mathrm{O}(\delta^{*2})$ deduced from the order-of-magnitude analysis of the x-momentum equation. As each inertia term on the left-hand side of Equation 4.5 is of the order of magnitude δ^*, and the diffusion term on the right-hand side is also of the order of magnitude δ^*, we conclude that in a boundary layer flow the static pressure gradient term $\partial p^* / \partial y^*$ is of the order of magnitude δ^*. Thus, we have obtained two important results from the order-of-magnitude analysis of the y-momentum equation. First, each term in this momentum equation is negligible in the boundary layer flow, allowing us to drop this equation altogether as a member of the boundary layer equations. Second, the static pressure gradient in the y direction is negligible. The static pressure, therefore, may be considered to be imposed entirely by the potential flow in the free stream, varying only along the flow direction. Thus, we can write: $\partial p^* / \partial x^* = \mathrm{d}p^* / \mathrm{d}x^*$.

Using the above order-of-magnitude analysis, we obtain the following boundary layer assumptions:

$$u >> v \qquad \frac{\partial u}{\partial y} >> \frac{\partial u}{\partial x} \qquad \frac{\partial p}{\partial y} \approx 0 \qquad \frac{\partial p}{\partial x} \approx \frac{\mathrm{d}p}{\mathrm{d}x}$$

In addition, in the analysis of a boundary layer flow, we may drop the momentum equation in the y direction. The continuity and x-momentum equations are sufficient to determine the velocity components u and v for the given pressure gradient term $\mathrm{d}p/\mathrm{d}x$. Thus, in addition to the continuity equation (Equation 4.1), the Navier–Stokes equations reduce to the following boundary layer equation:

$$\frac{\partial u}{\partial t}+u\frac{\partial u}{\partial x}+v\frac{\partial u}{\partial y}=-\frac{1}{\rho}\frac{\mathrm{d}p}{\mathrm{d}x}+\nu\frac{\partial^2 u}{\partial y^2} \tag{4.6}$$

As the total pressure remains constant in the potential flow outside the boundary layer, we write:

$$p_0 = p+\frac{1}{2}\rho u_\infty^2 = \text{constant}$$

whose derivative with respect to x yields:

$$\frac{1}{\rho}\frac{\mathrm{d}p}{\mathrm{d}x} = -u_\infty\frac{\mathrm{d}u_\infty}{\mathrm{d}x} \tag{4.7}$$

In addition, we can express the second term on the right-hand side of Equation 4.6 in terms of shear stress as:

$$\nu\frac{\partial^2 u}{\partial y^2}=\frac{1}{\rho}\frac{\partial \tau_{yx}}{\partial y} \tag{4.8}$$

Substituting Equations 4.7 and 4.8 in Equation 4.6 yields for the x-momentum equation:

$$\frac{\partial u}{\partial t}+u\frac{\partial u}{\partial x}+v\frac{\partial u}{\partial y}=u_\infty\frac{\mathrm{d}u_\infty}{\mathrm{d}x}+\frac{1}{\rho}\frac{\partial \tau_{yx}}{\partial y} \tag{4.9}$$

which is the second-order partial differential equation. This equation and the continuity equation (Equation 4.1) are sufficient to compute the distribution of u and v in the boundary layer for the given set of initial and boundary conditions.

4.4 THE VON KARMAN MOMENTUM INTEGRAL EQUATION

The primary motivation for the momentum equation (Equation 4.9) governing a boundary layer flow is to compute the shear stress distribution on a solid surface or its integrated value to obtain the drag force. Note that while making a wall adiabatic with zero heat transfer is possible, it is impossible to attain zero momentum transfer at the wall—an inevitable effect of fluid viscosity.

By integrating Equation 4.9 in the y direction over the boundary layer thickness at each value of x, we eliminate the variation in the y direction and render the equation an ordinary differential equation with quantities varying only in the x direction, the predominant flow direction. This ordinary differential equation is also known as the von Karman momentum integral equation, which we derive in the following sections in a couple of ways: first, by directly integrating Equation 4.9, and second, by using the control volume analysis presented in Chapter 1. Let us first introduce various integral quantities used in the momentum integral equation.

4.4.1 Boundary Layer Thickness

In a boundary layer flow, the boundary layer thickness δ, which varies in the flow direction, is the characteristic length scale showing the extent to which the viscous effects are considered significant. Theoretically, the boundary layer extends up to infinity in the crossflow direction; we will, however, neglect the impact of fluid viscosity beyond the boundary layer thickness δ, where the velocity in the flow direction reaches 99% of the local free-stream velocity. In our analysis of the boundary layer flow, shown in Figure 4.4a, the boundary layer thickness δ defines the edge of the boundary layer where $u = u_\infty(x)$ for all practical purposes. The shear stress is maximum at the wall, and we assume its value at the edge of the boundary layer to be zero. As we will see later in

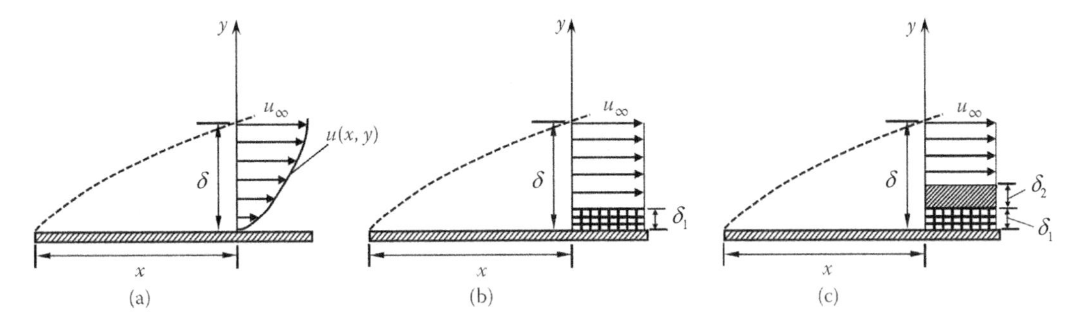

FIGURE 4.4 (a) Actual boundary layer flow showing $u(x, y)$ and boundary layer thickness δ, (b) hypothetical uniform velocity profile with displacement thickness δ_1, and (c) hypothetical uniform velocity profile with displacement thickness δ_1 and momentum thickness δ_2.

this section, the integration of Equation 4.9 in the y direction results in two characteristic integrals: displacement and momentum thickness.

4.4.2 Displacement Thickness

At each x location in a boundary layer, the velocity $u(x, y)$ varies in the y direction from zero at the wall to $u_\infty(x)$ at $y = \delta$. To model this velocity profile in terms of a uniform velocity u_∞ to yield the same mass flow rate, as shown in Figure 4.4b, we need a step function of zero velocity up to a distance δ_1 and a uniform velocity u_∞ over $(\delta - \delta_1)$. As the actual velocity profile $u(x, y)$ shown in Figure 4.4a and the modeled profile shown in Figure 4.4b must have equal mass flow rates, we obtain, for constant density and unit depth in the z direction:

$$\dot{m} = \rho \int_0^\delta u\,\mathrm{d}y = \rho \int_{\delta_1}^\delta u_\infty\,\mathrm{d}y = \rho \int_0^\delta u_\infty\,\mathrm{d}y - \rho \int_0^{\delta_1} u_\infty\,\mathrm{d}y = \rho \int_0^\delta u_\infty\,\mathrm{d}y - \rho\delta_1 u_\infty$$

$$\delta_1 = \int_0^\delta \left(1 - \frac{u}{u_\infty}\right)\mathrm{d}y \tag{4.10}$$

which defines the displacement thickness δ_1, shown in Figure 4.4b. In other words, if we displace the wall by δ_1, we can compute the same mass flow rate using the uniform velocity profile $u = u_\infty$ over the rest of the boundary layer thickness. Alternatively, for a given velocity profile $u(x, y)$ with known δ and δ_1, we can calculate the associated mass flow rate per unit depth as:

$$\dot{m} = \rho(\delta - \delta_1)u_\infty \tag{4.11}$$

4.4.3 Momentum Thickness

The step-function velocity profile shown in Figure 4.4b simulates the mass flow rate of the actual local velocity profile in the boundary layer but fails to simulate its x-momentum flow rate. Therefore, we need to further modify our step-function velocity profile with an additional step, called momentum thickness δ_2, shown in Figure 4.4c. Thus, in equation form, we write:

$$\dot{M}_x = \rho \int_0^\delta u^2\,\mathrm{d}y = \rho \int_{\delta_1+\delta_2}^\delta u_\infty^2\,\mathrm{d}y = \rho \int_0^\delta u_\infty^2\,\mathrm{d}y - \rho \int_0^{\delta_1} u_\infty^2\,\mathrm{d}y - \rho \int_0^{\delta_2} u_\infty^2\,\mathrm{d}y$$

$$\int_0^\delta u^2\,\mathrm{d}y = \int_0^\delta u_\infty^2\,\mathrm{d}y - u_\infty^2\delta_1 - u_\infty^2\delta_2$$

Substituting δ_1 from Equation 4.10 in this equation yields:

$$\int_0^\delta u^2\,\mathrm{d}y = \int_0^\delta u_\infty^2\,\mathrm{d}y - u_\infty^2 \int_0^\delta \left(1 - \frac{u}{u_\infty}\right)\mathrm{d}y - u_\infty^2 \delta_2$$

$$\int_0^\delta u^2\,\mathrm{d}y = u_\infty^2 \int_0^\delta \frac{u}{u_\infty}\,\mathrm{d}y - u_\infty^2 \delta_2 \tag{4.12}$$

$$\delta_2 = \int_0^\delta \frac{u}{u_\infty}\left(1 - \frac{u}{u_\infty}\right)\mathrm{d}y$$

For a given velocity profile $u(x, y)$ with known δ, δ_1, and δ_2, we can calculate the associated x-momentum flow rate per unit depth as:

$$\dot{M}_x = \rho(\delta - \delta_1 - \delta_2)u_\infty^2 \tag{4.13}$$

4.4.4 Derivation of the von Karman Momentum Integral Equation

4.4.4.1 Method 1: The Integration of Equation 4.9

For a steady incompressible boundary layer flow, the integration of Equation 4.9 in the y direction over the boundary layer thickness δ yields:

$$\int_0^\delta \left(u\frac{\partial u}{\partial x} + v\frac{\partial u}{\partial y}\right)\mathrm{d}y = \int_0^\delta u_\infty \frac{\mathrm{d}u_\infty}{\mathrm{d}x}\,\mathrm{d}y + \frac{1}{\rho}\int_0^\delta \frac{\partial \tau_{yx}}{\partial y}\,\mathrm{d}y$$

$$\int_0^\delta \left(u\frac{\partial u}{\partial x} + v\frac{\partial u}{\partial y} - u_\infty \frac{\mathrm{d}u_\infty}{\mathrm{d}x}\right)\mathrm{d}y = \frac{1}{\rho}\int_0^\delta \frac{\partial \tau_{yx}}{\partial y}\,\mathrm{d}y = \left.\frac{\tau_{yx}}{\rho}\right|_{y=\delta} - \left.\frac{\tau_{yx}}{\rho}\right|_{y=0}$$

which with $\tau_{yx}\big|_{y=\delta} = 0$ and $\tau_{yx}\big|_{y=0} = \tau_0$, where τ_o is the wall shear stress, reduces to:

$$\int_0^\delta \left(u\frac{\partial u}{\partial x} + v\frac{\partial u}{\partial y} - u_\infty \frac{\mathrm{d}u_\infty}{\mathrm{d}x}\right)\mathrm{d}y = -\frac{\tau_0}{\rho} \tag{4.14}$$

Integrating the second term on the left-hand side of this equation, we obtain:

$$\int_0^\delta v\frac{\partial u}{\partial y}\mathrm{d}y = vu\big|_{y=0}^{y=\delta} - \int_0^\delta u\frac{\partial v}{\partial y}\mathrm{d}y$$

which with $u = v = 0$ at $y = 0$ and $u = u_\infty$ at $y = \delta$ becomes:

$$\int_0^\delta v\frac{\partial u}{\partial y}\mathrm{d}y = v_\infty u_\infty - \int_0^\delta u\frac{\partial v}{\partial y}\mathrm{d}y$$

Substituting in this equation the following obtained from the continuity equation (Equation 4.1):

$$\frac{\partial v}{\partial y} = -\frac{\partial u}{\partial x} \text{ and } \int_0^\delta \frac{\partial v}{\partial y}\mathrm{d}y = -\int_0^\delta \frac{\partial u}{\partial x}\mathrm{d}y \text{ or } v_\infty = -\int_0^\delta \frac{\partial u}{\partial x}\mathrm{d}y$$

yields:

$$\int_0^\delta v\frac{\partial u}{\partial y}\mathrm{d}y = -\int_0^\delta u_\infty \frac{\partial u}{\partial x}\mathrm{d}y + \int_0^\delta u\frac{\partial u}{\partial x}\mathrm{d}y$$

whose substitution in Equation 4.14 further yields:

$$\int_0^\delta \left(2u\frac{\partial u}{\partial x} - u_\infty \frac{\partial u}{\partial x} - u_\infty \frac{\mathrm{d}u_\infty}{\mathrm{d}x} \right) \mathrm{d}y = -\frac{\tau_0}{\rho}$$

$$\int_0^\delta \left(\frac{\partial u^2}{\partial x} - \frac{\partial}{\partial x}(uu_\infty) + u\frac{\mathrm{d}u_\infty}{\mathrm{d}x} - u_\infty \frac{\mathrm{d}u_\infty}{\mathrm{d}x} \right) \mathrm{d}y = -\frac{\tau_0}{\rho}$$

$$\int_0^\delta \left(\frac{\partial}{\partial x}\{u(u_\infty - u)\} \right) \mathrm{d}y + \frac{\mathrm{d}u_\infty}{\mathrm{d}x} \int_0^\delta (u_\infty - u)\,\mathrm{d}y = \frac{\tau_0}{\rho}$$

We use the rule of Leibnitz (see Appendix A) to replace the partial derivative $\partial / \partial x$ in the first integral of this equation as an ordinary derivative $\mathrm{d} / \mathrm{d}x$ after the integration, giving:

$$\frac{\mathrm{d}}{\mathrm{d}x} \int_0^\delta u(u_\infty - u)\,\mathrm{d}y + \frac{\mathrm{d}u_\infty}{\mathrm{d}x} \int_0^\delta (u_\infty - u)\,\mathrm{d}y = \frac{\tau_0}{\rho}$$

which, using Equations 4.10 and 4.12, reduces to:

$$\frac{\mathrm{d}}{\mathrm{d}x}(u_\infty^2 \delta_2) + u_\infty \delta_1 \frac{\mathrm{d}u_\infty}{\mathrm{d}x} = \frac{\tau_0}{\rho} \tag{4.15}$$

which is the von Karman momentum integral equation for a steady incompressible boundary layer flow in the x–y plane. We can alternatively express this equation as:

$$\frac{\mathrm{d}\delta_2}{\mathrm{d}x} + \left(2 + \frac{\delta_1}{\delta_2} \right) \frac{\delta_2}{u_\infty} \frac{\mathrm{d}u_\infty}{\mathrm{d}x} = \frac{\tau_0}{\rho u_\infty^2} = \frac{C_\mathrm{f}}{2} \tag{4.16}$$

where δ_1 / δ_2 is known as the shape factor H, and we define the shear coefficient (also called the Fanning friction factor) C_f as:

$$C_\mathrm{f} = \frac{\tau_0}{\frac{1}{2}\rho u_\infty^2} \tag{4.17}$$

Note that Equation 4.16 features only the displacement thickness δ_1 and the momentum thickness δ_2 and is independent of the boundary layer thickness δ, which is somewhat arbitrarily defined, ideally extending up to infinity.

4.4.4.2 Method 2: Using the Control Volume Analysis

The approach we used in Method 1 is purely mathematical. The control volume analysis of the boundary layer flow provides a direct physic-based approach to deriving the momentum integral equation (Equation 4.15 or 4.16).

Figure 4.5a shows a control volume ABCD encompassing the boundary layer between x and $x + \Delta x$. At AB, the boundary layer thickness is δ, and at DC, it is $\delta + \Delta\delta$. In addition to changes in all boundary layer properties over Δx, we allow variation in the free-stream velocity from u_∞ to $u_\infty + \Delta u_\infty$, as shown in the figure.

Surface forces acting on the control volume in the x direction. For the x-momentum equation, Figure 4.5b shows the static pressures and wall shear stress, contributing to the surface force acting on the control volume. We have assumed that the static pressure acting on AD is the average of the pressures acting on AB and DC. Here, we have neglected the body force acting on the control volume. Assuming unit depth perpendicular to the plane of the boundary layer flow shown in Figure 4.5a, we can express the total surface force ΔF_s acting on the control volume in the x direction as:

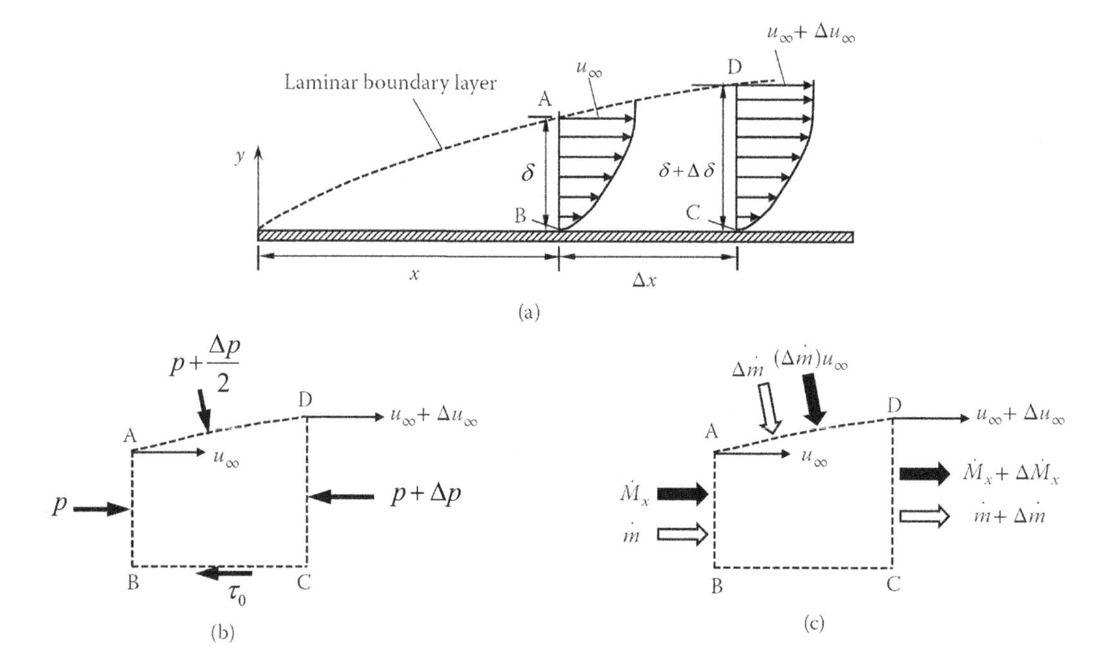

FIGURE 4.5 (a) A control volume encompassing the boundary layer flow between x and $x+\Delta x$, (b) surface forces acting on the boundary layer control volume, and (c) mass and x-momentum flow rates entering and leaving the boundary layer control volume.

$$\Delta F_s = -\tau_0 \Delta x + \Delta F_p$$

where ΔF_p is the net pressure force acting on the control surface, evaluated as:

$$\Delta F_p = p\delta + \left(p + \frac{\Delta p}{2}\right)\Delta\delta - (p + \Delta p)(\delta + \Delta\delta)$$

Simplifying this expression and neglecting higher-order term $\Delta p \Delta\delta$, we obtain:

$$\Delta F_P = -\Delta p \delta$$

giving:

$$\Delta F_s = -\tau_0 \Delta x - \Delta p \delta \tag{4.18}$$

Rates of mass and x-momentum inflow and outflow through the control surface. To satisfy the continuity equation for a steady boundary layer flow through the control volume, we obtain $\Delta\dot{m}$ entering at AD, as shown in Figure 4.5c. We can write the corresponding value of x-momentum as:

$$\Delta\dot{m}\left(u_\infty + \frac{\Delta u_\infty}{2}\right)$$

which, after neglecting the higher-order term $\Delta\dot{m}\Delta u_\infty$, reduces to $\Delta\dot{m}u_\infty$. Thus, we obtain the total rate of x-momentum inflow into the control volume as $\dot{M}_{\text{in}} = \dot{M}_x + \Delta\dot{m}u_\infty$ and the total rate of x-momentum outflow from the control volume as $\dot{M}_{\text{out}} = \dot{M}_x + \Delta\dot{M}_x$, giving the net rate of x-momentum outflow from the control volume as:

$$\begin{aligned}\dot{M}_{\text{out}} - \dot{M}_{\text{in}} &= \dot{M}_x + \Delta\dot{M}_x - \dot{M}_x - \Delta\dot{m}u_\infty \\ \dot{M}_{\text{out}} - \dot{M}_{\text{in}} &= \Delta\dot{M}_x - \Delta\dot{m}u_\infty\end{aligned} \tag{4.19}$$

Using Equations 4.18 and 4.19, the x-momentum equation for the control volume becomes:

$$-\tau_0\Delta x - \Delta p\delta = \Delta\dot{M}_x - \Delta\dot{m}u_\infty$$

Dividing this equation throughout by Δx and taking the limit $\Delta x \to 0$, after some rearrangement, we obtain:

$$u_\infty \frac{\text{d}\dot{m}}{\text{d}x} - \frac{\text{d}\dot{M}_x}{\text{d}x} - \frac{\text{d}p}{\text{d}x}\delta = \tau_0$$

$$\frac{\text{d}}{\text{d}x}\left(\dot{m}u_\infty - \dot{M}_x\right) - \dot{m}\frac{\text{d}u_\infty}{\text{d}x} - \frac{\text{d}p}{\text{d}x}\delta = \tau_0$$

Using Equations 4.7, 4.11, and 4.13 in this equation yields:

$$\frac{\text{d}}{\text{d}x}\left\{\rho(\delta-\delta_1)u_\infty^2 - \rho(\delta-\delta_1-\delta_2)u_\infty^2\right\} - \rho(\delta-\delta_1)u_\infty\frac{\text{d}u_\infty}{\text{d}x} + \rho u_\infty\frac{\text{d}u_\infty}{\text{d}x}\delta = \tau_0$$

$$\frac{\text{d}}{\text{d}x}(u_\infty^2\delta_2) + u_\infty\delta_1\frac{\text{d}u_\infty}{\text{d}x} = \frac{\tau_0}{\rho}$$

which is identical to Equation 4.15. In this derivation method, we have not assumed the laminar or turbulent flow. We can, therefore, use Equation 4.15 for both laminar and turbulent boundary layers. For a turbulent boundary layer, however, we determine the wall shear stress using an empirical correlation, as discussed in detail in Schlichting (1979). In this chapter, we have limited our discussion to laminar boundary layers.

4.5 LAMINAR BOUNDARY LAYER ON A FLAT PLATE

The free-stream velocity distribution on a flat plate is uniform with zero pressure gradient $(\partial p / \partial x = 0)$. Therefore, the momentum equation (Equation 4.9) for a steady boundary layer flow for this case reduces to:

$$u\frac{\partial u}{\partial x} + v\frac{\partial u}{\partial y} = \frac{1}{\rho}\frac{\partial \tau_{yx}}{\partial y} = \nu\frac{\partial^2 u}{\partial y^2} \tag{4.20}$$

subject to the boundary conditions: $y = 0: u = v = 0$; $y = \infty: u = u_\infty$, and the applicable continuity equation is Equation 4.1.

4.5.1 Exact Solution Using Similarity Variables

The order-of-magnitude analysis presented in the beginning of this chapter shows that $Re = 1 / \text{O}(\delta^{*2})$, which indicates that the boundary layer grows with the downstream distance x as follows:

$$\delta(x) \sim \left(\frac{\nu x}{u_\infty}\right)^{1/2} \tag{4.21}$$

Based on this physical insight, Blasius (a student of Prandtl) conjectured that the velocity profiles in the growing boundary layer on a flat plate are self-similar. Accordingly, he defined the following dimensionless similarity variable:

$$\eta = \frac{y}{\delta(x)} = y\sqrt{\frac{u_\infty}{\nu x}} \tag{4.22}$$

As the streamwise velocity u at each x varies from 0 at $y = 0$ to u_∞ at $y = \delta(x)$, for a self-similar boundary layer, we expect u velocity profiles at all x locations to collapse into one when we express each profile in terms of u / u_∞ as a function of the similarity variable η.

For the two-dimensional flow on a flat plate, we define a stream function Ψ as:

$$\Psi = \sqrt{\nu x u_\infty}\, f(\eta) = u_\infty \delta(x) f(\eta) \tag{4.23}$$

where we have Equation 4.21 for the dependence of $\delta(x)$ on x. In this equation, $f(\eta)$ is the dimensionless stream function. As $u = \partial\Psi / \partial y$ and $v = -\partial\Psi / \partial x$, the stream function Ψ automatically satisfies the continuity equation (Equation 4.1).

To transform the momentum equation (Equation 4.20) in terms of our new dimensionless stream function $f(\eta)$ and its derivatives with respect to the dimensionless distance η along the y direction, we evaluate various quantities as follows:

$\partial\eta / \partial x$ and $\partial\eta / \partial y$:

$$\frac{\partial\eta}{\partial x} = \frac{\partial}{\partial x}\left(y\sqrt{\frac{u_\infty}{\nu x}}\right) = -\frac{yx^{-3/2}}{2}\sqrt{\frac{u_\infty}{\nu}} \tag{4.24}$$

$$\frac{\partial\eta}{\partial y} = \frac{\partial}{\partial y}\left(y\sqrt{\frac{u_\infty}{\nu x}}\right) = \sqrt{\frac{u_\infty}{\nu x}} \tag{4.25}$$

Velocity in the x direction (u):

$$u = \frac{\partial\Psi}{\partial y} = \frac{\partial}{\partial y}\left(\sqrt{\nu x u_\infty}\, f(\eta)\right) = \sqrt{\nu x u_\infty}\,\frac{\partial f(\eta)}{\partial y} = \sqrt{\nu x u_\infty}\left(\frac{\partial f(\eta)}{\partial\eta}\right)\frac{\partial\eta}{\partial y}$$

Substituting $\partial\eta / \partial y$ from Equation 4.24 yields:

$$u = \frac{\partial\Psi}{\partial y} = \sqrt{\nu x u_\infty}\left(\frac{\partial f(\eta)}{\partial\eta}\right)\frac{\partial\eta}{\partial y}\sqrt{\nu x u_\infty} = u_\infty f' \tag{4.26}$$

which we use to define a dimensionless velocity in the x direction as:

$$U = \frac{u}{u_\infty} = f' \tag{4.27}$$

Velocity in the y direction (v):

$$v = -\frac{\partial\Psi}{\partial x} = -\frac{\partial}{\partial x}\left(\sqrt{\nu x u_\infty}\, f(\eta)\right) = -\sqrt{\nu x u_\infty}\,\frac{\partial f(\eta)}{\partial x} - f(\eta)\frac{\partial}{\partial x}\left(\sqrt{\nu x u_\infty}\right)$$

$$v = -\sqrt{\nu x u_\infty}\left(\frac{\partial f(\eta)}{\partial\eta}\right)\frac{\partial\eta}{\partial x} - \frac{1}{2}f(\eta)\sqrt{\frac{\nu u_\infty}{x}}$$

Substituting $\partial\eta/\partial x$ from Equation 4.25 yields:

$$v = -\sqrt{\nu x u_\infty}\, f'\left(-\frac{y x^{-3/2}}{2}\sqrt{\frac{u_\infty}{\nu}}\right) - \frac{1}{2} f(\eta)\sqrt{\frac{\nu u_\infty}{x}}$$
$$v = \frac{(\eta f' - f)}{2}\sqrt{\frac{\nu u_\infty}{x}} \tag{4.28}$$

which we use to define a dimensionless velocity in the y direction as:

$$V = \frac{v}{u_\infty}\sqrt{Re_x} = \frac{(\eta f' - f)}{2} \tag{4.29}$$

where $Re_x = u_\infty x/\nu$.

$\partial u/\partial x$:

$$\frac{\partial u}{\partial x} = \frac{\partial}{\partial x}(u_\infty f') = u_\infty \frac{\partial f'}{\partial x} = u_\infty \frac{\partial f'}{\partial \eta}\frac{\partial \eta}{\partial x} = u_\infty \frac{\partial \eta}{\partial x} f''$$
$$\frac{\partial u}{\partial x} = u_\infty\left(-\frac{y x^{-3/2}}{2}\sqrt{\frac{u_\infty}{\nu}}\right) f'' = -\frac{u_\infty \eta}{2x} f'' \tag{4.30}$$

$\partial u/\partial y$:

$$\frac{\partial u}{\partial y} = \frac{\partial}{\partial y}(u_\infty f') = u_\infty \frac{\partial f'}{\partial y} = u_\infty \frac{\partial f'}{\partial \eta}\frac{\partial \eta}{\partial y}$$
$$\frac{\partial u}{\partial y} = u_\infty f''\sqrt{\frac{u_\infty}{\nu x}} \tag{4.31}$$

$\partial^2 u/\partial y^2$:

$$\frac{\partial^2 u}{\partial y^2} = \frac{\partial}{\partial y}\left(u_\infty f''\sqrt{\frac{u_\infty}{\nu x}}\right) = u_\infty\sqrt{\frac{u_\infty}{\nu x}}\frac{\partial f''}{\partial y} = u_\infty\sqrt{\frac{u_\infty}{\nu x}}\frac{\partial f''}{\partial \eta}\frac{\partial \eta}{\partial y} = u_\infty\sqrt{\frac{u_\infty}{\nu x}}\frac{\partial f''}{\partial \eta}\sqrt{\frac{u_\infty}{\nu x}}$$
$$\frac{\partial^2 u}{\partial y^2} = \frac{u_\infty^2}{\nu x} f''' \tag{4.32}$$

Substituting all the terms in the momentum equation (Equation 4.20) from above, we finally obtain:

$$ff'' + 2f''' = 0 \tag{4.33}$$

which is a third-order nonlinear ordinary differential equation, also known as the Blasius equation. From Equations 4.26 and 4.28, the original boundary conditions of $u = v = 0$ at $y = 0$ translate into $f = f' = 0$ at $\eta = 0$ for this equation. Similarly, corresponding to $u = u_\infty$ at $y = \infty$, we obtain $f' = 1$ at $\eta = \infty$ from Equation 4.26. These three boundary conditions are sufficient to determine the complete solution of this equation, which can be expressed as a system of three first-order ordinary differential equations for the dependent variables f, f', and f'' as a function of the independent similarity variable η.

A closed-form analytical solution of Equation 4.33 is not known. As no boundary condition is known for f'', a numerical solution method must be iterative in nature. Accordingly, we update a guess value of f'' at $\eta = 0$ to satisfy the two-point boundary value problem for f'. In practice, the numerical integration of the system of equations need not be up to $\eta = \infty$, but should proceed up to $\eta = \eta_{max}$ such that the solution does not change significantly for $\eta > \eta_{max}$. In this case, $\eta_{max} = 9$ is found to be acceptable. As discussed in the work of Schlichting (1979), several numerical solutions have been published. The solutions presented in Figure 4.6 are from the results of Howarth (1938) reported in Table 7.1 of Schlichting (1979). Carnahan et al. (1969) presented a numerical solution method along with the related computer program. The method uses the fourth-order Runge–Kutta technique of integration for the system of three ordinary differential equations over successive steps, each of $\Delta\eta$. The results obtained from this method are in excellent agreement with those reported by Schlichting (1979).

4.5.1.1 Wall Shear Stress

As shown in Figure 4.6, an important result obtained from the numerical solution of Equation 4.33 is that $f'' = 0.332$ at $\eta = 0$. This result along with Equation 4.31 yields:

$$\tau_0(x) = \mu\left(\frac{\partial u}{\partial y}\right)_{y=0} = u_\infty f''(0)\sqrt{\frac{u_\infty}{\nu x}} = 0.332\, u_\infty\sqrt{\frac{u_\infty}{\nu x}}$$

$$C_f(x) = \frac{\tau_0(x)}{\frac{1}{2}\rho u_\infty^2} = \frac{0.664}{\sqrt{Re_x}} \tag{4.34}$$

4.5.1.2 Boundary Layer Thickness

Another important result shown in Figure 4.6 is that $u = 0.992\, u_\infty$ at $\eta = 5$. If we define the boundary layer thickness $\delta(x)$ at each x as the distance along the y direction where u asymptotically becomes within 1% of the free-stream velocity u_∞, Equation 4.22 yields:

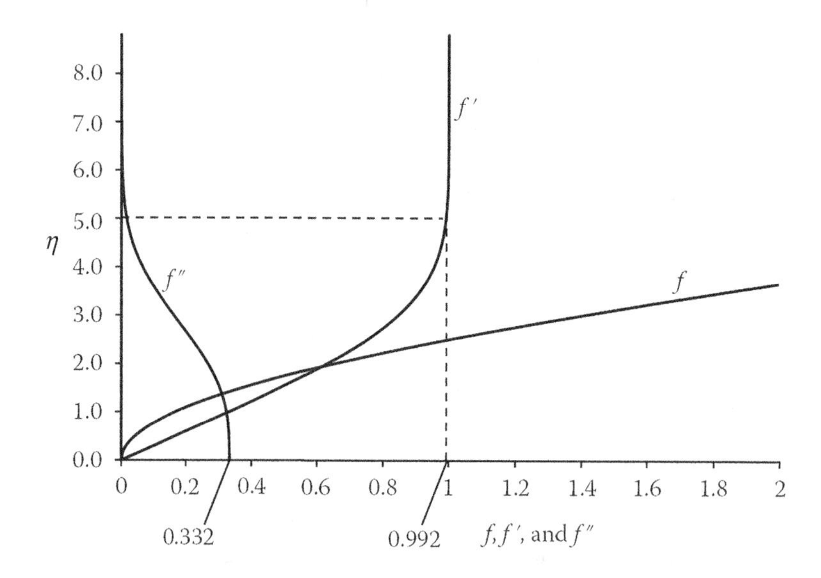

FIGURE 4.6 Solutions of the Blasius boundary layer equation.

$$5 = \delta(x)\sqrt{\frac{u_\infty}{\nu x}}$$

$$\delta(x) = 5\sqrt{\frac{\nu x}{u_\infty}} \tag{4.35}$$

which we accept as an exact solution of the Blasius boundary layer equation (Equation 4.33), for boundary layer thickness variation in the flow direction.

4.5.1.3 Displacement Thickness

From Equation 4.10, we express the displacement thickness as:

$$\delta_1 = \int_0^\delta \left(1 - \frac{u}{u_\infty}\right) \mathrm{d}y$$

In this equation, substituting u / u_∞ from Equation 4.27 and y from Equation 4.22 and replacing the outer limit of integration from δ to ∞, we obtain:

$$\delta_1 = \sqrt{\frac{\nu x}{u_\infty}} \int_0^\infty (1 - f')\,\mathrm{d}\eta$$

$$\delta_1 = \sqrt{\frac{\nu x}{u_\infty}} [\eta - f]_0^\infty = \sqrt{\frac{\nu x}{u_\infty}} [\eta - f]_{\eta\to\infty}$$

From Table 7.1 of Schlichting (1979), we obtain $[\eta - f]_{\eta=8.8} = 1.7208$. Note that for $\eta > 8.8$, the integrand $(1 - f')$ makes a negligible contribution to the integral. Thus, the displacement thickness of the laminar boundary layer on a flat plate varies as:

$$\delta_1 = 1.7208\sqrt{\frac{\nu x}{u_\infty}} \tag{4.36}$$

4.5.1.4 Momentum Thickness

From Equation 4.12, we express the momentum thickness as:

$$\delta_2 = \int_0^\delta \frac{u}{u_\infty}\left(1 - \frac{u}{u_\infty}\right) \mathrm{d}y$$

Again, substituting u / u_∞ from Equation 4.27 and y from Equation 4.22 and replacing the outer limit of integration from δ to ∞, we obtain:

$$\delta_2 = \sqrt{\frac{\nu x}{u_\infty}} \int_0^\infty f'(1 - f')\,\mathrm{d}\eta$$

As reported in the work of Schlichting (1979), the numerical evaluation of the integral in this equation yields a value of 0.664. Thus, the momentum thickness of the laminar boundary layer on a flat plate varies as:

$$\delta_2 = 0.664\sqrt{\frac{\nu x}{u_\infty}} \tag{4.37}$$

An alternate method to obtain the momentum thickness δ_2 is to use the momentum integral equation (Equation 4.15), which for the flow over a flat plate with constant u_∞ reduces to:

$$2\frac{d\delta_2}{dx} = \frac{\tau_0}{\frac{1}{2}\rho u_\infty^2}$$

Using Equation 4.34 to replace the right-hand side of this equation yields:

$$\frac{d\delta_2}{dx} = \frac{0.332}{\sqrt{Re_x}} = \frac{0.332}{\sqrt{\frac{u_\infty x}{\nu}}}$$

$$\delta_2 = 0.664\sqrt{\frac{\nu x}{u_\infty}}$$

which is identical to Equation 4.37. From the above results, we find that $\delta_2 < \delta_1 < \delta$.

4.5.1.5 Distributions of Velocities in the Boundary Layer

Figure 4.7 shows the streamwise and transverse velocities in their dimensionless forms obtained from the accurate numerical solution of the Blasius equation. The plot of U in this figure is identical to that of the first derivative of the dimensionless stream function f' shown in Figure 4.6. The plot of V in this figure shows that both its value and its derivative with η are positive. As $\partial v / \partial y = -\partial u / \partial x$ from the continuity equation, we conclude that everywhere within the boundary layer the u velocity locally decreases in the x direction. Also, using the continuity equation, we can extract the variation of U in the x direction from the variation of V with η shown in the figure.

Figure 4.7 further shows that $V = 0.837$ at $\eta = 5$, which we have selected to define the boundary layer thickness $\delta(x)$. However, the asymptotic value of V shown in the figure equals 0.8604. In terms of this asymptotic value at each x location, we can write from Equation 4.29:

$$V_{max} = \frac{v_{max}}{u_\infty}\sqrt{Re_x} = 0.8604$$

$$v_{max} = \frac{0.8604\, u_\infty}{\sqrt{Re_x}} \tag{4.38}$$

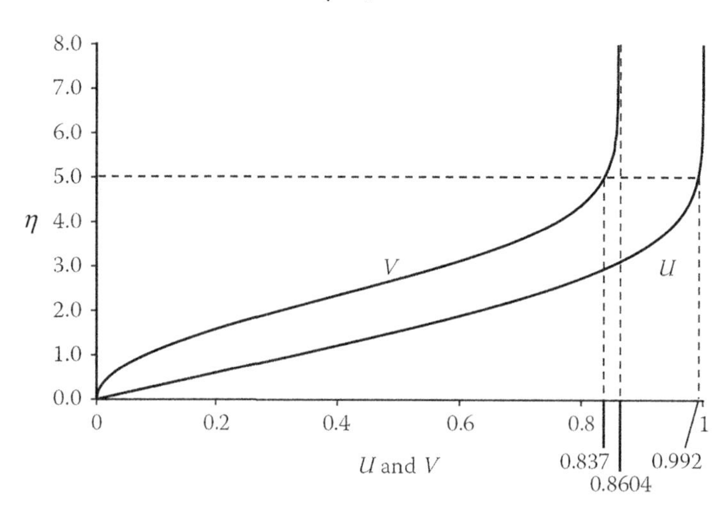

FIGURE 4.7 Variations of U and V in the laminar boundary layer on a flat plate.

At a far enough distance in the y direction, we may consider the flow to exit the boundary layer region with velocities $u = u_\infty$ and $v = v_{max}$, which, according to Equation 4.38, is much smaller than u_∞ for $Re_x >> 1$, a necessary condition for the existence of a boundary layer.

Let us consider a rectangular control volume bounded by $x = 0$ and $x = L$ along the flat plate and large enough in the transverse direction, where $u = u_\infty$ and $v = v_{max}$. For this two-dimensional control volume, assume unit length in the perpendicular direction. At $x = L$, the net deficit in volumetric flow rate for a constant-density flow equals $\delta_1(L)u_\infty$. To satisfy mass conservation for the control volume, this deficit must be equal to the volumetric flow rate exiting from the top, giving:

$$\int_0^L v_{max}\,dx = \delta_1(L)u_\infty$$

Using Equation 4.36 in this equation yields:

$$\int_0^L v_{max}\,dx = 1.7208\sqrt{\nu L u_\infty} \tag{4.39}$$

Using Equation 4.38 on the left-hand side of this equation, we obtain:

$$\int_0^L v_{max}\,dx = \int_0^L \frac{0.8604\,u_\infty}{\sqrt{Re_x}}\,dx$$

$$= 0.8604\sqrt{\nu u_\infty}\int_0^L \frac{dx}{\sqrt{x}}$$

$$= 1.7208\sqrt{\nu L u_\infty}$$

which verifies this equation, satisfying overall mass flow conservation for the boundary layer flow on a flat plate.

4.5.2 Approximate Solution Using the Momentum Integral Method

Engineers often strive for a quick method or an equation to obtain a reasonably accurate solution. In boundary layer analysis, one such equation is the momentum integral equation (Equation 4.15), which yields surprisingly good results even with a simple velocity profile, satisfying the no-slip boundary condition at the wall and free-stream velocity with zero shear stress at the edge of the boundary layer. Example 4.1 presents the momentum integral equation application methodology for a laminar boundary layer over a flat plate and compares the main results with those obtained from the exact solutions of the Blasius boundary layer equation, as discussed in Section 4.5.1.

Table 4.1 summarizes the results for various velocity profiles, including the Blasius exact solution.

4.6 LAMINAR BOUNDARY LAYER IN WEDGE FLOWS

The laminar boundary layer over a flat plate features a uniform free-stream velocity u_∞ in the potential flow, which results in zero pressure gradient throughout the flow. In this boundary layer, the velocity profiles in the flow direction are self-similar, leading to the Blasius equation (Equation 4.33), an ordinary differential equation in the similarity variables f and η. Thus, for a self-similar boundary layer on a flat plate, we could transform the momentum equation, a second-order partial differential, into an ordinary differential equation that is easier to solve numerically. Falkner and Skan (1931) found another class of flows, where the free-stream velocity obeys a power law ($u_\infty = Cx^m$), for whose boundary layer we can obtain the similarity solution, shown next along with the derivation of the ordinary differential equation called the Falkner–Skan equation. By solving this equation numerically, we can obtain nearly exact solutions of various boundary layer quantities in a wedge flow.

TABLE 4.1
Results of Laminar Boundary Layer over a Flat Plate from the Momentum Integral Methods along with the Exact Solution

Velocity Profile $U = u/u_\infty$; $\eta = y/\delta(x)$	$\delta_1\sqrt{\frac{u_\infty}{\nu x}}$	$\delta_2\sqrt{\frac{u_\infty}{\nu x}}$	$\frac{\tau_0}{\rho u_\infty^2}\sqrt{\frac{u_\infty x}{\nu}}$	$\bar{C}_f\sqrt{\frac{u_\infty L}{\nu}}$	$H = \delta_1/\delta_2$
$U = \eta$	1.732	0578	0.289	1.155	3.00
$U = \frac{3}{2}\eta - \frac{1}{2}\eta^3$	1.740	0.646	0.323	10.29	2.70
$U = 2\eta - 2\eta^3 + \eta^4$	1.752	0.686	0.343	1.372	2.55
$U = \sin\left(\frac{\pi\eta}{2}\right)$	1.741	0.654	0.327	1.310	2.66
Exact solution	1.721	0.664	0.332	1.328	259

4.6.1 Free-Stream Velocity in a Wedge Flow

Equation 2.33 gives the complex potential for a corner flow as:

$$F(z) = Az^n$$

which yields the following equations in cylindrical polar coordinates:

$$F(z) = \Phi + i\Psi = Ar^n e^{in\theta}$$

$$\Phi + i\Psi = Ar^n \cos(n\theta) + iAr^n \sin(n\theta)$$

which yields the stream function for a corner flow as:

$$\Psi = Ar^n \sin(n\theta)$$

giving $\Psi = 0$ for $\theta = 0$ and π / n. These two streamlines form a corner of included angle π / n, shown in Figure 4.8 as a concave corner (AOB). If we reflect this corner flow about AO, the convex corner angle becomes $2\pi / n$, and the streamline along AO stagnates at point O. Such a potential flow may be thought of as a flow over a wedge of included angle $(2\pi - 2\pi / n)$. From the geometry shown in the figure, we can write:

$$\beta\pi = \left(2\pi - \frac{2\pi}{n}\right)$$
$$\beta = \frac{2(n-1)}{n} \tag{4.40}$$

From the complex potential of the corner flow, we obtain the complex velocity as:

$$\bar{W}(z) = u - iv = \frac{dF(z)}{dz} = nAz^{n-1} = nA(x + iy)^{(n-1)}$$

For the flow along OB with $y = 0$ in Figure 4.8, this equation yields:

$$u_\infty = nAx^{(n-1)} = Cx^m$$

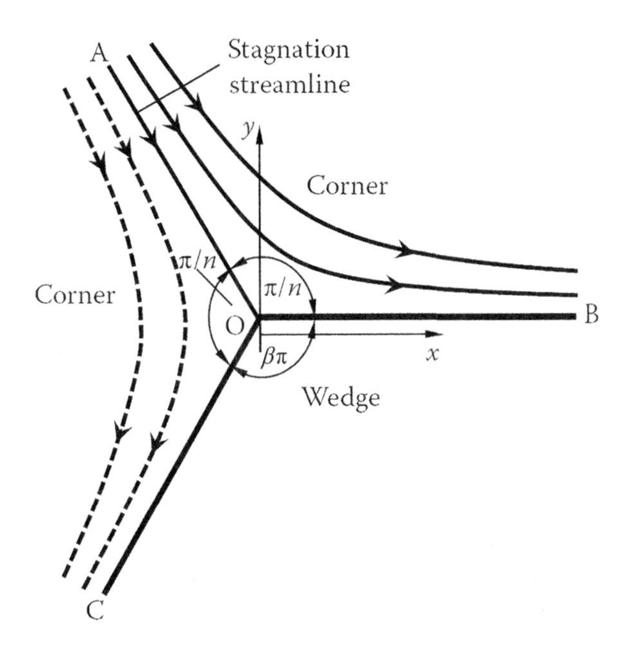

FIGURE 4.8 A wedge flow derived from a corner potential flow.

where C is a constant and $m = n - 1$. We can now write Equation 4.40 as:

$$\beta = \frac{2m}{m+1} \tag{4.41}$$

and

$$m = \frac{\beta}{2-\beta} \tag{4.42}$$

4.6.2 The Falkner–Skan Equation

Outside the boundary layer over a wedge, we can obtain the pressure gradient in the potential flow region governed by $u_\infty = Cx^m$ using the equation:

$$-\frac{1}{\rho}\frac{\mathrm{d}p}{\mathrm{d}x} = u_\infty \frac{\mathrm{d}u_\infty}{\mathrm{d}x} = Cx^m mCx^{m-1} = \frac{mu_\infty^2}{x}$$

whose substitution in the boundary layer equation (Equation 4.6) yields a steady laminar flow:

$$u\frac{\partial u}{\partial x} + v\frac{\partial u}{\partial y} = \frac{mu_\infty^2}{x} + \nu\frac{\partial^2 u}{\partial y^2} \tag{4.43}$$

Using the following dimensionless similarity variables η and $f(\eta)$ in Section 4.5.1, we derived the Blasius equation (Equation 4.33) for the boundary layer flow over a flat plate:

$$\eta = \frac{y}{\delta(x)} = y\sqrt{\frac{u_\infty(x)}{\nu x}} \text{ and } \Psi = \sqrt{\nu x u_\infty(x)}\, f(\eta) = u_\infty(x)\,\delta(x)\, f(\eta)$$

Using similar steps, we can transform Equation 4.43 into the following ordinary differential equation, known as the Falkner–Skan equation:

$$f''' + \frac{m+1}{2} ff'' + m(1 - f'^2) = 0 \tag{4.44}$$

with the following boundary conditions:

$$f(0) = 0, f'(0) = 0, f(\infty) = 1$$

The fact that we have succeeded in transforming the second-order partial differential equation (Equation 4.43) into an ordinary differential equation (Equation 4.44) indicates that, under the chosen similarity variables, wedge flows feature similar velocity profiles for the boundary layer. Note that for $m = 0$, Equation 4.44 becomes identical to the Blasius equation (Equation 4.33) for a flat plate. Thus, the Blasius boundary layer equation is a special case of the Falkner–Skan equation. Further note that, only for the flat plate case, the laminar velocity profiles in the boundary layer are self-similar, and we can collapse all velocity profiles in the flow direction into a single profile under the scaling of the similarity variables.

For the given boundary conditions, no closed-form solution of Equation 4.44 exists. As discussed in the work of Schlichting (1979), several nearly exact numerical solutions exist in the literature. More recently, for the Falkner–Skan equation, Warsi (2005) reported the results of various boundary layer parameters obtained from the numerical solution using a fourth-order Runge–Kutta scheme along with a shooting method. These results as a function of β adapted for Equation 4.44 and the associated definitions of the similarity variables are summarized in Table 4.2 and plotted in Figure 4.9. For $\beta = 0$, the results correspond to the laminar boundary layer flow over a flat plate with a uniform free-stream velocity u_∞. We obtain the results for a two-dimensional stagnation flow for $\beta = 1$. A limiting case of the wedge flow boundary layer occurs for $\beta = -0.1988$, which corresponds to $m = -0.0904$, when the separation occurs due to adverse pressure gradient. The wall shear stress becomes zero at the point of boundary layer separation. As shown in the figure, as *we* approaches $\beta = -0.1988$, both the displacement thickness δ_1 and the shape factor $H = \delta_1 / \delta_2$ increase rapidly.

For a flat plate boundary layer flow, Example 4.1 presents an approximate method of solution using the momentum integral equation (Equation 4.15). For various simple velocity profiles summarized in Table 4.1, the method yields reasonably accurate results compared to the nearly exact numerical solution of the Blasius equation. For laminar boundary layers in wedge flows, a similar approach to obtain an approximate solution of Equation 4.15 using the Karman–Pohlhausen fourth-degree polynomial is widely reported, for example, in the works of Schlichting (1979) and Currie (2013), to which interested readers may refer. We do not discuss this method here in favor of obtaining nearly exact numerical solutions using modern computing power.

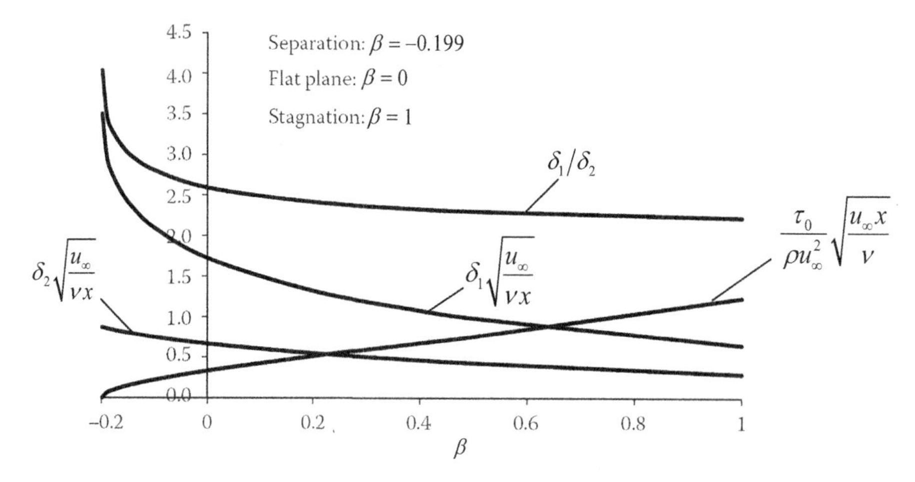

FIGURE 4.9 Results of exact solutions of the Falkner–Skan equation.

TABLE 4.2
Results of Exact Solution of the Falkner–Scan Boundary Layer Equation (Equation 4.40) ($u_\infty = Cx^m$)

β	m	$\delta_1\sqrt{\frac{u_\infty}{\nu x}}$	$\delta_2\sqrt{\frac{u_\infty}{\nu x}}$	$\frac{\tau_0}{\rho u_\infty^2}\sqrt{\frac{u_\infty x}{\nu}}$	$H=\delta_1/\delta_2$
–0.1988	–0.0904	3.498	0.867	0.000	4.032
–0.1900	–0.0868	2.970	0.854	0.058	3.478
–0.1800	–0.0826	2.762	0.839	0.087	3.294
–0.1600	–0.0741	2.510	0.811	0.130	3.094
–0.1400	–0.0654	2.336	0.788	0.164	2.963
–0.1000	–0.0476	2.093	0.746	0.220	2.804
0.0000	0.0000	1.721	0.665	0.332	2.589
0.2000	0.1111	1.320	0.547	0.512	2.412
0.4000	0.2500	1.079	0.464	0.675	2.324
0.5000	0.3333	0.985	0.429	0.757	2.297
1.0000	1.0000	0.648	0.292	1.233	2.219

4.7 BOUNDARY LAYER SEPARATION

Boundary layers feature the no-slip condition at the wall and a region significantly influenced by fluid viscosity. Outside a boundary layer, for all practical purposes, the flow may be treated as a potential flow with negligible vorticity. As part of the boundary layer assumptions, the transverse gradient of static pressure in the boundary layer is negligible, and the outer potential flow imposes its gradient in the flow direction.

There are two primary mechanisms for the separation of a wall boundary layer. First, a wall boundary layer separates under a change in wall geometry, as, for example, happens in a flow over a backward-facing step or a forward-facing step. Second, a wall boundary separates when subjected to an adverse pressure gradient in the flow direction. In this section, we will limit our discussion to the second mechanism of boundary layer separation.

The transverse gradient of streamwise velocity in a wall boundary layer results in skin friction, which contributes to the drag force on the body. As the wall shear stress is the product of viscosity and a transverse velocity gradient along the wall, the boundary layer is the main culprit responsible for the drag force, which we strive to reduce in an engineering design. For example, reducing skin friction on an airplane, automobile, or high-speed train will reduce fuel consumption. Similarly, reducing wall friction in long gas pipelines will significantly reduce the cost of transportation from the source point to the final distribution points. Thinking purely from the viewpoint of skin friction being the root cause of the drag force, we might consider the boundary layer separation a desirable flow feature. However, that is not the case. When a boundary layer separates from the wall, the static pressure drops in the separated region. As a result, the net pressure force acting on the body increases in the flow direction. This contribution to the drag force is called the form drag, which raises its ugly head whenever a boundary layer separates from the wall, thereby increasing the overall drag on the body. At a large angle of attack, the boundary layers on the lifting surfaces of an airplane separate. The airplane loses much of its lift force, creating a stall condition. Note that boundary layer separation occurs for both laminar and turbulent boundary layers, although it is harder to separate the latter than the former. One strategy to delay separation and reduce the form drag on a body is to trip the laminar boundary layer into a turbulent one. Another strategy is to use suction to continuously remove the retarded fluid near the wall.

Pressure gradients create the primary force field within a flow. Further, for a potential flow, Equation 4.7 relates the velocity gradient and the static pressure gradient as:

$$\frac{1}{\rho}\frac{\mathrm{d}p}{\mathrm{d}x} = -u_\infty \frac{\mathrm{d}u_\infty}{\mathrm{d}x}$$

which shows that a favorable pressure gradient ($\mathrm{d}p / \mathrm{d}x < 0$) results in an accelerating outer flow ($\mathrm{d}u_\infty / \mathrm{d}x > 0$), and the adverse pressure gradient ($\mathrm{d}p / \mathrm{d}x > 0$) leads to a decelerating outer flow ($\mathrm{d}u_\infty / \mathrm{d}x < 0$). A boundary layer over a flat plate features a uniform free-stream velocity with zero pressure gradients everywhere. Physically, in a boundary layer close to the wall, a favorable pressure gradient keeps pushing the fluid particles that are next to the wall. As a result, these particles maintain a positive velocity in the flow direction with no possibility of boundary layer separation. Conversely, an adverse pressure gradient creates an opposing force for the fluid particles next to the wall, slowing them down to eventually reach zero streamwise velocity. As a result, the boundary layer will no longer remain attached to the wall. Thus, the shear stress at the point of separation becomes zero, making it the formal criterion for boundary layer separation. Mathematically, for zero wall shear stress, the momentum integral equation (Equation 4.15) yields:

$$\frac{\mathrm{d}}{\mathrm{d}x}(u_\infty^2\delta_2) = -u_\infty\delta_1 \frac{\mathrm{d}u_\infty}{\mathrm{d}x} = \frac{\delta_1}{\rho}\frac{\mathrm{d}p}{\mathrm{d}x}$$

As the left-hand side of this equation is positive, we infer that the boundary layer separation (zero wall shear stress) can only occur when $\mathrm{d}u_\infty / \mathrm{d}x < 0$ or $\mathrm{d}p / \mathrm{d}x > 0$.

The development of a wall boundary layer on a convex surface, for example, the suction surface of an airfoil, under favorable and adverse pressure gradients is depicted in Figure 4.10. Starting with zero boundary layer thickness at the stagnation point located at the leading edge (not shown in the figure), the flow accelerates from Point A to Point B to Point M under a favorable pressure gradient. The streamwise velocity monotonically increases from zero at the wall to $u_\infty(x)$ at the edge of the boundary layer with a continuously decreasing slope. Beyond Point M, for example, at Point N, the flow decelerates under an adverse pressure gradient with increasing boundary layer thickness. The streamwise velocity shows an inflection point (PI) in the flow below which the slope of the velocity profile increases in the transverse direction, and above it, the slope decreases to zero at the edge of the boundary layer. Point S corresponds to the separation point at which we have $\partial u / \partial y = \tau_0 = 0$. Beyond Point S, a reverse flow at the wall occurs, as shown at Point R. The dividing streamline, which emanates from the separation point, encloses the reverse flow region in which the boundary layer assumptions are no longer valid. The curvature of the dividing streamline in the separated flow region implies lower static pressure in this region compared to its value in the outer flow. This reduction in static pressure beyond the point of separation increases the form drag on the body.

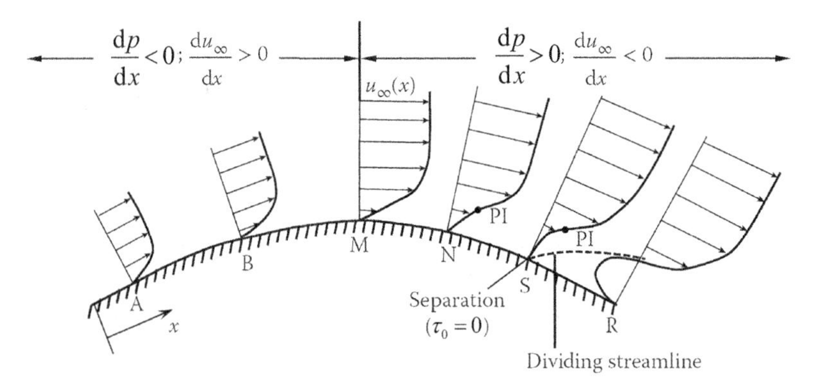

FIGURE 4.10 Boundary layer separation under adverse pressure gradient.

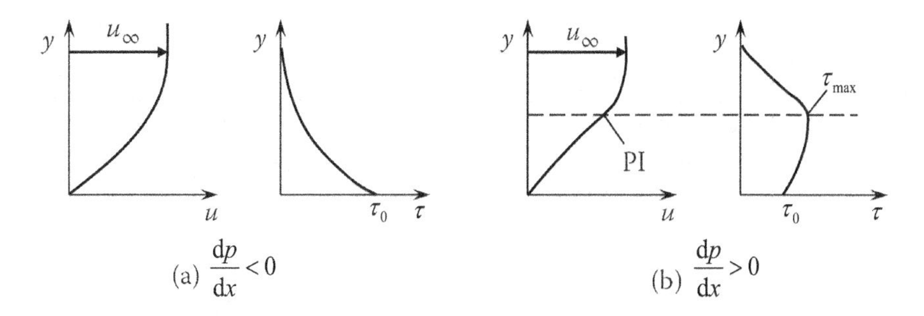

FIGURE 4.11 Shear stress distribution in a boundary layer: (a) favorable pressure gradient and (b) adverse pressure gradient.

On the one hand, for a boundary layer under a favorable pressure gradient, Figure 4.11a shows that the shear stress decreases monotonically from its maximum value at the wall to zero at the edge of the boundary layer. On the other hand, for a boundary layer under an adverse pressure gradient, Figure 4.11b shows that the shear stress initially increases from its value at the wall and then decreases to zero at the edge of the boundary layer. The maximum shear stress thus occurs within the boundary layer at the point of inflection in the velocity profile.

4.8 CONCLUDING REMARKS

Both as a matter of historical importance and their practical applications in design, we have introduced in this chapter some of the must-know concepts of wall boundary layers in two-dimensional laminar incompressible flows with constant viscosity. To fully understand the various methodologies used in the analysis of boundary layers, much of the chapter deals with the laminar boundary layer flow over a flat plate with a uniform velocity in the outer flow and zero pressure gradients everywhere. Using boundary layer assumptions, we have systematically derived the Prandtl boundary layer equation from the Navier–Stokes equations.

For the flat plate, we used the similarity variables to convert the Prandtl boundary layer equation, a second-order partial differential equation, and the continuity equation into a third-order ordinary differential equation called the Blasius equation. In this case, streamwise velocity profiles in the boundary layer developing along the flow direction are self-similar. These velocity profiles collapse into a single profile when expressed in terms of the similarity variables. Unlike a finite-difference method or a finite-element method needed for the numerical solution of the Prandtl boundary layer equation, we can solve the Blasius equation numerically using, for example, a fourth-order Runge–Kutta scheme combined with a shooting method. The von Karman momentum equation, a first-order ordinary differential equation involving integral properties of the boundary layer, such as displacement and momentum thickness, is derived using the control volume analysis, rendering it equally applicable to both laminar and turbulent flow boundary layers. As demonstrated for a flat plate, even using a crude velocity profile, one can obtain remarkably accurate solutions of the boundary layer quantities compared to their exact solutions obtained from the Blasius equation.

Wedge flows with streamwise velocity represented by a power law also admit similarity solutions under the similarity variables used for the flat plate. The corresponding ordinary differential equation is called the Falkner–Skan equation. Unlike the flat case, however, the velocity profiles in wedge flow boundary layers are not self-similar.

Boundary layer separation occurs under an adverse pressure gradient, and the wall shear stress vanishes at the point of separation. Beyond the point of separation, the flow changes its character from being parabolic to becoming elliptic. As a result, it is not easy to find the point of separation analytically using the boundary layer equations. One must, therefore, resort to the numerical solution of the full Navier–Stokes equations to accurately predict flow in the region of boundary layer separation. With state-of-the-art CFD technology and computing resources, we can routinely perform the numerical computations of such flows.

For comprehensive analyses of all kinds of boundary layers, including turbulent boundary layers, interested readers may want to study the classical book by Schlichting (1979).

To develop their skills in solving various fluid flow problems, readers may want to review numerous problems with solutions presented in Sultanian (2021, 2022).

WORKED EXAMPLES

Example 4.1

For a two-dimensional incompressible laminar flow on a flat plate, consider the following self-similar dimensionless profile in the boundary layer for the streamwise velocity:

$$U = C_1 + C_2\eta + C_3\eta^3$$

where $U = u/u_\infty$ and $\eta = y/\delta(x)$. The velocity profile must satisfy the following three boundary conditions: (1) $\eta = 0 : U = 0$, (2) $\eta = 1 : U = 1$, and (3) $\eta = 1 : \mathrm{d}U/\mathrm{d}\eta = 0$. Find the (a) boundary layer thickness $\delta(x)$, (b) displacement thickness $\delta_1(x)$, (c) momentum thickness $\delta_2(x)$, and (d) total drag force acting on a flat plate of length L and width b.

Solution

Let us first use the given boundary conditions to evaluate C_1, C_2, and C_3 in the given cubic polynomial for the velocity profile. The first boundary condition yields $C_1 = 0$. From the remaining two boundary conditions, we obtain $C_2 + C_3 = 1$ and $C_2 + 3C_3 = 0$, which together yield $C_2 = 3/2$ and $C_3 = -1/2$. The cubic polynomial for the velocity profile becomes:

$$U = \frac{3}{2}\eta - \frac{1}{2}\eta^3$$

a. **Boundary Layer Thickness $\delta(x)$**

The momentum integral equation (Equation 4.15) for the flat plate boundary layer flow with constant u_∞ reduces to:

$$\frac{\mathrm{d}\delta_2}{\mathrm{d}x} = \frac{\tau_0}{\rho u_\infty^2}$$

Wall shear stress τ_0:

$$\tau_0 = \mu\left(\frac{\partial u}{\partial y}\right)_{y=0} = \frac{\mu u_\infty}{\delta}\left(\frac{\partial U}{\partial \eta}\right)_{\eta=0} = \frac{3}{2}\frac{\mu u_\infty}{\delta}$$

Momentum thickness δ_2:

$$\frac{\delta_2}{\delta} = \int_0^1 U(1-U)\,\mathrm{d}\eta$$

$$= \int_0^1 \left(\frac{3}{2}\eta - \frac{1}{2}\eta^3\right)\left(1 - \frac{3}{2}\eta + \frac{1}{2}\eta^3\right)\mathrm{d}\eta$$

$$= \int_0^1 \left(\frac{3}{2}\eta - \frac{9}{4}\eta^2 - \frac{1}{2}\eta^3 + \frac{3}{2}\eta^4 - \frac{1}{4}\eta^6\right)\mathrm{d}\eta$$

$$= \left(\frac{3}{4} - \frac{9}{12} - \frac{1}{8} + \frac{3}{10} - \frac{1}{28}\right)$$

$$= \frac{39}{280}$$

Substituting τ_0 and δ_2 in the foregoing momentum integral equation yields:

$$\frac{39}{280}\frac{\mathrm{d}\delta}{\mathrm{d}x} = \frac{3\mu}{2\delta\rho u_\infty}$$

$$\frac{\delta\,\mathrm{d}\delta}{\mathrm{d}x} = \frac{140\ \mu}{13\,\rho u_\infty}$$

Separating the variables in this equation and integrating yields:

$$\delta = \sqrt{\frac{280\ \mu x}{13\rho u_\infty}} + C$$

which with $\delta = 0$ at $x = 0$ yields $C = 0$, giving:

$$\delta = \sqrt{\frac{280\ \mu x}{13\rho u_\infty}} = 4.641\sqrt{\frac{\nu x}{u_\infty}}$$

which is about 7% lower than that obtained from Equation 4.35.

b. Displacement Thickness $\delta_1(x)$

$$\frac{\delta_1}{\delta} = \int_0^1 (1-U)\,\mathrm{d}\eta == \int_0^1 \left(1 - \frac{3}{2}\eta + \frac{1}{2}\eta^3\right)\mathrm{d}\eta$$

$$\frac{\delta_1}{\delta} = \left(1 - \frac{1}{2} + \frac{1}{8}\right) = \frac{3}{8}$$

Substituting δ from (a) in this equation, we obtain:

$$\delta_1 = 1.740\sqrt{\frac{\nu x}{u_\infty}}$$

Compared to the exact solution given by Equation 4.36, this equation yields about 1% higher value of the displacement thickness, which is remarkably accurate for an approximate solution method.

c. Momentum Thickness $\delta_2(x)$

In (a), we obtained $\delta_2 / \delta = 39/280$, which with the substitution for δ yields:

$$\delta_2 = 0.646\sqrt{\frac{\nu x}{u_\infty}}$$

which is about 3% lower than the exact solution given by Equation 4.37.

d. Total Drag Force on the Flat Plate

Substituting δ in the expression for local wall shear stress τ_0 obtained in (a) yields:

$$\tau_0(x) = 0.323\rho u_\infty^2\sqrt{\frac{\nu}{u_\infty x}}$$

$$F_{\text{drag}} = b\int_0^L \tau_0(x)\,\mathrm{d}x = 0.323 b\rho u_\infty^2\sqrt{\frac{\nu}{u_\infty}}\int_0^L \sqrt{\frac{1}{x}}\,\mathrm{d}x$$

$$F_{\text{drag}} = 0.646(bL)\rho u_\infty^2 \sqrt{\frac{\nu}{u_\infty L}}$$

which is about 3% below the value computed from the exact solution of the wall shear stress from Equation 4.34.

This example demonstrates the power of the momentum integral method to quickly compute the key parameters of a laminar boundary layer on a flat plate using a cubic velocity profile.

Example 4.2

Verify the results given in Table 4.1 for the velocity profile.

$$U = \frac{u}{u_\infty} = \sin\left(\frac{\pi\eta}{2}\right)$$

Solution

Let us first evaluate the boundary layer thickness $\delta(x)$ for the given velocity profile. The momentum integral equation for the flat plate boundary layer flow with constant u_∞ reduces to:

$$\frac{\mathrm{d}\delta_2}{\mathrm{dx}} = \frac{\tau_0}{\rho u_\infty^2}$$

Wall shear stress τ_0:

$$\tau_0 = \mu\left(\frac{\partial u}{\partial y}\right)_{y=0} = \frac{\mu u_\infty}{\delta}\left(\frac{\partial U}{\partial \eta}\right)_{\eta=0} = \frac{\pi}{2}\frac{\mu u_\infty}{\delta}\cos\left(\frac{\pi\eta}{2}\right)_{\eta=0} = \frac{\pi}{2}\frac{\mu u_\infty}{\delta}$$

Momentum thickness δ_2:

$$\begin{aligned}
\frac{\delta_2}{\delta} &= \int_0^1 U(1-U)\mathrm{d}\eta \\
&= \int_0^1 \sin\left(\frac{\pi\eta}{2}\right)\left\{1-\sin\left(\frac{\pi\eta}{2}\right)\right\}\mathrm{d}\eta \\
&= \int_0^1 \left\{\sin\left(\frac{\pi\eta}{2}\right)-\sin^2\left(\frac{\pi\eta}{2}\right)\right\}\mathrm{d}\eta \\
&= \int_0^1 \sin\left(\frac{\pi\eta}{2}\right)d\eta - \int_0^1 \sin^2\left(\frac{\pi\eta}{2}\right)\mathrm{d}\eta \\
&= \frac{2}{\pi}\left[-\cos\left(\frac{\pi\eta}{2}\right)\right]_0^1 - \left[\frac{1}{2}\eta - \frac{1}{2\pi}\sin(\pi\eta)\right]_0^1 \\
&= \frac{2}{\pi} - \frac{1}{2} = 0.137
\end{aligned}$$

Substituting τ_0 and δ_2 in the momentum integral equation for the flat plate yields:

$$\left(\frac{2}{\pi}-\frac{1}{2}\right)\frac{\mathrm{d}\delta}{\mathrm{d}x} = \frac{\pi\mu}{2\delta\rho u_\infty}$$

$$\frac{\mathrm{d}\delta}{\mathrm{d}x} = \left(\frac{\pi^2}{4-\pi}\right)\frac{\nu}{\delta u_\infty}$$

Separating variables in this equation and integrating, we obtain:

$$\delta = \sqrt{\left(\frac{2\pi^2}{4-\pi}\right)\frac{\nu x}{u_\infty} + C}$$

As the boundary layer thickness is zero at $x = 0$, we obtain $C = 0$, giving:

$$\delta = \sqrt{\left(\frac{2\pi^2}{4-\pi}\right)\frac{\nu x}{u_\infty}} = 4.795\sqrt{\frac{\nu x}{u_\infty}}$$

Thus, the momentum thickness becomes:

$$\delta_2 = 0.137 \times 4.795\sqrt{\frac{\nu x}{u_\infty}} = 0.655\sqrt{\frac{\nu x}{u_\infty}}$$

$$\delta_2\sqrt{\frac{u_\infty}{\nu x}} = 0.655$$

Displacement thickness δ_1:

$$\frac{\delta_1}{\delta} = \int_0^1 (1-U)\,\mathrm{d}\eta == \int_0^1 \left\{1 - \sin\left(\frac{\pi\eta}{2}\right)\right\}\mathrm{d}\eta$$

$$\frac{\delta_1}{\delta} = \left[\eta + \frac{2}{\pi}\cos\left(\frac{\pi\eta}{2}\right)\right]_0^1 = 1 - \frac{2}{\pi} = 0.363$$

Substituting δ from the foregoing in this equation, we obtain:

$$\delta_1 = 1.743\sqrt{\frac{\nu x}{u_\infty}}$$

$$\delta_1\sqrt{\frac{u_\infty}{\nu x}} = 1.743$$

Substituting δ in this equation for the local wall shear stress τ_0 obtained in the foregoing yields:

$$\tau_0 = \frac{\pi}{2}\frac{\mu u_\infty}{\delta} = \frac{\pi}{2 \times 4.759}\frac{\mu u_\infty}{\delta}\sqrt{\frac{u_\infty}{\nu x}} = 0.328\rho u_\infty^2\sqrt{\frac{\nu}{u_\infty x}}$$

$$\frac{\tau_0}{\rho u_\infty^2}\sqrt{\frac{u_\infty x}{\nu}} = 0.328$$

For evaluating the average shear coefficient for a plate of length L and width b, we first determine the total frictional drag force on the plate as:

$$F_\mathrm{D} = b\int_0^L \tau_0(x)\,\mathrm{d}x = 0.328b\rho u_\infty^2\sqrt{\frac{\nu}{u_\infty}}\int_0^L \sqrt{\frac{1}{x}}\mathrm{d}x$$

$$F_\mathrm{D} = 0.656(bL)\rho u_\infty^2\sqrt{\frac{\nu}{u_\infty L}}$$

Therefore, we obtain:

$$\bar{C}_f = \frac{2F_D}{(bL)\rho u_\infty^2} = 2*0.656\sqrt{\frac{\nu}{u_\infty L}}$$

$$\bar{C}_f\sqrt{\frac{u_\infty L}{\nu}} = 1.321$$

Finally, we compute the shape factor as:

$$H = \frac{\delta_1}{\delta_2} = \frac{1.743}{0.655} = 2.640$$

$$H = 2.640$$

All the computed values for the given velocity profile are in good agreement with the values given in Table 4.1.

Problems

4.1 Consider a $0.5 \times 1.0\,\text{m}$ thin flat plate placed in an oncoming flow of water with a uniform velocity $u_\infty = 10\,\text{m/s}$. Using the exact solution of the Blasius boundary layer equation, compute the total drag on both the top and bottom surfaces of the plate for both orientations (long and short sides parallel to the flow). For the thin plate, neglect any contribution of the form drag.

4.2 For a laminar boundary layer, assume the following velocity profile:

$$\frac{u}{u_\infty} = a + b\eta - \eta^2$$

where $\eta = y/\delta$. What are the necessary values of the constants a and b? For the resulting velocity distribution, show that $\delta_2/\delta = 2/15$ and evaluate the boundary layer thickness as a function of x. In addition, find the friction drag coefficient (C_f) as a function of the Reynolds number.

4.3 For a laminar boundary layer on a flat plate, using the momentum integral method, compute various quantities given in Table 4.1 for the following approximate velocity profile:

$$U = \frac{u}{u_\infty} = 1 - e^{-\alpha\eta}$$

where α is a constant. Compare your results with the exact solutions given in the table.

4.4 For an incompressible turbulent boundary layer over a flat plate, the velocity profile may be approximated by the one-seventh power law $U = u/u_\infty = \eta^{1/7}$. From empirical data, the wall shear stress correlates to the following equation:

$$\tau_0 = 0.0228\rho u_\infty^2\left[\frac{\nu}{u_\infty \delta(x)}\right]^{1/4}$$

Using the von Karman momentum integral equation, find the expressions for $\delta(x)$, $\delta_1(x)$, $\delta_2(x)$, $C_f(x)$, and $\bar{C}_f(L)$.

4.5 For a flat plate in an incompressible laminar flow with a uniform velocity u_∞, assume a laminar boundary layer on the entire plate. At what fraction of the length from the leading edge would the drag force on the front portion equal half of the total drag force?

4.6 Consider a laminar flow of air in a circular pipe of diameter 1.0 cm. The density (ρ) and kinematic viscosity (ν) of air are constant at 1.0 kg/m^3 and 2.0×10^{-5} m^2/s, respectively. The uniform inlet velocity (V) is 4.0 m/s. As shown in Figure 4.2, at the end of the entrance length, the flow assumes a fully developed parabolic velocity profile. In the entrance region, the local shear coefficient on the pipe wall is assumed to vary as:

$$C_f = \frac{\tau_0}{\frac{1}{2}\rho V^2} = \frac{0.664}{\sqrt{Re_z}}$$

where τ_0 is the local wall shear stress and $Re_z = Vz/\nu$, z being the axial distance from the pipe inlet. Note that the above equation is similar to the exact solution for the laminar boundary layer over a flat plate where we have defined the local Reynolds number in terms of the average pipe flow velocity, which remains constant along the pipe. Determine the entrance length L_e. Comment on the accuracy of your result and suggest a method to improve it.

REFERENCES

Anderson, J.D. 2005. Ludwig Prandtl's Boundary Layer. *Phys. Today*. 58(12): 42–44.

Carnahan, B., H.A. Luther, and J.O. Wilkes. 1969. *Applied Numerical Methods*. New York: John Wiley.

Currie, I.G. 2013. *Fundamental Mechanics of Fluids*, 4th Edition. Boca Raton, FL: CRC Press.

Falkner, V.M. and S.W. Skan. 1931. Some approximate solutions of the boundary layer equations. *Phil. Mag.* 12: 865–896.

Howarth, L. 1934. On the solution of the laminar boundary layer equations. *Proc. Roy. Soc. London*. 164: 547–579.

Schlichting, H. 1979. *Boundary Layer Theory*, 7th Edition. New York: McGraw-Hill.

Sultanian, B.K. 2015. *Fluid Mechanics: An Intermediate Approach*, 1st Edition. Boca Raton, FL: Taylor & Francis.

Sultanian, B.K. 2021. *Fluid Mechanics and Turbomachinery: Problems and Solutions*. Boca Raton: Taylor & Francis.

Sultanian, B.K. 2022. *Thermal-Fluids Engineering: Problems with Solutions*. USA: Independently published by Kindle Direct Publishing (Amazon.com).

Warsi, Z.U.A. 2005. *Fluid Dynamics: Theoretical and Computational Approaches*, 3rd Edition. Boca Raton, FL: CRC Press.

BIBLIOGRAPHY

Babinsky, H. and J.K. Harvey. 2014. *Shock Wave-Boundary Layer Interactions*. Cambridge Aerospace Series #32. Cambridge: Cambridge University Press.

Cebeci, T. and P. Bradshaw. 1977. *Momentum Transfer in Boundary Layers*. Washington, DC: Hemisphere.

Cebeci, T. and J. Cousteix. 1998. *Modeling and Computation of Boundary-Layer Flows*. New York: Springer.

Durst, F. 2022. *Fluid* Mechanics*: An Introduction to the Theory of Fluid Flows*, 2nd Edition. Heidelberg: Springer Verlag GmbH.

Falkovich, G. 2018. *Fluid Mechanics*, 2nd Edition. Cambridge: Cambridge University Press.

Kundu, P.K., I.M. Cohen, D.R. Dowling, J. Capecilatro. 2024. *Fluid Mechanics*, 7th Edition. Burlington: Academic Press.

Panton, R.L. 2024. *Incompressible Flow*, 5th Edition. New York: Wiley.

Powers, J.M. 2023. *Mechanics of Fluids*. Cambridge: Cambridge University Press.

Schetz, J.A. 1984. *Foundations of Boundary Layer for Momentum, Heat, and Mass Transfer*, 1st Edition. New York: Prentice Hall.

Schlichting, H. and K. Gersten. 2017. *Boundary Layer Theory*, 9th Edition. New York: McGraw-Hill.

Sultanian, B.K. 2022. *Fluid Mechanics An Intermediate Approach*: Errata for *the First Edition Published in 2015*. USA: Independently published on KDP (Amazon.com).

Visconti, G. and P. Ruggieri. 2020. *Fluid Dynamics: Fundamental and Applications.* Cham, Switzerland: Spinger.

White, F.M and J. Majdalani. 2022. *Viscous Fluid Flow*, 4th Edition. New York: McGraw-Hill.

NOMENCLATURE

A	Coefficient in the complex potential function for the corner flows
C	Integration constant; constant coefficient in the power-law velocity variation in a wedge flow
C_f	Shear coefficient
$f(\eta)$	Dimensionless stream function used in similarity solution
f'	First derivative of f with η
f''	Second derivative of f with η
f'''	Third derivative of f with η
F	Complex potential function
H	Shape factor ($H = \delta_1 / \delta_2$)
i	Imaginary number ($\mathrm{i} = \sqrt{-1}$)
ΔF_s	Total surface force acting on the boundary layer control volume
ΔF_p	Total pressure force acting on the boundary layer control volume
L	Characteristic length scale
L_e	Entrance length in a pipe flow
m	Exponent in the power-law representation of free-stream velocity in a wedge flow
$\dot{m}$	Mass flow rate
$\dot{M}$	Momentum flow rate
n	Exponent in the power-law representation of the complex potential of corner flows ($F = Az^n$)
p	Static pressure
r	Coordinate r of the cylindrical polar coordinate system
R	Pipe radius
Re	Reynolds number
S	Stagnation point
t	Time
u, v	Velocity components in the x and y directions, respectively
U	Dimensional velocity in the x direction ($U = u / u_\infty$)
$\overline{W}$	Complex conjugate velocity ($\overline{W} = u - \mathrm{i}v$)
v_z	Axial (z direction) velocity
V	Average velocity of a pipe flow; dimensionless velocity in the y direction
x	Coordinate x of the Cartesian coordinate system
y	Coordinate y of the Cartesian coordinate system
z	Complex coordinate ($z = x + iy$)

SUBSCRIPTS AND SUPERSCRIPTS

in	Inflow
out	Outflow
x	Along x coordinate direction
$\left(^{*}\right)$	Nondimensional
$\left(\bar{\ }\right)$	Average value
∞	Infinity (free stream)

GREEK SYMBOLS

β	Wedge angle as a fraction of π
δ	Boundary layer thickness
δ_1	Displacement thickness
δ_2	Momentum thickness
η	Dimensionless similarity variable ($\eta = y / \delta$)
θ	Coordinate θ of the cylindrical polar coordinate system
μ	Dynamic viscosity
ν	Kinematic viscosity ($\nu = \mu / \rho$)
π	Ratio of the circumference of a circle to its diameter
ρ	Density
τ_{xy}	Local shear stress ($\tau_{xy} = \tau_{yx}$)
τ_0	Wall shear stress
Φ	Velocity potential function
Ψ	Stream function

5 Compressible Flow

5.1 INTRODUCTION

This chapter mainly provides an insightful review and reinforcement of the critical concepts of one-dimensional compressible flows with area change, friction, heat transfer, and rotation. The fluid density is no longer a constant, as in an incompressible flow, but changes with pressure and temperature. Mach number plays a crucial role in characterizing compressible flows: the higher the Mach number, the higher the flow compressibility with a two-way exchange between internal and external flow energies. Although the fluid in a compressible flow must always be compressible, like air or other gases, we may treat such a flow as incompressible for Mach numbers less than 0.3. In many compressible flows, the entropy considerations to satisfy the second law of thermodynamics become necessary to screen out physically impossible solutions. In addition to many one-dimensional flows, the chapter also discusses the key features of a couple of two-dimensional flows, such as oblique shock waves, which are nonisentropic, and Prandtl–Meyer expansion waves, which are isentropic.

Most of our day-to-day experience is limited to incompressible or subsonic compressible flows where the flow velocity is less than the fluid's sound speed. When the flow velocity exceeds the speed of sound, the compressible flow exhibits many fascinating phenomena that are, by and large, counterintuitive. Gas turbines, for example, feature nonisentropic compressible flows in their inlet systems, compressors, combustors, turbines, and exhaust systems. These flows are also prevalent in many other energy conversion devices. Normal and oblique shocks and their interactions with wall boundary layers are critical to designing supersonic wind tunnels and aircraft in commercial and military applications. This chapter provides a detailed understanding of the operation of a convergent–divergent nozzle under various back-pressure conditions. Understanding the operation of a supersonic wind tunnel, in which the downstream variable-throat-area diffuser swallows the starting normal shock in the test section, is essential for its optimal design and efficient performance.

This chapter features 19 worked examples and 45 chapter-end problems, and the original version of this chapter appears as Chapter 5 in the first edition of this book by Sultanian (2015).

5.2 BASICS OF COMPRESSIBLE FLOW

5.2.1 Speed of Sound

Figure 5.1a shows a disturbance wave moving at a constant speed C from right to left in a quiescent compressible fluid at the pressure and temperature conditions shown on the left of the wave front.

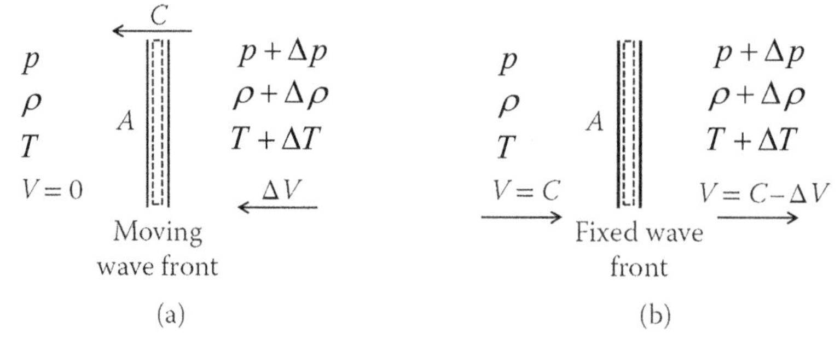

FIGURE 5.1 Speed of sound in a fluid medium. (a) Unsteady flow and (b) steady flow.

DOI: 10.1201/9781003323192-5

The changes in the fluid state as the wave passes through it are shown on the right. In a coordinate system fixed with the quiescent fluid, the problem is unsteady. We can make this problem steady by attaching the coordinate system to the wave front, and consistent with the Eulerian view point, using a control volume analysis across this front. In this coordinate system, shown in Figure 5.1b, the fluid flows across the wave front from left to right at a constant speed C.

The continuity equation across the wave front of area A, shown in Figure 5.1b, yields:

$$\rho AC = (\rho + \Delta\rho)A(C - \Delta V)$$

$$\Delta V = C\left(\frac{\Delta\rho}{\rho + \Delta\rho}\right) \tag{5.1}$$

Neglecting all surface and body forces acting on the control volume associated with the wave front, we can write the linear momentum equation as:

$$pA - (p + \Delta p)A = \rho AC(C - \Delta V - C)$$

$$\Delta p = \rho C \Delta V$$

which with the substitution for ΔV from Equation 5.1 yields:

$$\Delta p = \rho C^2\left(\frac{\Delta\rho}{\rho + \Delta\rho}\right)$$

$$C^2 = \frac{\Delta p}{\Delta\rho}\left(1 + \frac{\Delta\rho}{\rho}\right) \tag{5.2}$$

which shows that the larger the value of $\Delta\rho / \rho$, the faster the speed of sound. Accordingly, a powerful explosion wave will move much faster than the sound wave. In the limit of $\Delta\rho \to 0$, this equation yields the following expression for the speed of sound:

$$C^2 = \frac{\partial p}{\partial \rho} \tag{5.3}$$

Note that the value of C computed from this equation is isotropic, that is, equal in all directions. Let us now consider the speed of sound in air or in any other gaseous medium under the assumptions of isothermal and isentropic processes within the wave front.

5.2.1.1 Isothermal Process

If the process within the wave front is assumed to be isothermal (T = constant), we obtain the following expression for the speed of sound in air whose equation of state is given by $p = \rho RT$:

$$C^2_{\text{isothermal}} = \left(\frac{\partial p}{\partial \rho}\right)_{\text{isothermal}} = RT$$

$$C_{\text{isothermal}} = \sqrt{RT} \tag{5.4}$$

Newton first proposed Equation 5.4 in 1686. This equation underpredicted the measured value of the speed of sound in air. However, this erroneous equation remained in use for 150 years! In 1836, Laplace realized that the error in the equation resulted from the assumption of an isothermal process in the wavefront. He found that the assumption of the isentropic process within the wavefront was more realistic and proposed Equation 5.5, which is routinely in use, derived next.

5.2.1.2 Isentropic Process

For an isentropic process, air properties are related by the equation $p = K\rho^{\gamma}$ where K is a constant and γ is the ratio of specific heats. Using Equation 5.3, we obtain the following expression for the speed of sound:

$$C^2_{\text{isentropic}} = \left(\frac{\partial p}{\partial \rho}\right)_{\text{isentropic}} = K\gamma\rho^{\gamma-1} = \gamma\left(\frac{p}{\rho}\right) = \gamma RT$$
$$C_{\text{isentropic}} = \sqrt{\gamma RT} \tag{5.5}$$

Note that the speed of sound computed by this equation is about 18% higher than the value computed using Equation 5.4.

5.2.1.3 Motion of a Sound-Source

There is no known particle that travels faster than the speed of light, but it can travel faster than the speed of sound in a medium! For example, for a sound-source traveling in air, we have three possibilities: (a) the velocity of the sound-source is less than the speed of sound in air ($V_S < C$; subsonic velocity), (b) the velocity of the sound-source is equal to the speed of sound ($V_S = C$; sonic velocity), and (c) the velocity of the sound-source is greater than the speed of sound ($V_S > C$; supersonic velocity). Figure 5.2 shows these three cases.

For a sound-source traveling at subsonic velocity, shown in Figure 5.2a, the sound wave travels ahead of the sound-source, and the zone of sound spans 360 degrees. An observer standing far ahead of the sound-source can hear this source approaching. The limiting Mach wave divides the entire region into two halves for the sound-source traveling at sonic velocity, shown in Figure 5.2b. Ahead of the sound-source is the zone of silence, and behind it is the zone of sound. An observer standing in the zone of silence, shown left of the Mach wave in Figure 5.2b, will hear the sound-source

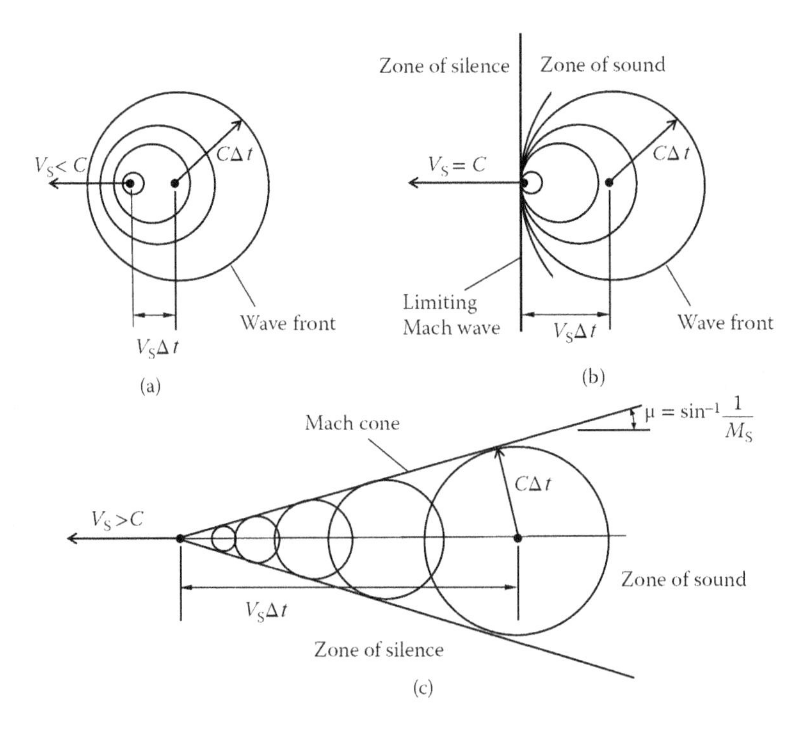

FIGURE 5.2 Motion of a sound-source. (a) Subsonic, (b) sonic, and (c) supersonic.

coming once it has just passed the observer. For the supersonic case, shown in Figure 5.2c, the sound zone travels in the form of a cone, also known as the Mach cone. We compute the half-angle of the Mach cone by the equation:

$$\mu = \sin^{-1}\frac{1}{M_S} \tag{5.6}$$

where $M_S = V_S / C$. An observer ahead of the sound-source will only hear the sound-source once it has passed the observer and the observer is within the Mach cone. Thus, standing on the ground, one cannot hear a supersonic plane approaching them. One realizes its flight only after it has flown past them and their ears are within the Mach cone.

5.2.2 Mach Number

The total or stagnation temperature at a point in a compressible flow of gas of constant c_p is a measure of its total energy. This total energy is the sum of its internal energy measured by its static temperature and its external energy associated with its flow velocity. Thus, we can write:

$$T_0 = T + \frac{V^2}{2c_p} \tag{5.7}$$

Since the equation $C = \sqrt{\gamma RT}$ computes the local speed of sound in a compressible flow, we can interpret the square of the Mach number, which is the ratio of the flow velocity and the local speed of sound, as the ratio of the external kinetic energy of the flow and its internal energy. For the compressible flow of constant total temperature, as the Mach number increases, more and more energy appears in the form of its external flow energy at the expense of its internal energy. When the static temperature of the gas theoretically equals zero, the external flow velocity becomes maximum.

We can write Equation 5.7 as:

$$\frac{C^2}{\gamma RT_0} + \frac{V^2}{2c_pT_0} = \tag{5.8}$$

which, for an adiabatic flow with constant T_0, yields the maximum value of $C = C_{max}$ for $V = 0$ and the maximum value of $V = V_{max}$ for $C = 0$ with $T = 0$. This equation thus becomes:

$$\frac{V^2}{V_{max}^2 \frac{C^2}{C_{max}^2}} \tag{5.9}$$

where $C_{max}\sqrt{\gamma RT_0}$ and $V_{max}\sqrt{2c_pT_0} = \sqrt{2\gamma RT_0/(\gamma-1)} = C_{max}\sqrt{2/(\gamma-1)}$, which is the equation of an ellipse with its major axis given by $2V_{max}$ and its minor axis by $2C_{max}$. For positive values of V and C, we plot this equation in Figure 5.3, which features different flow regimes characterized by Mach numbers as discussed below.

Subsonic Incompressible Flow ($M \leq 0.3$).

For $M \leq 0.3$, Figure 5.3 shows that V is very small compared with C. The change in C is negligible. Therefore, in a subsonic flow, the change in M is primarily due to the change in V. For most practical purposes, the flow in this regime may be assumed incompressible.

Subsonic Compressible Flow ($0.3 < M < 1.0$)

As shown in Figure 5.3, for $0.3 < M < 1.0$, while V is always less than C, they are of comparable magnitude. The changes in M occur mainly due to changes in V.

Transonic Flow ($0.8 < M < 1.2$)

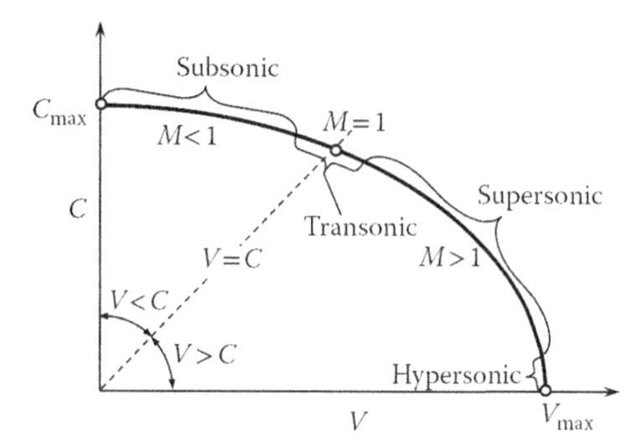

FIGURE 5.3 Variation of Mach number in an adiabatic compressible flow.

In this flow regime, not only V and C but also their changes are of comparable magnitude. The changes in M occur due to changes in both V and C. As the flows with Mach numbers greater than 1.0 are often counterintuitive, the transonic flows are generally more challenging to handle computationally than either subsonic or supersonic flows.

Supersonic Flow $(1 < M \leq 5)$

In this flow regime, V and C are of comparable magnitude, but V is larger than C. The changes in M occur due to large changes in both V and C.

Hypersonic Flow $(M > 5)$

In this flow regime, C is significantly smaller than V. The changes in M occur almost entirely due to changes in C while V has nearly reached its maximum value.

5.2.3 Isentropic Relations

In Chapter 1, we derived the expression to calculate the change in entropy of a fluid between two states. If the final state is the stagnation state (zero velocity) attained via an isentropic process, we can write:

$$\Delta s = 0 = c_p \ln\left(\frac{T_0}{T}\right) - R\ln\left(\frac{p_0}{p}\right)$$

$$\frac{p_0}{p} = \left(\frac{T_0}{T}\right)^{\frac{c_p}{R}} = \left(\frac{T_0}{T}\right)^{\frac{\gamma}{\gamma-1}}$$

Writing Equation 5.7 in terms of Mach number

$$\frac{T_0}{T} = 1 + \frac{\gamma-1}{2}M^2 \tag{5.10}$$

we can express the total-to-static pressure ratio as:

$$\frac{p_0}{p} = \left(\frac{T_0}{T}\right)^{\frac{c_p}{R}} = \left(\frac{T_0}{T}\right)^{\frac{\gamma}{\gamma-1}} = \left(1 + \frac{\gamma-1}{2}M^2\right)^{\frac{\gamma}{\gamma-1}} \tag{5.11}$$

Using the equation of state of a perfect gas at both static and stagnation states, given by $p = \rho RT$ and $p_0 = \rho_0 RT_0$, respectively, we obtain from Equations 5.10 and 5.11:

$$\frac{\rho_0}{\rho} = \left(\frac{T_0}{T}\right)^{\frac{1}{\gamma-1}} = \left(1 + \frac{\gamma-1}{2}M^2\right)^{\frac{1}{\gamma-1}} \tag{5.12}$$

The total temperature and total pressure of an isentropic compressible flow remain constant. When the flow velocity equals the speed of sound, it becomes sonic with $M = 1$. Using Equations 5.10–5.12, we obtain the following relations to compute the static properties (denoted by *) in a sonic flow:

$$\frac{T_0}{T^*} = \frac{\gamma+1}{2} \tag{5.13}$$

$$\frac{p_0}{p^*} = \left(\frac{\gamma+1}{2}\right)^{\frac{\gamma}{\gamma-1}} \tag{5.14}$$

$$\frac{\rho_0}{\rho^*} = \left(\frac{\gamma+1}{2}\right)^{\frac{1}{\gamma-1}} \tag{5.15}$$

The static properties at $M = 1$ are also known as the characteristic or critical properties of an isentropic flow.

5.2.4 Characteristic Mach Number

In Section 5.2.2, we defined the Mach number (M) in a compressible flow as the ratio of the local flow velocity and the speed of sound corresponding to the local static temperature. If the local velocity is divided instead by the speed of sound at the characteristic static temperature, which corresponds to $M = 1$, the resulting Mach number is called the characteristic Mach number M^* expressed as:

$$M^* = \frac{V}{C^*} = \frac{V}{\sqrt{\gamma R T^*}} \tag{5.16}$$

To find the relation between M and M^*, we write:

$$\frac{M^{*2}}{M^2} = \frac{C^2}{C^{*2}} = \frac{\gamma RT}{\gamma RT^*} = \frac{T}{T^*}$$

which for an adiabatic flow with constant total temperature, we can write as:

$$\frac{M^{*2}}{M^2} = \frac{T}{T^*}\frac{T_0}{T_0} = \frac{T_0}{T^*}\frac{T}{T_0} = \frac{\frac{\gamma+1}{2}}{1+\frac{\gamma-1}{2}M^2} = \frac{\gamma+1}{2+(\gamma-1)M^2}$$

$$M^* = \sqrt{\frac{(\gamma+1)M^2}{2+(\gamma-1)M^2}} \tag{5.17}$$

For $M \to \infty$, this equation yields $M^* = \sqrt{(\gamma+1)/(\gamma-1)}$, which for $\gamma = 1.4$ equals 2.449. Plotting this equation, see Figure 5.4, reveals that $M^* < 1$ for $M < 1$, $M^* = 1$ for $M = 1$, and $M^* > 1$ for $M > 1$.

For a normal shock in a compressible flow, Prandtl derived a simple relation between the characteristic Mach numbers before and after the shock, given by:

$$M_2^* = \frac{1}{M_1^*} \tag{5.18}$$

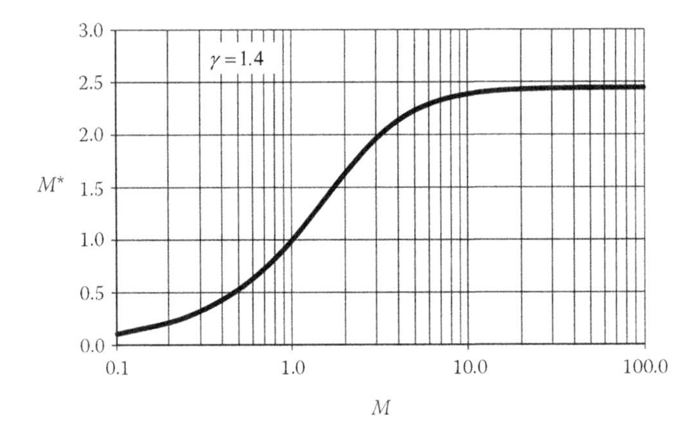

FIGURE 5.4 Relation between Mach number and characteristic Mach number.

As the upstream Mach number for a normal shock is always supersonic with $M_1^* > 1$, this equation shows that the characteristic Mach number M_2^* after the normal shock must always be subsonic and so will be the Mach number M_2. We will revisit these ideas in Section 5.11 on normal shocks later in this chapter.

5.3 COMPRESSIBLE FLOW FUNCTIONS

This section presents the concepts and applications of mass flow and impulse functions for one-dimensional compressible flows in a duct. For a given gas, these flow functions are Mach number functions at any duct section. The mass flow functions provide a convenient means of computing the mass flow rate at a duct section without explicitly using the gas density. They also offer a more intuitive understanding of the steady compressible flow that must satisfy the continuity equation within the duct. The impulse functions, on the other hand, are grounded in the linear momentum equation. They provide an easy way to compute stream thrust at a duct section. As discussed in Chapter 2, the change in stream thrust between any two sections of a duct equals the net surface force (in the absence of a body force and additional inflow or outflow) from the duct wall acting on the fluid control volume between these sections.

5.3.1 Mass Flow Functions

In a one-dimensional duct flow, we can compute the ideal mass flow rate at any section by the equation:

$$\dot{m} = \rho V A \tag{5.19}$$

where ρ is the fluid density computed at its static temperature and pressure, and V is the uniform flow velocity normal to the duct cross section A. In most design applications involving internal compressible flows, we often use total temperature, total pressure or static pressure, and Mach number to characterize the flow at any section. Expressing Equation 5.19 in terms of these quantities provides an improved intuitive understanding of how each parameter influences the mass flow rate at a duct cross section.

Let us rewrite Equation 5.19 as:

$$\frac{\dot{m}}{A} = \left(\frac{\rho}{\rho_0}\right)\rho_0 M C$$

where ρ_0 is the gas density at the stagnation (total) pressure and temperature. Using Equation 5.12 to substitute for the density ratio and the equation of state for the perfect gas to substitute for ρ_0, we obtain:

$$\frac{\dot{m}}{A} = \left(\frac{T}{T_0}\right)^{\frac{1}{\gamma-1}} \left(\frac{p_0}{RT_0}\right) M\sqrt{\gamma RT}$$

$$\frac{\dot{m}}{A} = \frac{p_0 M}{\sqrt{T_0}} \left(\sqrt{\frac{\gamma}{R}}\right) \left(\frac{T}{T_0}\right)^{\frac{1}{\gamma-1}} \sqrt{\frac{T}{T_0}}$$

$$\frac{\dot{m}}{A} = \frac{p_0 M}{\sqrt{T_0}} \left(\sqrt{\frac{\gamma}{R}}\right) \sqrt{\left(\frac{T}{T_0}\right)^{\frac{\gamma+1}{\gamma-1}}}$$

In this equation, substituting Equation 5.10 for the temperature ratio, we finally obtain after consolidating terms:

$$\dot{m} = \frac{p_0 AM}{\sqrt{T_0}} \sqrt{\frac{\gamma}{R\left(1+\frac{\gamma-1}{2}M^2\right)^{\frac{\gamma+1}{\gamma-1}}}} \tag{5.20}$$

Thus, through a duct section of area A, Equation 5.20 expresses the mass flow rate in terms of total pressure, total temperature, and Mach number, which are assumed uniform over the section, regardless of their values at any other section. This equation also reveals that sectional mass flow rate is directly proportional to the total pressure and inversely proportional to the square root of the total temperature at the section. The remaining terms in this equation depend on Mach number and gas properties R and γ.

5.3.1.1 Total Pressure Mass Flow Function

We rewrite Equation 5.20 as:

$$\dot{m} = \frac{F_{f0} A p_0}{\sqrt{T_0}} = \frac{\hat{F}_{f0} A p_0}{\sqrt{RT_0}} \tag{5.21}$$

where the total pressure mass flow function F_{f0} and its dimensionless counterpart $\hat{F}_{f0}$ are given as follows:

$$F_{f0} = M\sqrt{\frac{\gamma}{R\left(1+\frac{\gamma-1}{2}M^2\right)^{\frac{\gamma+1}{\gamma-1}}}} \tag{5.22}$$

$$\hat{F}_{f0} = M\sqrt{\frac{\gamma}{\left(1+\frac{\gamma-1}{2}M^2\right)^{\frac{\gamma+1}{\gamma-1}}}} \tag{5.23}$$

Equation 5.22 shows that the units of F_{f0} are those of $1/\sqrt{R}$. In combination with Equation 5.23, we can write $F_{f0} = \hat{F}_{f0}/\sqrt{R}$. The term "total pressure" in these mass flow functions reminds us that they are used together with the total pressure to calculate the mass flow rate at a section.

If the flow is choked ($M = 1$) at a section, we obtain the following expressions for F_{f0}^* and $\hat{F}_{f0}^*$:

$$F_{f0}^* = \sqrt{\frac{\gamma}{R\left(\frac{\gamma+1}{2}\right)^{\frac{\gamma+1}{\gamma-1}}}} \tag{5.24}$$

and

$$\hat{F}_{f0}^* = \sqrt{\frac{\gamma}{\left(\frac{\gamma+1}{2}\right)^{\frac{\gamma+1}{\gamma-1}}}} \tag{5.25}$$

For air with $\gamma = 1.4$ and $R = 287\,\text{J}/(\text{kg K})$, they yield $F_{f0}^* = 0.0404$ in units $\sqrt{(\text{kg K})/\text{J}}$ and the dimensionless $\hat{F}_{f0}^* = 0.6847$. Equation 5.21 shows that through a sonic section, for a given total pressure and temperature, the mass flow rate depends only on the section area.

For air with $\gamma = 1.4$, Figure 5.5 depicts the plot of Equation 5.23, which shows that, for subsonic flows with $M < 1$, $\hat{F}_{f0}$ increases monotonically with increasing Mach number. For supersonic flows with $M > 1$, $\hat{F}_{f0}$ decreases monotonically with increasing Mach number. For a given Mach number, there is a single value of $\hat{F}_{f0}$, but for a given value of $\hat{F}_{f0}$, we obtain two Mach numbers—one subsonic and the other supersonic. Equal values of $\hat{F}_{f0}$ at these Mach numbers ensure that, for an isentropic flow (constant total pressure and total temperature) in a convergent–divergent nozzle, equal areas on the convergent (subsonic flow) and divergent (supersonic flow) sections of the nozzle yield equal mass flow rates to satisfy the continuity equation.

5.3.1.2 Static Pressure Mass Flow Function

The static pressure is sometimes a known boundary condition at a duct cross section. For example, in a subsonic flow, the exit static pressure, not the total pressure, must equal the known ambient static pressure. In such cases, one can calculate the mass flow rate in the section as:

$$\dot{m} = \frac{F_f A p}{\sqrt{T_0}} = \frac{\hat{F}_f A p}{\sqrt{R T_0}} \tag{5.26}$$

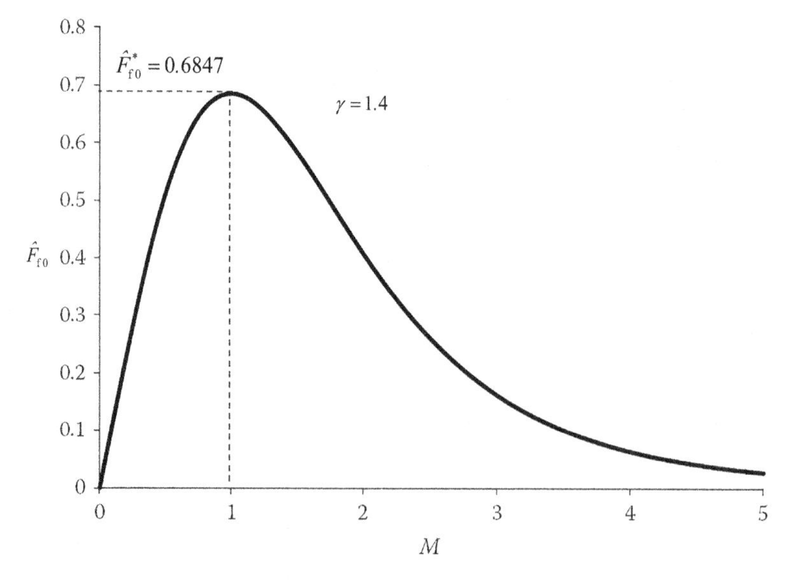

FIGURE 5.5 Variation of total pressure mass flow function $\hat{F}_{f0}$ with Mach number for $\gamma = 1.4$.

where the static pressure mass flow function F_f and its dimensionless counterpart $\hat{F}_f$ are related to the corresponding total pressure mass flow functions as follows:

$$F_f = F_{f0}\left(\frac{p_0}{p}\right) \tag{5.27}$$

and

$$\hat{F}_f = \hat{F}_{f0}\left(\frac{p_0}{p}\right) \tag{5.28}$$

Substituting F_{f0} from Equation 5.22 and p/p_0 from Equation 5.11 into Equation 5.27 and simplifying the resulting expression, we obtain:

$$F_f = M\sqrt{\frac{\gamma\left(1+\frac{\gamma-1}{2}M^2\right)}{R}} \tag{5.29}$$

and the corresponding expression for $\hat{F}_f$ as:

$$\hat{F}_f = M\sqrt{\gamma\left(1+\frac{\gamma-1}{2}M^2\right)} \tag{5.30}$$

At a section with choked flow ($M=1$), Equations 5.29 and 5.30 yield:

$$F_f^* = \sqrt{\frac{\gamma(\gamma+1)}{2R}} \tag{5.31}$$

and

$$\hat{F}_f^* = \sqrt{\frac{\gamma(\gamma+1)}{2}} \tag{5.32}$$

which, for air with $\gamma = 1.4$ and $R = 287\,\mathrm{J/(kg\ K)}$, yields $F_f^* = 0.07651$ in units $\sqrt{(\mathrm{kg\ K})/\mathrm{J}}$ and $\hat{F}_f^* = 1.2961$.

Figure 5.6 shows the variation of $\hat{F}_f$ with Mach number. Unlike $\hat{F}_{f0}$, which has a maximum at $M = 1.0$ and increases with Mach number in a subsonic flow and decreases with Mach number in a supersonic flow, $\hat{F}_f$ increases monotonically with Mach number. We can express $\hat{F}_{f0}$ as the product of $\hat{F}_f$, which increases with M, and p/p_0, which decreases with M. The shape of $\hat{F}_{f0}$ shown in Figure 5.5, therefore, results from the fact that for $M > 1.0$, p/p_0 decreases much more rapidly than the rate of increase in $\hat{F}_f$ with M.

For a given value of $\hat{F}_f$, we can use the following approach to directly compute M. Squaring both sides of Equation 5.30 yields:

$$\hat{F}_f^2 = \gamma M^2 + 0.5\gamma(\gamma-1)M^4$$

$$0.5\gamma(\gamma-1)M^4 + \gamma M^2 - \hat{F}_f^2 = 0$$

which is a quadratic equation in M^2 with the solution for M given as follows:

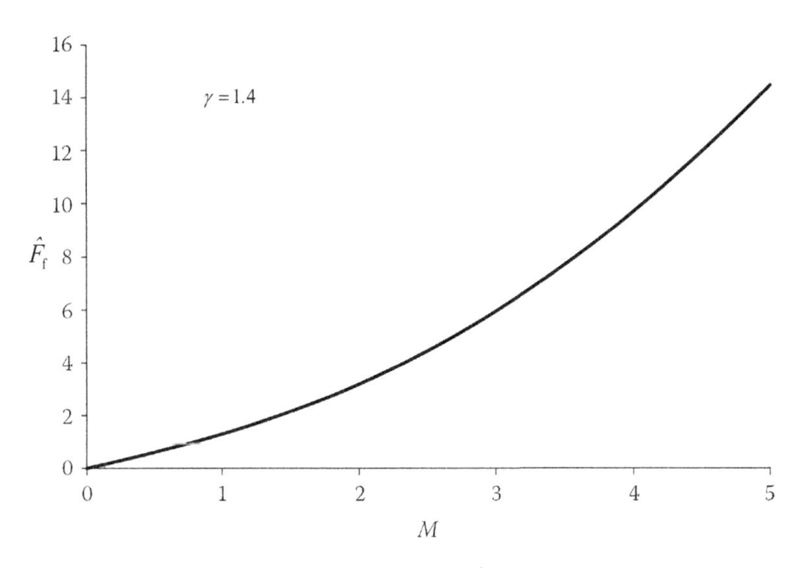

FIGURE 5.6 Variation of static pressure mass flow function $\hat{F}_f$ with Mach number for $\gamma = 1.4$.

$$M = \left(\frac{-\gamma + \sqrt{\gamma^2 + 2\gamma(\gamma - 1)\hat{F}_f^2}}{\gamma(\gamma - 1)} \right)^{\frac{1}{2}} \tag{5.33}$$

Similarly, for the given F_f, we obtain the following equation to compute M:

$$M = \left(\frac{-\gamma + \sqrt{\gamma^2 + 2\gamma(\gamma - 1)RF_f^2}}{\gamma(\gamma - 1)} \right)^{\frac{1}{2}} \tag{5.34}$$

We assumed that the entire velocity is normal to the flow cross section for the derivations of $\hat{F}_{f0}$ and $\hat{F}_f$. In many practical applications, this assumption may be false. For example, in the exhaust diffuser of a gas turbine, the exhaust gas has axial and swirl (tangential) velocities. If the axial direction is normal to the flow area, we can rewrite Equation 5.19 as:

$$\dot{m} = \rho \left(\frac{V_x}{V} \right) VA = C_V \rho VA \tag{5.35}$$

where the velocity coefficient $C_V = V_x / V$ is the fraction of the total flow velocity that contributes to the mass flow rate. Hence, we can rewrite Equations 5.21 and 5.26 as:

$$\dot{m} = \frac{C_V F_{f0} A p_0}{\sqrt{T_0}} = \frac{C_V \hat{F}_{f0} A p_0}{\sqrt{RT_0}} \tag{5.36}$$

$$\dot{m} = \frac{C_V F_f A p}{\sqrt{T_0}} = \frac{C_V \hat{F}_f A p}{\sqrt{RT_0}} \tag{5.37}$$

in which the Mach number used for calculating the mass flow functions is based on the total velocity V.

In Sultanian (2023), Table A.1 for $\gamma = 1.4$ and Table B.1 for $\gamma = 1.333$ provide values of $\hat{F}_f$ and $\hat{F}_{f0}$ as a function of Mach number.

5.3.2 Impulse Functions

In Chapter 2, we presented the concept of stream thrust (S_T) in the flow direction as the sum of the pressure force (pA) and the momentum flow ($\dot{m}V$), or the inertia force, in that direction. This concept is founded in the linear momentum equation and provides a convenient means to compute total thrust on a duct due to fluid flow simply from the change of stream thrust across the duct.

Figure 5.7 shows a rocket engine with a deflector plate that deflects its exhaust gases sideways. From the force-momentum balance on the control volume BCDE, we can compute the force of the rocket exhaust gases on the deflector plate as:

$$F = S_T = pA + \dot{m}V = pA + \rho AV^2 \tag{5.38}$$

If we divide the stream thrust by area, we obtain impulse pressure (p_i), given by:

$$p_i = p + \rho V^2 \tag{5.39}$$

We use Equation 5.39 to compute impulse pressure for both incompressible and compressible flows. However, the total pressure computed using $p_0 = p + 0.5\rho V^2$ is valid only for an incompressible flow.

5.3.2.1 Static Pressure Impulse Function

For a compressible flow in a duct, using the equation of state of a perfect gas, we can express the impulse pressure at a cross section in terms of Mach number as follows:

$$\begin{aligned} p_i &= p + \rho V^2 \\ &= p\left(1 + \frac{\rho V^2}{p}\right) \\ &= p\left(1 + \frac{\gamma V^2}{\gamma RT}\right) \\ p_i &= p\left(1 + \gamma M^2\right) = pI_f \end{aligned} \tag{5.40}$$

where I_f is the static pressure impulse function given by:

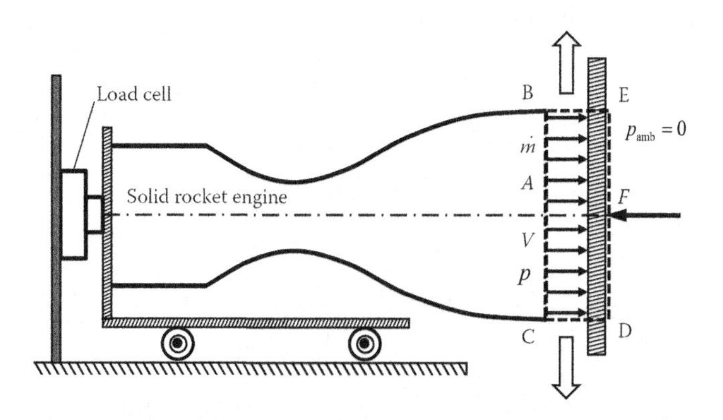

FIGURE 5.7 Stream thrust on a deflector plate.

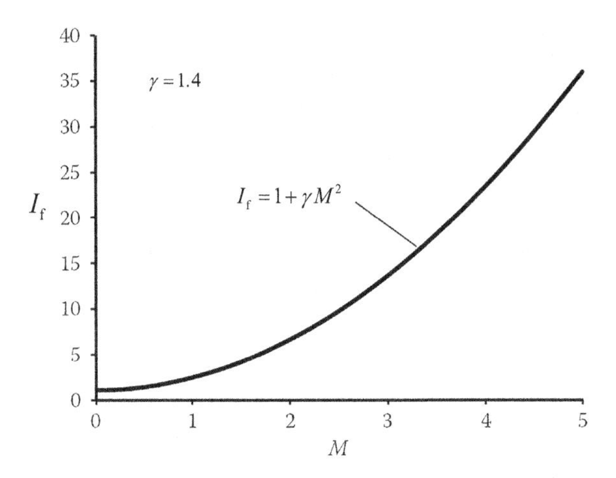

FIGURE 5.8 Variation of static pressure impulse function I_f with Mach number for $\gamma = 1.4$.

$$I_f = \left(1 + \gamma M^2\right) \tag{5.41}$$

which for $M = 1$ yields:

$$I_f^* = \left(1 + \gamma\right) \tag{5.42}$$

Figure 5.8 shows that I_f varies monotonically with M.

5.3.2.2 Total Pressure Impulse Function

We can also express the impulse pressure in terms of total pressure as:

$$p_i = p_0 \frac{p}{p_0}\left(1 + \gamma M^2\right)$$

which with the substitution for p / p_0 from Equation 5.11 becomes:

$$p_i = p_0 \frac{1 + \gamma M^2}{\left(1 + \frac{\gamma - 1}{2} M^2\right)^{\frac{\gamma}{\gamma-1}}} = p_0 I_{f0} \tag{5.43}$$

where I_{f0} is the total pressure impulse function given by:

$$I_{f0} = \frac{1 + \gamma M^2}{\left(1 + \frac{\gamma - 1}{2} M^2\right)^{\frac{\gamma}{\gamma-1}}} \tag{5.44}$$

which for $M = 1$ yields:

$$I_{f0}^* = \frac{1 + \gamma}{\left(\frac{\gamma + 1}{2}\right)^{\frac{\gamma}{\gamma-1}}} \tag{5.45}$$

The plot of Equation 5.44 in Figure 5.9 for $\gamma = 1.4$ shows that I_{f0}, which we can express as the product of I_f and p / p_0, increases for $M < 1$ and decreases for $M > 1$ and has a maximum value of 1.2679

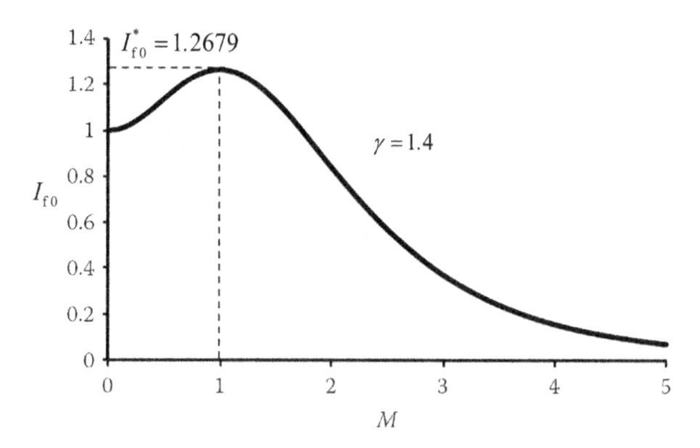

FIGURE 5.9 Variation of total pressure impulse function I_{f0} with Mach number for $\gamma = 1.4$.

at $M = 1$. The shape of the curve in this figure results from the fact that, for $M > 1$, p / p_0 decreases with M much more rapidly than the corresponding rate of increase in I_f.

Using impulse functions, we can express the stream thrust at any section of a duct flow as:

$$S_T = pAI_f = p_0 AI_{f0} \tag{5.46}$$

which, using mass flow functions, we can write as:

$$S_T = \frac{\dot{m}\sqrt{RT_0}}{\left(\hat{F}_f / I_f\right)} = \frac{\dot{m}\sqrt{RT_0}}{\left(\hat{F}_{f0} / I_{f0}\right)}$$

$$S_T = \frac{\dot{m}\sqrt{RT_0}}{N(M,\gamma)} \tag{5.47}$$

where $N(M,\gamma) = \hat{F}_f / I_f$.

In Sultanian (2023), Table A.1 for $\gamma = 1.4$ and Table B.1 for $\gamma = 1.333$ provide calculated values of I_f and I_{f0} as a function of Mach number.

5.3.3 Normal Shock Function

For a constant-area control volume simulating a normal shock, discussed in detail in Section 5.11, with no heat transfer and surface force, the mass flow rate, the stream thrust, and the total temperature remain constant over the control volume. Equation 5.47 reveals that $N(M,\gamma)$ must also remain constant across a normal shock. Accordingly, we call $N(M,\gamma)$ the normal shock function, given by:

$$N(M,\gamma) = \frac{\hat{F}_f}{I_f} = \frac{\hat{F}_{f0}}{I_{f0}} = \frac{M}{1+\gamma M^2}\sqrt{\gamma\left(1+\frac{\gamma-1}{2}M^2\right)} \tag{5.48}$$

which shows that, for a given gas, $N(M,\gamma)$ of M only. For $M < 1$, $\hat{F}_f$ increases with M faster than I_f does. For $M > 1$, $\hat{F}_f$ increases with M slower than I_f does. This explains the shape of the curve in Figure 5.10 showing the variation of N with M. For $M = 1$, this equation yields:

$$N^*(1,\gamma) = \sqrt{\frac{\gamma}{2(1+\gamma)}} \tag{5.49}$$

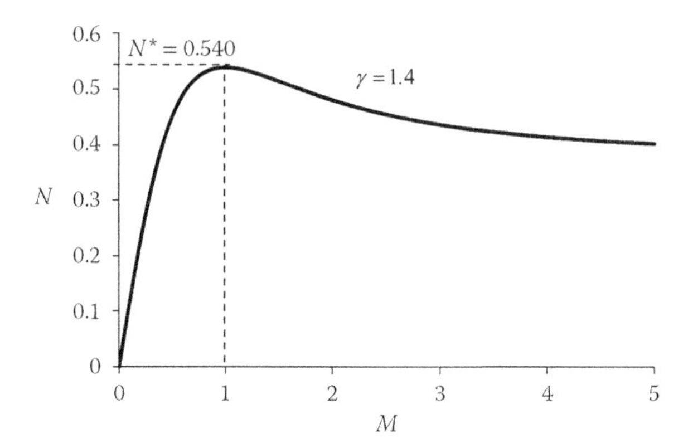

FIGURE 5.10 Variation of normal shock function N with Mach number for $\gamma = 1.4$.

which for $\gamma = 1.4$ equals 0.540. For a given mass flow rate and total temperature, Equation 5.47 yields the minimum value of the stream thrust for $M = 1$. For $M \to \infty$, Equation 5.48 yields the following asymptotic value of $N(M,\gamma)$:

$$N^{\infty}(\infty,\gamma) = \sqrt{\frac{\gamma - 1}{2\gamma}} \tag{5.50}$$

which for $\gamma = 1.4$ equals 0.378.

In Sultanian (2023), Table A.1 for $\gamma = 1.4$ and Table B.1 for $\gamma = 1.333$ provide values of normal shock function N as a function of Mach number.

5.4 VARIABLE-AREA DUCT FLOW WITH FRICTION, HEAT TRANSFER, AND ROTATION

This section presents the most general steady one-dimensional compressible flow in a duct whose cross section area varies arbitrarily in the flow direction. The duct can have arbitrary distributions of wall friction (shear force) and heat transfer. In addition, the duct rotates around an axis different from the flow direction, as shown in Figure 5.11. Among other design applications, such a duct typically forms the internal cooling passage of modern gas turbine blades (with rotation) and vanes (without rotation). This compressible duct flow may also feature choking ($M = 1$) and normal shocks.

Momentum and energy equations are fully coupled in a compressible flow. For a one-dimensional steady compressible flow shown in Figure 5.11, nonlinearly coupled effects of wall shear

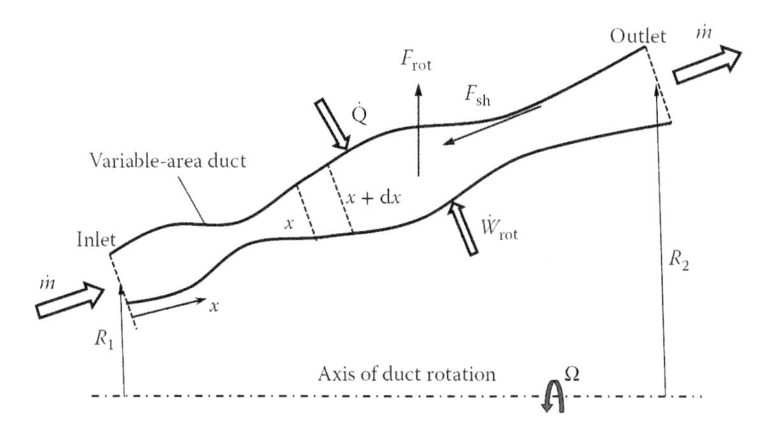

FIGURE 5.11 Variable-area duct flow with friction, heat transfer, and rotation.

force F_{sh} due to friction, heat transfer rate $\dot{Q}$, work transfer rate $\dot{W}_{rot}$ due to rotation, and rotational body force F_{rot} change the temperature and pressure from duct inlet to outlet. While F_{sh} and F_{rot} influence the flow properties through the momentum equation, $\dot{Q}$ and $\dot{W}_{rot}$ influence them through the energy equation. Note that the shear force F_{sh} always acts in the direction opposite to the flow direction and reduces the stream thrust. The rotational body force F_{rot}, on the other hand, always acts radially outward. Thus, for a radially outward flow in the duct, F_{rot} aids the flow; while for a radially inward flow, it opposes the flow. Similarly, the rotational work transfer $\dot{W}_{rot}$ is positive (into the fluid) for a radially outward flow and it is negative (out of the fluid) for a radially inward flow.

5.4.1 Differential Control Volume Analysis

For computing changes in flow properties along the duct, we need to express the conservation equations for a differential control volume delineated in Figure 5.11 between x and $x + dx$. Figure 5.12 shows the differential control volumes for the momentum and energy equations with changes in the primitive variables between their inlet and outlet. The static flow properties like p and T are independent of the reference frame, whether stationary or rotating. The flow velocity V shown in the figure is relative to the rotating duct and represents the through-flow velocity in the duct. The flow is in the state of solid-body rotation with the duct.

5.4.1.1 Continuity Equation

For a steady flow through the control volume, the continuity equation yields:

$$\mathrm{d}(\rho A V) = 0$$

$$A V \mathrm{d}\rho + \rho V \mathrm{d}A + \rho A \mathrm{d}V = 0$$

By dividing this equation throughout by $\rho A V$, we obtain:

$$\frac{\mathrm{d}\rho}{\rho} + \frac{\mathrm{d}A}{A} + \frac{\mathrm{d}V}{V} = 0 \tag{5.51}$$

5.4.1.2 Momentum Equation

For the control volume shown in Figure 5.12a, the linear momentum equation in the flow direction yields:

$$\dot{m}(V + \mathrm{d}V) - \dot{m}V = pA + \left(p + \frac{\mathrm{d}p}{2}\right)\mathrm{d}A - (p + \mathrm{d}p)(A + \mathrm{d}A) - \delta F_{sh\_x} + \delta F_{rot\_x}$$

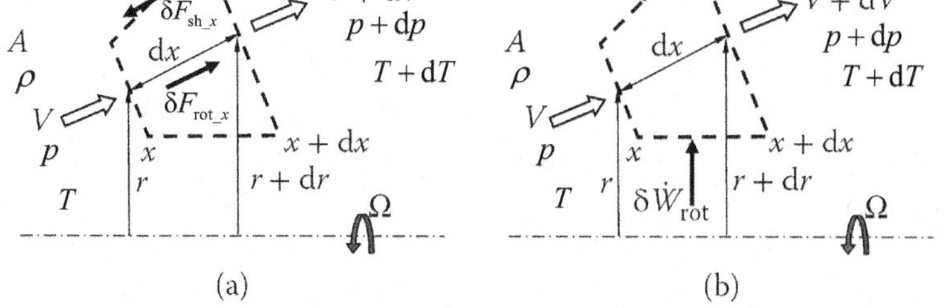

FIGURE 5.12 Differential control volumes; (a) Momentum and (b) Energy.

By substituting $\dot{m} = \rho AV$ in this equation and neglecting higher-order terms, we obtain:

$$\rho AV\mathrm{d}V + A\mathrm{d}p + \delta F_{\mathrm{sh}\_x} - \delta F_{\mathrm{rot}_x} = 0 \tag{5.52}$$

5.4.1.3 Energy Equation

For the control volume shown in Figure 5.12b, the energy equation yields:

$$\dot{m}\left(c_p\mathrm{d}T + V\mathrm{d}V\right) = \delta\dot{Q} + \delta\dot{W}_{\mathrm{rot}}$$

$$\mathrm{d}T + \frac{V\mathrm{d}V}{c_p} = \frac{\delta\dot{Q}}{\dot{m}c_p} + \frac{\delta\dot{W}_{\mathrm{rot}}}{\dot{m}c_p} \tag{5.53}$$

5.4.1.4 Equation of State

We write the equation of state of a perfect gas as:

$$\frac{p}{\rho} = RT$$

Taking the natural log of terms on each side of this equation and then differentiating them lead to the following equation:

$$\frac{\mathrm{d}\rho}{\rho} = \frac{\mathrm{d}p}{p} - \frac{\mathrm{d}T}{T} \tag{5.54}$$

Substituting for $\mathrm{d}\rho / \rho$ from this equation into Equation 5.51 results in the following equation:

$$\frac{\mathrm{d}A}{A} + \frac{\mathrm{d}V}{V} + \frac{\mathrm{d}p}{p} - \frac{\mathrm{d}T}{T} = 0 \tag{5.55}$$

5.4.2 Change in Relative Flow Velocity

Using the equation of state $p = \rho RT$ and Mach number $M = V / \sqrt{\gamma RT}$, we can rewrite Equation 5.52 as:

$$\frac{\mathrm{d}p}{p} = -\gamma M^2\frac{\mathrm{d}V}{V} - \left(\frac{\gamma M^2}{\rho AV^2}\right)\delta F_{\mathrm{sh}\_x} + \left(\frac{\gamma M^2}{\rho AV^2}\right)\delta F_{\mathrm{rot}\_x} \tag{5.56}$$

Using the Mach number $M = V / \sqrt{\gamma RT}$ and $c_p = \gamma R / (\gamma - 1)$, we can rewrite Equation 5.53 as:

$$\frac{\mathrm{d}T}{T} = -(\gamma - 1)M^2\frac{\mathrm{d}V}{V} + \frac{\delta\dot{Q}}{\dot{m}c_pT} + \frac{\delta\dot{W}_{\mathrm{rot}}}{\dot{m}c_pT} \tag{5.57}$$

Finally, substituting $\mathrm{d}T / T$ from this equation and $\mathrm{d}p / p$ from Equation 5.56 into Equation 5.55 and rearranging terms, we obtain the following equation:

$$\left(1 - M^2\right)\frac{\mathrm{d}V}{V} = -\frac{\mathrm{d}A}{A} - \left(\frac{\gamma M^2}{\rho AV^2}\right)\delta F_{\mathrm{rot}\_x} + \frac{\delta\dot{W}_{\mathrm{rot}}}{\dot{m}c_pT} + \left(\frac{\gamma M^2}{\rho AV^2}\right)\delta F_{\mathrm{sh}\_x} + \frac{\delta\dot{Q}}{\dot{m}c_pT} \tag{5.58}$$

which expresses changes in the flow velocity in a duct due to (a) change in flow area, (b) wall shear force due to friction, (c) heat transfer, and (d) duct rotation, which influences the change through the second and third terms on the right-hand side (RHS) of the equation. Let us further simplify these terms by expressing δF_{rot_x} and $\delta \dot{W}_{\text{rot}}$ in terms of the basic quantities.

Evaluation of δF_{rot_x}

For the differential control volume shown in Figure 5.14a, the centrifugal force in the radially outward direction can be expressed as:

$$\delta F_{\text{rot}} = (A\rho \text{d}x) r\Omega^2$$

We can express the component of this differential centrifugal force in the flow direction as:

$$\delta F_{\text{rot}\_x} = \delta F_{\text{rot}}\left(\frac{\text{d}r}{\text{d}x}\right) = A\rho r\Omega^2 \text{d}r$$

Thus, the second term on the RHS of Equation 5.58 simplifies to:

$$\left(\frac{\gamma M^2}{\rho A V^2}\right)\delta F_{\text{rot}\_x} = \left(\frac{\gamma M^2}{\rho A V^2}\right)\left(A\rho r\Omega^2 \text{d}r\right) = \frac{\Omega^2 r \text{d}r}{RT} \tag{5.59}$$

Evaluation of $\delta \dot{W}_{\text{rot}}$

For the differential control volume shown in Figure 5.14b, we can invoke the concept of rothalpy, presented in Chapter 2, to express the energy equation as:

$$\text{d}\dot{I} = \dot{m}\left(c_p \text{d}T + V\text{d}V - r\Omega^2 \text{d}r\right) = \delta \dot{Q}$$

Comparing this equation with Equation 5.53, we can write:

$$\delta \dot{W}_{\text{rot}} = \dot{m}\Omega^2 r \text{d}r$$

which simplifies the third term on the RHS of Equation 5.58 as:

$$\frac{\delta \dot{W}_{\text{rot}}}{\dot{m} c_p T} = \frac{\dot{m}\Omega^2 r \text{d}r}{\dot{m} c_p T} = \frac{\Omega^2 r \text{d}r}{c_p T} \tag{5.60}$$

Using this equation along with Equation 5.59, we can express the net effect of rotation in Equation 5.58 as:

$$-\left(\frac{\gamma M^2}{\rho A V^2}\right)\delta F_{\text{rot}\_x} + \frac{\delta \dot{W}_{\text{rot}}}{\dot{m} c_p T} = -\frac{r\Omega^2 \text{d}r}{RT} + \frac{r\Omega^2 \text{d}r}{c_p T} = -\frac{r\Omega^2 \text{d}r}{RT}\left(1 - \frac{R}{c_p}\right) = -\frac{r\Omega^2 \text{d}r}{\gamma RT}$$

which we can write as:

$$-\left(\frac{\gamma M^2}{\rho A V^2}\right)\delta F_{\text{rot}\_x} + \frac{\delta \dot{W}_{\text{rot}}}{\dot{m} c_p T} = -M_\theta^2 \frac{\text{d}r}{r} \tag{5.61}$$

where $M_\theta = r\Omega / \sqrt{\gamma RT}$ is the rotational Mach number. This equation shows that the effects of rotation on the control volume—the body force in the momentum equation and the rotational work transfer in the energy equation—collapse into a single term that depends on the rotational Mach number. This term becomes singular at $r = 0$ and is nonzero only when the change in radius along the duct is nonzero. With the substitution of this equation in Equation 5.58, we finally obtain:

$$\left(1-M^2\right)\frac{dV}{V}=-\frac{dA}{A}-M_\theta^2\frac{dr}{r}+\left(\frac{\gamma M^2}{\rho AV^2}\right)\delta F_{sh\_x}+\frac{\delta\dot{Q}}{\dot{m}c_pT} \quad (5.62)$$

We note the coefficient $\left(1-M^2\right)$ multiplying dV/V on the LHSmakes this equation nonlinear. When all effects on the RHS are present, this equation will generally require a numerical solution. We will discuss the analytical solution for each effect in Sections 5.5–5.10. We will also discuss a few analytical solutions when two effects are coupled, for example, an isothermal flow in a constant-area nonrotating duct where both friction and heat transfer are present. Nevertheless, this equation provides valuable insight into how a compressible flow responds to various effects of area change, rotation, friction, and heat transfer.

With no area change, rotation, friction, or heat transfer, Equation 5.62 reduces to:

$$\left(1-M^2\right)dV=0$$

which has two solutions: first, the trivial isentropic solution with no change in velocity ($dV = 0$), and the second, a nonequilibrium solution (normal shock) with an abrupt change in entropy, like what happens in a "hydraulic jump" in an open-channel water flow. We will discuss normal and oblique shocks later in Sections 5.11 and 5.12. Because the property changes across a shock are discontinuous, we cannot use the differential equation (Equation 5.62) to compute such changes.

Because the coefficient $\left(1-M^2\right)$ is positive for a subsonic flow and negative for a supersonic flow, the terms on the RHS of Equation 5.62 will have opposite effects in subsonic and supersonic flow regimes, highlighting a critical feature that results from the compressibility in a compressible flow, unlike an incompressible flow.

From Equation 5.62, we can make the following qualitative observations for subsonic and supersonic flows in a duct:

5.4.2.1 Subsonic Flow

- The flow will accelerate ($dV > 0$) if the duct converges ($dA < 0$), the flow is radially inward ($dr < 0$) under rotation, or we heat the gas ($\delta\dot{Q} > 0$).
- Conversely, the flow will decelerate if the duct diverges, flows radially outward under rotation, or we cool it.
- The friction always accelerates a subsonic flow.

5.4.2.2 Supersonic Flow

- The flow will accelerate ($dV > 0$) if the duct diverges ($dA > 0$), the flow is radially outward ($dr > 0$) under rotation, or we cool it ($\delta\dot{Q} < 0$).
- Conversely, the flow will decelerate if the duct converges, flows radially inward under rotation, or we heat it.
- Wall friction always decelerates a supersonic flow.

5.5 ISENTROPIC FLOW IN A VARIABLE-AREA DUCT

The isentropic compressible flows in a variable-area duct represent the most ideal one-dimensional flows without the effects of duct rotation, friction, and heat transfer. They represent the reference flows in the design of a lot of engineering equipment, for example, the flow in a converging–diverging (C–D) nozzle used in rocket propulsion. The assumption of constant entropy (isentropic) implies that total pressure and temperature remain constant in such flows. Accordingly, Equation 5.62 reduces to:

$$\left(1-M^2\right)\frac{dV}{V}=-\frac{dA}{A} \quad (5.63)$$

In addition, the linear momentum equation (Equation 5.52) becomes:

$$\frac{dp}{\rho V^2} = -\frac{dV}{V} \tag{5.64}$$

According to Equation 5.63, the decrease or increase in flow velocity with duct area depends on the Mach number, whether subsonic or supersonic. On the other hand, Equation 5.64 shows that the change in the static pressure is always opposite to that in the velocity, a well-known tenet of the Bernoulli equation—high velocity, low pressure. Using Equations 5.63 and 5.64, Figure 5.13 shows the changes in velocity and static pressure with the changes in the duct area for both subsonic and supersonic flows. This figure shows that the behavior of a subsonic flow in a converging duct is identical to that of a supersonic flow in a diverging duct. Similarly, the behavior of a subsonic flow in a diverging duct is identical to that of a supersonic flow in a converging duct.

For an initially subsonic flow to become supersonic or vice versa, it must go through the sonic condition with $M = 1$, which, according to Equation 5.63, calls for $dA = 0$. In the variable-area duct shown in Figure 5.14, $dA = 0$ at both sections 1 and 2. In section 1, the duct first diverges and then converges. Based on the flow physics presented in Figure 5.13, going through section 1, a subsonic flow will remain subsonic, and a supersonic flow will remain supersonic. The flow cannot transition from subsonic to supersonic or vice versa at this section with $dA = 0$. On the other hand, at section 2, a subsonic flow can become supersonic or vice versa when $M = 1$ in this section. This explains the throat section in the converging–diverging (C–D) nozzle of a rocket motor. When this C–D nozzle operates with $M = 1$ at the throat section, a choke-flow condition, the upstream flow in the converging section is subsonic, and the downstream flow in the diverging section is supersonic. Similarly, for a subsonic flow at its outlet, a diffuser with a supersonic flow at its inlet must be a converging–diverging duct with a throat section operating at $M = 1$.

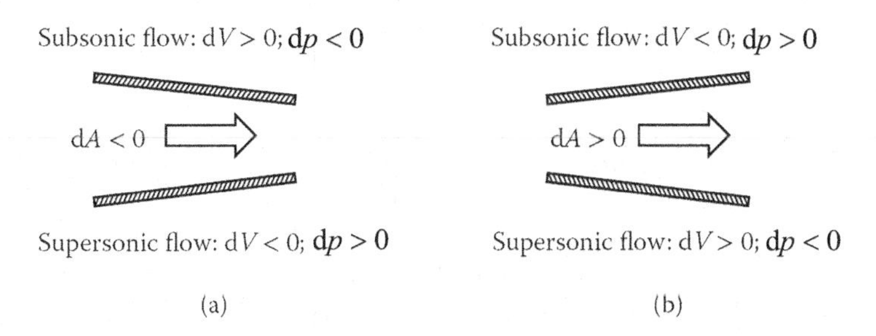

FIGURE 5.13 Changes in velocity and static pressure in a variable-area duct: (a) converging duct and (b) diverging duct.

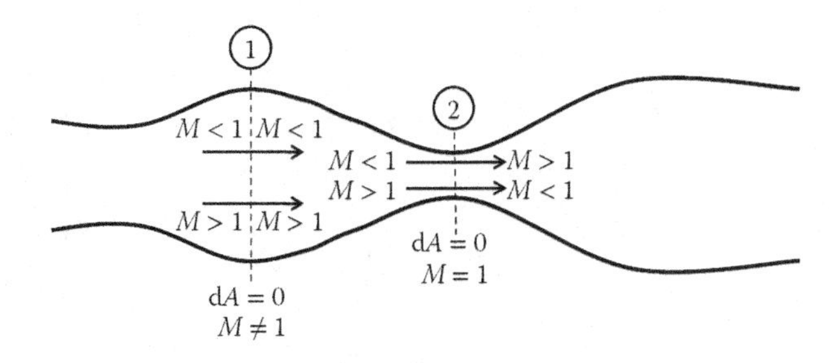

FIGURE 5.14 Compressible flow through a variable-area duct with sections where $dA = 0$.

5.5.1 Stream Thrust in Converging, Straight, and Diverging Ducts

Figure 5.15 shows fluid control volumes in converging, straight, and diverging sections of a variable-area duct with isentropic flow. Without friction, we have no wall shear stress acting on the control volume. The shown static pressure distribution is normal to the lateral fluid control volume surface in contact with the duct wall. For the converging duct shown in Figure 5.15a, the lateral pressure distribution creates a pressure force that acts on the control volume in the direction opposite to the flow direction. As a result, the fluid stream thrust decreases in the flow direction. For the straight duct shown in Figure 5.15b, the lateral pressure distribution is normal to the flow direction. Thus, the stream thrust for an isentropic flow in a straight duct remains constant across the duct. For the diverging duct shown in Figure 5.15c, a component of the lateral pressure force acts in the flow direction. As a result, the fluid stream thrust increases in a diverging duct from its inlet to its outlet. From this discussion, it should be clear why we need a sizable divergent section to maximize the thrust generated by a rocket nozzle.

5.5.2 Critical Throat Area

At a duct cross section of area A, if we know the Mach number, total pressure, and total temperature, we know the ideal mass flow rate of a gas through this section. We compute this mass flow rate using Equation 5.21 as:

$$\dot{m} = \frac{\hat{F}_{f0} A p_0}{\sqrt{RT_0}}$$

where we compute $\hat{F}_{f0}$ using Equation 5.23 as:

$$\hat{F}_{f0} = M \sqrt{\frac{\gamma}{\left(1 + \frac{\gamma - 1}{2} M^2\right)^{\frac{\gamma+1}{\gamma-1}}}}$$

For the given total pressure and temperature, there is a minimum area A^*, which is called the critical throat area or the sonic area with $M = 1$. The computed mass flow rate will not pass through an area smaller than this critical area. Note that A^* may not physically exist in the duct; under isentropic flow conditions, it acts as a reference critical throat area for all sections in the duct. In a C–D nozzle, the throat area equals A^* only for a choked nozzle with isentropic flow.

Equating the mass flow rates at A and A^*, we obtain:

$$\dot{m} = \frac{\hat{F}_{f0} A p_0}{\sqrt{RT_0}} = \frac{\hat{F}^*_{f0} A^* p_0}{\sqrt{RT_0}}$$

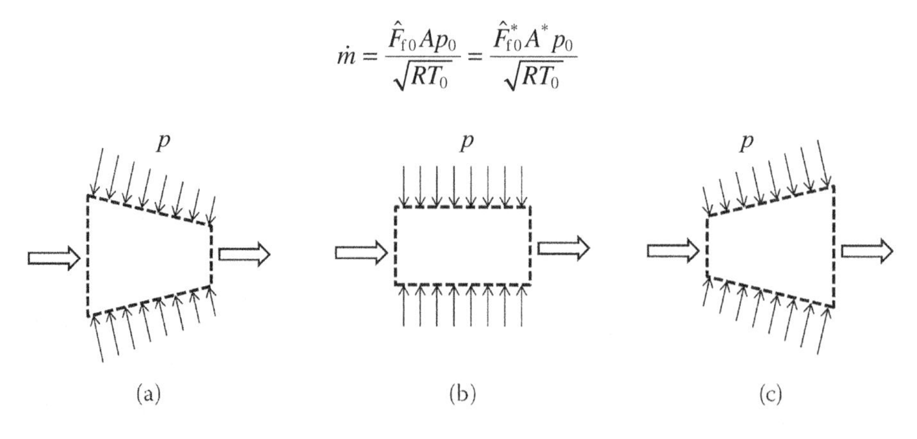

FIGURE 5.15 Fluid control volumes: (a) converging duct, (b) straight duct, and (c) diverging duct.

giving

$$\frac{A}{A^*} = \frac{\hat{F}_{f0}^*}{\hat{F}_{f0}} \tag{5.65}$$

Substituting $\hat{F}_{f0}^*$ from Equation 5.25 and $\hat{F}_{f0}$ from Equation 5.23 in this equation and simplifying the resulting expression, we obtain:

$$\frac{A}{A^*} = \frac{1}{M}\sqrt{\left(\frac{2+(\gamma-1)M^2}{\gamma+1}\right)^{\frac{\gamma+1}{\gamma-1}}} \tag{5.66}$$

Figure 5.16 shows the plots A/A^* and $\hat{F}_{f0}$ as a function of Mach number for $\gamma = 1.4$. For a given Mach number at a duct cross section, we have one set of values for A/A^* and $\hat{F}_{f0}$. For a given value of A/A^*, however, we obtain two Mach numbers—one subsonic and the other supersonic. Isentropic flow tables generally tabulate A/A^* as a function of Mach number; see, for example, Tables A.2 and B.2 in Sultanian (2023). Knowing $\hat{F}_{f0}^*$, we can use the isentropic flow tables and Equation 5.65 to compute $\hat{F}_{f0}$ for a given Mach number.

The critical area ratio A/A^* provides a useful measure of the area margin available to pass a desired mass flow rate. For example, for $M = 0.8$, $A/A^* = 1.0382$, which drops to 1.0088 for $M = 0.9$. As shown in Figure 5.16, with only small changes in the flow area, a transonic flow (near $M = 1.0$) can exhibit significant variations in Mach number.

5.5.3 Key Characteristics of Isentropic Flows in a Variable-Area Duct

Based on various concepts discussed in this section, we can summarize the following key characteristics of an isentropic compressible flow in a variable-area duct:

- The total pressure p_0 and the total temperature T_0 remain constant.
- The static pressure p and the static temperature T decrease monotonically with increasing Mach number.
- The minimum flow area corresponds to $M = 1$ when the flow becomes choked.

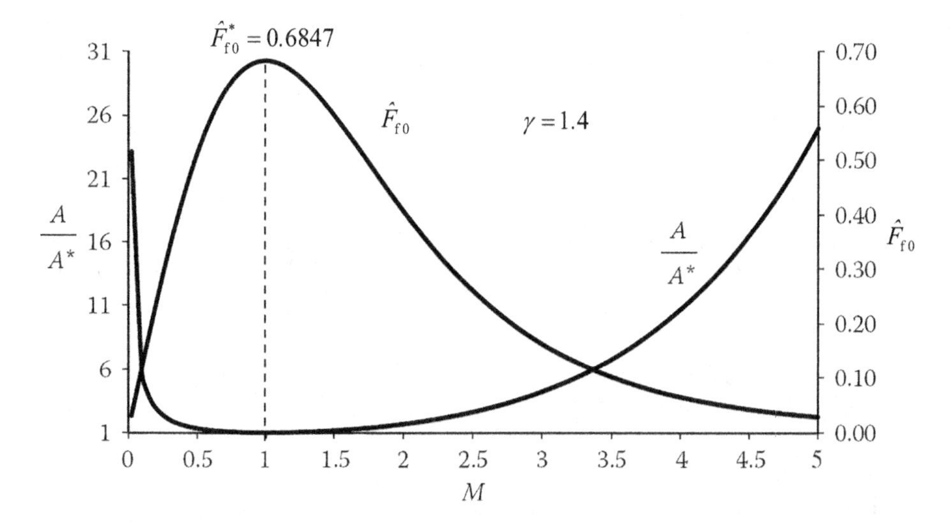

FIGURE 5.16 Variations of A/A^* and $\hat{F}_{f0}$ with Mach number in an isentropic duct flow for $\gamma = 1.4$.

- For a choked flow, any change in the downstream conditions will not change the mass flow rate.
- For a supersonic flow, both the velocity and the Mach number increase as the area increases, which are opposite to the behavior found in subsonic or incompressible flow.
- For both subsonic and supersonic flows, the change in velocity is opposite to the change in static pressure, like what happens in a potential flow governed by the Bernoulli equation.

5.6 ISENTROPIC FLOW IN A CONSTANT-AREA DUCT WITH ROTATION

With no area change, friction, and heat transfer, Equation 5.62 reduces to:

$$\left(1-M^2\right)\frac{\mathrm{d}V}{V}=-M_\theta^2\frac{\mathrm{d}r}{r} \tag{5.67}$$

Comparing this equation with Equation 5.63 shows that a change in flow velocity due to area change is analogous to a change in radius in a rotating constant-area duct. For a subsonic flow, velocity will decrease in the flow direction for a radially outward flow and increase for a radially inward flow. Just the opposite will happen in a supersonic flow. That is, the velocity will increase in the flow direction for a radially outward flow and decrease for a radially inward flow.

For no area change, friction, and heat transfer, the momentum Equation 5.52 reduces to:

$$\rho AV\mathrm{d}V+A\mathrm{d}p-\delta F_{\mathrm{rot}\_x}=0$$

Substituting $\delta F_{\mathrm{rot}\_x}=A\rho r\Omega^2 dr$ in this equation yields:

$$\frac{\mathrm{d}p}{\rho V^2}=-\frac{\mathrm{d}V}{V}+\frac{M_\theta^2}{M^2}\frac{\mathrm{d}r}{r}$$

Using Equation 5.67 to substitute $\mathrm{d}V/V$ in this equation and simplifying the resulting equation, we finally obtain:

$$\left(1-M^2\right)\frac{\mathrm{d}p}{\rho V^2}=\frac{M_\theta^2}{M^2}\frac{\mathrm{d}r}{r} \tag{5.68}$$

Comparing this equation with Equation 5.67, we find that the change in static pressure is opposite to the change in velocity for both subsonic and supersonic flows in a constant-area rotating duct. This trend is identical to that found in the duct with area change and a potential flow governed by the Bernoulli equation.

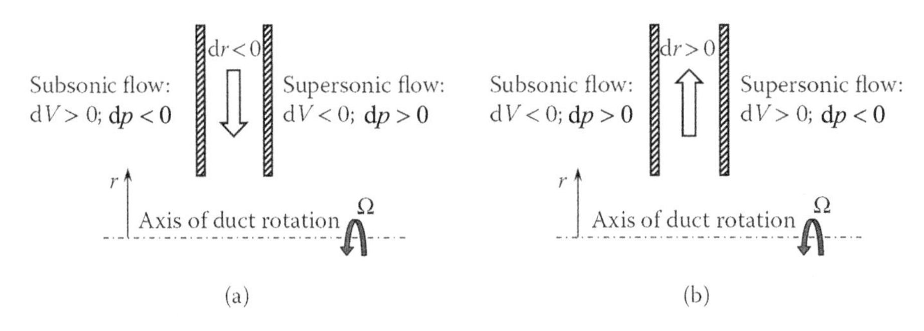

FIGURE 5.17 Changes in velocity and static pressure in a rotating constant-area duct. (a) Radially inward flow and (b) radially outward flow.

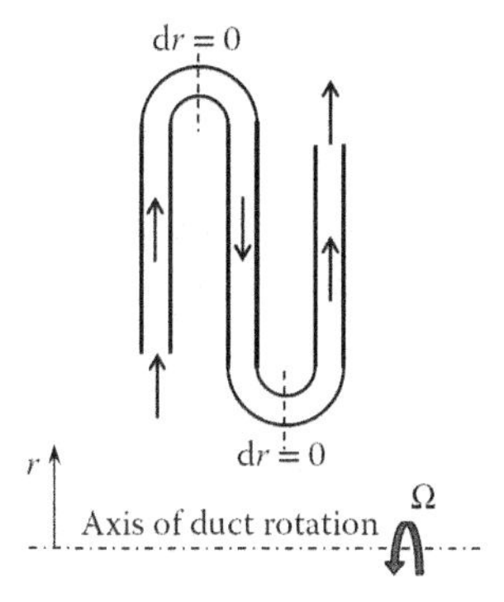

FIGURE 5.18 Choking in a rotating serpentine duct of constant area.

Figure 5.17 depicts changes in velocity and static pressure in subsonic and supersonic flows in a rotating constant-area duct flowing radially inward and outward. It may be instructive to compare these changes to those shown in Figure 5.13 for a stationary duct with area change.

According to Equation 5.67, the choking ($M = 1$) in a rotating constant-area duct can occur only at a section where $\mathrm{d}r = 0$, needed for a subsonic flow to become supersonic or vice versa, just as we need a throat in a nonrotating variable-area duct for the transition between subsonic and supersonic flows. According to the flow physics presented in Figure 5.17, we can conclude that the transition between subsonic and supersonic flows in a rotating constant-area serpentine duct shown in Figure 5.18 can occur at the lower U-turn but not at the upper U-turn, although both have sections where $\mathrm{d}r = 0$. At the upper U-turn, the duct radius first increases and then decreases around the turn for either flow direction. For the lower U-turn, however, the duct radius first decreases and then increases around the turn for either flow direction, a condition essential for the transition of subsonic flow to become supersonic or vice versa for the choked flow at $\mathrm{d}r = 0$.

5.7 ISENTROPIC FLOW IN A VARIABLE-AREA DUCT WITH ROTATION

With no friction and heat transfer, the velocity change given by Equation 5.62 reduces to:

$$\left(1 - M^2\right)\frac{\mathrm{d}V}{V} = -\frac{\mathrm{d}A}{A} - M_\theta^2\,\frac{\mathrm{d}r}{r} \tag{5.69}$$

which reveals that if $\mathrm{d}A$ and $\mathrm{d}r$ are both negative or both positive, the trend for the change in velocity in subsonic ($M < 1$) and supersonic ($M > 1$) flows is identical to those shown in Figure 5.13 for a variable-area duct without rotation or those in Figure 5.17 for a rotating duct of constant area. If $\mathrm{d}A$ is positive and $\mathrm{d}r$ is negative or vice versa, the change in velocity for subsonic and supersonic flows will depend on the sign of the resultant of both terms on the RHS of Equation 5.69.

Equation 5.69 further reveals that the isentropic flow in a variable-area duct with rotation will not choke at a physical throat where $\mathrm{d}A = 0$ or where $\mathrm{d}r = 0$ but at a section where $\mathrm{d}A / \mathrm{d}r = -\left(AM_\theta^2 / r\right)$. Because AM_θ^2 / r is positive definite, we can conclude that for a radially inward flow ($\mathrm{d}r < 0$), the choking will occur at a section downstream of the physical throat. Similarly, for a radially outward flow ($\mathrm{d}r > 0$), only a section upstream of the physical throat can choke. These interesting features of a compressible flow are not intuitive and widely known.

The momentum equation (Equation 5.52) with no frictional force at the duct wall reduces to:

$$\rho AV\mathrm{d}V + A\mathrm{d}p - \delta F_{\mathrm{rot}\_x} = 0$$

Substituting $\delta F_{\mathrm{rot}\_x} = A\rho r\Omega^2\mathrm{d}r$ and $\mathrm{d}V / V$ from Equation 5.69 in this equation yields:

$$\left(1-M^2\right)\frac{\mathrm{d}p}{\rho V^2}=\frac{\mathrm{d}A}{A}+\frac{M_\theta^2}{M^2}\frac{\mathrm{d}r}{r} \tag{5.70}$$

From Equations 5.69 and 5.70, we conclude that the change in the static pressure is always opposite to the change in velocity when both dA and dr are either negative or positive. Because the coefficient of the second term on the RHS of Equation 5.69 is different from that in Equation 5.70, it is possible to have flows in a rotating duct with area change where velocity and static pressure changes have the same sign.

5.8 FANNO FLOW

In this section, we will discuss the effect of friction on a one-dimensional steady compressible flow in a constant-area duct without heat transfer and rotation. Because of no heat transfer and work transfer, the fluid total temperature remains constant in this flow, widely known as the Fanno flow. Due to wall friction, the entropy gradually increases, and total pressure decreases downstream in the duct for both subsonic and supersonic flows. An exciting and counterintuitive feature of a subsonic Fanno flow is that the friction increases the Mach number in the flow direction and, if the duct is long enough, chokes the flow at the exit. This phenomenon is called frictional choking. A further extension of the duct after it has choked will decrease the mass flow rate with the sonic condition prevailing at the new exit. For a supersonic flow, however, the Mach number decreases downstream to the sonic condition at the duct exit if the duct is long enough. A further duct extension will result in a normal shock with an abrupt change in entropy and other properties inside the duct. The flow will now choke at the new exit at the same mass flow rate. Using the conservation laws of mass, momentum, energy, and entropy along with the equation of state of a perfect gas, we present physical and mathematical arguments in this section to develop a more intuitive understanding of a Fanno flow.

Let us use some of the flow functions introduced at the beginning of this chapter to develop a good understanding of subsonic and supersonic Fanno flows. Figure 5.19 shows the control volume between sections 1 and 2 of a Fanno flow. Whether the flow is subsonic or supersonic, the shear force F_{sh} acts on the control volume surface to oppose the flow; as a result, the stream thrust S_{T2} at section 2 will be lower than the stream thrust S_{T1} at section 1.

From the force and linear momentum balance on the control volume, we can write:

$$S_{T1}=S_{T2}+F_{\text{sh}}$$

which implies that $S_{T1}>S_{T2}$. Using Equation 5.47, we can express the stream thrust at sections 1 and 2 as:

$$S_{T1}=\frac{\dot{m}\sqrt{RT_{01}}}{N_1\left(M_1,\gamma\right)} \text{ and } S_{T2}=\frac{\dot{m}\sqrt{RT_{02}}}{N_2\left(M_2,\gamma\right)}$$

Because the total temperature remains constant in a Fanno flow, these expressions yield:

$$\frac{S_{T2}}{S_{T1}}=\frac{N_1\left(M_1,\gamma\right)}{N_2\left(M_2,\gamma\right)} \tag{5.71}$$

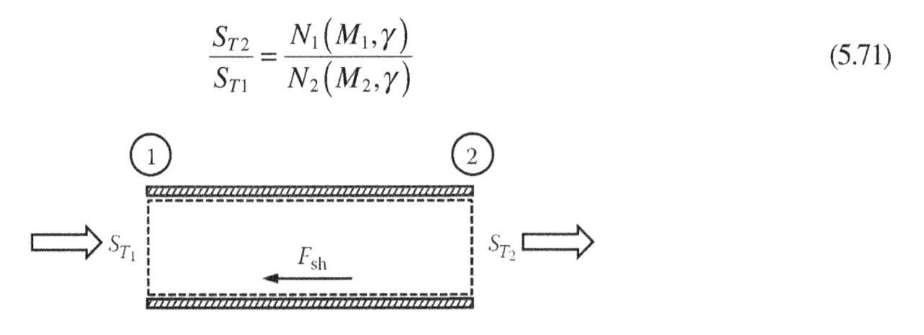

FIGURE 5.19 Control volume of a Fanno flow in a duct.

Between any two sections of a Fanno flow, this equation reveals that the ratio of outlet S_{T2} to inlet S_{T1} equals the ratio of the flow function N_1 at the inlet to the flow function N_2 at the outlet. Since $S_{T1} > S_{T2}$, we infer that $N_2 > N_1$. According to the variation of the flow function N with Mach number shown in Figure 5.10, we further conclude that the downstream Mach number in a Fanno flow changes to increase N toward its maximum at $M = 1$. In other words, if the inlet flow is subsonic, the downstream Mach number will increase; if it is supersonic, the downstream Mach number will decrease. Thus, in a Fanno flow, friction raises a subsonic Mach number and lowers a supersonic Mach number toward the sonic condition ($M = 1$), known as frictional choking. There is a maximum stream thrust loss (due to high friction factor and/or long duct) before the flow becomes sonic for the given inlet conditions.

Let us examine what we can conclude from the continuity equation when we express it in terms of the total pressure mass flow function. Using Equation 5.21, we write:

$$\dot{m} = \frac{\hat{F}_{f01} A_1 p_{01}}{\sqrt{RT_{01}}} = \frac{\hat{F}_{f02} A_2 p_{02}}{\sqrt{RT_{02}}}$$

Because $A_1 = A_2$ and $T_{01} = T_{02}$ in a Fanno flow, this equation simplifies to:

$$\hat{F}_{f01} p_{01} = \hat{F}_{f02} p_{02} \tag{5.72}$$

Because the variation of $\hat{F}_{f0}$ with Mach number, shown in Figure 5.5, is similar to that of the flow function N, and we have shown that $N_2 > N_1$ in a Fanno flow, we conclude that $\hat{F}_{f02} > \hat{F}_{f01}$, which yields $p_{02} < p_{01}$. That is, in a Fanno flow, the total pressure decreases downstream for both subsonic and supersonic inlet flows.

We can compute the change in entropy between two sections of a Fanno flow using the following equation:

$$s_2 - s_1 = c_p \ln\left(\frac{T_{02}}{T_{01}}\right) - R \ln\left(\frac{p_{02}}{p_{01}}\right)$$

Because the Fanno flow is adiabatic ($T_{01} = T_{02}$), this equation reduces to:

$$s_2 - s_1 = -R \ln\left(\frac{p_{02}}{p_{01}}\right) \tag{5.73}$$

From the requirement that the entropy increases downstream of an adiabatic flow with friction, we again conclude from this equation that $p_{02} < p_{01}$ for both subsonic and supersonic flows.

In the above, from the requirement that the stream thrust decreases downstream of a Fanno flow due to friction, we concluded that the Mach number must increase downstream for a subsonic flow and decrease downstream for a supersonic flow toward the sonic state. Since the Fanno flow is adiabatic, the entropy consideration leads us to conclude that the total pressure must decrease downstream of the flow. When we combine this requirement with the continuity equation expressed in terms of total pressure mass flow function, we again conclude that the friction in a Fanno flow must raise the Mach number for a subsonic flow and lower it for a supersonic flow toward the sonic state.

Section 5.8.2 presents equations for quantitatively evaluating changes in flow properties like static pressure, static temperature, density, and velocity in a Fanno flow. Based on the conservation principles, let us first use some simple reasoning to understand whether these quantities will increase or decrease in the flow direction in a Fanno flow. For example, for a subsonic Fanno flow, because $M_2 > M_1$, we obtain $V_2 > V_1$. The continuity equation ($\rho_2 V_2 = \rho_1 V_1$) leads to $\rho_2 < \rho_1$, and from the constancy of total temperature ($T_0 = T + 0.5V^2/c_p$), we obtain $T_2 < T_1$. The substitution $\rho_2 < \rho_1$ and $T_2 < T_1$ in the equation of state for a perfect gas ($p = \rho RT$) leads to $p_2 < p_1$. We can use a similar argument to forecast opposite trends of variation in the static flow properties in a supersonic Fanno flow.

5.8.1 Fanno Line

A Fanno line represents the variation of static temperature with entropy on a *T-s* plane. The total temperature and mass flux (mass flow rate per unit area) remain constant along each Fanno line. By combining the second law of thermodynamics with the first law expressed in terms of enthalpy, we can write the change in entropy as:

$$T\mathrm{d}s = \mathrm{d}h - \frac{\mathrm{d}p}{\rho}$$

which for a perfect gas ($h = c_p T$ and $p = \rho RT$) becomes:

$$\mathrm{d}s = c_p \frac{\mathrm{d}T}{T} - R\frac{\mathrm{d}p}{p}$$

$$\frac{\mathrm{d}s}{R} = \frac{c_p}{R}\frac{\mathrm{d}T}{T} - \frac{\mathrm{d}p}{p} \tag{5.74}$$

Substituting $\mathrm{d}p / p$ in this equation from the differential form of the equation of state (Equation 5.54), we obtain:

$$\frac{\mathrm{d}s}{R} = \left(\frac{\gamma}{\gamma - 1}\right)\frac{\mathrm{d}T}{T} - \frac{\mathrm{d}T}{T} - \frac{\mathrm{d}\rho}{\rho}$$

Substituting $\mathrm{d}\rho / \rho$ in this equation from the differential form of the continuity equation (Equation 5.51) for a constant-area duct yields:

$$\frac{\mathrm{d}s}{R} = \left(\frac{1}{\gamma - 1}\right)\frac{\mathrm{d}T}{T} + \frac{\mathrm{d}V}{V} \tag{5.75}$$

Integrating this equation between initial State 1 to an arbitrary state yields:

$$\frac{s - s_1}{R} = \frac{1}{\gamma - 1}\ln\left(\frac{T}{T_1}\right) + \ln\left(\frac{V}{V_1}\right)$$

With the constant total temperature, substituting $V = \sqrt{2c_p(T_0 - T)}$ and $V_1 = \sqrt{2c_p(T_0 - T_1)}$ in this equation yields:

$$\frac{s - s_1}{R} = \hat{s} = \frac{1}{\gamma - 1}\ln\left(\frac{T}{T_1}\right) + \frac{1}{2}\ln\left(\frac{T_0 - T}{T_0 - T_1}\right) \tag{5.76}$$

where $\hat{s}$ is the dimensionless change in entropy. For a given gas (constant γ and R), T_0, and T_1, this equation generates the corresponding Fanno line, shown in Figure 5.20.

In Figure 5.20, the horizontal line corresponding to the critical temperature T^* with $M = 1$ divides the Fanno line into two branches. The flow is subsonic along the upper branch and supersonic along the lower branch. This figure further shows that, for both subsonic and supersonic flows, the

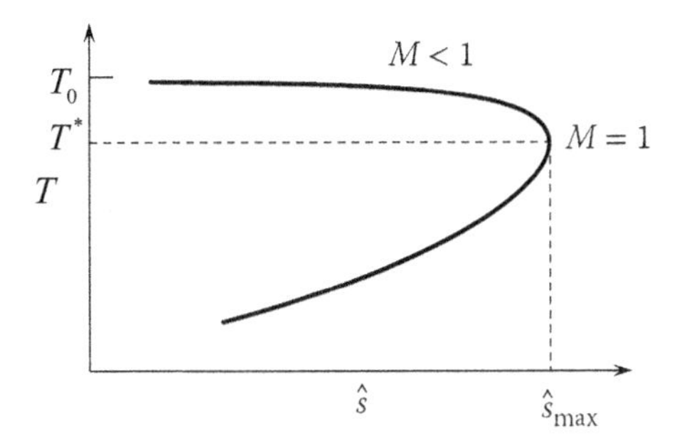

FIGURE 5.20 A Fanno line in the $T-\hat{s}$ plane.

maximum value $\hat{s} = \hat{s}_{mas}$ occurs at $M = 1$. We can also demonstrate this by taking the derivative of Equation 5.76 with respect to T and setting it to zero, as follows:

$$\frac{d\hat{s}}{dT} = \left(\frac{1}{\gamma-1}\right)\frac{1}{T} - \frac{1}{2}\left(\frac{1}{T_0 - T}\right) = 0$$

which simplifies to $T_0 / T = (\gamma+1)/2$ and leads to $M = 1$.

The occurrence of maximum entropy at the sonic point has an important ramification. For a subsonic Fanno flow with a continuous increase in entropy by friction, the Mach number can, at most, reach the sonic point. The flow cannot cross the sonic point to become supersonic, as it would imply a decrease in entropy with friction, which is physically impossible. Similarly, a supersonic Fanno flow cannot become subsonic by going through the sonic point with a continuous increase in entropy by friction. Equation 5.76 is, however, mathematically continuous and analytic at the sonic point.

Let us rewrite Equation 5.76 to express the entropy variation in a Fanno flow as:

$$s = s_1 + \frac{R}{\gamma-1}\ln\left(\frac{T}{T_1}\right) + \frac{R}{2}\ln\left(\frac{T_0 - T}{T_0 - T_1}\right) \tag{5.77}$$

where s_1 is evaluated relative to an assumed reference state of zero entropy with static pressure p_{ref} and static temperature T_{ref}. For a fixed total temperature T_0 at state 1 having static temperature T_1 and static pressure p_1, we can compute s_1 and the mass flux using the following equations:

$$s_1 = c_p \ln\left(\frac{T_1}{T_{ref}}\right) - R\ln\left(\frac{p_1}{p_{ref}}\right),\ M_1 = \sqrt{\frac{2}{\gamma-1}\left(\frac{T_0}{T_1} - 1\right)},$$

$$\hat{F}_{f1} = M_1\sqrt{\gamma\left(1 + \frac{\gamma-1}{2}M_1^2\right)},\ \text{and}\ \dot{m}/A = \hat{F}_{f1}p_1/\sqrt{RT_0}.$$

Figure 5.21 shows the plot of Equation 5.77 for three values of mass flux. We have $T_{0A} = T_{0B} = T_{0C}$ at the sonic points of the three Fanno lines in this figure. For $s_A^* < s_B^* < s_C^*$, we obtain $p_{0A}^* > p_{0B}^* > p_{0C}^*$. By using the total pressure mass flow function, we can express the mass flux in a Fanno flow at the sonic point as $\dot{m}/A = \hat{F}_{f0}^* p_0^* / \sqrt{RT_0}$, from which we conclude that $(\dot{m}/A)_A > (\dot{m}/A)_B > (\dot{m}/A)_C$ as shown in the figure.

5.8.2 Quantitative Evaluation of Properties in a Fanno Flow

In this section, we will use the conservation equations of mass, momentum, energy, and entropy to derive relevant equations to compute changes in static temperature, static pressure, static density,

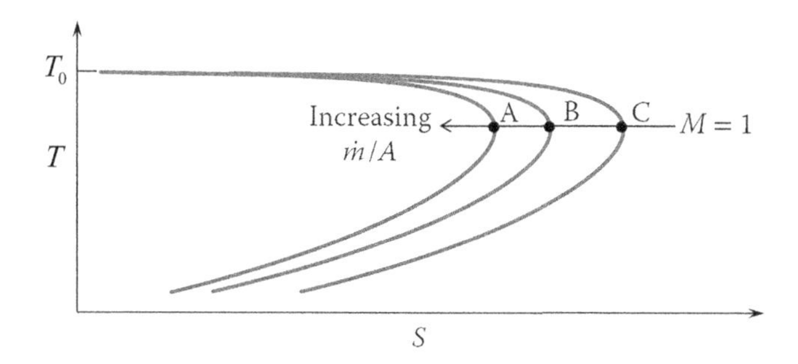

FIGURE 5.21 Fanno lines in the ***T-s*** plane with varying mass flux.

and total pressure in terms of their ratios with the corresponding values at the sonic point as the reference state. In the following derivations, we will use the facts that the total temperature and the mass flux remain constant in a Fanno flow.

5.8.2.1 Static Temperature Ratio

At a duct cross section, we compute the total-to-static temperature ratio as:

$$\frac{T_0}{T} = 1 + \frac{\gamma - 1}{2} M^2$$

which at the sonic section becomes:

$$\frac{T_0}{T^*} = \frac{\gamma + 1}{2}$$

giving

$$\frac{T}{T^*} = \frac{\gamma + 1}{2 + (\gamma - 1) M^2} \tag{5.78}$$

Note that this equation, which relates the local static temperature to its critical value at the sonic point, is valid even in an isentropic flow.

5.8.2.2 Static Density Ratio

For the mass flux remaining constant in a Fanno flow, we can write:

$$\frac{\rho}{\rho^*} = \frac{V^*}{V} = \frac{1}{M}\sqrt{\frac{T^*}{T}}$$

which, using Equation 5.78, becomes:

$$\frac{\rho}{\rho^*} = \frac{1}{M}\sqrt{\frac{2 + (\gamma - 1) M^2}{\gamma + 1}} \tag{5.79}$$

5.8.2.3 Static Pressure Ratio

For a perfect gas, the equation of state yields:

$$\frac{p}{p^*} = \frac{\rho}{\rho^*}\frac{T}{T^*}$$

In this equation, substituting ρ / ρ^* from Equation 5.78 and T / T^* from Equation 5.79, we obtain:

$$\frac{p}{p^*} = \left(\frac{\gamma+1}{2+(\gamma-1)M^2}\right) \times \left(\frac{1}{M}\sqrt{\frac{2+(\gamma-1)M^2}{\gamma+1}}\right)$$

$$\frac{p}{p^*} = \frac{1}{M}\sqrt{\frac{\gamma+1}{2+(\gamma-1)M^2}} \tag{5.80}$$

5.8.2.4 Total Pressure Ratio

At a duct cross section, we compute the total-to-static pressure ratio from the total-to-static temperature ratio using the isentropic relation:

$$\frac{p_0}{p} = \left(\frac{T_0}{T}\right)^{\frac{\gamma}{\gamma-1}}$$

which yields:

$$\frac{p_0}{p_0^*} = \frac{p}{p^*}\frac{\left(\frac{T_0}{T}\right)^{\frac{\gamma}{\gamma-1}}}{\left(\frac{T_0}{T^*}\right)^{\frac{\gamma}{\gamma-1}}} = \frac{p}{p^*}\left(\frac{T^*}{T}\right)^{\frac{\gamma}{\gamma-1}}$$

In this equation, substituting $\frac{p^*}{p}$ from Equation 5.80 and $\frac{T^*}{T}$ from Equation 5.78, we obtain:

$$\frac{p_0}{p_0^*} = \frac{1}{M}\left(\sqrt{\frac{\gamma+1}{2+(\gamma-1)M^2}}\right)\left(\frac{2+(\gamma-1)M^2}{\gamma+1}\right)^{\frac{\gamma}{\gamma-1}}$$

which simplifies to:

$$\frac{p_0}{p_0^*} = \frac{1}{M}\left(\frac{2+(\gamma-1)M^2}{\gamma+1}\right)^{\frac{\gamma+1}{2(\gamma-1)}} \tag{5.81}$$

Figure 5.22 shows how various properties vary with Mach number in a Fanno flow. Although the plotted lines show a smooth transition in properties across the sonic point, as we have discussed in Section 5.8.1, we have two distinct Fanno flow regimes, one subsonic with increasing Mach number up to the sonic point and the other supersonic with decreasing Mach number up to the sonic point. As indicated by the arrows on the Mach number axis in this figure, a subsonic Fanno flow will be from left to right, and a supersonic Fanno flow will be from right to left without ever crossing over the sonic point in either direction. For example, the figure shows that the total pressure ratio decreases in the flow direction for both subsonic and supersonic flows.

5.8.2.5 Variation of Mach number Due to Friction

In the above, we have derived equations to compute various flow properties as a function of Mach number, which depends on friction in a Fanno flow. We have found how the Mach number changes in an isentropic duct flow due to area change and rotation. In the following, we present how wall friction changes the Mach number in a Fanno flow duct of constant wetted perimeter P_w and the Darcy friction factor f, which equals four times the shear coefficient C_f.

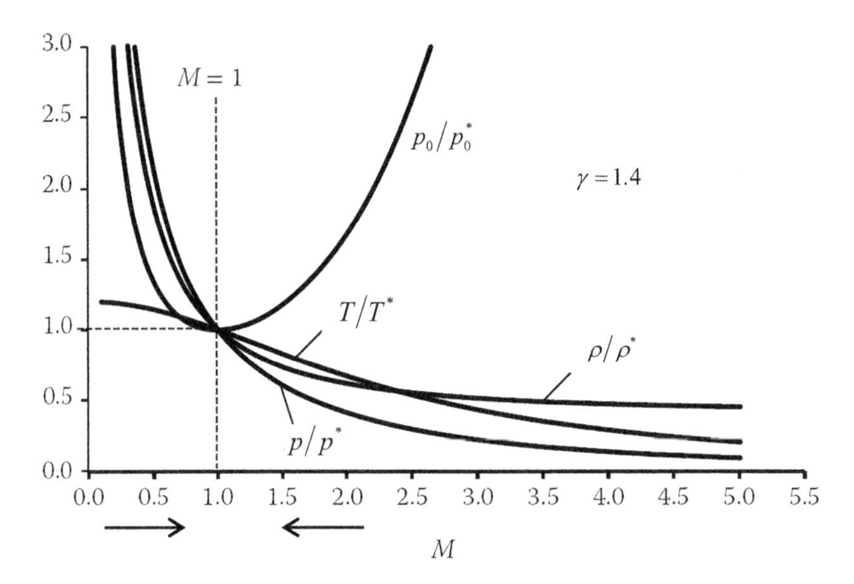

FIGURE 5.22 Variations of properties with Mach number in a Fanno flow.

The momentum equation (Equation 5.52) without the rotational body force term δF_{rot_x} reduces to:

$$\rho A V \mathrm{d}V + A\mathrm{d}p + \delta F_{\text{sh}\_x} = 0 \tag{5.82}$$

Over a fluid control volume of length dx in the Fanno flow duct, we can express the shear force δF_{sh_x} as:

$$\delta F_{\text{sh}\_x} = \tau_{\text{w}} p_{\text{w}} \mathrm{d}x = C_{\text{f}} \frac{\rho V^2}{2} P_{\text{w}} \mathrm{d}x$$

whose substitution in Equation 5.82, after rearranging terms, yields:

$$\begin{aligned} \frac{\delta F_{\text{sh}\_x}}{A} &= 4C_{\text{f}} \frac{\rho V^2}{2}\left(\frac{P_{\text{w}}}{4A}\right)\mathrm{d}x = -\mathrm{d}p - \rho V \mathrm{d}V \\ \frac{f\mathrm{d}x}{D_{\text{h}}} &= -2\frac{\mathrm{d}p}{\rho V^2} - 2\frac{\mathrm{d}V}{V} \end{aligned} \tag{5.83}$$

where D_{h} is the mean hydraulic diameter of the duct. Using Equation 5.55 for a constant-area duct to substitute $\mathrm{d}V / V$ in this equation yields:

$$\frac{f\mathrm{d}x}{D_{\text{h}}} = -2\frac{\mathrm{d}p}{\rho V^2} + 2\frac{\mathrm{d}p}{p} - 2\frac{\mathrm{d}T}{T}$$

which with $\rho V^2 = \gamma p M^2$ becomes:

$$\frac{f\mathrm{d}x}{D_{\text{h}}} = -2\left(\frac{1}{\gamma M^2} - 1\right)\frac{\mathrm{d}p}{p} - 2\frac{\mathrm{d}T}{T} \tag{5.84}$$

As ρV is proportional to $pM / \sqrt{T}$ and remains constant in a Fanno flow, we can write:

$$\frac{pM}{\sqrt{T}} = \text{const.}$$

whose differentiation yields:

$$\frac{dp}{p} = \frac{1}{2}\frac{dT}{T} - \frac{dM}{M} \tag{5.85}$$

Similarly, for the total temperature remaining constant in a Fanno flow, we can write:

$$\frac{dT}{T} = -\frac{d\{2+(\gamma-1)M^2\}}{\{2+(\gamma-1)M^2\}} \tag{5.86}$$

whose substitution in Equation 5.85 yields:

$$\frac{dp}{p} = -\frac{1}{2}\frac{d\{2+(\gamma-1)M^2\}}{\{2+(\gamma-1)M^2\}} - \frac{dM}{M} \tag{5.87}$$

Finally, substituting this equation and Equation 5.86 in Equation 5.84, we obtain:

$$\frac{f dx}{D_h} = \frac{1}{\gamma M^2}\frac{d\{2+(\gamma-1)M^2\}}{\{2+(\gamma-1)M^2\}} + \frac{d\{2+(\gamma-1)M^2\}}{\{2+(\gamma-1)M^2\}} - 2\frac{dM}{M} - \frac{2}{\gamma M^2}\frac{dM}{M} \tag{5.88}$$

whose integration from the duct section at x with Mach number M to the duct section at x_1 with Mach number M_1, after simplifying the algebra, yields:

$$\frac{f(x_1 - x)}{D_h} = \frac{\gamma+1}{2\gamma}\ln\left[\frac{M^2\{2+(\gamma-1)M_1^2\}}{M_1^2\{2+(\gamma-1)M^2\}}\right] + \frac{1}{\gamma}\left(\frac{1}{M^2} - \frac{1}{M_1^2}\right) \tag{5.89}$$

In this equation, if x_1 corresponds to the sonic point with $M_1 = 1$, we obtain the maximum duct length $(x_1 - x)$ for the Mach number specified at x, giving:

$$\frac{fL_{max}}{D_h} = \frac{\gamma+1}{2\gamma}\ln\left(\frac{(\gamma+1)M^2}{\{2+(\gamma-1)M^2\}}\right)\frac{1}{\gamma}\left(\frac{1}{M^2} - 1\right) \tag{5.90}$$

For a given Mach number at a Fanno flow duct inlet, we can use this equation to directly compute fL_{max}/D_h. For a known value of fL_{max}/D_h, however, we need to use an iterative method (for example, "Goal Seek" in MS Excel) to find the inlet Mach number. For $M \to \infty$, this equation has an asymptotic value given by:

$$\left(\frac{fL_{max}}{D_h}\right)_{M\to\infty} = \frac{\gamma+1}{2\gamma}\ln\left(\frac{\gamma+1}{\gamma-1}\right) - \frac{1}{\gamma} \tag{5.91}$$

which for $\gamma = 1.4$ equals 0.8215.

Figure 5.23 depicts the asymptotic behavior of Equation 5.90 for supersonic Fanno flows, including some exciting features on how the maximum duct length of a Fanno flow changes with the inlet Mach number. Compared to the case of a supersonic inlet flow, for a subsonic inlet flow, the required maximum length exhibits steep dependence on Mach number.

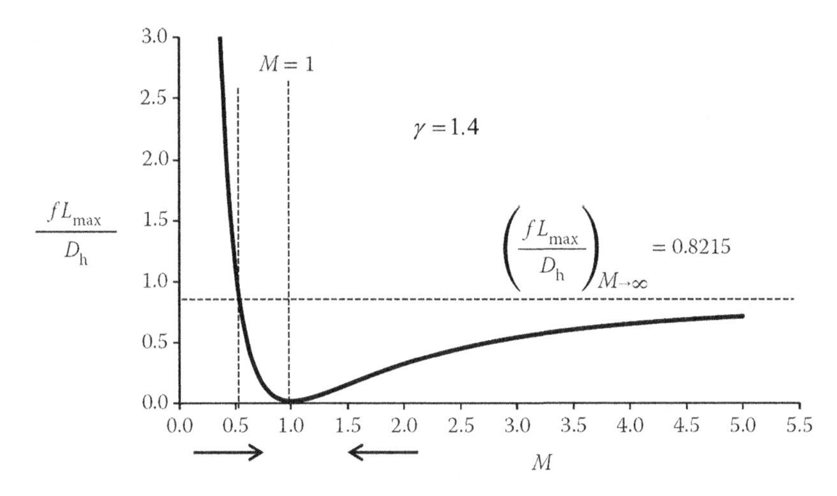

FIGURE 5.23 Variations in maximum Fanno flow duct length with inlet Mach number for $\gamma = 1.4$.

For a subsonic inlet flow, let us consider a maximum length Fanno flow duct (sonic exit). If we extend this duct by attaching a short piece of the duct, the sonic point will shift to the extended exit plane, requiring a lower Mach number at the inlet. With the inlet total pressure and temperature remaining constant, the mass flow rate through the duct will reduce, and the flow will move to a new Fanno line to the right (see Figure 5.21). We should expect this behavior because the added friction from the duct extension will increase the overall pressure loss across the duct with the original mass flow rate.

If the inlet flow is supersonic, the duct extension will result in a very different Fanno flow behavior. Because the downstream changes are not transmitted upstream in a supersonic flow, the flow will feature a normal shock (abrupt and nonequilibrium changes in flow properties) at an intermediate section, becoming subsonic and then choking at the new duct exit without requiring a change in duct inlet conditions and mass flow rate.

5.8.3 Fanno Flow Tables with the Reference State at $M = 1$

In Sultanian (2023), Table A.3 for $\gamma = 1.4$ and Table B.3 for $\gamma = 1.333$ present calculated properties of subsonic and supersonic Fanno flows. Although each value in the table is accurate to five decimal places, a linear interpolation between two adjacent values may yield an inaccurate result if the actual variation is linear. In that case, unless interested in an approximate solution, one should use the actual equations for the computation. In addition to yielding a quick and approximate solution, tables are helpful where the original equations require an iterative solution. One can enter each table through any known variable and quickly obtain the values of all variables along the corresponding row.

5.9 RAYLEIGH FLOW

In this section, we will discuss the effect of heat transfer (both heating and cooling) on a one-dimensional steady compressible flow in a constant-area duct without friction and rotation. In this flow, known as the Rayleigh flow, the stream thrust or the corresponding impulse pressure remains constant. The total temperature remains constant in the duct in a Fanno flow, as presented in the previous section. Although the effect of friction in a Fanno flow cannot be assumed to be reversible, we will assume that the heat transfer process in a Rayleigh flow is reversible. Thus, we can reverse all the changes in a Rayleigh flow due to the heating of the gas by an equal amount of cooling of the gas. According to the second law of thermodynamics, $\delta q_{rev} = Tds$ relates the entropy change to the amount of reversible heat transfer, which shows how heating will increase and cooling will decrease gas entropy, regardless of the flow Mach number, a feature absent in a Fanno flow. Note here that it is much more impractical to eliminate friction in a duct than to make it adiabatic. As a result, Rayleigh flows are an excellent idealization for a short duct only. Nevertheless, developing an improved intuitive understanding of all the features of a Rayleigh flow is essential.

Continuous increase in entropy with a concurrent decrease in the total pressure in a Rayleigh flow due to heating increases the Mach number for a subsonic inlet and decreases it for a supersonic inlet. Entropy reaches its maximum value at the sonic condition. Thermal choking allows a subsonic gas flow to become sonic ($M=1$) at the duct exit by heating the flow. For the given duct inlet conditions, thermal choking requires a certain amount of heat transfer. Increasing heat transfer beyond this value will demand a reduction in the mass flow rate through the duct, keeping the flow at the duct exit sonic. It is not possible to go from subsonic flow to supersonic flow or vice versa by continuous heating in a Rayleigh flow beyond the point of thermal choking. However, it is possible to achieve it by cooling the flow preceded by a short adiabatic section. This behavior in a Rayleigh flow is analogous to area change in an isentropic C–D nozzle to convert a subsonic flow into a supersonic flow or vice versa using a throat section with zero area change. Using the conservation laws of mass, momentum, energy, and entropy along with the equation of state of a perfect gas, we present physical and mathematical arguments in this section to develop a better intuitive understanding of a Rayleigh flow.

Figure 5.24 shows the control volume between sections 1 and 2 of a Rayleigh flow with the heat transfer rate $\dot{Q}$ that increases the total temperature from T_{01} to T_{02}. Without the wall friction in the constant-area duct of a Rayleigh flow, the stream thrust or the impulse pressure, which equals the stream thrust per unit cross section area, remains constant over the duct ($S_{T1} = S_{T2}$). The heat transfer rate $\dot{Q}$ and the associated rise in total temperature are related as follows:

$$T_{02} - T_{01} = \frac{\dot{Q}}{\dot{m}c_p}$$

$$\frac{T_{02}}{T_{01}} = 1 + \frac{\dot{Q}}{\dot{m}c_p T_{01}}$$

Using Equation 5.47, we can express the stream thrust at sections 1 and 2 as:

$$S_{T1} = \frac{\dot{m}\sqrt{RT_{01}}}{N_1(M_1, \gamma)}$$

and

$$S_{T2} = \frac{\dot{m}\sqrt{RT_{02}}}{N_2(M_2, \gamma)}$$

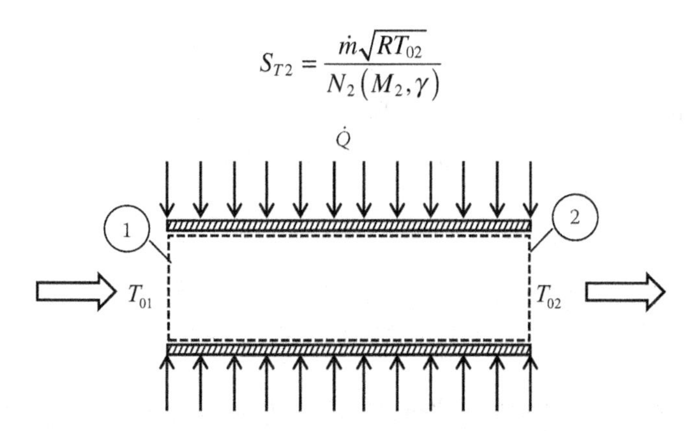

FIGURE 5.24 Control volume of a Rayleigh flow with heating

As the stream thrust remains constant in a Rayleigh flow, we obtain from the above expressions:

$$\frac{N_2(M_2,\kappa)}{N_1(M_1,\kappa)} = \sqrt{\frac{T_{02}}{T_{01}}} \tag{5.92}$$

which shows that for $T_{02} > T_{01}$ (heating the gas flow), we have $N_2 > N_1$. From the variation of flow function N with M, shown in Figure 5.10, we conclude that M_2 should be closer to $M = 1$ than to M_1. In other words, if the inlet flow is subsonic, the Mach number will increase downstream; if the flow is supersonic at the inlet, the Mach number will decrease downstream. Thus, heating in a Rayleigh flow raises a subsonic Mach number and lowers a supersonic Mach number toward the sonic condition ($M = 1$), called thermal choking. That is, for given inlet conditions, there is a maximum value of $\dot{Q}$ that makes the flow sonic.

To see how heating in a Rayleigh flow reduces the total pressure downstream, let use Equation 5.43 to express the impulse pressure p_i. For the constant impulse pressure between any two sections of a Rayleigh flow, we can write:

$$p_{01} I_{f01} = p_{02} I_{f02}$$

$$\frac{p_{02}}{p_{01}} = \frac{I_{f01}}{I_{f02}} \tag{5.93}$$

As M_2 is on the same side as M_1 and closer to $M = 1$, based on the variation I_{f1} with M, shown in Figure 5.9, we infer that $I_{f01} < I_{f02}$. Thus, Equation 5.93 yields $p_{02} < p_{01}$ for both subsonic and supersonic Rayleigh flows, which confirms that the increase in entropy due to heating in a Rayleigh flow results in a decrease in total pressure in the flow direction. This phenomenon in a Rayleigh flow is not as intuitive as the role friction plays in a Fanno flow. Conversely, unlike in a Fanno flow, the cooling in a Rayleigh flow increases total pressure downstream.

For a Rayleigh flow, Equation 5.62 reduces to:

$$\left(1 - M^2\right)\frac{dV}{V} = \frac{\delta \dot{Q}}{\dot{m} c_p T} \tag{5.94}$$

which shows that heating will increase flow velocity for a subsonic Rayleigh flow ($M < 1$) and will decrease it for a supersonic Rayleigh flow ($M > 1$). In addition, the equation also shows that cooling of the flow will have the opposite trends in velocity changes in the flow direction. Using the second law of thermodynamics and assuming reversible heat transfer, we can rewrite this equation as:

$$\left(1 - M^2\right)\frac{dV}{V} = \frac{ds}{c_p} \tag{5.95}$$

For the constant-area duct of a Rayleigh flow, Equation 5.55 reduces to:

$$\frac{dV}{V} + \frac{dp}{p} - \frac{dT}{T} = 0 \tag{5.96}$$

and Equation 5.56 reduces to:

$$\frac{dp}{p} = -\gamma M^2 \frac{dV}{V} \tag{5.97}$$

Eliminating dp / p between Equations 5.96 and 5.97 yields:

$$\frac{dV}{V} = \frac{1}{\left(1 - \gamma M^2\right)} \frac{dT}{T} \tag{5.98}$$

Finally, eliminating dV / V between Equations 5.95 and 5.98 yields:

$$\frac{ds}{c_p} = \frac{\left(1 - M^2\right)}{\left(1 - \gamma M^2\right)} \frac{dT}{T} \tag{5.99}$$

which reveals additional nonintuitive features of a Rayleigh flow. For a supersonic flow ($M > 1$), the coefficient of dT / T is always positive, resulting in monotonically increasing (for heating) and decreasing (for cooling) relationship between entropy and changes in static temperature. For a subsonic flow, however, the coefficient becomes negative for $1 / \sqrt{\gamma} < M < 1$. This implies that, in the Mach number range $1 / \sqrt{\gamma} < M < 1$, the static temperature decreases with heating and increases with cooling, an unexpectedly interesting behavior found in a Rayleigh flow. In this Mach number range for the case of heating, although the total temperature increases, the flow accelerates rapidly to satisfy the continuity equation with a concomitant reduction in the static temperature.

Further, we can find the points of maximum static temperature and maximum entropy from Equation 5.99. For the first, we set $dT / ds = 0$ and obtain $M = 1 / \sqrt{\gamma}$. For the second, we set $ds / dT = 0$ and obtain $M = 1$. Thus, in a Rayleigh flow, the maximum static temperature occurs at $M = 1 / \sqrt{\gamma}$ and the maximum entropy occurs at the sonic condition, corresponding to thermal choking. A Rayleigh line, which we present in Section 5.9.2, graphically shows these features deduced from Equation 5.99.

5.9.1 Quantitative Evaluation of Properties in a Rayleigh Flow

This section uses conservation equations of mass, momentum, energy, and entropy to derive relevant equations to compute changes in static pressure, static temperature, static density, total temperature, total pressure, and entropy in terms of their ratios between two sections of a Rayleigh flow with its impulse pressure and mass flux remaining constant.

5.9.1.1 Static Pressure Ratio

For the constant impulse pressure between any two sections of a Rayleigh flow, Equation 5.40 yields:

$$\frac{p_2}{p_1} = \frac{1 + \gamma M_1^2}{1 + \gamma M_2^2} \tag{5.100}$$

5.9.1.2 Static Temperature Ratio

For the mass flux remaining constant in a Rayleigh flow, we can write:

$$\rho_1 V_1 = \rho_2 V_2$$

which, using the equation of state of a perfect gas and the definition of Mach number, becomes:

$$\frac{p_1 M_1 \sqrt{\gamma R T_1}}{R T_1} = \frac{p_2 M_2 \sqrt{\gamma R T_2}}{R T_2}$$

$$\frac{T_2}{T_1} = \left(\frac{p_2 M_2}{p_1 M_1}\right)^2$$

which, using Equation 5.100, finally yields the static temperature ratio as:

$$\frac{T_2}{T_1} = \frac{M_2^2}{M_1^2}\left(\frac{1+\gamma M_1^2}{1+\gamma M_2^2}\right)^2 \tag{5.101}$$

5.9.1.3 Static Density Ratio

For the mass flux remaining constant in a Rayleigh flow, we can write:

$$\frac{\rho_2}{\rho_1} = \frac{V_1}{V_2} = \left(\frac{p_2}{p_1}\right)\left(\frac{T_1}{T_2}\right)$$

which, using Equations 5.100 and 5.101, becomes:

$$\frac{\rho_2}{\rho_1} = \frac{V_1}{V_2} = \frac{M_1^2}{M_2^2}\left(\frac{1+\gamma M_2^2}{1+\gamma M_1^2}\right) \tag{5.102}$$

5.9.1.4 Total Temperature Ratio

Using the isentropic relationship between the static temperature and the total temperatures at any section of the Rayleigh flow, we write:

$$\frac{T_{02}}{T_{01}} = \frac{T_2}{T_1}\left(\frac{1+\dfrac{\gamma-1}{2}M_2^2}{1+\dfrac{\gamma-1}{2}M_1^2}\right) = \frac{T_2}{T_1}\left(\frac{2+(\gamma-1)M_2^2}{2+(\gamma-1)M_1^2}\right)$$

which, using Equation 5.101, finally yields the total temperature ratio as:

$$\frac{T_{02}}{T_{01}} = \frac{M_2^2}{M_1^2}\left(\frac{1+\gamma M_1^2}{1+\gamma M_2^2}\right)^2\left(\frac{2+(\gamma-1)M_2^2}{2+(\gamma-1)M_1^2}\right) \tag{5.103}$$

5.9.1.5 Total Pressure Ratio

At a duct cross section, we rewrite the following isentropic relation between the total-to-static pressure ratio and the total-to-static temperature ratio:

$$\frac{p_0}{p} = \left(\frac{T_0}{T}\right)^{\frac{\gamma}{\gamma-1}}$$

as

$$\frac{p_{02}}{p_{01}} = \frac{p_2}{p_1}\frac{\left(\dfrac{T_{02}}{T_2}\right)^{\frac{\gamma}{\gamma-1}}}{\left(\dfrac{T_{01}}{T_1}\right)^{\frac{\gamma}{\gamma-1}}}$$

which, using Equation 5.100 and each total-to-static temperature ratio in terms of Mach number, becomes:

$$\frac{p_2}{p_1} = \left(\frac{1+\gamma M_1^2}{1+\gamma M_2^2}\right)\left(\frac{2+(\gamma-1)M_2^2}{2+(\gamma-1)M_1^2}\right)^{\frac{\gamma}{\gamma-1}} \tag{5.104}$$

5.9.1.6 Entropy Change

Combining the first and second laws of thermodynamics, we can write the entropy change between two states of a perfect gas (see Chapter 1) as:

$$ds = c_p \frac{dT}{T} - R\frac{dp}{p}$$

which upon integration between two sections of a Rayleigh flow yields:

$$\frac{s_2 - s_1}{R} = \left(\frac{\gamma}{\gamma - 1}\right)\ln\left(\frac{T_2}{T_1}\right) - \ln\left(\frac{p_2}{p_1}\right)$$

In this equation, using Equations 5.100 and 5.101 and simplifying the resulting expression, we finally obtain:

$$\frac{s_2 - s_1}{R} = \left(\frac{2\gamma}{\gamma - 1}\right)\ln\left(\frac{M_2}{M_1}\right) + \left(\frac{\gamma + 1}{\gamma - 1}\right)\ln\left(\frac{1 + \gamma M_1^2}{1 + \gamma M_2^2}\right) \tag{5.105}$$

5.9.2 Rayleigh Line

A Rayleigh line represents the variation of static temperature with entropy on a T-s plane. The impulse pressure and mass flux (mass flow rate per unit area) remain constant along a Rayleigh line. Using the Mach number as an intermediate independent variable, we can use Equations 5.101 and 5.105 to plot a Rayleigh line easily. In these equations, subscript 1 represents the fixed initial state, and we eliminate subscript 2 to describe a general point on the Rayleigh line. These equations then become:

$$\frac{T}{T_1} = \frac{M^2}{M_1^2}\left(\frac{1 + \gamma M_1^2}{1 + \gamma M^2}\right)^2 \tag{5.106}$$

and

$$\frac{s - s_1}{R} = \hat{s} = \left(\frac{2\gamma}{\gamma - 1}\right)\ln\left(\frac{M}{M_1}\right) + \left(\frac{\gamma + 1}{\gamma - 1}\right)\ln\left(\frac{1 + \gamma M_1^2}{1 + \gamma M^2}\right) \tag{5.107}$$

where $\hat{s}$ represents a dimensionless change in entropy from the fixed value at section 1 with known T_1 and M_1.

Figure 5.25 depicts some crucial features of a Rayleigh line. The heating increases the Mach number in a subsonic flow (upper branch) and decreases it in a supersonic flow (lower branch) to $M = 1$, corresponding to the maximum change in entropy. The point of maximum static temperature

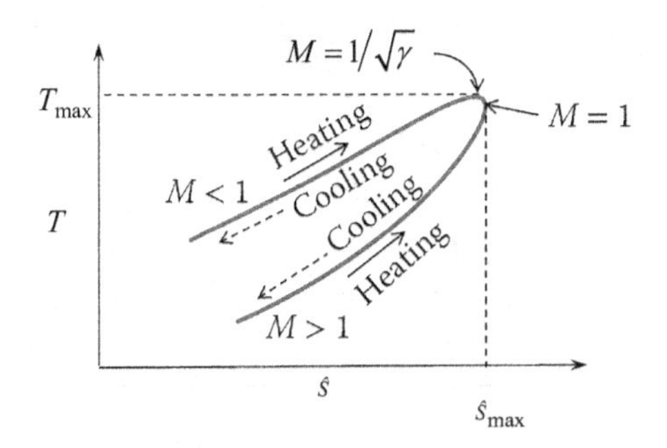

FIGURE 5.25 A Rayleigh line in the $T - \hat{s}$ plane.

occurs on the subsonic branch at $M = 1/\sqrt{\gamma}$. Beyond this point, the continued heating decreases static temperature until $M = 1$ (thermal choking), where the flow attains the maximum total temperature and entropy. The figure also shows that all the trends reverse for the cooling case in both subsonic and supersonic flows.

Let us write Equation 5.107 for entropy variation in a Rayleigh flow as:

$$s = s_1 + \left(\frac{2\gamma R}{\gamma - 1}\right)\ln\left(\frac{M}{M_1}\right) + R\left(\frac{\gamma+1}{\gamma-1}\right)\ln\left(\frac{1+\gamma M_1^2}{1+\gamma M^2}\right) \tag{5.108}$$

in which we evaluate s_1 relative to a reference state with the static pressure p_{ref} and static temperature T_{ref} where we assume the entropy to be zero. For a Rayleigh line, M_1 and p_1 at state 1 determine the impulse pressure $p_{i1} = p_1\left(1+\gamma M_1^2\right)$. Each value of T_{01} at state 1 yields a different value of mass flux, T_1, and s_1, computed by the following equations:

$$s_1 = c_p \ln\left(\frac{T_1}{T_{ref}}\right) - R\ln\left(\frac{p_1}{p_{ref}}\right)$$

$$T_1 = \frac{T_{01}}{\left(1+\frac{\gamma-1}{2}M_1^2\right)}$$

$$\hat{F}_{f1} = M_1\sqrt{\gamma\left(1+\frac{\gamma-1}{2}M_1^2\right)}$$

$$\dot{m}/A = \hat{F}_{f1}p_1/\sqrt{RT_{01}}.$$

Figure 5.26 shows three Rayleigh lines plotted using Equation 5.108 and the abovementioned approach. These Rayleigh lines have equal impulse pressure. At Mach number M_1, the outer Rayleigh line has a higher total temperature than the inner one. Because the outer Rayleigh line at the sonic point has higher entropy than the inner one, its total pressure at this point must be lower than for the inner one. Using the equation $\dot{m}/A = \hat{F}_{f0}^* p_0^* / \sqrt{RT_0}$ to compute the mass flux at the sonic point for each Rayleigh line shown in Figure 5.26 leads us to conclude that the mass flux of the outer Rayleigh line must be lower than that of the inner one.

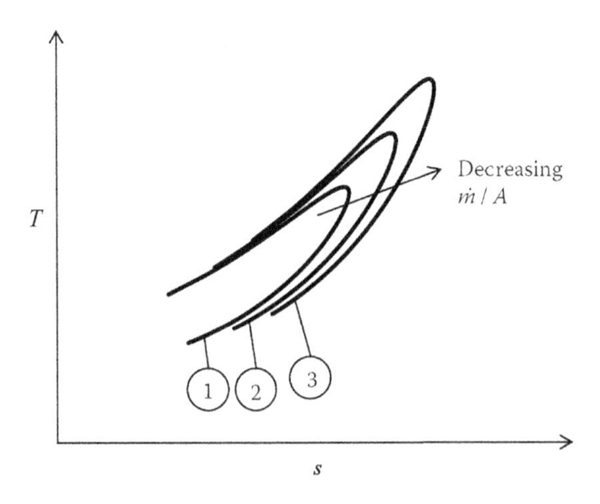

FIGURE 5.26 Rayleigh lines in the $T - s$ plane with varying mass flux.

5.9.3 Trends of Changes in Rayleigh Flow Properties

In Section 5.9.1, we have derived equations to evaluate ratios of various flow properties between any two sections of a Rayleigh flow. If section 1 corresponds to the reference section with thermal choking ($M = 1$) represented by quantities with a superscript* and section 2 corresponds to a general section without a subscript, we obtain the following equations to compute various property ratios:

$$\frac{p}{p^*} = \frac{1+\gamma}{1+\gamma M^2} \tag{5.109}$$

$$\frac{T}{T^*} = M^2\left(\frac{1+\gamma}{1+\gamma M^2}\right)^2 \tag{5.110}$$

$$\frac{\rho}{\rho^*} = \frac{V^*}{V} = \frac{1}{M^2}\left(\frac{1+\gamma M^2}{1+\gamma}\right) \tag{5.111}$$

$$\frac{T_0}{T^*} = M^2\left(\frac{1+\gamma}{1+\gamma M^2}\right)^2\left(\frac{2+(\gamma-1)M^2}{\gamma+1}\right) \tag{5.112}$$

$$\frac{p_0}{p^*} = \left(\frac{1+\gamma}{1+\gamma M^2}\right)\left(\frac{2+(\gamma-1)M^2}{\gamma+1}\right)^{\frac{\gamma}{\gamma-1}} \tag{5.113}$$

Figure 5.27 shows Equations 5.109–5.113 plots, depicting how various property ratios vary in a Rayleigh flow with Mach number for both heating and cooling. This figure does show that the maximum static temperature in a subsonic Rayleigh flow occurs at $M = 1/\sqrt{\gamma}$, an interesting behavior of the flow. Figure 5.28 summarizes the trends of property changes in a Rayleigh flow shown in Figure 5.27 and the entropy change shown in Figure 5.25 for a ready reference.

5.9.4 Loss in Total Pressure in a Rayleigh Flow with Heating

Loss in total pressure in a gas flow system is a critical engineering consideration with the design intent to reduce it as much as possible. In a Fanno flow, we can reduce the loss by lowering the wall friction factor. In a Rayleigh flow with heating, even with an assumed frictionless wall, the total

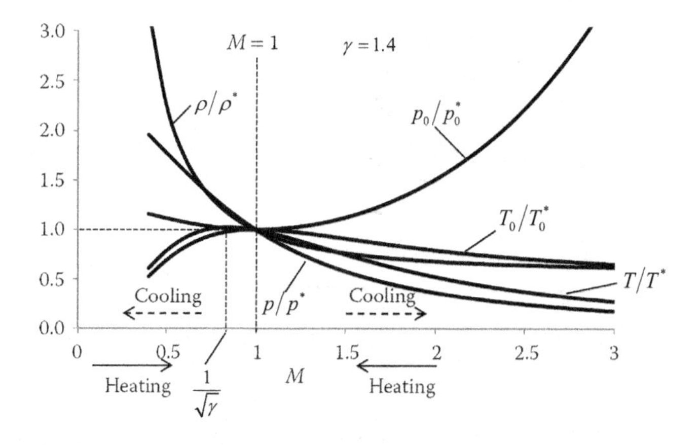

FIGURE 5.27 Variations of properties with Mach number in a Rayleigh flow for $\gamma = 1.4$.

	Heating		Cooling	
Property	$M < 1$	$M > 1$	$M < 1$	$M > 1$
M	↑	↓	↓	↑
s	↑	↑	↓	↓
p	↓	↑	↑	↓
V	↑	↓	↓	↑
ρ	↓	↑	↑	↓
p_0	↓	↓	↑	↑
T_0	↑	↑	↓	↓
T	$M < 1/\sqrt{\gamma}$ ↑ $1/\sqrt{\gamma} < M < 1$ ↓	↑	$M < 1/\sqrt{\gamma}$ ↓ $1/\sqrt{\gamma} < M < 1$ ↑	↓

FIGURE 5.28 Trends of property changes in a Rayleigh flow.

pressure loss is inevitable. In this section, let us understand how to reduce the total pressure loss in a Rayleigh flow resulting from a desired amount of gas heating.

Let us define the total pressure loss coefficient K_R between two sections of a Rayleigh flow as:

$$K_R = \frac{p_{01} - p_{02}}{p_{01} - p_1} \tag{5.114}$$

where the subscripts 1 and 2 correspond to duct inlet and outlet, respectively. For values of T_{02} / T_{01} ranging from 2 to 8, Figure 5.29 shows the variation of K_R in a Rayleigh flow with inlet Mach number M_1. The dash line shown in the figure corresponds to the limit of thermal choking with $M_2 = 1$. We leave the method of generating the plot in Figure 5.29 as an exercise for the reader.

Figure 5.29 highlights some important features of heating in a Rayleigh flow. For each value of inlet Mach number, K_R increases with T_{02} / T_{01} to the maximum value limited by thermal choking. For each value of T_{02} / T_{01}, the minimum value of K_R asymptotes to

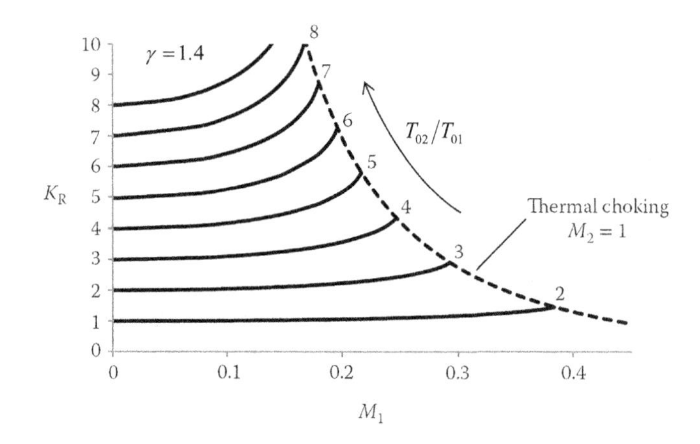

FIGURE 5.29 Variation of total pressure loss coefficient K_R with inlet Mach number in a Rayleigh flow for $\gamma = 1.4$.

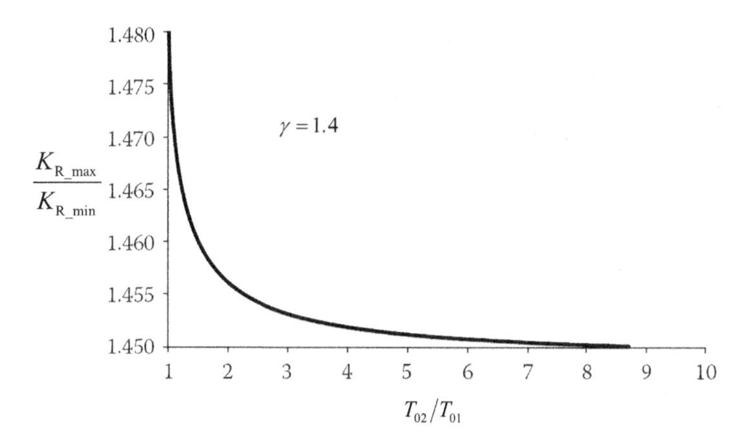

FIGURE 5.30 Variation of K_{R_max}/K_{R_min} with $T_{02}\,/\,T_{01}$ in a Rayleigh flow for $\gamma = 1.4$.

$$K_{R\_min} = \frac{T_{02}}{T_{01}} - 1 \tag{5.115}$$

in the incompressible flow regime. Thus, to minimize the loss in total pressure for a given value of $T_{02}\,/\,T_{01}$ in a Rayleigh flow, we should heat it at as low a Mach number as possible. The corresponding maximum value K_{R_max} occurs at the inlet Mach number M_1 that results in thermal choking ($M_2 = 1$), as shown in the figure.

Figure 5.30 shows the variation of the ratio K_{R_max}/K_{R_min} with $T_{02}\,/\,T_{01}$ for air with $\gamma = 1.4$. We obtain the following simple relation that fits the curve in this figure within an error of less than 0.4%:

$$\frac{K_{R\_max}}{K_{R\_min}} = \frac{1.450}{\tanh(2.3T_{02}\,/\,T_{01})} \tag{5.116}$$

Using the influence coefficients presented in Table 8.2 of Shapiro (1953), we obtain the following simple relation between the fractional change in total temperature and the corresponding fractional change in total pressure as a function of Mach number:

$$\frac{dp_0}{p_0} = -\gamma M^2 \frac{dT_0}{T_0}$$
$$\frac{d(\ln p_0)}{d(\ln T_0)} = -\gamma M^2 \tag{5.117}$$

which is true also for a variable-area duct and shows that the total pressure in a Rayleigh flow will decrease with an increase in total temperature (heating) and will increase with a decrease in total temperature (cooling). For a given change in total temperature, the lowest change in total pressure occurs at the lowest possible Mach number.

5.9.5 Rayleigh Flow Table with the Reference State at $M = 1$

In Sultanian (2023), Table A.4 for $\gamma = 1.4$ and Table B.4 for $\gamma = 1.333$ present properties of subsonic and supersonic Rayleigh flows using Mach number as the independent variable. Although each value in the table is accurate to five decimal places, a linear interpolation between two adjacent values may yield an inaccurate result if the actual variation is not linear. In that case, one should use the actual equations unless one is interested in an approximate solution. Besides yielding a quick and approximate solution, these tables are helpful where the original equations require an iterative solution. One can enter each table through any known variable and quickly obtain the values of all other variables along the corresponding row.

5.10 ISOTHERMAL CONSTANT-AREA FLOW WITH FRICTION

From the behavior of a Fanno flow, we can conclude that an isothermal compressible flow in a constant-area duct with friction will require simultaneous heat transfer (heating or cooling depending on the flow Mach number) to maintain a constant static temperature of the gas. Due to the nonlinear coupling between the friction and heat transfer effects in an isothermal flow, we cannot model them by a linear superposition of the Fanno and Rayleigh flows. We can model long underground gas pipelines as an isothermal flow.

While a general compressible flow in a constant-area duct with friction and heat transfer will require a numerical solution, an isothermal compressible flow allows us to obtain closed-form analytical solutions to compute various property changes. For an isothermal flow, using the definition of the Mach number and the perfect gas law, we get the following equations:

$$\frac{dV}{V} = \frac{dM}{M} \tag{5.118}$$

$$\frac{dp}{p} = \frac{d\rho}{\rho} \tag{5.119}$$

From the continuity equation in a constant-area duct, we obtain $dV / V = -d\rho / \rho$. Combining this relation with Equations 5.118 and 5.119 yields:

$$\frac{dp}{p} = -\frac{dM}{M} \tag{5.120}$$

whose integration between two sections of the duct results in the following equation for the static pressure ratio:

$$\ln\left(\frac{p_2}{p_1}\right) = \ln\left(\frac{M_1}{M_2}\right)$$

$$\frac{p_2}{p_1} = \frac{M_1}{M_2} \tag{5.121}$$

Using isentropic relations between static and total properties, we obtain the total pressure ratio as:

$$\frac{p_{02}}{p_{01}} = \frac{p_2\left(1+\frac{\gamma-1}{2}M_2^2\right)^{\frac{\gamma}{\gamma-1}}}{p_1\left(1+\frac{\gamma-1}{2}M_1^2\right)^{\frac{\gamma}{\gamma-1}}} = \frac{M_1}{M_2}\left(\frac{2+(\gamma-1)M_2^2}{2+(\gamma-1)M_1^2}\right)^{\frac{\gamma}{\gamma-1}} \tag{5.122}$$

and the total temperature ratio as:

$$\frac{T_{02}}{T_{01}} = \frac{T_2\left(1+\frac{\gamma-1}{2}M_2^2\right)}{T_1\left(1+\frac{\gamma-1}{2}M_1^2\right)} = \frac{2+(\gamma-1)M_2^2}{2+(\gamma-1)M_1^2} \tag{5.123}$$

With the substitution of $\mathrm{d}T = 0$ and $\mathrm{d}p / p$ from Equation 5.120, Equation 5.84 becomes:

$$\frac{f\mathrm{d}x}{D_h} = 2\left(\frac{1}{\gamma M^2} - 1\right)\frac{\mathrm{d}M}{M} \tag{5.124}$$

which illustrates some interesting features of an isothermal compressible flow with friction. Unlike a Fanno flow, where the critical point corresponds to frictional choking at $M = 1$, the critical point in the isothermal flow occurs at $M = 1/\sqrt{\gamma}$ where the coefficient of $\mathrm{d}M / M$ in this equation becomes zero. For $M < 1/\sqrt{\gamma}$, the Mach number increases in the flow direction, and for $M > 1/\sqrt{\gamma}$, the Mach number decreases downstream. Equation 5.123 reveals that for $M < 1/\sqrt{\gamma}$, the flow needs to be heated ($T_{02} > T_{01}$), and for $M > 1/\sqrt{\gamma}$, the flow needs to be cooled ($T_{02} < T_{01}$). Because the wall friction effect is irreversible, it is physically impossible to isothermally accelerate or decelerate the flow across the critical Mach number of $M = 1/\sqrt{\gamma}$. Thus, a supersonic isothermal flow can proceed to cross the choking point ($M = 1$) to become subsonic up to the critical Mach number of $1/\sqrt{\gamma}$. Figure 5.31 shows some of these interesting and counterintuitive features of an isothermal flow.

We can easily integrate Equation 5.124 between two sections of an isothermal flow duct to yield:

$$\frac{fL}{D_\mathrm{h}} = 2\ln\left(\frac{M_1}{M_2}\right) + \frac{1}{\gamma M_1^2} - \frac{1}{\gamma M_2^2} \tag{5.125}$$

which relates the length of an isothermal flow duct of hydraulic mean diameter D_h and wall with the Darcy friction factor f to the Mach numbers at the duct inlet and outlet. The gas viscosity also remains constant because the static temperature remains constant in an isothermal flow. As a result, for a given mass flow rate, the flow Reynolds number remains constant and so does the Darcy friction f.

Let us assume the reference state at the critical point with $M_1 = 1/\sqrt{\gamma}$. Denoting the properties at this state with a star, we obtain the following equations from Equations 5.121 through 5.123 and 5.125. We have dropped the subscript 2 in these equations to represent a general section.

Static pressure ratio:

$$\frac{p}{p^*} = \frac{\rho}{\rho^*} = \frac{1}{M\sqrt{\gamma}} \tag{5.126}$$

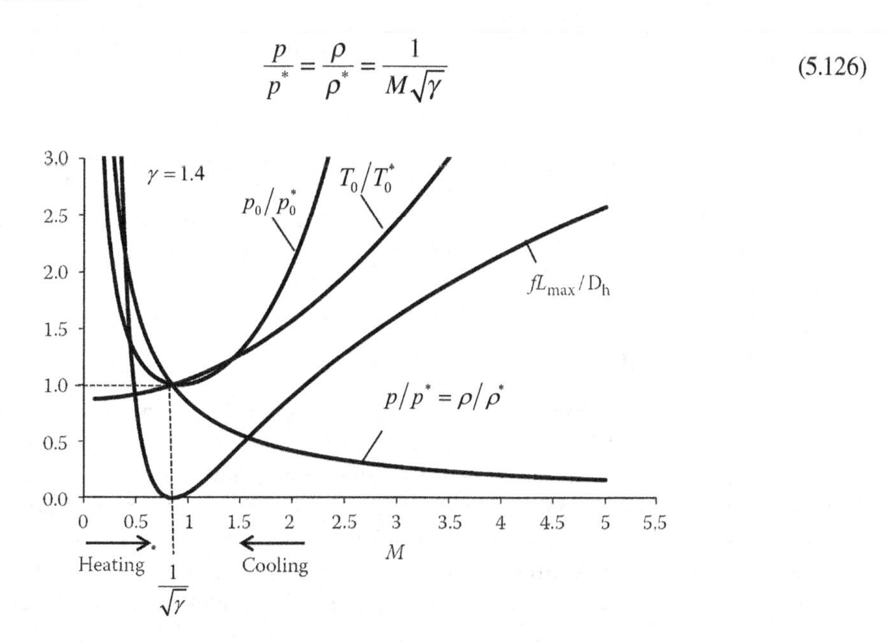

FIGURE 5.31 Variations of properties with Mach number in an isothermal flow for $\gamma = 1.4$

Total pressure ratio:

$$\frac{p_0}{p_0^*} = \frac{1}{M\sqrt{\gamma}}\left(\frac{\gamma(2+(\gamma-1)M^2)}{(3\gamma-1)}\right)^{\frac{\gamma}{\gamma-1}} \tag{5.127}$$

Total temperature ratio:

$$\frac{T_0}{T_0^*} = \frac{\gamma\left(2+(\gamma-1)M^2\right)}{(3\gamma-1)} \tag{5.128}$$

Maximum duct length:

$$\frac{fL_{max}}{D_h} = \ln\left(\gamma M^2\right) + \frac{1}{\gamma M^2} - 1 \tag{5.129}$$

To show the trend of variations of various flow properties with Mach number, the above equations are plotted in Figure 5.31. As the flow cannot cross the critical point $M = 1/\sqrt{\gamma}$, we should read the plots in the figure from left to right for flows with $M < 1/\sqrt{\gamma}$ and right to left for flows with $M > 1/\sqrt{\gamma}$. A few features of the isothermal flow shown in the figure are worth noting. Because heating and friction reduce total pressure and cooling increases the total pressure in the flow direction, the total pressure variation shown in the figure for $M < 1/\sqrt{\gamma}$ is much steeper than that for $M > 1/\sqrt{\gamma}$. In addition, the heating rate for $M < 1/\sqrt{\gamma}$ is much lower than the cooling rate needed for $M > 1/\sqrt{\gamma}$ to maintain an isothermal flow with friction. Note that in the range $1/\sqrt{\gamma} < M < 1.5$, the rate of total pressure decrease is nearly equal to the rate of total temperature decrease in the flow direction.

5.10.1 Table for Isothermal Constant-Area Flow with Friction Using the Reference State at $M = 1/\sqrt{\gamma}$

In Sultanian (2023), Table A.5 for $\gamma = 1.4$ and Table B.5 for $\gamma = 1.333$ present properties of isothermal flows in a constant-area duct with friction as a function of Mach number. Although each value in the table is accurate to five decimal places, a linear interpolation between two adjacent values may yield an inaccurate result if the variation is nonlinear. In that case, one should use the actual equations unless one requires a quick and approximate solution. Besides yielding such a solution, each table is helpful in cases where the original equations require an iterative solution. One can enter the table through any known variable and quickly obtain the values of all other variables along the corresponding row.

5.11 NORMAL SHOCK

A normal shock is a compression wave normal to the direction of a compressible flow. Even with no area change, friction, heat transfer, and rotation, variations of flow properties across a normal shock are abrupt. These changes primarily occur due to the nonequilibrium and nonisentropic changes the flow undergoes to adjust to the specified higher downstream static pressure boundary condition. Because the flow across a normal shock is adiabatic without friction, the entropy must increase through the shock in the flow direction: a condition often used to screen out a physically unrealistic shock wave solution. While a normal shock wave has a finite thickness, we will neglect it in all our analyses in this book. Because property changes across a normal shock are discontinuous, we cannot use differential equations to model a normal shock region. Instead, we will use a control volume analysis of the governing conservation laws to develop simple algebraic equations that relate flow properties before and after the shock.

Two essential properties of a normal shock are worth noting. First, a normal shock can occur only in a supersonic compressible flow. Second, regardless of how strong the upstream supersonic flow is, as measured by its Mach number, the flow downstream of the normal shock is always subsonic. While the total temperature and impulse pressure remain constant across a normal shock, the total pressure decreases, and the static pressure increases.

5.11.1 Governing Conservation and Auxiliary Equations

Figure 5.32 shows a normal shock enclosed by a constant-area control volume, illustrating various flow properties at the inlet (section 1) and the exit (section 2). The control volume is thin enough, and the flow is fast enough to render the normal shock adiabatic (not isentropic) with negligible effect of wall friction. Under these assumptions, we can summarize the conservation and auxiliary equations governing a normal shock as follows:

5.11.1.1 Continuity Equation

With equal flow area at inlet and exit of the control volume around the normal shock, the mass conservation yields:

$$\rho_1 V_1 = \rho_2 V_2 \tag{5.130}$$

5.11.1.2 Linear Momentum Equation

With no surface and body force acting on the control volume, the linear momentum equation reduces to:

$$p_1 + \rho_1 V_1^2 = p_2 + \rho_2 V_2^2 \tag{5.131}$$

5.11.1.3 Energy Equation

With no heat transfer and work transfer, the energy conservation over the control volume yields:

$$h_1 + \frac{V_1^2}{2} = h_2 + \frac{V_2^2}{2} = h_0 = \text{constant} \tag{5.132}$$

In addition to Equations 5.130 through 5.132, the assumptions of a perfect gas and constant c_p lead to two more equations: $p = \rho RT$ and $h = c_p T$. Thus, we have the system of five equations for finding all five quantities in section 2 when we know them in section 1.

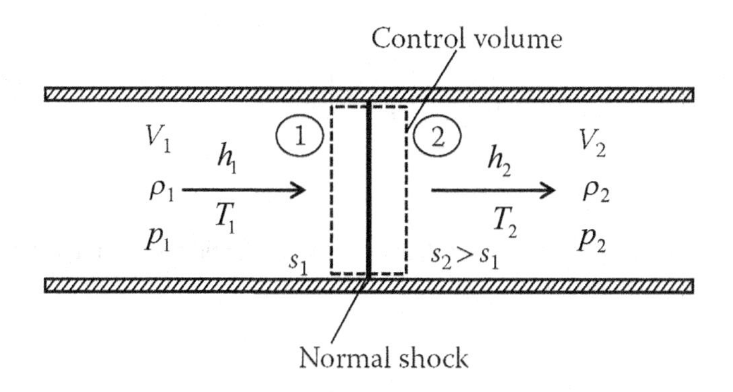

FIGURE 5.32 Control volume analysis of a normal shock.

5.11.2 Rankine–Hugoniot Equations

For a normal shock being a compression wave, the static pressure ratio across it is a measure of its strength—the higher the static pressure rises across the shock, the stronger it is. The total pressure decreases across a normal shock because the entropy increases. In the following, we derive Rankine–Hugoniot equations that express ratios of densities, velocities, and temperatures in terms of the static pressure ratio across a normal shock.

Let us rewrite the energy equation (Equation 5.132) in terms of static temperature as:

$$c_p T_1 + \frac{V_1^2}{2} = c_p T_2 + \frac{V_2^2}{2}$$

$$V_1^2 - V_2^2 = 2c_p\left(T_2 - T_1\right) = \frac{2c_p}{R}\left(\frac{p_2}{\rho_2} - \frac{p_1}{\rho_1}\right) \tag{5.133}$$

where we have used the equation of state $p = \rho RT$ to replace the static temperature in terms of static pressure and density.

Combining the continuity equation (Equation 5.130) with the linear momentum equation (Equation 5.131) yields:

$$V_1 - V_2 = \frac{p_2 - p_1}{\rho_1 V_1}$$

Multiplying both sides of this equation by $(V_1 + V_2)$ yields:

$$V_1^2 - V_2^2 = \left(\frac{p_2 - p_1}{\rho_1 V_1}\right)(V_1 + V_2) = (p_2 - p_1)\left(\frac{1}{\rho_1} + \frac{1}{\rho_2}\right) \tag{5.134}$$

where we have used the continuity equation to replace V_2 / V_1 by ρ_1 / ρ_2. From this equation and Equation 5.133, we obtain:

$$\frac{2c_p}{R}\left(\frac{p_2}{\rho_2} - \frac{p_1}{\rho_1}\right) = (p_2 - p_1)\left(\frac{1}{\rho_1} + \frac{1}{\rho_2}\right)$$

Replacing c_p / R by $\gamma / (\gamma - 1)$ in this equation and simplifying the algebra, we finally obtain the following Rankine–Hugoniot equation that relates the density ratio to the static pressure ratio across a normal shock:

$$\frac{\rho_2}{\rho_1} = \frac{1 + \left(\dfrac{\gamma+1}{\gamma-1}\right)\dfrac{p_2}{p_1}}{\left(\dfrac{\gamma+1}{\gamma-1}\right) + \dfrac{p_2}{p_1}} = \frac{V_1}{V_2} \tag{5.135}$$

If the flow were isentropic across a normal shock, the density ratio will be related to the static pressure ratio by the following equation:

$$\left(\frac{\rho_2}{\rho_1}\right)_{\text{isentopic}} = \left(\frac{p_2}{p_1}\right)^{\frac{1}{\gamma}} \tag{5.136}$$

Using the equation of state along with Equation 135, we obtain the static temperature ratio across a normal shock as:

$$\frac{T_2}{T_1}=\left(\frac{p_2}{p_1}\right)\left(\frac{\rho_1}{\rho_2}\right)=\frac{\left(\frac{\gamma+1}{\gamma-1}\right)\frac{p_2}{p_1}+\left(\frac{p_2}{p_1}\right)^2}{1+\left(\frac{\gamma+1}{\gamma-1}\right)\frac{p_2}{p_1}} \tag{5.137}$$

We write the corresponding isentropic relation between the static temperature ratio and the static pressure ratio as:

$$\left(\frac{T_2}{T_1}\right)_{\text{isentropic}}=\left(\frac{p_2}{p_1}\right)^{\frac{\gamma-1}{\gamma}} \tag{5.138}$$

Finally, we can compute entropy change across a normal shock as:

$$\Delta s = c_p \ln\left(\frac{T_2}{T_1}\right)-R\ln\left(\frac{p_2}{p_1}\right)$$

$$\frac{s_2-s_1}{R}=\left(\frac{\gamma}{\gamma-1}\right)\ln\left(\frac{T_2}{T_1}\right)-\ln\left(\frac{p_2}{p_1}\right) \tag{5.139}$$

where we obtain T_2/T_1 from Equation 5.137.

Figure 5.33 illustrates variations of various properties (for a gas with $\gamma=1.4$) computed from Rankine–Hugoniot equations derived in this section. As shown in this figure, only for $p_2/p_1 \geq 1$, we obtain $(s_2-s_1)\geq 0$, which means that the static pressure must always increase across a normal shock, and so should the static density and static temperature. Thus, the results plotted in the third quadrant of this figure are physically unrealizable. Because the static density increases across a normal shock, from the continuity equation (Equation 5.130), we can conclude that the velocity must decrease across it. The figure further shows that the weak normal shocks with $p_2/p_1 \leq 2$ are nearly isentropic. For strong normal shocks with $p_2/p_1 > 2$, the increase in the static density is lower than, and the increase in the static temperature is higher than, their corresponding isentropic changes.

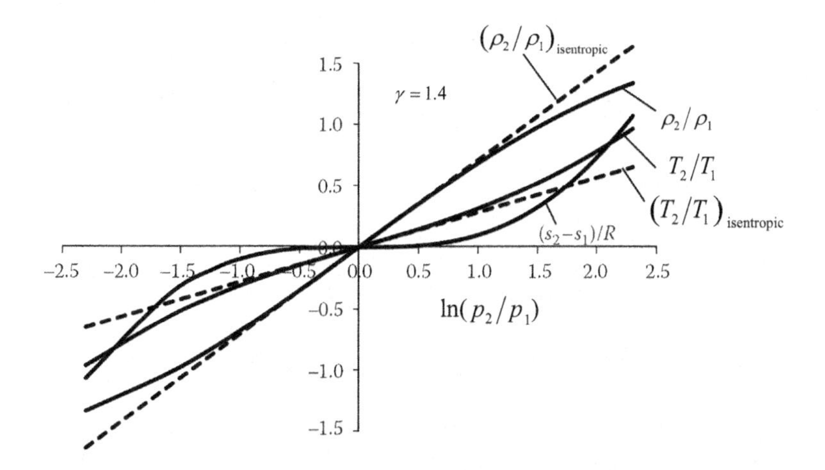

FIGURE 5.33 Variations of properties across a normal shock from Rankine–Hugoniot equations for $\gamma = 1.4$.

The increase in static temperature downstream of a normal shock indicates how the external kinetic energy associated with the flow is scrambled into random molecular motions over a very short distance (thickness of the normal shock) through intense irreversible dissipation processes driven by fluid viscosity. This leads to an abrupt increase in entropy across a normal shock even when the overall process is adiabatic, and the impulse pressure remains constant.

5.11.3 Constancy of Normal Shock Function

In Section 5.3 on compressible flow functions, we introduced the normal shock function $N(M,\gamma)$ as the ratio of the mass flow function and impulse function, given by:

$$N(M,\gamma)=\frac{\hat{F}_{\mathrm{f}}}{I_{\mathrm{f}}}=\frac{\hat{F}_{\mathrm{f0}}}{I_{\mathrm{f0}}}=\frac{M}{1+\gamma M^2}\sqrt{\gamma\left(1+\frac{\gamma-1}{2}M^2\right)}$$

As the mass flow rate, total temperature, and impulse pressure remain constant across a normal shock, so should the normal shock function $N(M,\gamma)$. We can use this constraint to derive a unique relationship between the Mach numbers across a normal shock. If M_1 and M_2 are the Mach numbers upstream and downstream of a normal shock, this equation for $N(M,\gamma)$ yields:

$$\frac{M_1}{1+\gamma M_1^2}\sqrt{\gamma\left(1+\frac{\gamma-1}{2}M_1^2\right)}=\frac{M_2}{1+\gamma M_2^2}\sqrt{\gamma\left(1+\frac{\gamma-1}{2}M_2^2\right)}$$

By squaring both sides of this equation, cross-multiplying terms, and simplifying, we obtain:

$$M_1^2\left\{2+(\gamma-1)M_1^2\right\}\left(1+\gamma M_2^2\right)^2=M_2^2\left\{2+(\gamma-1)M_2^2\right\}\left(1+\gamma M_1^2\right)^2$$

Note that we obtain the RHS of this equation from the left-hand side (LHS) by replacing M_1 by M_2 and M_2 by M_1.

LHS:

$$\begin{aligned}
&M_1^2\left\{2+(\gamma-1)M_1^2\right\}\left(1+\gamma M_2^2\right)^2\\
&=M_1^2\left\{2+(\gamma-1)M_1^2\right\}\left(1+2\gamma M_2^2+\gamma^2 M_2^4\right)\\
&=2M_1^2+4\gamma M_1^2M_2^2+2\gamma^2M_1^2M_2^4+(\gamma-1)M_1^4\\
&+2\gamma(\gamma-1)M_1^4M_2^2+\gamma^2(\gamma-1)M_1^4M_2^4
\end{aligned}$$

Similarly, we can write the RHS by inspection as:

$$\begin{aligned}
&M_2^2\left\{2+(\gamma-1)M_2^2\right\}\left(1+\gamma M_1^2\right)^2\\
&=M_2^2\left\{2+(\gamma-1)M_2^2\right\}\left(1+2\gamma M_1^2+\gamma^2 M_1^4\right)\\
&=2M_2^2+4\gamma M_2^2M_1^2+2\gamma^2M_2^2M_1^4+(\gamma-1)M_2^4\\
&+2\gamma(\gamma-1)M_2^4M_1^2+\gamma^2(\gamma-1)M_2^4M_1^4
\end{aligned}$$

Subtracting RHS from LHS results in the following equation:

$$2\left(M_1^2-M_2^2\right)-2\gamma^2M_1^2M_2^2\left(M_1^2-M_2^2\right)+(\gamma-1)\left(M_1^4-M_2^4\right)+2\gamma(\gamma-1)M_1^2M_2^2\left(M_1^2-M_2^2\right)=0$$

As $\left(M_1^2 - M_2^2\right) \neq 0$, factoring it out reduces this equation to:

$$2 - 2\gamma^2 M_1^2 M_2^2 + (\gamma - 1)\left(M_1^2 + M_2^2\right) + 2\gamma(\gamma - 1) M_1^2 M_2^2 = 0$$

$$2 + (\gamma - 1) M_1^2 + (\gamma - 1) M_2^2 - 2\gamma M_1^2 M_2^2 = 0$$

which finally leads to the following equation that allows us to compute M_2 in terms of M_1:

$$M_2^2 = \frac{2 + (\gamma - 1) M_1^2}{2\gamma M_1^2 - (\gamma - 1)} \tag{5.140}$$

which indicates a one-to-one relationship between the Mach numbers across a normal shock. For $M_1 \to \infty$, this equation has an asymptotic value $M_2 = \sqrt{(\gamma - 1)/(2\gamma)}$, which for $\gamma = 1.4$ yields $M_2 = 0.378$.

5.11.4 Prandtl–Meyer Equation

Using the conservation and auxiliary equations presented at the beginning of this section, we will now derive a powerful equation that relates the characteristic Mach numbers across a normal shock. This equation, known as the Prandtl–Meyer equation, states that the product of characteristic Mach numbers across a normal shock is always unity; that is, one is the reciprocal of the other.

Dividing the momentum equation (Equation 5.131) by the continuity equation (Equation 5.130) yields:

$$\frac{p_1}{\rho_1 V_1} + V_1 = \frac{p_2}{\rho_2 V_2} + V_2$$

Substituting $p_1 / \rho_1 = RT_1$ and $p_2 / \rho_2 = RT_2$ from the perfect gas law in this equation, we obtain:

$$\frac{RT_1}{V_1} + V_1 = \frac{RT_2}{V_2} + V_2 \tag{5.141}$$

Because the total temperature remains constant in a normal shock, we can substitute $T_1 = T_0 - \frac{V_1^2}{2c_p}$ and $T_2 = T_0 - \frac{V_2^2}{2c_p}$ in this equation to yield:

$$\frac{R}{V_1}\left(T_0 - \frac{V_1^2}{2c_p}\right) + V_1 = \frac{R}{V_2}\left(T_0 - \frac{V_2^2}{2c_p}\right) + V_2$$

$$RT_0\left(\frac{V_2 - V_1}{V_1 V_2}\right) + (V_2 - V_1)\left(\frac{R}{2c_p} - 1\right) = 0$$

$$(V_2 - V_1)\left(\frac{\gamma R T_0}{V_1 V_2} - \frac{\gamma + 1}{2}\right) = 0$$

which has two solutions. The first solution $V_2 - V_1 = 0$ is a trivial one. The second solution yields:

$$\frac{\gamma R T_0}{V_1 V_2} - \frac{\gamma + 1}{2} = 0$$

$$V_1V_2 = \gamma R\left(\frac{2T_0}{\gamma+1}\right) = \gamma RT^* = C^{*2} \tag{5.142}$$

where we have used Equation 5.13, which gives the ratio between the total temperature T_0 and the static temperature T^* at $M = 1$ equal to $(\gamma+1)/2$. We can further write this equation in terms of characteristic Mach numbers as:

$$M_1^* M_2^* = 1 \tag{5.143}$$

which is a powerful simple equation, known as the Prandtl–Meyer equation, which relates the characteristic Mach numbers across a normal shock. From our discussion of the Rankine–Hugoniot equations, we concluded that a normal shock is a compression wave with $\rho_2/\rho_1 = V_1/V_2 > 1$. Combining this fact with this equation, we can write:

$$\frac{\rho_2}{\rho_1} = \frac{V_1}{V_2} = \frac{M_1^*}{M_2^*} = M_1^{*2} > 1 \tag{5.144}$$

which reveals an important feature of a normal shock that its upstream Mach number must be supersonic. In other words, we cannot have a normal shock in a subsonic flow. Writing $M_2^* = 1/M_1^*$ from Equation 5.143 reveals another important feature of a normal shock: its downstream Mach number must always be subsonic.

5.11.5 Changes in Properties with Upstream Mach Number

The Rankine–Hugoniot equations presented in Section 5.11.2 provide the equations to compute the density ratio and static temperature ratio in terms of static pressure ratio across a normal shock. In the following, we use the Prandtl–Meyer equation (Equation 5.143) to conveniently derive various equations to compute changes in all flow properties across a normal shock as a function of the upstream Mach number M_1, uniquely determining the downstream Mach number.

Downstream Mach Number (M_2)

Using Equation 5.17 to express the characteristic Mach number in terms of the regular Mach number in Equation 5.143, we obtain:

$$\left\{\frac{(\gamma+1)M_1^2}{2+(\gamma-1)M_1^2}\right\}\left\{\frac{(\gamma+1)M_2^2}{2+(\gamma-1)M_2^2}\right\} = 1$$

$$(\gamma+1)^2 M_1^2 M_2^2 = \left\{2+(\gamma-1)M_1^2\right\}\left\{2+(\gamma-1)M_2^2\right\}$$

$$(\gamma+1)^2 M_1^2 M_2^2 = 4 + 2(\gamma-1)M_1^2 + 2(\gamma-1)M_2^2 + (\gamma-1)^2 M_1^2 M_2^2$$

$$2\gamma M_1^2 M_2^2 - (\gamma-1)M_2^2 = 2 + (\gamma-1)M_1^2$$

$$M_2^2 = \frac{2+(\gamma-1)M_1^2}{2\gamma M_1^2 - (\gamma-1)}$$

which is identical to Equation 5.140, resulting from $N_2 = N_1$ in a normal shock.

Density Ratio (ρ_2 / ρ_1)

Substituting Equation 5.17 in Equation 5.144 yields:

$$\frac{\rho_2}{\rho_1} = M_1^{*2} = \frac{(\gamma+1)M_1^2}{2+(\gamma-1)M_1^2} \tag{5.145}$$

which reveals that, for $M_1 \to \infty$, the asymptotic value of ρ_2 / ρ_1 becomes $(\gamma+1)/(\gamma-1)$, which equals 6 for $\gamma = 1.4$.

Static Pressure Ratio (p_2 / p_1)

Combining the continuity equation (Equation 5.130) with the momentum equation (Equation 5.131), we obtain:

$$p_2 - p_1 = \rho_1 V_1 (V_1 - V_2)$$

Dividing this equation by p_1 yields:

$$\frac{p_2}{p_1} - 1 = \frac{\rho_1 V_1^2}{p_1}\left(1 - \frac{V_2}{V_1}\right)$$

in which, substituting the equation of state for a perfect gas and Equation 5.144, we obtain:

$$\frac{p_2}{p_1} - 1 = \gamma M_1^2\left(1 - \frac{1}{M_1^{*2}}\right)$$

which, after substituting M_1^* from Equation 5.17 and simplifying the resulting equation, yields:

$$\frac{p_2}{p_1} = \frac{2\gamma M_1^2 - (\gamma-1)}{(\gamma+1)} \tag{5.146}$$

which asymptotes to $p_2 / p_1 = 2\gamma M_1^2 / (\gamma+1)$ for large values of M_1.

Static Temperature Ratio (T_2 / T_1)

Using the equation of state of a perfect gas, we can write:

$$\frac{T_2}{T_1} = \left(\frac{p_2}{p_1}\right)\left(\frac{\rho_1}{\rho_2}\right)$$

which, with the substitutions of Equations 5.145 and 5.146, yields:

$$\frac{T_2}{T_1} = \left\{\frac{2\gamma M_1^2 - (\gamma-1)}{(\gamma+1)}\right\}\left\{\frac{2+(\gamma-1)M_1^2}{(\gamma+1)M_1^2}\right\}$$

$$\frac{T_2}{T_1} = \frac{\{2\gamma M_1^2 - (\gamma-1)\}\{2+(\gamma-1)M_1^2\}}{(\gamma+1)^2 M_1^2} \tag{5.147}$$

which reduces to the parabolic dependence given by $T_2 / T_1 = 2\gamma(\gamma-1)M_1^2 / (\gamma+1)^2$ for large values of M_1.

Total Pressure Ratio (p_{02} / p_{01})

We can write the total pressure ratio across a normal shock as:

$$\frac{p_{02}}{p_{01}} = \frac{\left(\dfrac{p_{02}}{p_2}\right)}{\left(\dfrac{p_{01}}{p_1}\right)}\left(\frac{p_2}{p_1}\right)$$

in which expressing the total-to-static pressure ratio in terms of total-to-static temperature ratio from the isentropic relation on either side of the normal shock yields:

$$\frac{p_{02}}{p_{01}} = \frac{\left(\dfrac{T_{02}}{T_2}\right)^{\frac{\gamma}{\gamma-1}}}{\left(\dfrac{T_{01}}{T_1}\right)^{\frac{\gamma}{\gamma-1}}}\left(\frac{p_2}{p_1}\right)$$

For $T_{02} = T_{01}$ in a normal shock, this equation reduces to:

$$\frac{p_{02}}{p_{01}} = \left(\frac{p_2}{p_1}\right)\left(\frac{T_1}{T_2}\right)^{\frac{\gamma}{\gamma-1}}$$

Using Equations 5.146 and 5.147 in this equation yields:

$$\frac{p_{02}}{p_{01}} = \left\{\frac{2\gamma M_1^2-(\gamma-1)}{(\gamma+1)}\right\}\left[\frac{(\gamma+1)^2 M_1^2}{\{2\gamma M_1^2-(\gamma-1)\}\{2+(\gamma-1)M_1^2\}}\right]^{\frac{\gamma}{\gamma-1}}$$

$$\frac{p_{02}}{p_{01}} = \left\{\frac{(\gamma+1)}{2\gamma M_1^2-(\gamma-1)}\right\}^{\frac{1}{\gamma-1}}\left[\frac{(\gamma+1)M_1^2}{\{2+(\gamma-1)M_1^2\}}\right]^{\frac{\gamma}{\gamma-1}} \tag{5.148}$$

Given the mass flow rate of gas, its total temperature T_0, and total pressure p_0 at a section, we can associate a unique critical area A^* where the Mach number is unity (sonic condition). At the prevailing total pressure and temperature conditions, an area lower than this critical area will not allow the specified mass flow rate to pass. For the isentropic gas flow through a converging–diverging nozzle, the throat area will correspond to the critical area A^* under the choked nozzle flow. The total pressure decreases across a normal shock for the given mass flow rate and constant total temperature. The critical areas associated with the isentropic flow on either side of a normal shock will differ. In fact, from the continuity equation, we easily deduce that the product of the total pressure and the critical area must remain constant across a normal shock, giving:

$$p_{02}A_2^* = p_{01}A_1^*$$

$$\frac{A_1^*}{A_2^*} = \frac{p_{02}}{p_{01}} \tag{5.149}$$

which shows that, for $p_{02} < p_{01}$ across a normal shock, we need a larger downstream critical area ($A_2^* > A_1^*$) to pass the upstream mass flow rate.

Total-to-Static Pressure Ratio (p_{02}/p_1)

We can write the ratio of total pressure downstream of a normal shock to its upstream static pressure as:

$$\frac{p_{02}}{p_1} = \left(\frac{p_{02}}{p_{01}}\right)\left(\frac{p_{01}}{p_1}\right)$$

In this equation, substituting $\frac{p_{02}}{p_{01}}$ from Equation 5.148 and $\frac{p_{01}}{p_1}$ from the isentropic relation yields:

$$\frac{p_{02}}{p_1} = \left\{\frac{2\gamma M_1^2 - (\gamma - 1)}{(\gamma + 1)}\right\}^{\frac{-1}{\gamma-1}} \left[\frac{(\gamma+1)M_1^2}{\{2 + (\gamma - 1)M_1^2\}}\right]^{\frac{\gamma}{\gamma-1}} \left(1 + \frac{\gamma - 1}{2} M_1^2\right)^{\frac{\gamma}{\gamma-1}}$$

which further simplifies to

$$\frac{p_{02}}{p_1} = \left\{\frac{2\gamma M_1^2 - (\gamma - 1)}{(\gamma + 1)}\right\}^{\frac{-1}{\gamma-1}} \left(\frac{\gamma + 1}{2} M_1^2\right)^{\frac{\gamma}{\gamma-1}} \quad (5.150)$$

Equation 5.150 is also known as the Rayleigh supersonic pitot tube equation.

Entropy Change ($(s_2 - s_1)/R$)

In terms of total pressure and total temperature upstream and downstream of a normal shock, we can compute the change in entropy by the equation:

$$s_2 - s_1 = c_p \ln\left(\frac{T_{02}}{T_{01}}\right) - R \ln\left(\frac{p_{02}}{p_{01}}\right)$$

With $T_{02} = T_{01}$ in a normal shock, this equation reduces to

$$\frac{(s_2 - s_1)}{R} = -\ln\left(\frac{p_{02}}{p_{01}}\right)$$

which, with the substitution of Equation 5.148, finally yields:

$$\frac{(s_2 - s_1)}{R} = -\ln\left[\left\{\frac{(\gamma + 1)}{2\gamma M_1^2 - (\gamma - 1)}\right\}^{\frac{1}{\gamma-1}} \left[\frac{(\gamma+1)M_1^2}{\{2 + (\gamma - 1)M_1^2\}}\right]^{\frac{\gamma}{\gamma-1}}\right] \quad (5.151)$$

Using the equations derived in the preceding, for a gas with $\gamma = 1.4$, Figure 5.34 illustrates variations of various flow properties with upstream Mach number. While both downstream Mach number and total pressure decrease across a normal shock, all other flow properties increase with upstream Mach number. The figure further shows that the changes across the shock are nearly isentropic for a weak normal shock with $M_1 < 1.5$.

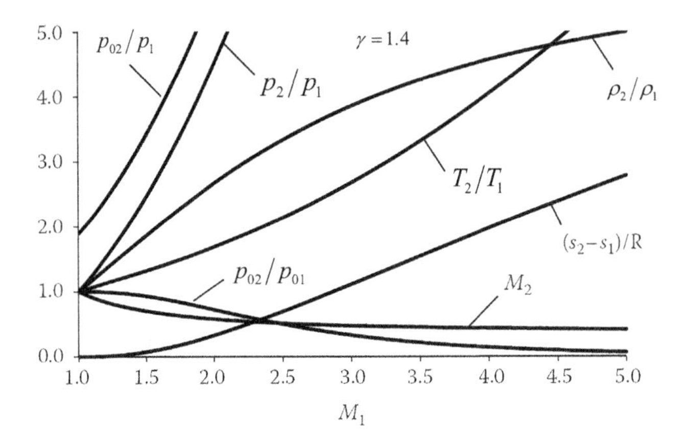

FIGURE 5.34 Variations of properties across a normal shock with upstream Mach number for $\gamma = 1.4$.

5.11.6 Normal Shock Table

In Sultanian (2023), Table A.6 for $\gamma = 1.4$ and Table B.6 for $\gamma = 1.333$ present variations of properties across a normal shock as a function of upstream Mach number greater than or equal to unity (sonic). Besides yielding a quick and approximate solution, these tables are helpful in cases where the original equations require an iterative solution. One can enter each table through any known variable and easily obtain the values of all other variables along the corresponding row.

5.11.7 Intersections of Fanno and Rayleigh Lines

As we have discussed earlier in Section 5.8.1, a Fanno line is a locus of points on the $h-s$ or $T-s$ (for gas with constant c_p) diagram for the given constant mass flux and stagnation enthalpy. The upper branch corresponds to a subsonic flow, and the lower branch represents a supersonic flow. Along the upper branch, the Mach number increases, and the impulse pressure decreases with increasing entropy, becoming maximum at the sonic point. Along the lower branch, the Mach number and the impulse pressure decrease with increasing entropy, again becoming maximum at the sonic point. The Fanno line models a steady adiabatic gas flow in a constant-area duct with frictional shear force at its wall. This shear force changes the Mach number along the subsonic and supersonic branches, leading to frictional choking at the sonic point. As the normal shock process is both adiabatic and frictionless (constant impulse pressure), for a point on either branch of a Fanno line, we can find a corresponding point on the other branch with equal impulse pressure. These two points will thus represent the states across a normal shock. To satisfy the requirement of entropy increase, we find that the point on the supersonic branch must be upstream, and the one on the subsonic branch must be downstream of the normal shock.

Just as the states across a normal shock lie on a Fanno line, they are also obtainable from a Rayleigh line, a locus of points on the $h-s$ or $T-s$ (for gas with constant c_p) diagram for the given constant mass flux and impulse pressure. For a Rayleigh line, the flow is subsonic along the upper branch and supersonic along the lower branch. Both branches converge at the sonic point with maximum entropy. The Rayleigh line models the heating or cooling of a steady gas flow in a constant-area duct with frictionless walls, an idealization that is difficult to achieve in the real world. The heating moves the flow toward the sonic point (thermal choking) along both branches and cooling moves it away from the sonic point. For a point on either branch of a Rayleigh line, we can find a corresponding point on the other branch with equal stagnation enthalpy (total temperature). These two points will thus represent the states across a normal shock. From entropy consideration, the shock process must occur from the point on the supersonic branch to the one on the subsonic branch.

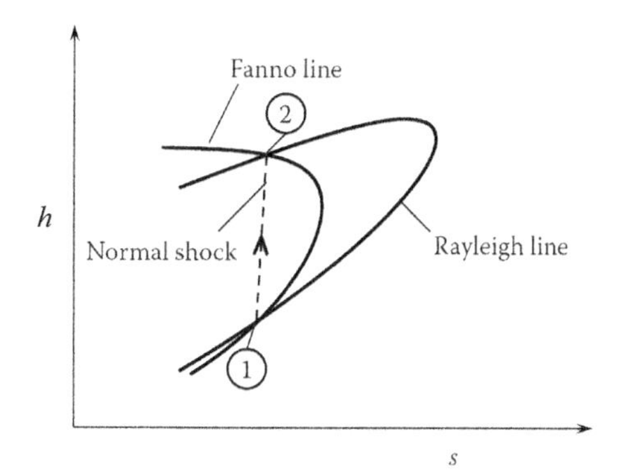

FIGURE 5.35 Intersection points of Fanno and Rayleigh lines.

Considering the above discussion, as shown in Figure 5.35, the intersection points of Fanno and Rayleigh lines must represent the states across a normal shock. From entropy consideration, the direction of the normal shock must be from point 1 to point 2, which is to the right of point 1. Note that point 1 lies on the supersonic branches and point 2 on the subsonic branches of the Fanno and Rayleigh lines. This figure offers a graphical technique to obtain normal shock solutions for gases where closed-form analytical solutions are not possible for lack of a simple equation of state and the fact that the gases may not be assumed to be calorically perfect.

5.12 OBLIQUE SHOCK

In the preceding, we found that the gas flow upstream of a normal shock must be supersonic, becoming subsonic downstream. The flow direction before and after a normal shock remains unchanged. The normal shock in an internal duct flow arises as the oncoming supersonic flow must adjust to a downstream elevated static pressure. When we have a geometric protrusion, like a bump on the duct wall, or an obstruction in an external flow, like a wedge sitting in an oncoming supersonic flow, an oblique shock wave occurs and provides the necessary mechanism for the flow to negotiate these geometric features. In an oblique shock, the velocity component normal to the shock wave undergoes a normal shock, and the velocity component along the shock wave passes unchanged. As a result, the flow downstream of the oblique shock turns, flowing along the wall of the obstruction. Like a normal shock, an oblique shock is also compressive.

Figure 5.36a shows that a subsonic flow goes around the wedge placed in cross flow with no discontinuity in the flow properties. Figure 5.36b shows that the oblique shock waves appear when the gas flow is supersonic, allowing the incoming flow to turn to negotiate the wedge. As the wedge half-angle is less than the maximum turning angle ($\theta < \theta_{max}$), based on the Mach number of the incoming supersonic flow, the oblique shock waves remain attached to the wedge at its leading edge. As we will see later in this section, for air with $\gamma = 1.4$, we obtain $\theta_{max} = 45.58^{\circ}$ for $M_1 \rightarrow \infty$.

When $\theta > \theta_{max}$ for the wedge, as shown in Figure 5.36c, the oblique shock waves do not remain attached to the wedge and instead become a bow shock ahead of the wedge. In this figure, the streamline going through the center of the bow shock undergoes a normal shock. As we move away from the center along the bow shock, the shock wave gets weaker than the normal shock. All incoming streamlines go through strong oblique shocks within the sonic line, resulting in subsonic flow behind the shock. The shock strength of streamline B is less than that of streamline A, but strong enough to yield a subsonic flow. For streamline C, and others toward the outer edge of the bow shock and away from the wedge, the flow after the shock remains supersonic. For each incoming flow Mach number, the weakest oblique shock is the Mach wave at an angle of $\sin^{-1}(1/M_1)$ and causes no turning of the flow. It is clear from this discussion that a normal shock is a particular case

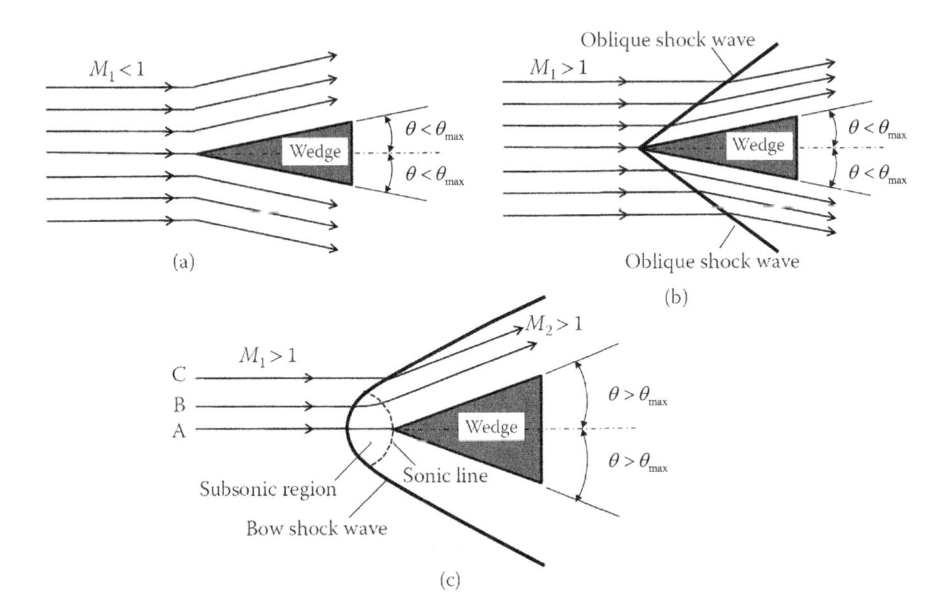

FIGURE 5.36 A wedge in uniform gas flow: (a) subsonic flow, (b) supersonic flow with an attached oblique shock wave, and (c) supersonic flow with a bow shock wave.

of an oblique shock that is the strongest for the given upstream Mach number, and the flow direction before and after the shock remains the same. On the other hand, the Mach wave represents the weakest oblique shock, introducing no change in the flow direction and entropy.

5.12.1 Analysis of an Oblique Shock

Figure 5.37a illustrates the flow geometry of an oblique shock at a wave angle β with the incoming flow direction. The upstream velocity V_1 has two orthogonal components V_{n1} and $V_{\beta 1}$: the first is normal to the wave, and the second is tangent to the wave. Across the oblique shock wave, only V_{n1} undergoes a normal shock, decreasing to V_{n2} downstream. The tangential component $V_{\beta 1}$ simply passes through unchanged ($V_{\beta 1} = V_{\beta 2} = V_\beta$); as a result, the resultant downstream velocity V_2 gets deflected by an angle θ toward the shock wave. Thus, under an inertial transformation of the flow relative to an observer moving with velocity V_β, the flow will appear to undergo a normal shock. Using V_β as the common base, Figure 5.37b depicts the upstream and downstream velocity triangles. Note that, for a shock to occur, we must have $V_{n1} > C_1$ where C_1 is the local speed of sound. This implies that V_1 must be supersonic. On the downstream side, we must have $V_{n2} < C_2$; as a result, V_2 may become subsonic for a strong shock or remain supersonic for a weak shock.

From the velocity vectors shown in Figure 5.37b, we easily obtain the following relations:

$$V_{n1} = V_1 \sin\beta \text{ or } M_{n1} = M_1 \sin\beta \tag{5.152}$$

$$V_{n2} = V_2 \sin(\beta - \theta) \text{ or } M_{n2} = M_2 \sin(\beta - \theta) \tag{5.153}$$

As mentioned earlier, in an oblique shock, one would observe the same flow pattern by running along a normal shock with a constant velocity V_β. Thus, the normal and oblique shocks are related by an inertial velocity transformation. Accordingly, the equations we derived earlier for a normal shock are also applicable for an oblique shock by replacing M_1 and M_2 in these equations by M_{n1} and M_{n2}, respectively.

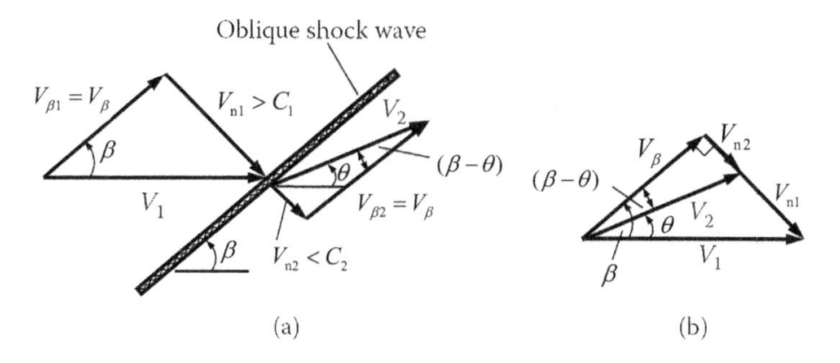

FIGURE 5.37 (a) Flow geometry of an oblique shock wave and (b) upstream and downstream velocity triangles with V_β as the common base.

5.12.2 Changes in Properties with Upstream Mach Number

In this section, starting from the equations derived for a normal shock in the preceding, we derive equations for changes in various flow properties across an oblique shock.

Downstream Mach Number (M_2)

For an oblique shock, we can write Equation 5.140 as:

$$M_{n2}^2 = \frac{2 + (\gamma - 1) M_{n1}^2}{2\gamma M_{n1}^2 - (\gamma - 1)} \tag{5.154}$$

Substituting Equations 5.152 and 5.153 in this equation yields:

$$M_2^2 = \frac{1}{\sin^2(\beta - \theta)} \left\{ \frac{2 + (\gamma - 1) M_1^2 \sin^2 \beta}{2\gamma M_1^2 \sin^2 \beta - (\gamma - 1)} \right\} \tag{5.155}$$

Density Ratio (ρ_2 / ρ_1)

For an oblique shock, Equation 5.145 for the density ratio becomes:

$$\frac{\rho_2}{\rho_1} = \frac{(\gamma + 1) M_1^2 \sin^2 \beta}{2 + (\gamma - 1) M_1^2 \sin^2 \beta} \tag{5.156}$$

Static Pressure Ratio (p_2 / p_1)

For an oblique shock, Equation 5.146 for the static pressure ratio becomes:

$$\frac{p_2}{p_1} = \frac{2\gamma M_1^2 \sin^2 \beta - (\gamma - 1)}{(\gamma + 1)} \tag{5.157}$$

Static Temperature Ratio (T_2 / T_1)

For an oblique shock, Equation 5.147 for the static temperature ratio becomes:

$$\frac{T_2}{T_1} = \frac{\{2\gamma M_1^2 \sin^2 \beta - (\gamma - 1)\}\{2 + (\gamma - 1) M_1^2 \sin^2 \beta\}}{(\gamma + 1)^2 M_1^2 \sin^2 \beta} \tag{5.158}$$

Total Pressure Ratio (p_{02} / p_{01})

Equation 5.148, developed for a normal shock, yields the following equation for an oblique shock:

$$\frac{p_{02}}{p_{01}} = \left\{ \frac{(\gamma+1)}{2\gamma M_1^2 \sin^2\beta - (\gamma-1)} \right\}^{\frac{1}{\gamma-1}} \left[\frac{(\gamma+1) M_1^2 \sin^2\beta}{\{2+(\gamma-1)M_1^2 \sin^2\beta\}} \right]^{\frac{\gamma}{\gamma-1}} \tag{5.159}$$

Entropy Change ($(s_2 - s_1)/R$)

In terms of the total pressure and temperature upstream and downstream of an oblique shock, we can express the change in entropy as:

$$s_2 - s_1 = c_p \ln\left(\frac{T_{02}}{T_{01}}\right) - R \ln\left(\frac{p_{02}}{p_{01}}\right)$$

Because $T_{02} = T_{01}$ in an oblique shock, this equation reduces to:

$$\frac{(s_2 - s_1)}{R} = -\ln\left(\frac{p_{02}}{p_{01}}\right)$$

in which substituting Equation 5.159 finally yields:

$$\frac{(s_2 - s_1)}{R} = -\ln\left[\left\{ \frac{(\gamma+1)}{2\gamma M_1^2 \sin^2\beta - (\gamma-1)} \right\}^{\frac{1}{\gamma-1}} \left[\frac{(\gamma+1) M_1^2 \sin^2\beta}{\{2+(\gamma-1)M_1^2 \sin^2\beta\}} \right]^{\frac{\gamma}{\gamma-1}} \right] \tag{5.160}$$

5.12.3 Equations Relating Wave Angle, Deflection Angle, and Upstream Mach Number

Section 5.12.2 presents equations to calculate property changes across an oblique shock. For a perfect gas of a known value of γ, these equations require the knowledge of the upstream Mach number M_1 and wave angle β. In this section, we derive equations to determine θ when we know β and M_1 and find β when we know θ and M_1. In addition, we present equations for a given M_1 to find the maximum value of the deflection angle (θ_{max}), determining whether the shock wave remains attached to the body or becomes detached as a bow shock with no analytical solution.

Finding θ from β and M_1

The continuity equation across an oblique shock yields:

$$V_{n1}\rho_1 = V_{n2}\rho_2$$

$$\frac{\rho_2}{\rho_1} = \frac{V_{n1}}{V_{n2}} = \frac{V_{\beta 1}\tan\beta}{V_{\beta 2}\tan(\beta-\theta)} = \frac{\tan\beta}{\tan(\beta-\theta)}$$

where we used the fact that $V_{\beta 1} = V_{\beta 2}$ in an oblique shock. Using Equation 5.156 in this equation, we obtain:

$$\frac{\tan\beta}{\tan(\beta-\theta)} = \frac{(\gamma+1)M_1^2\sin^2\beta}{2+(\gamma-1)M_1^2\sin^2\beta}$$

which, using the trigonometric identity for $\tan(\beta-\theta)$ and some algebra, finally yields:

$$\tan\theta = 2\cot\beta\left[\frac{M_1^2\sin^2\beta-1}{M_1^2(\gamma+\cos 2\beta)+2}\right] \tag{5.161}$$

Knowing the supersonic Mach number of an incoming gas flow, we use this equation to compute θ for an oblique shock with its wave angle ranging from 0 to 90°. This equation shows two limiting cases to yield $\theta = 0$. First, when $\beta = 90°$ (because $\cot\beta = 0$) regardless of M_1, and the second when $\beta = \mu = \sin^{-1}(1/M_1)$. The first limit corresponds to a normal shock, representing the strongest oblique shock, and the second limit corresponds to the Mach wave, representing the weakest oblique shock. Note that $\mu = 90°$ for $M_1 = 1$ and $\mu = 0$ for $M_1 = \infty$.

For each upstream Mach number, we can differentiate Equation 5.161with respect to β and set the resulting expression to zero (i.e., $\mathrm{d}(\tan\theta)/\mathrm{d}\beta = 0$). After some algebra, as reported in Oosthuizen and Carscallen (2013), we obtain the following equation for β_{max} that corresponds to θ_{max}, which we can compute using Equation 5.161. Figure 5.38 shows the variation of β_{max} and θ_{max} with M_1. As M_1 increases from 1.0 to ∞, θ_{max} increases from 0 to 45.58° while β_{max} initially decreases from 90° to 64.67° (at $M_1 = 2$) and then increases to 67.79°. This figure also includes the variation of β^* and θ^* with M_1, where $M_2 = 1$. It is interesting to note in this figure that the variations of θ_{max} and θ^* with M_1 are practically indistinguishable. On the other hand, β^* is always less than β_{max}; the difference is significant for $1 < M_1 < 4$.

$$\sin^2\beta_{max} = \frac{\gamma+1}{4\gamma}\frac{1}{\gamma M_1^2}\left[1-\sqrt{(\gamma+1)\left(1+\frac{\gamma-1}{2}M_1^2+\frac{\gamma+1}{16}M_1^4\right)}\right] \tag{5.162}$$

Using Equation 5.161, Figure 5.39 shows how θ varies with β for various values of M_1 ranging from 1.2 to ∞. The curve for each M_1 looks like an inverted parabola skewed to the right. Along a constant M_1 curve, each value of β yields a single value of θ, but for each value of θ, we obtain two values of β; the lower value corresponds to a weak oblique shock, and the higher value corresponds to a strong shock, which renders the downstream flow subsonic. This figure also includes two lines: the solid line, obtained using Equations 5.161 and 5.162, corresponds to $\theta = \theta_{max}$ for each value of M_1, and the dotted line corresponds to $M_2 = 1$ for each value of M_1. All values of β on the right of the

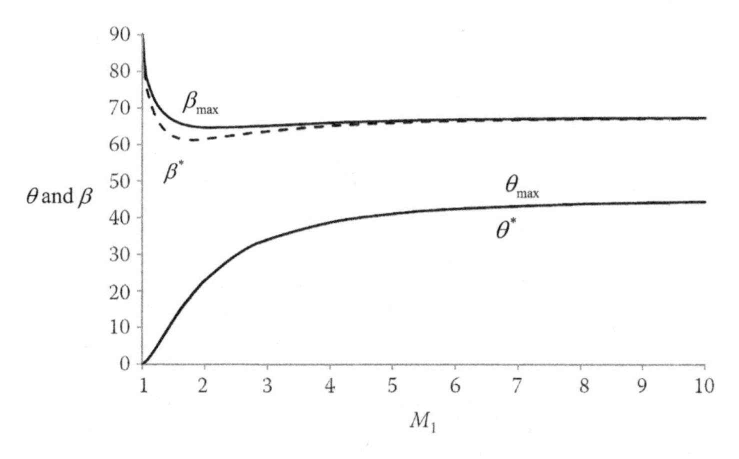

FIGURE 5.38 Variation of deflection and wave angles with upstream Mach number in an oblique shock.

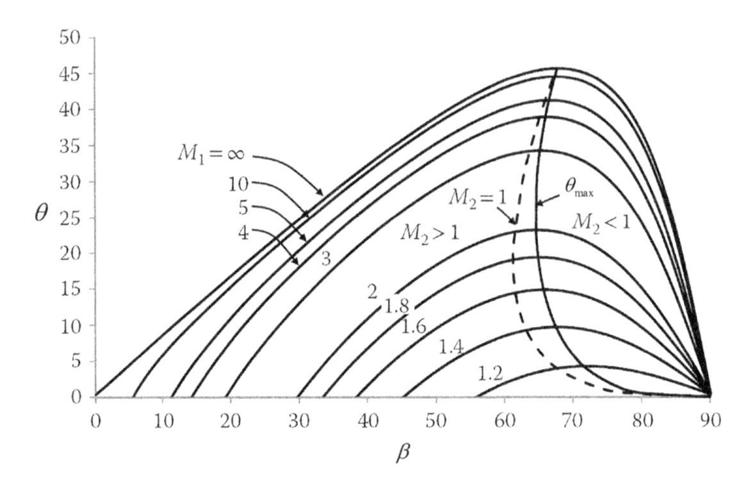

FIGURE 5.39 Variation of deflection angle (θ) with wave angle (β) for various upstream Mach numbers in an oblique shock.

dotted line in this figure represent strong shocks (subsonic downstream flows), and on the left, they represent weak shocks (supersonic downstream flows). We obtain oblique solid shocks that result in subsonic downstream flows within the narrow region bounded by these two lines.

Finding θ from β and M_1

The preceding discussion shows that except when $\theta = \theta_{max}$, we obtain two values of β for each incoming supersonic gas flow: the smaller value represents the weak oblique shock, and the larger value corresponds to the strong oblique shock with a subsonic downstream flow. As shown in Figure 5.39, except when both solutions fall on the right of the dotted line with $M_2 = 1$, the downstream flow remains supersonic for the smaller β. Practically, whether a weak or strong shock solution prevails depends upon the downstream static pressure boundary condition to which the incoming flow needs to adjust past the obstruction.

While we can use an iterative method to obtain both weak and strong solutions of β from Equation 5.161, it is more convenient to use equations to directly compute β for given M_1 and θ (half wedge angle). Such equations from Anderson (2021) are summarized below:

$$\tan\beta = \frac{M_1^2 - 1 + 2\lambda\cos\left(\dfrac{4\pi\delta + \cos^{-1}\chi}{3}\right)}{3\left(1 + \dfrac{\gamma - 1}{2}M_1^2\right)\tan\theta} \tag{5.163}$$

in which $\delta = 0$ yields the strong shock solution and $\delta = 1$ corresponds to the weak shock solution. In both cases, we compute λ and χ as follows:

$$\lambda = \sqrt{\left(M_1^2 - 1\right)^2 - 3\left(1 + \frac{\gamma - 1}{2}M_1^2\right)\left(1 + \frac{\gamma + 1}{2}M_1^2\right)\tan^2\theta} \tag{5.164}$$

$$\chi = \frac{\left(M_1^2 - 1\right)^3 - 9\left(1 + \dfrac{\gamma - 1}{2}M_1^2\right)\left(1 + \dfrac{\gamma - 1}{2}M_1^2 + \dfrac{\gamma + 1}{4}M_1^4\right)\tan^2\theta}{\lambda^3} \tag{5.165}$$

5.12.4 Oblique Shock Polar Hodograph

Figure 5.37 presents upstream and downstream velocity diagrams of an oblique shock on a physical plane with the flow velocity resolved into components along and normal to the oblique shock wave. This section presents these velocities in a hodograph plane where the x and y velocity components are the coordinates. As the total temperature remains constant across a shock, it is convenient to normalize all velocities in a hodograph plane by the speed of sound C^* that corresponds to the static temperature of the sonic state ($M = 1$). Accordingly, the dimensionless coordinates in the hodograph plane become $V_x^* = V_x / C^*$ and $V_y^* = V_y / C^*$ where $C^* = \sqrt{2\gamma R T_0 / (\gamma + 1)}$, which is constant for a given gas and its total temperature. For a given upstream velocity V_1^*, which equals the characteristic Mach number M_1^*, Figure 5.40 shows a shock polar in the hodograph plane. This shock polar, with deflection angle θ as a parameter, represents the locus of all feasible oblique shock solutions. Shapiro (1953) provides the following equation to plot a shock polar on a hodograph plane:

$$V_y^{*2} = \frac{(M_1^* - V_x^*)^2 \left(V_x^* M_1^* - 1\right)}{\left(\dfrac{2}{\gamma + 1}\right) M_1^{*2} - \left(V_x^* M_1^* - 1\right)} \tag{5.166}$$

As shown in Figure 5.40, a shock polar provides two solutions for $\theta = 0$: Point S represents the strongest shock solution (the normal shock) and W represents the weakest solution (isentropic) corresponding to the Mach wave whose wave angle β equals the Mach angle $\mu = \sin^{-1}(1/M_1)$. Thus, for a shock polar, the wave angle varies from $\beta = \mu$ to $\beta = 90°$. The part of the shock polar shown by the dotted line corresponds to the lower symmetric half of the wedge. As shown in the figure, line WA is orthogonal to the Mach line OA. The equations of these lines in the hodograph plane are given by:

Line OA:

$$V_y^* = (\tan\mu) V_x^* = \frac{V_x^*}{\sqrt{M_1^2 - 1}}$$

Line WA:

$$V_y^* = \left(\sqrt{M_1^2 - 1}\right)\left(M_1^* - V_x^*\right)$$

Further note that line WA is tangent to the shock polar at W. From the Prandtl–Meyer relation, we obtain $M_2^* = 1/M_1^*$.

Figure 5.41 shows that for a general deflection angle θ, a radial line represented by the equation $V_y^* = (\tan\theta) V_x^*$ intersects the shock polar at two points W and S, which correspond to the weak and strong solutions, respectively. In this figure, line W_1W is orthogonal to line OA having a wave angle β_W. We can write the equation of line WA in the hodograph plane as $V_y^* = (\cot\beta_W)\left(M_1^* - V_x^*\right)$,

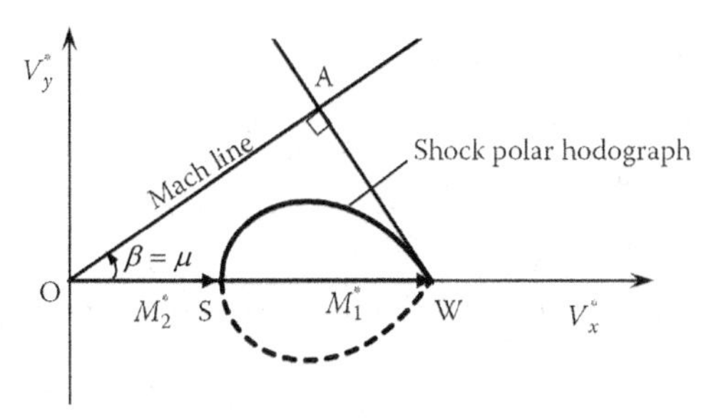

FIGURE 5.40 Strongest and weakest shocks on a shock polar hodograph.

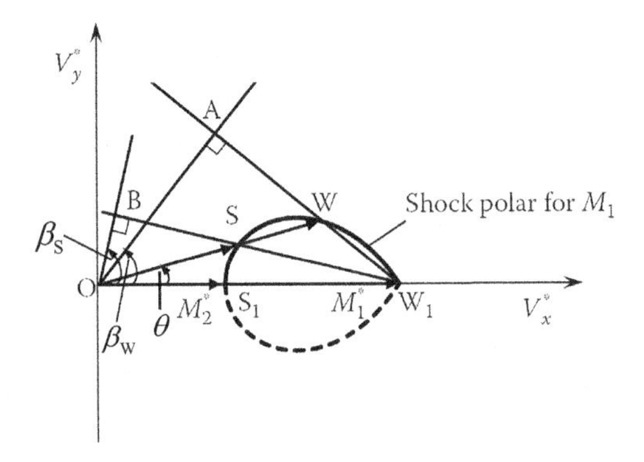

FIGURE 5.41 Weak and strong shock waves on a shock polar hodograph for given deflection angle and upstream Mach number.

which yields the coordinates of point W as $\left\{M_1^* / \left(1+\tan\theta\tan\beta_W\right), M_1^* \tan\theta / \left(1+\tan\theta\tan\beta_W\right)\right\}$. In this figure, while OW represents the total velocity V_2^* downstream of the oblique shock, the line segments AW_1 and AW, respectively, correspond to the upstream and downstream velocity components normal to the shock wave. Note that the velocity triangles OW_1A and OWA shown in this figure are identical to the velocity triangles shown in Figure 5.37b in the physical plane.

Similar to the details of the geometric construction for the weak solution point Won, the shock polar that we discussed in the foregoing, Figure 5.41 also features the geometric construction for the strong solution point S with the corresponding wave angle β_S. We can write the coordinates of S in the hodograph plane as $\left\{M_1^* / \left(1+\tan\theta\tan\beta_S\right), M_1^* \tan\theta / \left(1+\tan\theta\tan\beta_S\right)\right\}$.

Based on the above discussion, we can use the following alternate procedure, instead of Equation 5.166, to create a shock polar in a hodograph plane for an upstream Mach number M_1. For this Mach number, we compute the wave angle $\mu = \sin^{-1}(1/M_1)$. The shock polar intersects the V_x^* axis at two points: $(M_1^*,0)$ and $(M_2^*,0)$ where

$$M_1^* = \sqrt{\frac{(\gamma+1)M_1^2}{2+(\gamma-1)M_1^2}}$$

$$M_2^* = \frac{1}{M_1^*}$$

The point $(M_1^*,0)$ corresponds to the weak solution with $\beta = \mu$ and $\theta = 0$, and the point $(M_2^*,0)$ corresponds to the normal shock solution with $\beta = 90°$ and $\theta = 0$. For a general point $P\left(V_x^*, V_y^*\right)$ with β between μ and 90°, we use the following two-step procedure to compute θ and the coordinates V_x^* and V_y^*:

Step1: Using Equation 5.161, compute θ as:

$$\tan\theta = 2\cot\beta\left[\frac{M_1^2\sin^2\beta - 1}{M_1^2\left(\gamma+\cos 2\beta\right)+2}\right]$$

Step 2: With the known values of β and θ, compute the coordinates of point $P\left(V_x^*, V_y^*\right)$ as:

$$V_x^* = \frac{M_1^*}{1+\tan\theta\tan\beta}$$

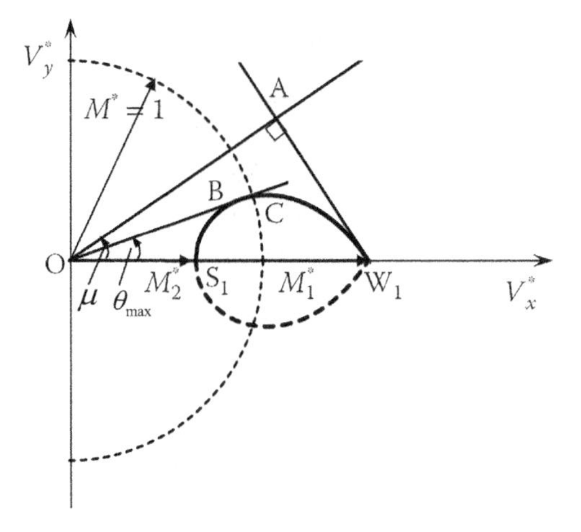

FIGURE 5.42 Points corresponding to θ_{max} and $M_2^* = 1$ on a shock polar.

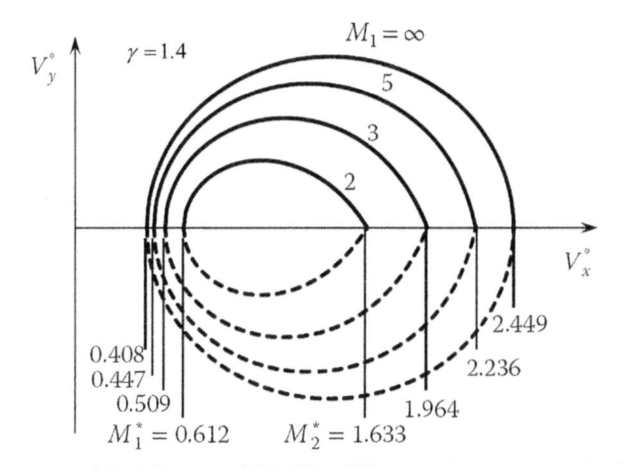

FIGURE 5.43 Shock polars for different upstream Mach numbers.

$$V_y^* = \frac{M_1^* \tan\theta}{1 + \tan\theta \tan\beta}$$

Varying β from μ to 90° in small increments and repeating Steps 1 and 2 above, we first compute points that form the upper half of the shock polar and then reflect them to generate the lower half.

Figure 5.42 shows that the line OB with maximum deflection angle θ_{max} is tangent to the shock polar at point B, within the sonic semicircle, intersecting the shock polar at point C. For $\theta > \theta_{max}$, there is no oblique shock solution, and the shock wave cannot remain attached to the wedge of half-angle θ.

Figure 5.43 presents a family of shock polars for different upstream Mach numbers where the shock polar is a circle for $M_1 = \infty$. The figure also shows the upstream and downstream characteristic Mach numbers for each upstream Mach number. These values correspond to $\theta = 0$.

5.13 PRANDTL–MEYER FLOW

Oblique shocks are compression waves; except for an infinitesimal deflection angle, they are non-isentropic. When an oncoming supersonic flow needs to negotiate a convex corner, it happens through a series of expansion waves, which are isentropic. Such a supersonic flow is called the Prandtl–Meyer flow. Figure 5.44a shows a Prandtl–Meyer flow expanding from $M = 1$ to $M_1 > 1$ and turning by an angle ϕ_1. Similarly, Figure 5.44b shows a Prandtl–Meyer turn of ϕ_2 while expanding from $M = 1$ to $M_2 > 1$. As shown in Figure 5.44c, if the flow needs to expand isentropically from $M_1 > 1$ to $M_2 > M_1$, it will turn by an angle $\phi_2 - \phi_1$. We compute all the flow properties before and after a Prandtl–Meyer turn using isentropic flow equations.

Figure 5.45 shows a Prandtl–Meyer turn $\mathrm{d}\phi$ for an incoming supersonic flow of Mach number M. After the turn, the flow velocity increases by $\mathrm{d}V$. We compute the Mach wave angle $\mu = \sin^{-1}(1/M)$. From triangles OBC and OAC, we can write:

$$V_\mu = V \cos \mu$$

$$V_\mu = (V + \mathrm{d}V)\cos(\mu + \mathrm{d}\phi)$$

Both these equations yield:

$$V \cos \mu = (V + \mathrm{d}V)\cos(\mu + \mathrm{d}\phi)$$

$$V \cos \mu = (V + \mathrm{d}V)(\cos \mu \cos(\mathrm{d}\phi) - \sin \mu \sin(\mathrm{d}\phi))$$

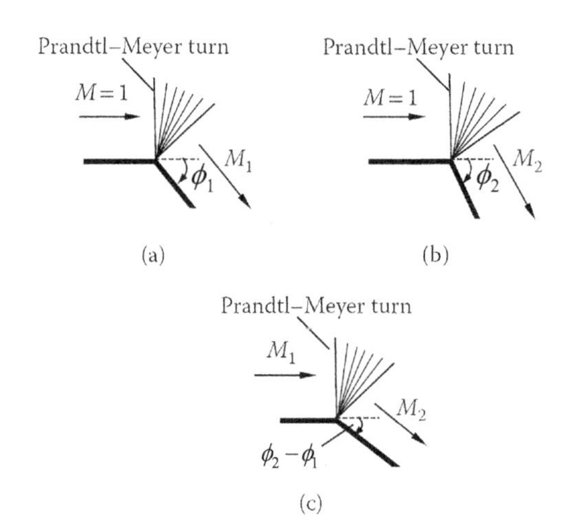

FIGURE 5.44 Differential Prandtl–Meyer flow: (a) Expansion turn from $M = 1$ to $M_1 > 1$, (b) expansion turn form $M = 1$ to $M_2 > 1$, and (c) expansion turn $M_1 > 1$ to $M_2 > M_1$.

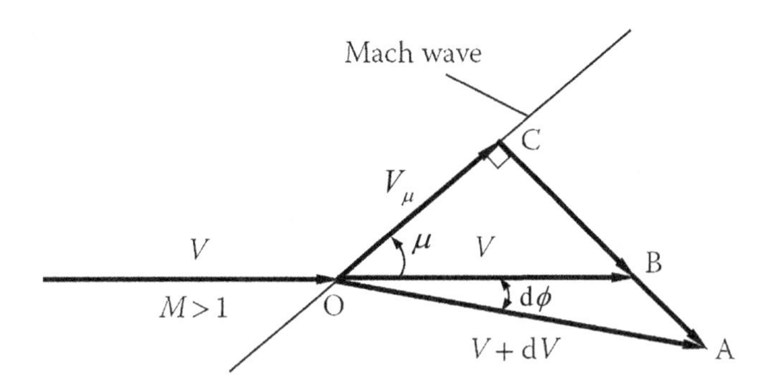

FIGURE 5.45 Differential changes in velocity and turning angle in a Prandtl–Meyer flow.

Substituting $\sin(d\phi) = d\phi$ and $\cos(d\phi) = 1$ for small $d\phi$ in this equation yields:

$$V\cos\mu = (V + dV)(\cos\mu - d\phi\sin\mu)$$

In this equation, neglecting higher-order terms and simplifying the resulting equation, we obtain:

$$d\phi = \sqrt{M^2 - 1}\,\frac{dV}{V} \tag{5.167}$$

where we have used the relation $\tan\mu = 1/\sqrt{M^2 - 1}$.

For an isentropic flow, we can either use the momentum or the energy equation to recast the RHS of Equation 5.167 in a form that we can easily integrate. We can write the energy equation in the form:

$$\frac{C^{*2}}{V^2} = b^2\cos^2\mu + \sin^2\mu \tag{5.168}$$

where

$$C^{*2} = \frac{2\gamma RT_0}{\gamma + 1} \text{ and } b^2 = \frac{\gamma - 1}{\gamma + 1}$$

Taking log and differentiating both sides of Equation 5.168, after further simplification, yields:

$$\sqrt{M^2 - 1}\,\frac{dV}{V} = \left(\frac{b^2 - 1}{b^2 + \tan^2\mu}\right)d\mu \tag{5.169}$$

From Equations 5.167 and 5.169, we write:

$$d\phi = \left(\frac{b^2 - 1}{b^2 + \tan^2\mu}\right)d\mu$$

which we integrate to finally obtain:

$$\phi = \sqrt{\frac{\gamma + 1}{\gamma - 1}}\tan^{-1}\sqrt{\frac{\gamma - 1}{\gamma + 1}\left(M^2 - 1\right)} - \tan^{-1}\sqrt{M^2 - 1} \tag{5.170}$$

in which we have arbitrarily assumed that $\phi = 0$ for $M = 1$. For a given gas with known γ, the equation shows that ϕ is a function of Mach number only. This function is also called the Prandtl–Meyer function. Note that, for $M \to \infty$, the equation asymptotes to:

$$\phi_{max} = \frac{\pi}{2}\left(\sqrt{\frac{\gamma + 1}{\gamma - 1}} - 1\right) \tag{5.171}$$

which for $\gamma = 1.4$ yields $\phi_{max} = 130.45^\circ$.

In Sultanian (2023), Table A.7 for $\gamma = 1.4$ and Table B.7 for $\gamma = 1.333$ present computed values of the Prandtl–Meyer function ϕ and Mach wave angle μ as a function of Mach number.

5.14 OPERATION OF NOZZLES AND DIFFUSERS

This section discusses the applications of various concepts of compressible flow presented in the preceding to understand, for example, the operation of a converging nozzle, a C–D nozzle, and a supersonic wind tunnel, which includes a C–D diffuser. Nozzles, where we need to accelerate the flow velocity, are part of numerous engineering equipment, including rockets in spacecraft, aircraft engines, and missiles. Diffusers, where we need to slow the flow to convert its dynamic pressure into static pressure, are commonly found in power generation gas turbines in their exhaust systems and steam turbines to exhaust into condensers. Various operational characteristics of nozzles and diffusers discussed here involve several numerical experiments, which the readers can reproduce to develop skills needed to perform engineering calculations on such flows for their improved physical understanding and design applications.

5.14.1 Operation of a Converging Nozzle

Figure 5.46 shows an axisymmetric converging nozzle used here for numerical experiments using air as a perfect gas with $\gamma = 1.4$ and $R = 287\ \text{J}/(\text{kg K})$. The following two parabolas express the axial variation of the nozzle diameter D in meter.

$$D = -2.048x^2 + 0.31 \quad 0 \le x \le 0.25$$

$$D = 6(x - 0.35)^2 + 0.122 \quad 0.25 \le x \le 0.35$$

This figure shows that the nozzle is 0.35 m long, and its minimum (throat) diameter is 0.122 m.

In all our numerical experiments, we assume that the flow is steady and isentropic with constant total pressure and temperature throughout the nozzle. These pressure and temperature conditions are set by the source plenum, which is maintained at $p_0 = 5$ bar and $T_0 = 500$ K. The sink plenum sets the back pressure for each flow situation. Because the dynamic pressure is zero in each plenum, its static pressure equals the total pressure. For $p_b = p_0$, the nozzle will have no flow. For $p_b < p_0$, the nozzle will flow from the source plenum to the sink plenum where the nozzle exit static pressure p_e must equal the back pressure p_b until the nozzle is choked. When the flow chokes at the nozzle throat with $M_e = 1$, we obtain the maximum mass flow rate of 11.219 kg/s for the given source plenum conditions.

Six cases chosen for the numerical simulation of air flow in the converging nozzle are summarized in Table 5.1. In each case, the nozzle outlet boundary condition is specified by the sink plenum pressure p_b. For example, Case e_1 corresponds to $p_b / p_0 = 0.9$. Assuming that $p_e = p_b$, which is true

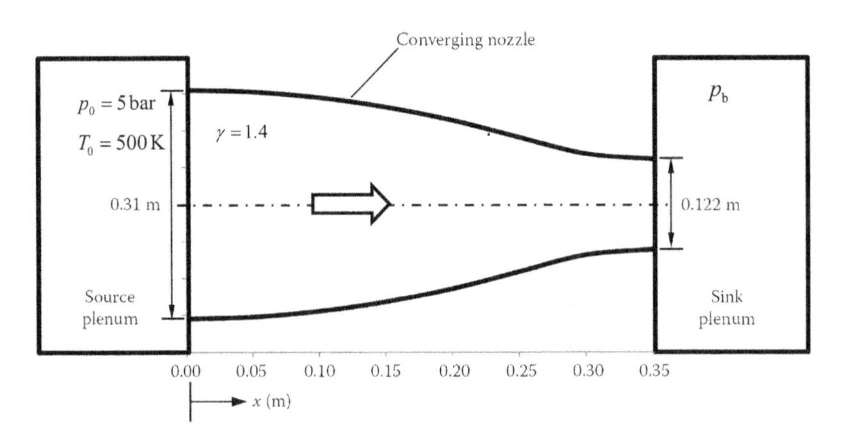

FIGURE 5.46 Geometry and operating conditions of an axisymmetric converging nozzle.

TABLE 5.1
Converging nozzle operating conditions for each numerical experiment

Case	$\frac{p_b}{p_0}$	M_e	$\dot{m}/\dot{m}_{max}$
e_1	0.9	0.391	0.617
e_2	0.8	0.574	0.819
e_3	0.7	0.732	0.932
e_4	0.528	1.0	1.0
e_5	0.4	1.0	1.0
e_6	0.3	1.0	1.0

until the flow is choked at the nozzle exit, we perform the initial calculations of M_e and A^* using the following isentropic relations:

$$M_e = \sqrt{\frac{2}{\gamma-1}\left\{\left(\frac{p_0}{p_e}\right)^{\frac{\gamma-1}{\gamma}} - 1\right\}} \text{ and } A^* = \frac{A_e}{\frac{1}{M_e}\sqrt{\left(\frac{2+(\gamma-1)M_e^2}{\gamma+1}\right)^{\frac{\gamma+1}{\gamma-1}}}}$$

Note that for each isentropic flow in the nozzle, there exists a unique value of the critical area A^*, which equals the nozzle throat area with $M = 1$. Once we know A^*, we obtain the distribution of A/A^* along the nozzle, which allows us to compute the Mach number distribution using an iterative method. Knowing the Mach number at each nozzle section, we use the isentropic flow equations to compute other flow properties.

Figure 5.47 presents the results of numerical experiments on the converging nozzle, showing the axial variation of the nozzle flow area in Figure 5.47a and that of p / p_0 for each case in Figure 5.47b. For Cases e_1, e_2, and e_3, the flow within the nozzle remains subsonic, as shown by the Mach number variations for these cases in Figure 5.47c. Figure 5.47d shows that the mass flow rate in the nozzle increases as we decrease the back pressure in the sink plenum. For Case e_4, the nozzle operation assumes an exciting character, becoming choked with $M_e = 1$, $A_{min} = A^*$ and $\dot{m} = \dot{m}_{max}$. Upon further reducing the sink plenum pressure for Cases e_5 and e_6, the flow field within the nozzle remains unchanged. At the nozzle exit, the flow expands like a Prandtl–Meyer flow with a local increase in Mach number while adjusting to the lower static pressure in the plenum. The flow entering the sink plenum loses all its dynamic pressure through complex shear flow interactions and viscous dissipation, the numerical simulation of which is outside the present scope. Within a converging nozzle, a supersonic flow is impossible regardless of how much we lower the back pressure in the sink plenum.

5.14.2 Operation of a Converging–Diverging Nozzle

Figure 5.48 shows an axisymmetric converging–diverging (C–D) nozzle. The following three parabolas express the axial variation of the nozzle diameter D in meter.

$$D = -0.8x^2 + 0.31 \quad 0 \le x \le 0.2$$

$$D = 1.244(x - 0.35)^2 + 0.25 \quad 0.2 \le x \le 0.45$$

$$D = -0.2896(x - 1.0)^2 + 0.35 \quad 0.45 \le x \le 1.0$$

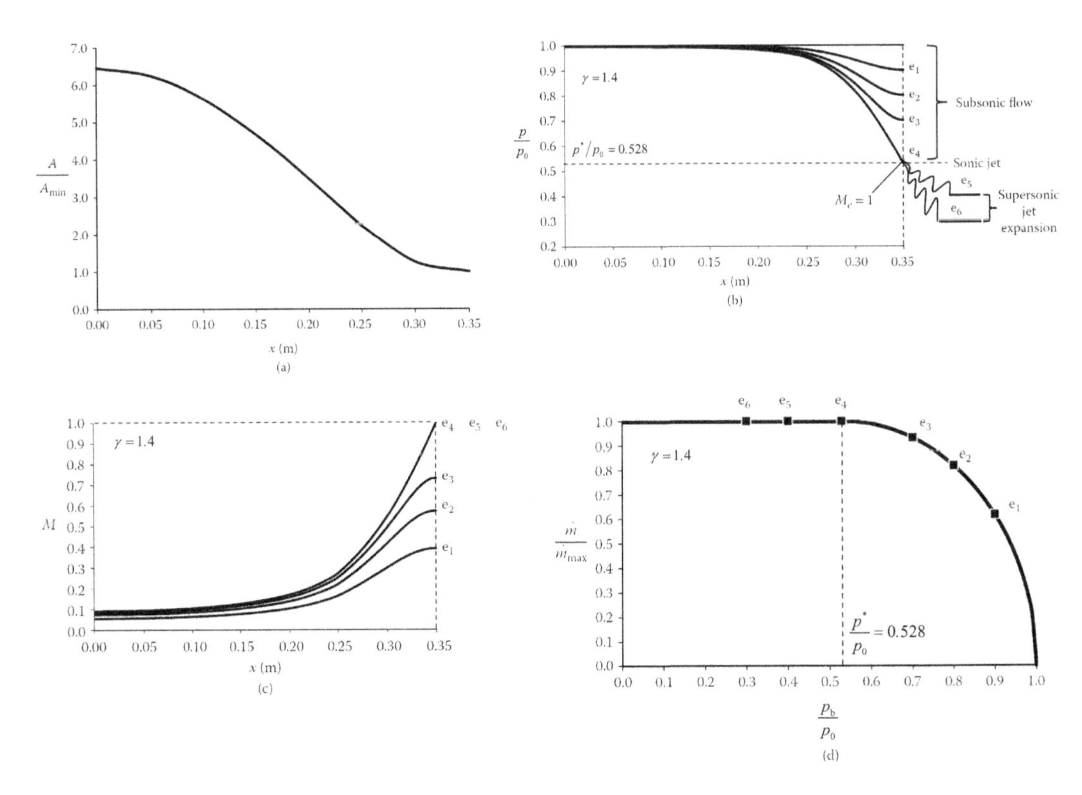

FIGURE 5.47 (a) Axial variation of the converging nozzle area, (b) axial variation of pressure ratio in the converging nozzle, (c) axial variation of Mach number in the converging nozzle, and (d) variation of mass flow rate with overall pressure ratio for the converging nozzle.

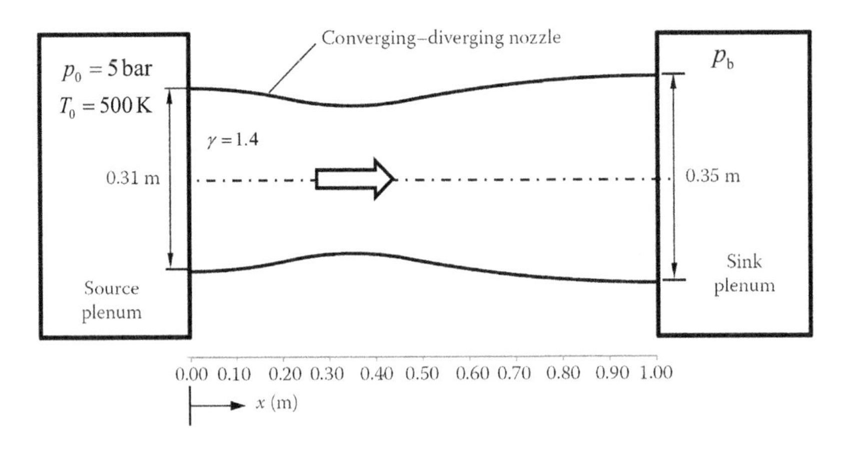

FIGURE 5.48 Geometry and operating conditions of an axisymmetric converging–diverging nozzle.

The figure shows that the C–D nozzle is 1.0 m long, and its minimum (throat) diameter is 0.25 m.

For all eight numerical experiments, the total pressure and temperature in the source plenum remain constant at 5 bar and 500 K, respectively. Table 5.2 summarizes the back-pressure boundary condition set in the sink plenum for each numerical experiment. The table also includes the calculated values of the exit Mach number and nozzle mass flow rate. For the given source plenum boundary conditions, the maximum mass flow rate computed for the C–D nozzle is 44.344 kg/s.

TABLE 5.2
C–D nozzle operating conditions for each numerical experiment

Case	p_b/p_0	M_e	$\dot{m}/\dot{m}_{max}$
E_1	0.9600	0.2422	0.7921
E_2	0.9450	0.2854	0.9210
E_3	0.9344	0.3129	1.0
E_4	0.6248	0.4627	1.0
E_5	0.5207	0.5505	1.0
E_6	0.1500	2.1746	1.0
E_7	0.0973	2.1746	1.0
E_8	0.0250	2.1746	1.0

5.14.2.1 Subsonic Flow

Figure 5.49a shows the axial variation of C–D nozzle flow area. The minimum (throat) area is 0.0491 m^2, occurring at an axial distance of 0.35 m. When the back pressure in the sink plenum equals the pressure in the source plenum, the nozzle has no flow. When we lower the back pressure, a subsonic flow prevails throughout the C–D nozzle as for Cases E_1 and E_2. Each case features a single isentropic flow where the converging section acts like a nozzle and the diverging section acts like a diffuser. The critical area computed in each case is lower than the throat area of the C–D nozzle. The mass flow rate for Case E_2 is higher than that for Case E_2. For both cases, Figure 5.49b

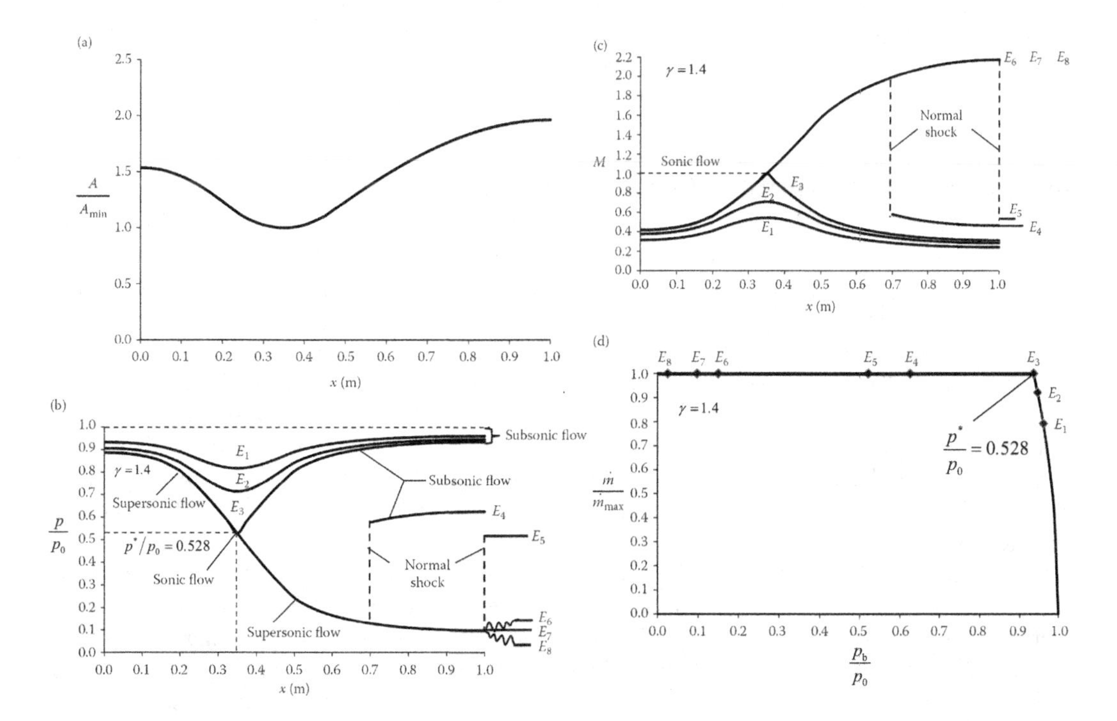

FIGURE 5.49 (a) Axial variation of C–D nozzle area, (b) axial variation of pressure ratio in the C–D nozzle, (c) axial variation of Mach number in the C–D nozzle, and (d) variation of mass flow rate with overall pressure ratio for the C–D nozzle.

shows the axial variation of the pressure ratio, and Figure 5.49c shows that of the Mach number. Because of the diffusion in the diverging section, the pressure ratio computed at the throat is much lower, and the Mach number is much higher than their values calculated at the nozzle exit plane. Figure 5.49d shows that the mass flow rate increases with the overall C–D nozzle pressure ratio at a much higher rate than that for the converging nozzle, as shown in Figure 5.47d.

5.14.2.2 Choked Flow

When the back pressure is lowered further, the C–D nozzle becomes choked for Case E_3. In this case, the Mach number reaches unity at the throat with the critical pressure ratio of 0.528. At this operating point, the C–D nozzle attains the maximum mass flow rate, which remains constant for all remaining cases with lower back pressures. For this limiting case, the critical area for the isentropic flow in both the converging and diverging sections equals the throat area. Even with the sonic condition at the nozzle throat, the imposed back pressure in the sink plenum will not generate a supersonic flow in the diverging section. In this case, we may think of the flow as having a nearly isentropic normal shock at the throat, rendering the downstream flow subsonic and the diverging section behaving like a diffuser.

5.14.2.3 Normal Shock in the Diverging Section

As we lower the back pressure below the value for Case E_3, the normal shock location in the diverging section moves downstream from the throat. For a shock location between the throat and the nozzle exit plane, as for Case E_4, we obtain two isentropic flows, one upstream and the other downstream of the normal shock. For the upstream flow, the critical area equals the nozzle throat area, and for the downstream flow, the total pressure decreases. The critical area increases to keep the mass flow rate and the total temperature unchanged. Because the supersonic Mach number in the diverging section rises as we move away from the nozzle throat, the normal shock gets stronger with a higher increase in static pressure and a higher reduction in total pressure. Case E_5, thus, yields the strongest normal shock, occurring at the nozzle exit plane at a Mach number of 2.175. As the flow downstream of a normal shock is subsonic, the part of the diverging section with this subsonic flow acts like a diffuser. For all operating points from E_4 through E_5, the C–D nozzle features a normal shock in its diverging section.

5.14.2.4 Shock-Free Supersonic Flow in the Diverging Section

Further lowering the back pressure below the value for Case E_5, the supersonic flow in the diverging section becomes shock-free, as it happens for Cases E_6, E_7, and E_8, and the entire flow field in the C–D nozzle is no longer influenced by the sink plenum pressure. For Case E_6, the nozzle operates under-expanded with its exit static pressure higher than that in the sink plenum. The initial adjustment of the nozzle flow with this sink condition happens via an oblique shock, shown in Figure 5.49b by a wiggly line. For Case E_7, the nozzle exit static pressure equals the back pressure in the plenum, requiring no compression or expansion wave for further adjustment of the flow with the ambient, representing the optimal operating point for the C–D nozzle. For Case E_8, the plenum pressure is below the nozzle exit static pressure, requiring an initial Prandtl–Meyer expansion for the flow to adjust to the plenum conditions, and the C–D nozzle operates over-expanded.

We typically design a rocket engine as a C–D nozzle to operate at altitude for missile applications and in vacuum for space applications. When we test such a rocket on the ground, it works over-expanded, exhibiting a series of compression and expansion waves appearing in diamond shapes in the rocket exhaust. An enormous plume associated with a space rocket at a high altitude indicates that the rocket is operating under-expanded. Note that the area ratio of the space rocket can never be large enough to yield zero exit pressure (vacuum).

5.14.3 Operation of a Supersonic Wind Tunnel

Figure 5.50 schematically shows a supersonic wind tunnel where a converging–diverging nozzle accelerates the airflow to the desired supersonic Mach number in the working section, and the downstream converging–diverging diffuser turns the flow subsonic in the discharge. For the wind tunnel configuration shown in the figure, the airflow is pulled by a compressor (not shown in the figure) at the discharge side while the inlet corresponds to the ambient conditions. For an isentropic flow through an ideal wind tunnel where the test section is between an identical C–D nozzle and C–D diffuser, one would expect the flow conditions at the discharge to be the same as at the inlet. Once started, we will need zero power to run such a tunnel. For the most efficient operation of a supersonic wind tunnel, we must achieve the slightest loss in total pressure through the tunnel.

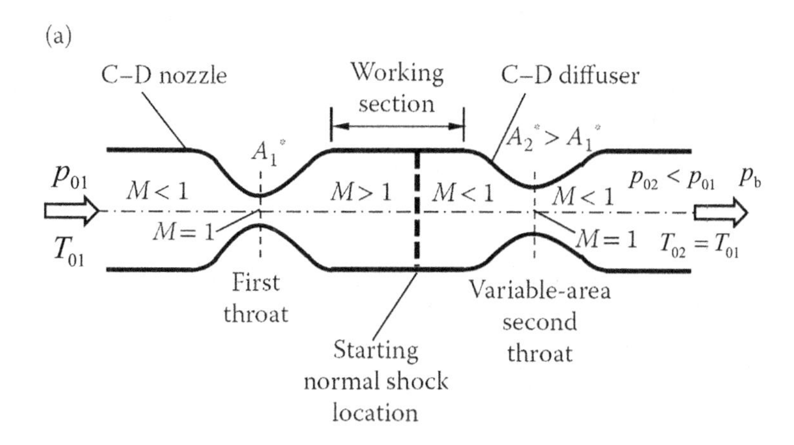

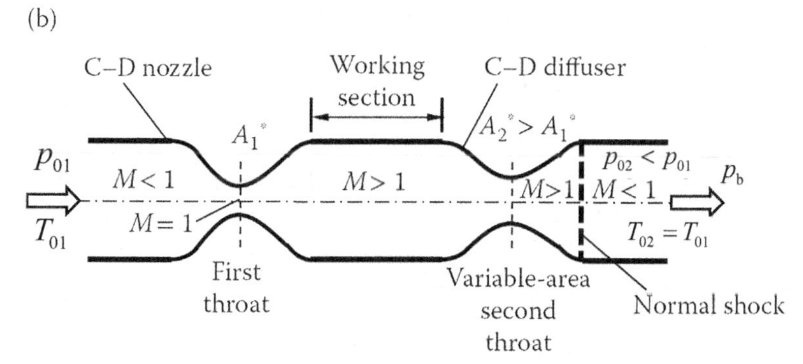

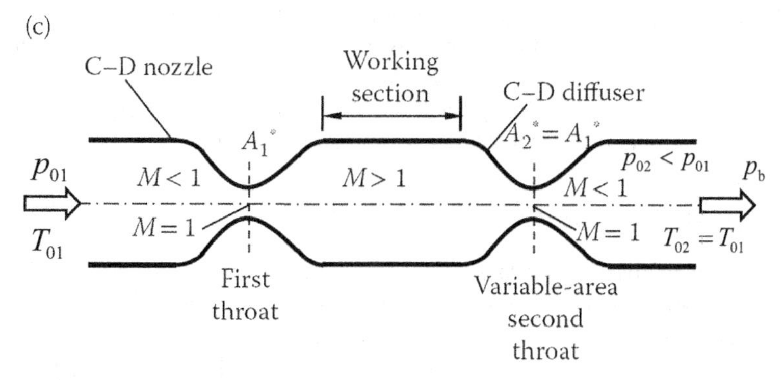

FIGURE 5.50 Supersonic wind tunnel: (a) Starting operation, (b) normal shock swallowed by the second throat, and (c) shock-free operation.

Notwithstanding the pressure loss due to friction and shock boundary layer interactions, running a supersonic wind tunnel shock-free yields a significant reduction in total pressure loss.

To understand how we can run a supersonic wind tunnel shock-free, let us consider an ideal one, shown in Figure 5.50, without friction and boundary layer, where the airflow is isentropic everywhere except across the normal shock, if present. As shown in this figure, when we gradually lower the back pressure of the wind tunnel, a normal shock will appear in the working section during start-up, similar to Case E_5 shown in Figure 5.49b for a C–D nozzle where the normal shock stands at the nozzle exit plane. As the upstream nozzle is flowing choked, the mass flow rate and the total temperature remain constant for all downstream changes in the tunnel flow conditions. The total pressure and temperature for the upstream isentropic flow are those set at the inlet. For the present numerical experiments in the tunnel, we assume $p_{01} = 1$ bar and $T_{01} = 293$ K. The C–D nozzle throat area (A_1^*) and exit area (A_e), which equals the flow area of the working section, are 2.3704 and 6.25 m^2, respectively. The flow Mach number in the working section is computed as 2.5.

The normal shock in the working section results in a subsonic downstream flow at Mach number 0.513 at a reduced total pressure $p_{02} = 0.499$ bar. To allow the mass flow rate (559.453 kg/s) set by the upstream sonic throat to pass downstream at the reduced total pressure, the diffuser throat area (A_2^*) must be larger than A_1^*, related by the equation:

$$p_{02} A_2^* = p_{01} A_1^*$$

giving $A_2^* = 4.7501$ m^2. As shown in Figure 5.50a, with $A_2^* = 4.7501$ m^2, both throat sections are choked with $M = 1$. The converging section of the diffuser operates like a nozzle, and the diverging section functions as a subsonic diffuser.

When we slightly increase the second throat area beyond the value calculated above, the normal shock jumps to the downstream diffuser diverging section, having its area equal to the working section area. The second throat is said to have swallowed the shock from the working section. As shown in Figure 5.50b, the entire upstream flow after the first throat remains supersonic at the inlet total pressure (p_{01}). The Mach number at the second throat becomes 2.2. Thus, we achieve the first design objective of a shock-free supersonic working section of the wind tunnel without reducing its overall total pressure loss, which now occurs across the normal shock in the diffuser diverging section.

Note that a normal shock in the converging section of the diffuser is not stable. To illustrate this, let's assume a normal shock in the converging section occurring at a Mach number of 2.4. For the subsonic flow downstream of this hypothetical normal shock, the static pressure at the diffuser exit is calculated as 0.4672 bar, which is higher than the static back pressure of 0.4170 bar set for the tunnel, and therefore, it is unacceptable for a subsonic flow condition at the exit.

When the normal shock is at the diffuser exit, the flow upstream is supersonic, and it is no longer sensitive to back-pressure changes. To reduce the total pressure loss across a normal shock, it must occur at as low a supersonic Mach number as possible. We achieve this by decreasing the second throat area and increasing the back pressure. As a result, as shown in Figure 5.50, the steady operation of the wind tunnel becomes shock-free with $A_2^* = A_1^*$. Only one isentropic flow prevails in the entire wind tunnel at the specified inlet total pressure and temperature.

5.15 CONCLUDING REMARKS

With the introduction of the mass flow function, impulse function, and normal shock function, we have discussed in this chapter the isentropic and nonisentropic behavior of compressible flows. Considering the wide industrial applications, the chapter presents the physics-based modeling of one-dimensional compressible flow in an arbitrary duct with area change, friction, heat transfer, and rotation. The main thrust of this chapter has been to develop an improved intuitive understanding of the following features of various one-dimensional compressible flows:

- For an isentropic compressible flow in a choked convergent–divergent nozzle, both area and Mach number increase in the divergent section, and the nozzle exit static pressure may differ from the ambient static pressure.
- In a constant-area duct with wall friction but no heat transfer and no rotation (Fanno flow), the Mach number for a subsonic inlet continuously increases downstream in the duct. The reverse is true if the inlet flow is supersonic. In both cases, the friction choking ($M = 1$) is the limit at the duct exit.
- In a constant-area duct with heat transfer (heating of the fluid) but no friction and no rotation (Rayleigh flow), the total pressure for a subsonic flow at the inlet decreases, and the Mach number increases continuously downstream. The thermal choking ($M = 1$) is the limit at the duct exit.
- For a Rayleigh flow with heating, the static temperature for a subsonic inlet continuously decreases when the Mach number exceeds $M = \dfrac{1}{\sqrt{\gamma}}$.
- For a constant-area duct, we cannot change a subsonic flow at the inlet into a supersonic flow or a supersonic flow at the inlet into a subsonic flow by any amount of wall friction and heat transfer.
- The upstream flow must always be supersonic for a normal shock, and the flow immediately after the shock becomes subsonic.
- In a constant-area duct with friction, we can never accelerate an isothermal subsonic flow or decelerate an isothermal supersonic flow beyond $M = \dfrac{1}{\sqrt{\gamma}}$.

The chapter ends with a detailed discussion on the operations of both converging and C–D nozzles under various back-pressure conditions, including how to run a supersonic wind tunnel where the starting normal shock in the test section gets swallowed by the downstream variable-throat-area diffuser for an efficient shock-free operation.

To develop their skills in solving various fluid flow problems, readers may want to review numerous problems with solutions presented in Sultanian (2021, 2022).

WORKED EXAMPLES

Example 5.1

The total temperature and pressure at the inlet to a three-stage axial flow turbine are 1500°C and 10 bar, respectively. The flow conditions at the inlet to the annular exhaust diffuser correspond to the total velocity Mach number of 0.6, the static pressure of 0.85 bar, and the total temperature of 400°C. The swirl velocity (tangential velocity) and the radial velocity at the diffuser inlet are 20% and 5% of the total velocity, respectively. The flow exits the diffuser fully axially. In the diffuser, from inlet to exit, the total temperature drops by 15°C due to heat transfer and the total pressure reduces by 5500 Pa due to wall friction and secondary flows. The diffuser design calls for a static pressure of 1.013 bar at the diffuser exit to allow the exhaust gases to discharge into the ambient air. Calculate the annular diffuser exit-to-inlet flow area ratio. Assume $R = 287\ \mathrm{J/(kg\ K)}$ and $\gamma = 1.4$ for the exhaust gases.

Solution

In the solution of this example, the given total temperature and pressure conditions at the inlet to turbine are not relevant. We need to work only with the data given at the inlet and exit sections of the annular diffuser.

ANNULAR DIFFUSER INLET

Velocity coefficient:

$$V^2 = V_x^2 + V_\theta^2 + V_r^2$$

$$C_V = \frac{V_x}{V} = \sqrt{1-(0.2)^2-(0.05)^2} = 0.9785$$

Static pressure mass flow function: Equation 5.30 yields"

$$\hat{F}_{\text{f\_inlet}} = M_{\text{inlet}}\sqrt{\gamma\left(1+\frac{\gamma-1}{2}M_{\text{inlet}}^2\right)} = 0.6\sqrt{1.4\left(1+\frac{1.4-1}{2}(0.6)^2\right)} = 0.7350$$

Mass flow rate per unit area: Using Equation 5.37, we obtain:

$$\frac{\dot{m}_{\text{inlet}}}{A_{\text{inlet}}} = \frac{C_V\hat{F}_{\text{f\_inlet}}p_{\text{inlet}}}{\sqrt{RT_{0\_\text{inlet}}}} = \frac{0.9785\times0.7350\times0.85\times10^5}{\sqrt{287\times(400+273)}} = 139.108\ \text{kg}/\left(\text{sm}^2\right)$$

ANNULAR DIFFUSER EXIT

Total pressure at inlet: Equation 5.11 yields:

$$p_{0\_\text{inlet}} = p_{\text{inlet}}\left(1+\frac{\gamma-1}{2}M_{\text{inlet}}^2\right)^{\frac{\gamma}{\gamma-1}} = 0.85\times10^5\left(1+\frac{1.4-1}{2}(0.6)^2\right)^{\frac{1.4}{1.4-1}} = 108{,}418\ \text{Pa}$$

Total pressure at exit:

$$p_{0\_\text{exit}} = 108418 - 5500 = 102{,}918\ \text{Pa}$$

Total temperature at exit:

$$T_{0\_\text{exit}} = 673 - 15 = 658\ \text{K}$$

Exit Mach number: Equation 5.11 yields:

$$\frac{p_{0\_\text{exit}}}{p_{\text{exit}}} = \left(1+\frac{\gamma-1}{2}M_{\text{exit}}^2\right)^{\frac{\gamma}{\gamma-1}}$$

$$M_{\text{exit}} = \sqrt{\frac{2}{\gamma-1}\left\{\left(\frac{p_{0\_\text{exit}}}{p_{\text{exit}}}\right)^{\frac{\gamma-1}{\gamma}}-1\right\}} = \sqrt{\frac{2}{1.4-1}\left\{\left(\frac{102918}{1.013\times10^5}\right)^{\frac{1.4-1}{1.4}}-1\right\}} = 0.1506$$

Static pressure mass flow function: Using Equation 5.30, we obtain:

$$\hat{F}_{\text{f\_exit}} = M_{\text{exit}}\sqrt{\gamma\left(1+\frac{\gamma-1}{2}M_{\text{exit}}^2\right)} = 0.1506\sqrt{1.4\left(1+\frac{1.4-1}{2}(0.1506)^2\right)} = 0.1786$$

Mass flow rate per unit area: Using Equation 5.26, we obtain:

$$\frac{\dot{m}_{\text{exit}}}{A_{\text{exit}}} = \frac{F_{\text{f\_exit}}p_{\text{exit}}}{\sqrt{T_{0\_\text{exit}}}} = \frac{0.1786\times1.013\times10^5}{\sqrt{287\times658}} = 41.637\ \text{kg}/\left(\text{sm}^2\right)$$

As the mass flow rates at annular diffuser inlet and exit are equal, we write:

$$\frac{A_{\text{exit}}}{A_{\text{inlet}}} = \frac{139.108}{41.637} = 3.341$$

Thus, the computed exit-to-inlet flow area ratio of the annular exhaust diffuser is 3.341.

Example 5.2

Figure 5.51 shows an air compressor operating in a duct of constant area $A = 0.0645\ \text{m}^2$. For the one-dimensional flow in the duct, the total pressure and the Mach number at sections 1 and 2 are given as $p_{01} = 1.5\ \text{bar}$, $M_1 = 0.3$, $p_{02} = 2.0\ \text{bar}$, and $M_2 = 0.4$. Assuming $\gamma = 1.4$ for air, find (a) the force acting on the compressor, (b) the ratio of total temperatures T_{02} / T_{01}, and (c) the entropy change $(s_2 - s_1) / R$.

Solution

In the solution of this example, we further assume that the frictional force acting on the duct wall is negligible compared to the drag force of the compressor body.

FORCE ACTING ON THE COMPRESSOR IN THE FLOW DIRECTION

If F is the force acting on the compressor in the flow direction, the drag force acting on the flow will be $-F$, and it should equal the reduction in stream thrust between sections 1 and 2.

$$-F = S_{T2} - S_{T1}$$

$$F = S_{T1} - S_{T2} = p_{01} A I_{f01} - p_{02} A I_{f02}$$

$$F = 2.0 \times 10^5 \times 0.0645 \times 1.0962 - 1.5 \times 10^5 \times 0.0645 \times 1.0578$$

$$F = 14141.4 - 10234.6 = 3906.8\ N$$

RATIO OF TOTAL TEMPERATURES T_{02} / T_{01}

As the mass flow rates at sections 1 and 2 are equal, we write:

$$\dot{m} = \frac{A \hat{F}_{f01} p_{01}}{\sqrt{RT_{01}}} = \frac{A \hat{F}_{f02} p_{02}}{\sqrt{RT_{01}}}$$

which yields:

$$\frac{T_{02}}{T_{01}} = \left(\frac{\hat{F}_{f02} p_{02}}{\hat{F}_{f01} p_{01}} \right)^2$$

$$\frac{T_{02}}{T_{01}} = \left(\frac{0.4306 \times 3 \times 10^5}{0.3365 \times 1.5 \times 10^5} \right)^2 = 2.912$$

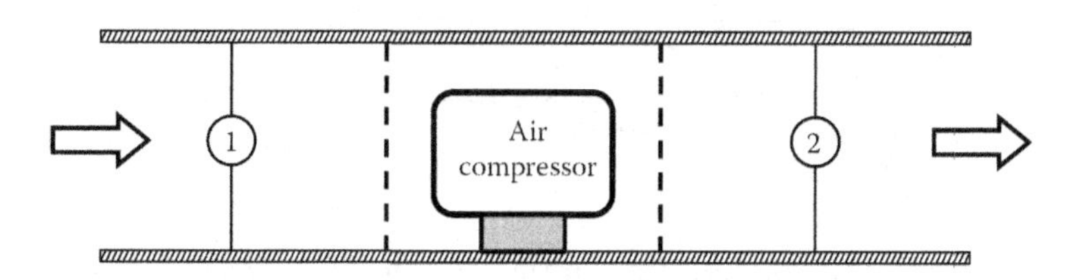

FIGURE 5.51 An air compressor operating in a constant-area duct (Example 5.2).

Entropy Change $(s_2 - s_1)/R$

In terms of total properties, we can compute entropy changes between sections 1 and 2 as:

$$s_2 - s_1 = c_p \ln\left(\frac{T_{02}}{T_{01}}\right) - R\ln\left(\frac{p_{02}}{p_{01}}\right)$$

$$\frac{s_2 - s_1}{R} = \frac{c_p}{R}\ln\left(\frac{T_{02}}{T_{01}}\right) - \ln\left(\frac{p_{02}}{p_{01}}\right) = \frac{\gamma}{\gamma - 1}\ln\left(\frac{T_{02}}{T_{01}}\right) - \ln\left(\frac{p_{02}}{p_{01}}\right)$$

$$\frac{s_2 - s_1}{R} = \frac{1.4}{1.4 - 1}\ln(2.912) - \ln\left(\frac{2\times10^5}{1.5\times10^5}\right) = 3.453$$

Therefore, using the mass flow function and the impulse function in this example, we readily obtain: (a) force acting on the compressor = 3906.8 N, (b) total temperature ratio $T_{02}/T_{01} = 2.912$, and (c) entropy change $(s_2 - s_1)/R = 3.453$.

Example 5.3

Figure 5.52 shows two air streams of different properties entering a long duct with a cross-sectional area of $0.002\ \text{m}^2$. At section 1 (inlet), the hotter stream forms the central core, surrounded by the colder stream. Each stream occupies an equal area (neglect the wall thickness of the inner duct). The uniform static pressure in section 1 equals 895614 Pa, and in section 2 (outlet) 10 bar. Total pressure and temperature of each stream at section 1 (inlet) are given below. Assume both streams mix thoroughly with uniform properties before exiting the duct at section 2.

	Stream 1	*Stream 2*
Total temperature (K)	1173	1473
Total pressure (bar)	10	15

Assuming $\gamma = 1.4$ and $R = 287\ \text{J/(kg K)}$, calculate the following:

a. Mass flow rate of each stream
b. Section-average total temperature and static temperature at inlet
c. Section-average total pressure at inlet
d. Percentage drop in total pressure from inlet to outlet
e. Total wall shear force acting on the flow between inlet and outlet

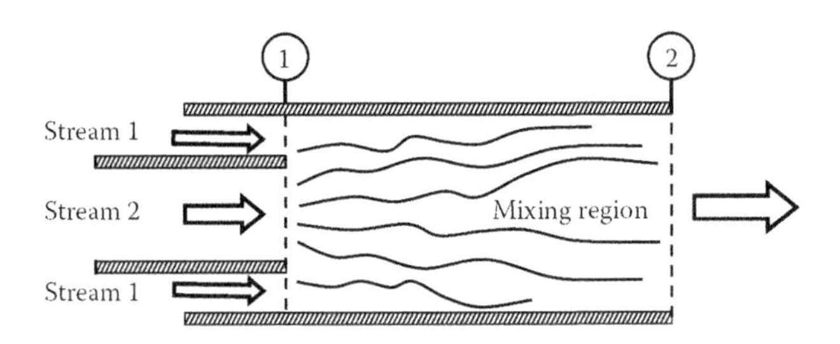

FIGURE 5.52 Mixing of two compressible flows in a duct (Example 5.3).

Solution

In this example, stream 1 forms the annular flow and stream 2 forms the central flow. Both streams start mixing at section 1 (inlet), where the uniform static pressure is given as 895614 Pa. Both streams become thoroughly mixed when the flow reaches section 2 (outlet). The flow exits the duct with uniform velocity and temperature at a uniform static pressure of 10 bar.

(A) MASS FLOW RATE OF EACH STREAM

Stream 1 at inlet:

$$A_1 = 0.001\ \text{m}^2; p_{01} = 10\times10^5\ \text{Pa}; T_{01} = 1173\ \text{K; and } p_1 = 895614\ \text{Pa}$$

$$\frac{p_{01}}{p_1} = \frac{10\times10^5}{895614} = 1.117$$

which yields $\frac{T_{01}}{T_1} = 1.032$, $M_1 = 0.400$, and $\hat{F}_{f1} = 0.481$. We express the mass flow rate at this section as:

$$\dot{m}_1 = \frac{\hat{F}_{f1} A_1 p_1}{\sqrt{RT_{01}}} = \frac{0.481\times0.001\times895614}{\sqrt{287\times1173}} = 0.742\ \text{kg/s}$$

Stream 2 at inlet:

$$A_2 = 0.001\ \text{m}^2; p_{02} = 15\times10^5\ \text{Pa}; T_{01} = 1473\ \text{K; and } p_2 = 895614\ \text{Pa}$$

$$\frac{p_{02}}{p_2} = \frac{15\times10^5}{895614} = 1.675$$

which yields $\frac{T_{02}}{T_2} = 1.159$, $M_2 = 0.891$, and $\hat{F}_{f2} = 1.135$. We compute the mass flow rate at this section as:

$$\dot{m}_2 = \frac{\hat{F}_{f2} A_2 p_2}{\sqrt{RT_{02}}} = \frac{1.135\times0.001\times895614}{\sqrt{287\times1473}} = 1.563\ \text{kg/s}$$

(B) SECTION-AVERAGE TOTAL TEMPERATURE AND STATIC TEMPERATURE AT INLET

As the static and total temperatures represent scalar properties of the fluid flow, consistent with the steady flow energy equation, we compute their section-average values at the inlet using the mass-weighted averaging of their values for each stream. Thus,

$$\dot{m}_{\text{inlet}} = \dot{m}_1 + \dot{m}_2 = 0.742 + 1.563 = 2.305\ \text{kg/s}$$

$$T_{0\_\text{inlet}} = \frac{\dot{m}_1 T_{01} + \dot{m}_2 T_{02}}{\dot{m}_{\text{inlet}}} = \frac{0.742\times1173 + 1.563\times1473}{2.305} = 1376.4\ \text{K}$$

$$T_1 = \frac{T_{01}}{T_{01}/T_1} = \frac{1173}{1.032} = 1136.6\ \text{K and } T_2 = \frac{T_{02}}{T_{02}/T_2} = \frac{1473}{1.159} = 1271.2\ \text{K}$$

$$T_{\text{inlet}} = \frac{\dot{m}_1 T_1 + \dot{m}_2 T_2}{\dot{m}_{\text{inlet}}} = \frac{0.742\times1136.6 + 1.563\times1271.2}{2.305} = 1227.9\ \text{K}$$

(C) SECTION-AVERAGE TOTAL PRESSURE AT INLET

The following technique to compute section-average value of total pressure at the inlet does not invoke either the area-weighted averaging or the mass-weighted averaging:

$$\frac{T_{0\_inlet}}{T_{inlet}} = \frac{1376.4}{1227.9} = 1.121$$

which yields:

$$\frac{p_{0\_inlet}}{p_{inlet}} = \left(\frac{T_{0\_inlet}}{T_{inlet}}\right)^{\frac{\gamma}{\gamma-1}} = 1.491$$

giving

$$p_{0\_inlet} = \frac{p_{0\_inlet}}{p_{inlet}} \times p_{inlet} = 1.491 \times 895614 = 1335712 \text{ Pa}$$

(D) PERCENTAGE DROP IN TOTAL PRESSURE FROM INLET TO OUTLET

Total pressure at outlet:

$$A_{outlet} = 0.002 \text{ m}^2; \dot{m}_{outlet} = \dot{m}_{inlet} = 2.305 \text{ kg/s};$$

$$T_{0\_outlet} = T_{0\_inlet} = 1376.4 \text{ K; and } p_{outlet} = 10 \times 10^5 \text{ Pa}$$

$$\hat{F}_{f\_outlet} = \frac{\dot{m}_{outlet}\sqrt{RT_{0\_outlet}}}{A_{outlet}\, p_{outlet}} = \frac{2.305 \times \sqrt{287 \times 1376.4}}{0.002 \times 10 \times 10^5} = 0.724$$

We use Equation 5.33 to obtain the flow Mach number at the outlet as:

$$M_{outlet} = \left(\frac{-\gamma + \sqrt{\gamma^2 + 2\gamma(\gamma-1)\hat{F}_{f\_outlet}^2}}{\gamma(\gamma-1)}\right)^{\frac{1}{2}} = 0.592$$

which further yields $\frac{p_{0\_outlet}}{p_{outlet}} = 1.2675$ and $p_{0\_outlet} = 1267465$ Pa. Thus,

$$\Delta p_0 = \left(\frac{p_{0\_inlet} - p_{0\_outlet}}{p_{0\_inlet}}\right) \times 100 = \left(\frac{1335712 - 1267465}{1335712}\right) \times 100 = 5.109\%$$

(E) TOTAL WALL SHEAR FORCE ACTING ON THE FLOW BETWEEN INLET AND OUTLET

As the duct cross section between inlet and outlet remains constant, from the linear momentum equation, we can write:

$$F_{shear} = S_{T\_inlet} - S_{T\_outlet} = S_{T1} + S_{T2} - S_{T\_outlet}$$

where we evaluate the stream thrusts at inlet and outlet as follows:

$$S_{T1} = p_1 A_1 \left(1 + \gamma M_1^2\right) = 895614 \times 0.001 \times (1 + 1.4 \times 0.4 \times 0.4) = 1096.232 \text{ N}$$

$$S_{T2} = p_2 A_2 \left(1 + \gamma M_2^2\right) = 895614 \times 0.001 \times (1 + 1.4 \times 0.891 \times 0.891) = 1890.895 \text{ N}$$

$$S_{T\_outlet} = p_{outlet} A_{outlet} \left(1 + \gamma M_{outlet}^2\right) = 1000000 \times 0.002 \times (1 + 1.4 \times 0.592 \times 0.592) = 2980.913 \text{ N}$$

giving

$$F_{\text{shear}} = 1096.232 + 1890.895 - 2980.913 = 6.214 \text{ N}$$

We summarize the results obtained in this example as follows:

Mass flow rate of each stream: $\dot{m}_1 = 0.742$ kg/s and $\dot{m}_2 = 1.563$ kg/s

Section-average total temperature and static temperature at inlet: $T_{0\_\text{inlet}} = 1376.4$ K and $T_{\text{inlet}} = 1227.9$ K

Section-average total pressure at inlet: $p_{0\_\text{inlet}} = 1335712$ Pa

Percentage drop in total pressure from inlet to outlet: $\Delta p_0 = 5.109\%$

Total wall shear force acting on the flow between inlet and outlet: $F_{\text{shear}} = 6.214$ N

Example 5.4

Figure 5.53 shows an airflow through a duct fitted with a sharp-edge orifice of area $A_o = 0.001$ m^2. Constant pressure and temperature conditions upstream of the orifice are $p_0 = 5$ bar and $T_0 = 500$ K. Calculate the mass flow rate through the duct for three values of back pressure: (a) $p_b = 4$ bar, (b) $p_b = 3$ bar, and (c) $p_b = 2$ bar. Assume no loss in total pressure and temperature across the duct. Use $\gamma = 1.4$ and $R = 287 \text{ J}/(\text{kg K})$.

Solution

In this example, we first need to establish the critical static pressure (p_o^*) at the orifice. If the static back pressure is above this critical value, the mass flow rate through the orifice will change; if it is below this value, the orifice operates under a choked flow condition with maximum mass flow rate, becoming independent of the downstream back pressure. Note that for $p_b > p_o^*$, the orifice exit static pressure p_o equals the downstream back pressure p_b.

Critical pressure at orifice: Using Equation 5.14, we obtain:

$$\frac{p_0}{p_o^*} = \left(\frac{\gamma+1}{2}\right)^{\frac{\gamma}{\gamma-1}} = 1.893$$

$$p_o^* = \frac{5}{1.893} = 2.641 \text{ bar}$$

a. $p_b = p_o = 4$ bar

Because the back pressure of 4 bar is above the critical pressure of 2.641 bar, the orifice will flow subsonic with Mach number computed from Equation 5.11 as:

$$\frac{p_0}{p_o} = \left(1 + \frac{\gamma-1}{2}M^2\right)^{\frac{\gamma}{\gamma-1}}$$

$$M = \sqrt{\frac{2}{\gamma-1}\left[\left(\frac{p_0}{p_b}\right)^{\frac{\gamma-1}{\gamma}} - 1\right]} = 0.574$$

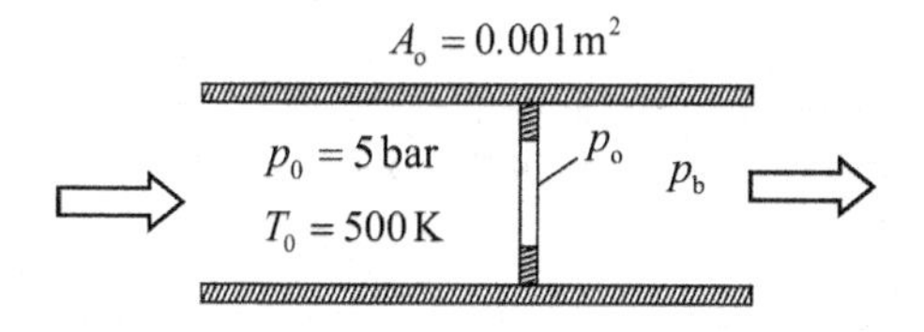

FIGURE 5.53 Air mass flow rate through an orifice for various back-pressure values (Example 5.4).

Equation 5.30 yields:

$$\hat{F}_f = M\sqrt{\gamma\left(1+\frac{\gamma-1}{2}M^2\right)} = 0.701$$

We calculate the mass flow rate using Equation 5.26 as:

$$\dot{m} = \frac{\hat{F}_f A_o p_o}{\sqrt{RT_0}} = \frac{0.701\times 0.001\times 4.0\times 10^5}{\sqrt{287\times 500}} = 0.740 \text{ kg/s}$$

b. $p_b = p_0 = 3$ bar

Because the back pressure of 3 bar is above the critical pressure of 2.641 bar, the orifice will also flow subsonic in this case. Repeating the steps used in (a), we obtain the following values:

$M = 0.886, \hat{F}_f = 1.128$, and

$$\dot{m} = \frac{\hat{F}_f A_o p_o}{\sqrt{RT_0}} = \frac{1.128\times 0.001\times 3.0\times 10^5}{\sqrt{287\times 500}} = 0.893 \text{ kg/s}$$

c. $p_b = 2$ bar

Because the back pressure of 2 bar is below the critical pressure of 2.641 bar, the orifice will flow sonic $(M = 1)$ with $p_o = p_o^* = 2.641$ bar and $\hat{F}_f^* = 1.296$. We compute the mass flow rate in this case as:

$$\dot{m} = \frac{\hat{F}_f^* A_o p_o^*}{\sqrt{RT_0}} = \frac{1.296\times 0.001\times 2.641\times 10^5}{\sqrt{287\times 500}} = 0.904 \text{ kg/s}$$

For the given upstream conditions, the computed mass flow rate corresponds to the maximum value. Note that if in this case we had assumed the static pressure at the orifice to be equal to the back pressure of 2 bar, we had erroneously computed a supersonic flow at the orifice and a lower mass flow rate.

Example 5.5

Figure 5.54 shows a constant-area duct used for air impingement cooling of a cylindrical surface. At the outer edge, the duct is fitted with a nozzle to produce a jet of diameter $d_j = 0.01$ m. The air jet exits the nozzle at a radius $R_1 = 1.0$ m into the ambient static pressure of 1 bar. At the origin, the air enters the duct at a total pressure and total temperature of $p_{0o} = 10$ bar and $T_{0o} = 373$ K, respectively. Assuming the duct airflow to be isentropic (adiabatic and frictionless), calculate its mass flow rate when (a) $\Omega = 0$ and (b) $\Omega = 3000$ rpm. Use $\gamma = 1.4$ and $R = 287\ \text{J}/(\text{kg K})$ for the coolant air.

Solution

In this example, the given total pressure and temperature at the origin corresponds to zero radius. When the duct is stationary, the isentropic flow assumption implies no change in total pressure and temperature in the duct. When the duct rotates, we consider the total pressure and temperature to be in the rotating reference frame. In both cases, as the pressure ratio across the nozzle is greater than the critical value of 1.8929, the nozzle operates under the choked flow condition with no influence from the ambient pressure of 1 bar in the cylinder.

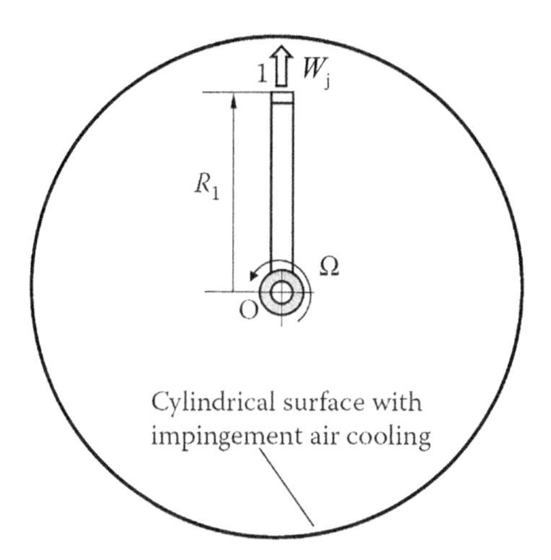

FIGURE 5.54 Impingement cooling of a cylindrical surface with a rotating air jet (Example 5.5).

(A) MASS FLOW RATE THROUGH THE DUCT WHEN $\Omega = 0$

Nozzle throat area:

$$A_j = \frac{\pi d_j^2}{4} = \frac{\pi \times (0.01)^2}{4} = 7.854 \times 10^{-5}\ \text{m}^2$$

Mass flow rate:

$$\dot{m}_{(a)} = \frac{\hat{F}_{f0}^{*} A_j p_{0o}}{\sqrt{RT_{0o}}} = \frac{0.6847 \times 7.854 \times 10^{-5} \times 10^6}{\sqrt{287 \times 373}} = 0.164\ \text{kg/s}$$

(B) MASS FLOW RATE THROUGH THE DUCT WHEN $\Omega = 3000$ rpm

Duct rotational speed:

$$\Omega = \frac{2\pi \times 3000}{60} = 314.159\ \text{rad/s}$$

Air specific heat at constant pressure:

$$c_p = \frac{R\gamma}{\gamma - 1} = 1004.5\ \text{J}/(\text{kg K})$$

Increase in relative total temperature due to rotational work transfer from origin to nozzle throat:

$$\Delta T_0 = \frac{\Omega^2 \left(R_1^2 - R_o^2\right)}{2c_p} = \frac{(314.159)^2 \times 1}{2 \times 1004.5} = 49.127\ \text{K}$$

Relative total temperature at nozzle throat:

$$T_{01} = T_{0o} + \Delta T_0 = 373 + 49.127 = 422.127\ \text{K}$$

Relative total pressure at nozzle throat:

As the flow remains isentropic, we can obtain the relative total pressure at the nozzle throat using the equation:

$$\frac{p_{01}}{p_{0o}} = \left(\frac{T_{01}}{T_{0o}}\right)^{\frac{\gamma}{\gamma-1}} = \left(\frac{422.127}{373}\right)^{\frac{1.4}{1.4-1}} = 1.542$$

$$p_{01} = 1.542 \times 10^6 \text{ Pa}$$

Mass flow rate:

$$\dot{m}_{(b)} = \frac{\hat{F}^*_{f01} A_j p_{01}}{\sqrt{RT_{01}}} = \frac{0.6847 \times 7.854 \times 10^{-5} \times 1.542 \times 10^6}{\sqrt{287 \times 422.127}} = 0.238 \text{ kg/s}$$

The results show that the mass flow rate for the rotating duct increases by about 45% due to the pumping effect.

Example 5.6

Consider a Fanno flow in a pipe of 0.0508 m diameter and average Darcy friction factor of 0.022. The total pressure and temperature at the pipe inlet are 11 bar and 459 K, respectively. Assuming $\gamma = 1.4$ and $R = 287$ J/(kg K) for air, calculate the maximum pipe length to deliver a mass flow rate of 1.75 kg/s for both subsonic and supersonic inlet conditions. In each case, determine the total pressure at the pipe exit.

Solution

In this example, for the given total pressure and temperature at the pipe inlet, two flow solutions are possible for the specified mass flow rate through the pipe: one subsonic and the other supersonic. In both cases, the flow becomes sonic at the exit with different pipe lengths.

(A) SUBSONIC FLOW AT INLET

Pipe flow area:

$$A = \frac{\pi D^2}{4} = \frac{\pi \times (0.0508)^2}{4} = 2.027 \times 10^{-3} \text{ m}^2$$

Inlet Mach number: Using Equation 5.21, we express the mass flow rate at the pipe inlet as:

$$\dot{m} = \frac{\hat{F}_{f0(a)} A p_{0\_inlet}}{\sqrt{RT_{0\_inlet}}}$$

where

$$\hat{F}_{f0(a)} = M_{inlet(a)} \sqrt{\frac{\gamma}{\left(1 + \frac{\gamma - 1}{2} M^2_{inlet(a)}\right)^{\frac{\gamma+1}{\gamma-1}}}}$$

The above equations show that we can compute pipe inlet Mach number using an iterative method only. Using the "Goal Seek" function in Excel with a subsonic Mach number as the initial guess, we obtain $M_{inlet(a)} = 0.25$.

Maximum pipe length: Using the inlet Mach number, we compute $fL_{max(a)} / D$ by Equation 5.90 as:

$$\frac{fL_{max(a)}}{D} = \frac{1.4+1}{2 \times 1.4} \ln\left(\frac{(1.4+1)(0.25)^2}{\{2 + (1.4-1)(0.25)^2\}}\right) + \frac{1}{1.4}\left(\frac{1}{(0.25)^2} - 1\right) = 8.491$$

giving

$$L_{max(a)} = \frac{8.491 \times 0.0508}{0.022} = 19.606 \text{ m}$$

Total pressure at pipe exit: Using Equation 5.81, we compute the total pressure ratio as:

$$\frac{p_{0\_inlet}}{p^*_{0(a)}} = \frac{1}{M_{inlet(a)}}\left(\frac{2+(\gamma-1)M^2_{inlet(a)}}{\gamma+1}\right)^{\frac{\gamma+1}{2(\gamma-1)}} = \frac{1}{0.25}\left(\frac{2+(1.4-1)(0.25)^2}{1.4+1}\right)^{\frac{1.4+1}{2(1.4-1)}} = 2.404$$

giving

$$p^*_{0(a)} = \frac{p_{0\_inlet}}{2.404} = \frac{11\times 10^5}{2.404} = 457670 \text{ Pa}$$

(B) SUPERSONIC FLOW AT INLET

Inlet Mach number: We again use an iterative method (the "Goal Seek" function in Excel with a supersonic Mach number as the initial guess) to solve the equations:

$$\dot{m} = \frac{\hat{F}_{f0(b)} A p_{0\_inlet}}{\sqrt{R T_{0\_inlet}}}$$

and

$$\hat{F}_{f0(b)} = M_{inlet(b)} \sqrt{\frac{\gamma}{\left(1+\frac{\gamma-1}{2} M^2_{inlet(b)}\right)^{\frac{\gamma+1}{\gamma-1}}}}$$

giving

$$M_{inlet(b)} = 2.4$$

Maximum pipe length: Using $M_{inlet(b)} = 2.4$, we compute $fL_{max(b)} / D$ by Equation 5.90 as:

$$\frac{fL_{max(b)}}{D} = \frac{1.4+1}{2\times 1.4}\ln\left(\frac{(1.4+1)(2.4)^2}{\{2+(1.4-1)(2.4)^2\}}\right) + \frac{1}{1.4}\left(\frac{1}{(2.4)^2} - 1\right) = 0.410$$

giving

$$L_{max(b)} = \frac{0.410 \times 0.0508}{0.022} = 0.947 \text{ m}$$

Total pressure at pipe exit: We use Equation 5.81 to compute the total pressure ratio as:

$$\frac{p_{0\_inlet}}{p^*_{0(b)}} = \frac{1}{M_{inlet(b)}}\left(\frac{2+(\gamma-1)M^2_{inlet(b)}}{\gamma+1}\right)^{\frac{\gamma+1}{2(\gamma-1)}} = \frac{1}{2.4}\left(\frac{2+(1.4-1)(2.4)^2}{1.4+1}\right)^{\frac{1.4+1}{2(1.4-1)}} = 2.404$$

giving

$$p^*_{0(b)} = \frac{p_{0\_inlet}}{2.404} = \frac{11\times 10^5}{2.404} = 457670 \text{ Pa}$$

The results obtained in this example show that the maximum duct length with the subsonic inlet flow is 19.606 m and that for the supersonic inlet flow is 0.947 m, which provides a much shorter delivery system. As the computed total pressure at the pipe exit is the same in both cases, they have the same overall drop in total pressure over two different pipe lengths. We can use an isentropic convergent nozzle for a subsonic Fanno flow to create the subsonic inlet conditions. We can use an isentropic convergent–divergent nozzle for the supersonic Fanno flow to generate the required inlet conditions. In this case, however, the flow system will have two sonic sections, one at the physical throat of the nozzle and the other at the pipe exit.

Example 5.7

Calculate and describe the properties of the Fanno flow of air presented in Example 5.6 if the maximum pipe lengths with sonic exit plane computed for the subsonic and supersonic inlet conditions are extended by10% without changing the specified total pressure and temperature at the pipe inlet. In both cases, the flow remains choked at the extended pipe exit.

Solution

Figure 5.55 schematically (not to scale) shows pipe extensions in initially choked Fanno flows with subsonic and supersonic inlet conditions. As shown in Figure 5.55a, when we extend a choked Fanno flow with a subsonic inlet, the sonic section moves downstream to the extended pipe exit with a reduction in both mass flow rate and exit total pressure. A choked Fanno flow with a supersonic inlet, shown in Figure 5.55b, features a different flow behavior when we extend the pipe. The pipe extension results in a normal shock at an intermediate section because the downstream changes are not communicated upstream in a supersonic flow. The flow is supersonic upstream of the normal shock, becoming subsonic downstream and reaching the sonic condition at the extended pipe exit. In this case, the mass flow rate and the total pressure at the new exit remain unchanged.

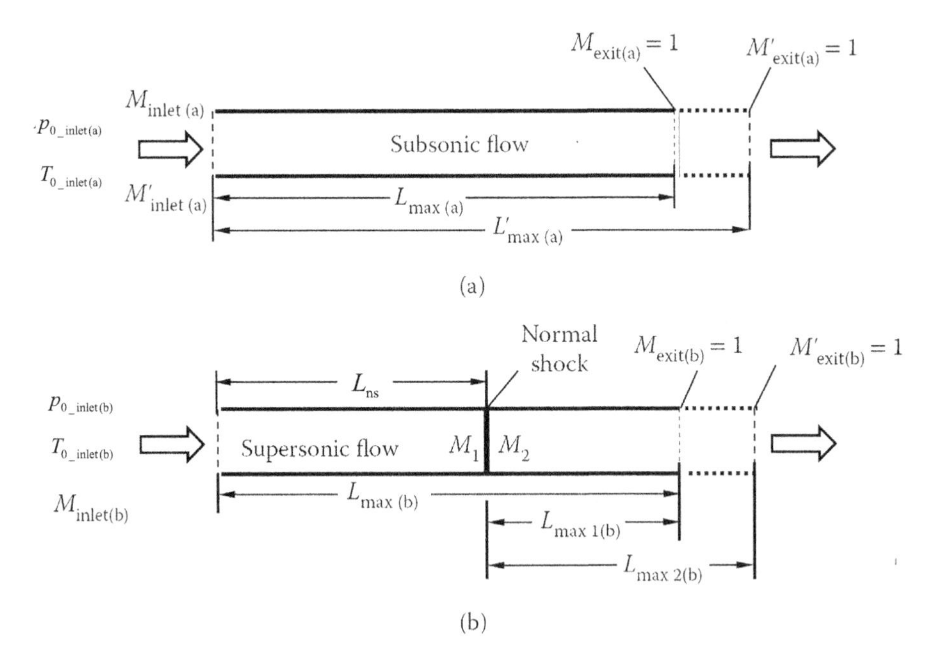

FIGURE 5.55 (a) Pipe extension in an initially choked Fanno flow with subsonic inlet and (b) pipe extension in an initially choked Fanno flow with supersonic inlet (Example 5.7).

(A) SUBSONIC FLOW AT INLET

With reference to Figure 5.55a, the inlet conditions and calculated properties from Example 5.6 for the case of subsonic flow are summarized below.

$$p_{0\_inlet(a)} = 11\times10^5 \text{ Pa}, T_{0\_inlet(a)} = 459 \text{ K}, M_{inlet(a)} = 0.25, L_{max(a)} = 19.606 \text{ m, and } p_{0\_exit(a)} = 457670 \text{ Pa}$$

With 10% extension of the pipe, the new maximum pipe length becomes:

$$L'_{max(a)} = 1.1\times19.606 = 21.566 \text{ m}$$

giving

$$\frac{fL'_{max(a)}}{D} = 0.022\times21.566/0.054 = 9.34$$

Using the "Goal Seek" iterative method in Excel, we obtain the new inlet Mach number from Equation 5.90 as $M'_{inlet(a)} = 0.241$. With this inlet Mach number, we calculate the new mass flow rate from Equation 5.21 as $\dot{m}'_{(a)} = 1.689$ kg/s, which is 3.51% lower than the initial value of 1.75 kg/s without the pipe extension. Similarly, at $M'_{inlet(a)} = 0.241$, we compute the new exit total pressure from Equation 5.81 as $p'^{*}_{0\_exit(a)} = 441616$ Pa, which is also 3.51% lower than its value calculated before the pipe extension.

(B) SUPERSONIC FLOW AT INLET

With reference to Figure 5.55b, the inlet conditions and calculated properties from Example 5.6 for this case of supersonic inlet are summarized below.

$$p_{0\_inlet(b)} = 11\times10^5 \text{ Pa}, T_{0\_inlet(b)} = 459 \text{ K}, M_{inlet(b)} = 2.4, L_{max(b)} = 0.947 \text{ m, and } p_{0\_exit(b)} = 457670 \text{ Pa}$$

With 10% extension of the pipe, the new maximum pipe length becomes:

$$L'_{max(b)} = 1.1\times0.947 = 1.0417 \text{ m}$$

From the flow physics presented above for this case, the pipe inlet and new exit conditions remain the same as before the pipe extension. We only need to find the distance L_{ns} from the pipe inlet where the normal shock occurs.

From Figure 5.55b, we can write:

$$L'_{max(b)} = L_{ns} + L_{max\,2(b)} \text{ and } L_{ns} = L_{max(b)} - L_{max\,1(b)}$$

where $L_{max(b)}$ is the maximum pipe length corresponding to the supersonic Mach number at the pipe inlet, $L_{max\,1(b)}$ is the maximum pipe length computed at M_1, and $L_{max\,2(b)}$ is the maximum pipe length computed at M_2. Note that M_1 is the supersonic Mach number upstream of the normal shock, and M_2 is the downstream subsonic Mach number. Next, we present the steps of an iterative method to calculate L_{ns}.

1. Assume M_1.
2. Calculate M_2. with the following equation, which we present in the section on normal shocks:

$$M_2^2 = \frac{(\gamma-1)M_1^2 + 2}{2\gamma M_1^2 - (\gamma-1)}$$

3. Use Equation 5.90 to calculate $fL_{max\,1(b)}/D$ at M_1, hence $L_{max\,1(b)}$.
4. Calculate $L_{ns} = L_{max(b)} - L_{max\,1(b)}$.

5. Use Equation 5.90 to calculate $fL_{max\,2(b)} / D$ at M_2, hence $L_{max\,2(b)}$.
6. Repeat steps from 1 to 6 until $L'_{max(b)} = L_{ns} + L_{max\,2(b)}$.

Using the "Goal Seek" function in Excel, this iterative method yields the following values for various quantities:

$$M_1 = 1.3967, M_2 = 0.7411, L_{max\,1(b)} = 0.228 \text{ m}, L_{ns} = 0.719 \text{ m, and } L_{max\,2(b)} = 0.322 \text{ m}$$

These results show that the normal shock in the supersonic Fanno flow makes the downstream flow subsonic in such a way as to keep the overall loss in the total pressure constant when extending the initially choked pipe. This feature of a choked supersonic Fanno flow makes it an ideal constant mass flow delivery system robust to variations in downstream conditions.

Example 5.8

Using Fanno flow tables, calculate the exit Mach number M_2 in the Fanno flow of air shown in Figure 5.56. The inlet Mach number M_1 and the parameter fL / D_h for the duct are given as 0.2 and 10.0, respectively. Assume $\gamma = 1.4$ for air.

Solution

As the duct hydraulic mean diameter and Darcy friction factor remain constant, we can write from Figure 5.56:

$$\frac{fL_{max\,2}}{D_h} = \frac{fL_{max\,1}}{D_h} - \frac{fL}{D_h}$$

where $L_{max\,1}$. and $L_{max\,2}$ are the maximum lengths of the duct required for choking with inlet Mach numbers M_1 and M_2, respectively. From Table A.3 in Sultanian (2015, 2023), we obtain $fL_{max\,1} / D_h = 14.53328$ for $M_1 = 0.2$, giving:

$$\frac{fL_{max\,2}}{D_h} = \frac{fL_{max\,1}}{D_h} - \frac{fL}{D_h} = 14.53328 - 10.0 = 4.53328$$

Entering this table for $fL_{max\,2} / D_h = 4.53328$ with linear interpolation, we obtain $M_2 = 0.317$.

Example 5.9

Consider a Rayleigh flow of air in a 5 m long pipe of area 0.002 m^2 at a steady mass flow rate of 2.0 kg/m^2. The pipe wall is maintained at a constant temperature of 1000 K and the heat transfer coefficient between the pipe wall and the air is constant at $1000 \text{ W}/(\text{m}^2\text{K})$. Total pressure and temperature at the pipe inlet are 7×10^5 Pa and 500 K, respectively. At the pipe inlet and outlet, compute the Mach number, static pressure, and temperature. Also, compute the total temperature and pressure at the pipe outlet. Assume $\gamma = 1.4$ and $R = 287 \text{ J}/(\text{kg K})$ for air.

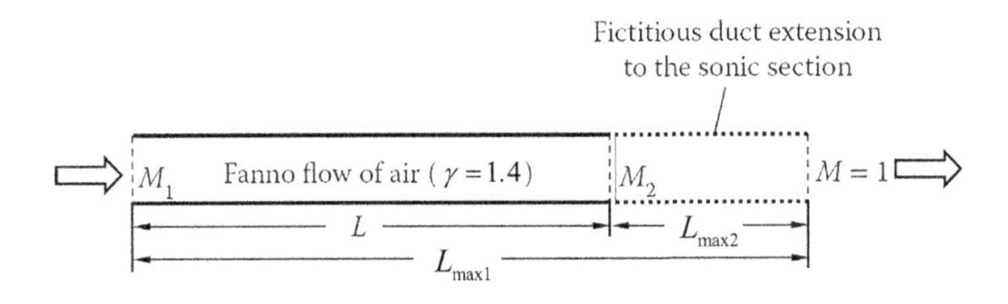

FIGURE 5.56 Fanno flow of air in a duct with given inlet Mach number (Example 5.8).

Solution

In this example, the airflow in the pipe is heated by forced convection with an isothermal pipe wall and a constant heat transfer coefficient. Under this condition, we know that the difference between the wall and inlet air temperatures will decay exponentially toward the pipe outlet. We calculate the air temperature rise in the pipe as follows:

Pipe wall area A_w:

$$D = \sqrt{\frac{4A}{\pi}} = \sqrt{\frac{4 \times 0.002}{\pi}} = 0.0505 \text{ m}$$

$$A_w = \pi DL = \pi \times 0.0505 \times 5 = 0.793 \text{ m}^2$$

Number of transfer units (η):

$$c_p = \frac{R\gamma}{\gamma - 1} = \frac{287 \times 1.4}{(1.4 - 1)} = 1004.5 \text{ J/(kg K)}$$

$$\eta = \frac{hA_w}{\dot{m}c_p} = \frac{1000 \times 0.793}{2 \times 1004.5} = 0.395$$

Total temperature at pipe outlet:

$$T_w - T_{02} = (T_w - T_{01})\,e^{-\eta}$$

$$T_{02} = T_w - (T_w - T_{01})e^{-\eta}$$

$$T_{02} = 1000 - (1000 - 500)e^{-0.395} = 663 \text{ K}$$

Total temperature ratio:

$$\frac{T_{02}}{T_{01}} = \frac{663}{500} = 1.326$$

Inlet Mach number, static pressure, and static temperature: At the pipe inlet, we are given the total pressure, temperature, and the mass flow rate. Using Equations 5.21 and 5.22, we can write:

$$\dot{m} = \frac{\hat{F}_{f01} A p_{01}}{\sqrt{RT_{01}}} \text{ and } \hat{F}_{f01} = M_1 \sqrt{\frac{\gamma}{\left(1 + \frac{\gamma - 1}{2} M_1^2\right)^{\frac{\gamma+1}{\gamma-1}}}}$$

Using an iterative method, such as "Goal Seek" in Excel, we solve the above two equations to yield $M_1 = 0.543$.

Using the isentropic relations between the total and static properties, we obtain the static pressure and temperature as follows:

$$p_1 = \frac{p_{01}}{\left(1 + \frac{\gamma - 1}{2} M_1^2\right)^{\frac{\gamma}{\gamma-1}}} = \frac{7 \times 10^5}{\left(1 + \frac{1.4 - 1}{2} 0.543^2\right)^{\frac{1.4}{1.4-1}}} = 5.727 \times 10^5 \text{ Pa}$$

$$T_1 = \frac{T_{01}}{\left(1 + \frac{\gamma - 1}{2} M_1^2\right)^{\frac{\gamma}{\gamma-1}}} = \frac{500}{\left(1 + \frac{1.4 - 1}{2} 0.543^2\right)} = 472.14 \text{ K}$$

Flow properties at pipe outlet: Knowing the temperature ratio across the pipe (as calculated above) and the inlet Mach number, we iteratively solve Equation 5.103 to obtain $M_2 = 0.924$, giving total pressure ratio using Equation 5.104 as $p_{02} / p_{01} = 0.9144$. Thus, we obtain $p_{02} = 0.9144 \times p_{01} = 0.9144 \times 7 \times 10^5 = 6.401 \times 10^5$ Pa. Again, using the isentropic relations between the total and static properties, we obtain the static pressure and temperature at the pipe outlet as follows:

$$p_2 = \frac{p_{02}}{\left(1 + \frac{\gamma - 1}{2} M_2^2\right)^{\frac{\gamma}{\gamma-1}}} = \frac{6.401 \times 10^5}{\left(1 + \frac{1.4 - 1}{2} 0.924^2\right)^{\frac{1.4}{1.4-1}}} = 3.686 \times 10^5 \text{ Pa}$$

$$T_2 = \frac{T_{02}}{\left(1 + \frac{\gamma - 1}{2} M_2^2\right)^{\frac{\gamma}{\gamma-1}}} = \frac{663}{\left(1 + \frac{1.4 - 1}{2} 0.924^2\right)} = 566.3 \text{ K}$$

Example 5.10

For a Rayleigh flow, you are given M_1, p_{01} and T_{01} at section 1 and T_{02} at section 2. Making use of the Rayleigh flow table, for example, Table A.4 in Sultanian (2015, 2023), outline the steps needed to find M_2 and p_{02} at section 2.

Solution

Knowing the Mach number at a section, we can obtain static properties from corresponding known total (stagnation) properties or vice versa using the isentropic relations. In this example, we are given the total properties and Mach number at section 1. For the case of heating, the total temperature at section 2 will be higher than that at section 1. For the cooling case, the total temperature at section 2 will be lower than at section 1. The solution method outlined below is valid for both cases. While reading the Rayleigh flow table, we may need to use linear interpolation between two adjacent values.

1. For the given M_1, read p_{01} / p_0^* and T_{01} / T_0^* from a Rayleigh flow, for example, Table A.4 in Sultanian (2015, 2023).
2. Knowing T_{02} / T_{01}, compute $T_{02} / T_0^* = \left(T_{01} / T_0^*\right)\left(T_{02} / T_{01}\right)$.
3. For T_{02} / T_0^*, read M_2 and p_{02} / p_0^* from the Rayleigh flow table.
4. Evaluate $p_{02} = p_{01}\left(p_{02} / p_0^*\right) / \left(p_{01} / p_0^*\right)$.

Example 5.11

Consider an isothermal flow in an underground constant-area gas pipeline with inlet conditions: $M_1 = 0.1$, $p_{01} = 15$ bar, and $T_{01} = 300$ K. Assuming the gas to behave like air with $\gamma = 1.4$, find the maximum value of fL_{max} / D_h possible for this pipeline with the mass flow rate computed at the specified inlet conditions. Also, compute the total pressure, the static pressure, and the total temperature at the pipeline exit (critical point). What is the change in impulse pressure between the pipeline inlet and exit?

Solution

The solution to this problem involves directly applying the abovementioned equations for an isothermal flow in a constant-area duct with friction and the isentropic flow equations that relate total and static properties.

Maximum dimensionless pipeline length: We use Equation 5.129 to obtain fL_{max} / D_h as:

$$\frac{fL_{max}}{D_h} = \ln\left(\gamma M^2\right) + \frac{1}{\gamma M^2} - 1 = \ln(1.4 \times 0.1 \times 0.1) - \frac{1}{1.4 \times 0.1 \times 0.1} - 1 = 66.16$$

Total pressure at pipeline exit: Using Equation 5.127, we obtain $p_{02} = p_0^*$ as:

$$\frac{p_{01}}{p_{02}} = \frac{p_{01}}{p_0^*} = \frac{1}{M\sqrt{\gamma}}\left(\frac{\gamma(2+(\gamma-1)M^2}{(3\gamma-1)}\right)^{\frac{\gamma}{\gamma-1}} = \frac{1}{0.1\sqrt{1.4}}\left(\frac{1.4(2+(1.4-1)\times 0.1\times 0.1}{(3\times 1.4-1)}\right)^{\frac{1.4}{1.4-1}} = 5.333$$

$$p_{02} = p_0^* = \frac{15}{5.333} = 2.812 \text{ bar}$$

Static pressure at pipeline exit:

$$p_2 = \frac{p_{02}}{\left(1+\frac{\gamma-1}{2}M_2^2\right)^{\frac{\gamma}{\gamma-1}}} = \frac{2.812}{\left(1+\frac{(1.4-1)}{2}\times 0.845\times 0.845\right)^{\frac{1.4}{1.4-1}}} = 1.763 \text{ bar}$$

Total temperature at pipeline exit: Using Equation 5.128, we obtain $T_{02} = T_0^*$ as:

$$\frac{T_{01}}{T_{02}} = \frac{T_{01}}{T_0^*} = \frac{\gamma\left(2+(\gamma-1)M^2\right)}{(3\times 1.4-1)} = \frac{1.4\left(2+(1.4-1)\times 0.1\times 0.1\right)}{(3\times 1.4-1)} = 0.877$$

$$T_{02} = T_0^* = \frac{300}{0.877} = 342.2 \text{ K}$$

Change in impulse pressure between pipeline inlet and exit:

$$p_1 = \frac{p_{01}}{\left(1+\frac{\gamma-1}{2}M_1^2\right)^{\frac{\gamma}{\gamma-1}}} = \frac{15}{\left(1+\frac{(1.4-1)}{2}0.1\times 0.1\right)^{\frac{1.4}{1.4-1}}} = 14.895 \text{ bar}$$

$$p_{i1} = p_1\left(1+\gamma M_1^2\right) = 14.895\times(1+1.4\times 0.1\times 0.1) = 15.104 \text{ bar}$$

$$p_{i2} = p_2\left(1+\gamma M_2^2\right) = 1.762\times(1+1.4\times 0.845\times 0.845) = 3.525 \text{ bar}$$

$$\Delta p_i = p_{i2} - p_{i1} = 3.525 - 15.104 = -11.579 \text{ bar}$$

The negative sign for the change in impulse pressure indicates that it decreases (due to pipe wall friction) from the inlet to the exit. For the given gas pipeline, the maximum dimensionless length is 66.16. From the pipeline inlet to exit, the total pressure decreases from 15 to 2.812 bar, the total temperature increases from 300 to 342.2 K due to heat transfer needed to keep the flow isothermal in the pipeline, and the fluid impulse pressure decreases by 11.579 bar. We calculate the static pressure at the exit as 1.762 bar.

Example 5.12

Knowing the given inlet conditions, the net increase in total temperature, and $fL/D_h = 66.16$ from Example 5.11, a student decides to compute the pipeline exit flow properties by the linear superposition of the solutions of the Fanno and Rayleigh flows. He first calculates the loss in total pressure across the pipeline by simply adding the total pressure losses obtained from the Fanno and Rayleigh flow solutions, neglecting the nonlinear coupling between the two flows. He then computes the final exit Mach number to ensure that the mass flux at the exit equals that at the inlet. Thus, knowing the total temperature, pressure, and the Mach number at the pipeline exit, he computes the remaining quantities at this section. Find the error in the total pressure, the static pressure, and the inlet-to-exit change in impulse pressure calculated by the student at the pipeline exit using his new approach.

Solution

In his solution approach, the student uses the following quantities from Example 5.11:

$$M_1 = 0.1,\ p_{01} = 15\ \text{bar},\ T_{01} = 300\ \text{K},\ \frac{fL}{D_h} = 66.16,\ \text{and}\ T_{02} = 342.2\ \text{K}$$

Mass flux at the pipeline inlet: The student uses Equation 5.22 to compute the total pressure mass flow function as:

$$\hat{F}_{f01} = M_1\sqrt{\frac{\gamma}{\left(1+\frac{\gamma-1}{2}M_1^2\right)^{\frac{\gamma+1}{\gamma-1}}}} = 0.1\sqrt{\frac{1.4}{\left(1+\frac{1.4-1}{2}M_1^2\right)^{\frac{1.4+1}{1.4-1}}}} = 0.118$$

Equation 5.21 then yields the mass flux value at the inlet as:

$$\dot{m}/A = \frac{\hat{F}_{f01}\,p_{01}}{\sqrt{RT_{01}}} = \frac{0.118\times 15\times 10^5}{\sqrt{287\times 300}} = 601.244\ \text{kg}/\left(\text{s m}^2\right)$$

Rayleigh flow solution: Knowing the total temperature ratio

$$\frac{T_{02}}{T_{01}} = \frac{342.2}{300} = 1.141$$

the student uses Equation 5.103 to compute M_2 using an iterative method, such as "Goal Seek" in Excel, to calculate:

$$\frac{T_{02}}{T_{01}} = \frac{M_2^2}{M_1^2}\left(\frac{1+\gamma M_1^2}{1+\gamma M_2^2}\right)^2\left(\frac{2+(\gamma-1)M_2^2}{2+(\gamma-1)M_1^2}\right)$$

$$\frac{M_2^2}{0.1\times 0.1}\left(\frac{1+1.4\times 0.1\times 0.1}{1+1.4M_2^2}\right)^2\left(\frac{2+(\gamma-1)M_2^2}{2+(1.4-1)\times 0.1\times 0.1}\right) = 1.141$$

$$M_2 = 0.107$$

He uses Equation 5.104 to compute the total pressure ratio as:

$$\frac{p_{02}}{p_{01}} = \left(\frac{1+\gamma M_1^2}{1+\gamma M_2^2}\right)\left(\frac{2+(\gamma-1)M_2^2}{2+(\gamma-1)M_1^2}\right)^{\frac{\gamma}{\gamma-1}}$$

$$\frac{p_{02}}{p_{01}} = \left(\frac{1+1.4\times 0.1\times 0.1}{1+1.4\times 0.1068\times 0.1068}\right)\left(\frac{2+(1.4-1)\times 0.1068\times 0.1068}{2+(1.4-1)\times 0.1\times 0.1}\right)^{\frac{1.4}{1.4-1}} = 0.999$$

which yields $\Delta p_{0\_\text{Raleigh}} = (p_{02} - p_{01})_{\text{Rayleigh}} = -0.0148\ \text{bar}$

Fanno flow solution: To calculate M_2, the student first expresses Equation 5.89 as:

$$\frac{f(x_2-x_1)}{D_h}=\frac{fL}{D_h}=\frac{\gamma+1}{2\gamma}\ln\left(\frac{M_1^2\{2+(\gamma-1)M_2^2\}}{M_2^2\{2+(\gamma-1)M_1^2\}}\right)+\frac{1}{\gamma}\left(\frac{1}{M_1^2}-\frac{1}{M_2^2}\right)$$

Substituting $fL/D_h = 66.16$ and $M_1 = 0.1$ in this equation, he uses an iterative solution method, such as "Goal Seek" in Excel, to obtain $M_2 = 0.544$. Using Equation 5.81, he then calculates:

$$\frac{p_{01}}{p_0^*}=\frac{1}{M_1}\left(\frac{2+(\gamma-1)M_1^2}{\gamma+1}\right)^{\frac{\gamma+1}{2(\gamma-1)}}=\frac{1}{0.1}\left(\frac{2+(1.4-1)\times 0.1\times 0.1}{1.4+1}\right)^{\frac{1.4+1}{2(1.4-1)}}=5.822$$

$$\frac{p_{02}}{p_0^*}=\frac{1}{M_2}\left(\frac{2+(\gamma-1)M_2^2}{\gamma+1}\right)^{\frac{\gamma+1}{2(\gamma-1)}}=\frac{1}{0.544}\left(\frac{2+(1.4-1)\times 0.544\times 0.544}{1.4+1}\right)^{\frac{1.4+1}{2(1.4-1)}}=1.264$$

giving $p_{02}/p_{01} = 0.217$ and $\Delta p_{0\_\text{Fanno}} = (p_{02}-p_{01})_{\text{Fanno}} = -11.744$ bar.

Linearly combining the losses in total pressure from Rayleigh and Fanno flow solutions, the student calculates the total pressure at the pipeline exit as:

$$p_{02}=p_{01}+\Delta p_{0\_\text{Rayleigh}}+\Delta P_{0\_\text{Fanno}}=15-0.0148-11.744=3.242\text{ bar}$$

Mach number at pipeline exit: Equating the mass flux at the inlet and exit yields:

$$\frac{\dot{m}}{A}=\frac{\hat{F}_{f02}\,p_{02}}{\sqrt{RT_{02}}}=\frac{\hat{F}_{f02}\times 3.242\times 10^5}{\sqrt{287\times 342.2}}=601.244\text{ kg}/\left(\text{s m}^2\right)$$

$$\hat{F}_{f02}=0.581$$

Using an iterative method, such as "Goal Seek" in Excel, the student solves the equation:

$$\hat{F}_{f02}=M_2\sqrt{\frac{\gamma}{\left(1+\frac{\gamma-1}{2}M_2^2\right)^{\frac{\gamma+1}{\gamma-1}}}}=M_2\sqrt{\frac{1.4}{\left(1+\frac{1.4-1}{2}M_2^2\right)^{\frac{1.4+1}{1.4-1}}}}=0.581$$

and obtains $M_2 = 0.609$.

Impulse pressure at exit:

$$p_2=\frac{p_{02}}{\left(1+\frac{\gamma-1}{2}M_2^2\right)^{\frac{\gamma}{\gamma-1}}}=\frac{3.242}{\left(1+\frac{(1.4-1)}{2}0.609\times 0.609\right)^{\frac{1.4}{1.4-1}}}=2.524\text{ bar}$$

$$p_{i2}=p_2\left(1+\gamma M_2^2\right)=2.524\times(1+1.4\times 0.609\times 0.609)=3.833\text{ bar}$$

Inlet-to-exit change in impulse pressure:

$$\Delta p_i = p_{i2}-p_{i1}=3.833-15.104=-11.271\text{ bar}$$

$$\text{Error in computed exit total pressure}=\frac{3.242-2.812}{2.812}=15.3\%$$

$$\text{Error in computed exit static pressure}=\frac{2.524-1.762}{1.762}=43.2\%$$

$$\text{Error in computed inlet-to-exit change in impulse pressure} = \frac{-11.271-(-11.579)}{-11.579} = -2.7\%$$

This example, which solves the isothermal flow of Example 5.11 using the linear superposition of Fanno and Rayleigh flows, brings out some key features of the nonlinear coupling between heat transfer and friction in a constant-area pipe. It is easy to see that the small amount of heating required to convert a Fanno flow into an isothermal flow in a constant-area pipe with friction dramatically changes the variation in flow total pressure, static pressure, and Mach number along the pipe. For the same inlet conditions and mass flux, the isothermal flow yields the maximum Mach number of 0.845. In contrast, the corresponding Fanno flow reaches the exit Mach number of 0.544, which increases to 0.609 when the exit total temperature equals that of the isothermal flow. For the subsonic pipe flow, while the friction increases the downstream Mach number and decreases the total pressure without changing the flow total temperature, the heat transfer increases the total temperature and Mach number, reducing the total pressure along the pipe. The net effect of simultaneous heat transfer and friction leads to a higher increase in Mach number and a higher decrease in total pressure than obtained from simply adding them from Rayleigh and Fanno flow solutions.

Results show that the new approach used by the student may be acceptable for estimating the total fluid drag force acting on the pipeline. However, this approach needs to be revised in computing the total and static pressure at the pipeline exit and avoided for a meaningful engineering calculation.

Example 5.13

A supersonic plane is cruising at an altitude of 10000 m where the atmospheric pressure and temperature are $p_{amb} = 2.65\times10^4$ Pa and $T_{amb} = -49.90^\circ$C. A pitot tube fitted at the nose of the plane measures a total (stagnation) pressure of 8.15×10^4 Pa. Calculate the plane's speed if the airflow undergoes a normal shock ahead of the pitot tube. Assume $\gamma = 1.4$ and $R = 287\ \mathrm{J/(kg\ K)}$ for air.

Solution

In this example, knowing the total pressure downstream of the shock (measured by the pitot tube) and the static atmospheric pressure upstream, we can directly use the Rayleigh supersonic pitot equation, Equation 5.150, to compute the Mach number upstream of the normal shock as follows:

$$\frac{p_{02}}{p_1} = \frac{p_{02}}{p_{amb}} = \frac{8.15\times10^4}{2.65\times10^4} = 3.075$$

whose substitution in Equation 5.150 yields using an iterative method, such as "Goal Seek" in Excel, $M_1 = 1.407$.

Speed of sound at the cruising altitude of 10000 m:

$$C_1 = \sqrt{1.4\times287\times(-49.9+273)} = 299.402\ \mathrm{m/s}$$

Speed of the plane:

$$V_{plane} = C_1\times M_1 = 299.402\times1.407 = 421.4\ \mathrm{m/s} = 1517\ \mathrm{km/h}$$

Therefore, the supersonic plane is cruising at a speed of 1517 km/h.

Example 5.14

A designer is considering two alternative designs of a supersonic airflow diffuser, shown in Figure 5.57. The first design, shown in Figure 5.57a, uses a convergent–divergent diffuser where the incoming flow passes through a choked throat to transition from the supersonic to the subsonic flow at the exit. In the second design, shown in Figure 5.57b, the normal shock ahead of the

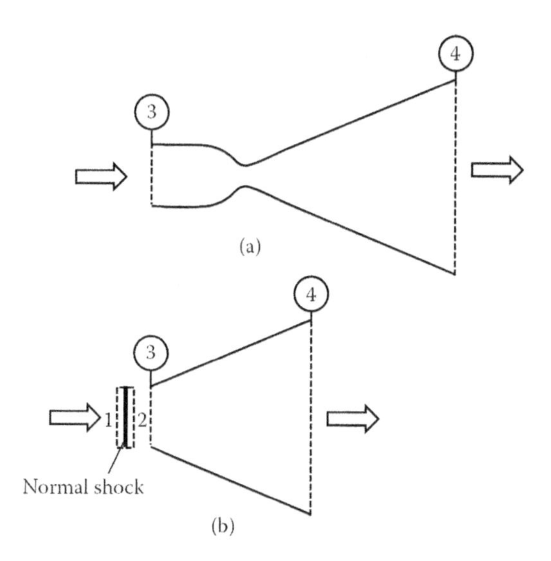

FIGURE 5.57 Supersonic diffuser designs: (a) A convergent–divergent diffuser and (b) a divergent diffuser with a normal shock at its inlet (Example 5.14).

diffuser converts the supersonic flow into a subsonic flow, further diffusing in the divergent duct. For both designs, the incoming flow Mach number is 1.1 at a static pressure of 0.476 bar, and the diffuser discharges at the ambient static pressure of 1.0 bar. Perform the necessary calculations for both designs and evaluate their advantages and disadvantages. Consider the loss in total pressure across the normal shock in the second design but neglect all other losses in both designs. Assume $\gamma = 1.4$ and $R = 287\ \mathrm{J/(kg\ K)}$ for air.

Solution

As shown in Figure 5.57, in both designs, section 3 represents the diffuser inlet, and section 4 the diffuser exit. Within each diffuser, we assume the airflow to be isentropic.

(A) CONVERGENT–DIVERGENT DIFFUSER

Given: $M_{1(a)} = M_{3(a)} = 1.1$, $p_{1(a)} = p_{3(a)} = 0.476$ bar, and $p_{4(a)} = 1.0$ bar

Total pressure at diffuser inlet: From Equation 5.11, we obtain:

$$\frac{p_{03(a)}}{p_{3(a)}} = \left(1 + \frac{\gamma - 1}{2} M_{3(a)}^2\right)^{\frac{\gamma}{\gamma-1}} = \left(1 + \frac{1.4 - 1}{2} \times 1.1 \times 1.1\right)^{\frac{1.4}{1.4-1}} = 2.135$$

$$p_{03(a)} = p_{04(a)} = 2.135 \times 0.476 = 1.016 \text{ bar}$$

Critical area ratio at diffuser inlet ($A_{3(a)} / A^*_{(a)}$): Using Equation 5.66, we obtain:

$$\frac{A_{3(a)}}{A^*_{(a)}} = \frac{1}{M_{3(a)}} \sqrt{\left(\frac{2 + (\gamma - 1) M_{3(a)}^2}{\gamma + 1}\right)^{\frac{\gamma+1}{\gamma-1}}} = \frac{1}{1.1} \sqrt{\left(\frac{2 + (1.4 - 1) \times 1.1 \times 1.1}{1.4 + 1}\right)^{\frac{1.4+1}{1.4-1}}} = 1.008$$

Mach number at diffuser exit: Rewriting Equation 5.11 at section 4, we obtain the Mach number as:

$$M_{4(a)} = \sqrt{\frac{2}{\gamma-1}\left\{\left(\frac{p_{04(a)}}{p_{4(a)}}\right)^{\frac{\gamma-1}{\gamma}} - 1\right\}} = \sqrt{\frac{2}{1.4-1}\left\{(1.016)^{\frac{1.4-1}{1.4}} - 1\right\}} = 0.152$$

Critical area ratio at diffuser exit ($A_{4(a)} / A^*_{(a)}$): Using Equation 5.66, we obtain:

$$\frac{A_{4(a)}}{A^*_{(a)}} = \frac{1}{M_{4(a)}}\sqrt{\left(\frac{2+(\gamma-1)M^2_{4(a)}}{\gamma+1}\right)^{\frac{\gamma+1}{\gamma-1}}} = \frac{1}{0.152}\sqrt{\left(\frac{2+(1.4-1)\times 0.152\times 0.152}{1.4+1}\right)^{\frac{1.4+1}{1.4-1}}} = 3.854$$

Overall diffuser area ratio: The overall diffuser area ratio for this design works out to be:

$$\frac{A_{4(a)}}{A_{3(a)}} = \frac{A_{4(a)}}{A^*_{(a)}} \times \frac{A^*_{(a)}}{A_{3(a)}} = \frac{3.854}{1.008} = 3.823$$

(B) DIVERGENT DIFFUSER WITH A NORMAL SHOCK AT THE INLET

Given: $M_{1(b)} = 1.1$ and $p_{1(b)} = 0.476$ bar

Conditions downstream of the normal shock: Using Equation 5.140, we obtain the Mach number downstream of the normal shock as:

$$M^2_{2(b)} = \frac{2+(\gamma-1)M^2_{1(b)}}{2\gamma M^2_{1(b)} - (\gamma-1)} = \frac{2+(1.4-1)\times 1.1\times 1.1}{2\times 1.4\times 1.1\times 1.1-(1.4-1)} = 0.831$$

$$M_{2(b)} = \sqrt{0.831} = 0.912$$

Using Equation 5.150, we obtain:

$$\frac{p_{02(b)}}{p_{1(b)}} = \left\{\frac{2\gamma M^2_{1(b)} - (\gamma-1)}{(\gamma+1)}\right\}^{\frac{-1}{\gamma-1}} \left\{\frac{(\gamma+1)}{2} M^2_{1(b)}\right\}^{\frac{\gamma}{\gamma-1}}$$

$$\frac{p_{02(b)}}{P_{1(b)}} = \left\{\frac{2\times 1.4\times 1.1\times 1.1-(1.4-1)}{(1.4+1)}\right\}^{\frac{-1}{1.4-1}} \left\{\frac{(1.4+1)}{2}\times 1.1\times 1.1\right\}^{\frac{1.1}{1.1-1}} = 2.133$$

which yields:

$$p_{03(b)} = p_{02(b)} = 2.133\times 0.476 = 1.015 \text{ bar}$$

Critical area ratio at diffuser inlet ($A_{3(b)} / A^*_{(b)}$): The conditions downstream of the normal shock prevail at the diffuser inlet (section 3). Using Equation 5.66, we obtain:

$$\frac{A_{3(b)}}{A^*_{(b)}} = \frac{1}{M_{3(b)}}\sqrt{\left(\frac{2+(\gamma-1)M^2_{3(b)}}{\gamma+1}\right)^{\frac{\gamma+1}{\gamma-1}}} = \frac{1}{0.912}\sqrt{\left(\frac{2+(1.4-1)\times 0.912\times 0.912}{1.4+1}\right)^{\frac{1.4+1}{1.4-1}}} = 1.007$$

Mach number at diffuser exit: Rewriting Equation 5.11 at section 4, we obtain the Mach number as:

$$M_{4(b)} = \sqrt{\frac{2}{\gamma-1}\left\{\left(\frac{p_{04(b)}}{p_{4(b)}}\right)^{\frac{\gamma-1}{\gamma}} - 1\right\}} = \sqrt{\frac{2}{1.4-1}\left\{(1.015)^{\frac{1.4-1}{1.4}} - 1\right\}} = 0.147$$

Critical area ratio at diffuser exit ($A_{4(b)} / A^*_{(b)}$): Using Equation 5.66, we obtain:

$$\frac{A_{4(b)}}{A^*_{(b)}} = \frac{1}{M_{4(b)}}\sqrt{\left(\frac{2+(\gamma-1)M^2_{4(b)}}{\gamma+1}\right)^{\frac{\gamma+1}{\gamma-1}}} = \frac{1}{0.147}\sqrt{\left(\frac{2+(1.4-1)\times 0.147\times 0.147}{1.4+1}\right)^{\frac{1.4+1}{1.4-1}}} = 3.985$$

Overall diffuser area ratio: The overall diffuser area ratio for this design works out to be:

$$\frac{A_{4(b)}}{A_{3(b)}} = \frac{A_{4(b)}}{A^*_{(b)}} \times \frac{A^*_{(b)}}{A_{3(b)}} = \frac{3.985}{1.007} = 3.957$$

The calculations in this example show that the convergent–divergent diffuser design shown in Figure 5.57a requires an overall area ratio of 3.823, and the airflow exits the diffuser at a Mach number of 0.152. The divergent diffuser design, shown in Figure 5.57b, has an overall area ratio of 3.985 with an exit Mach number of 0.147. Thus, with only a 3.5% increase in the area ratio, the divergent diffuser design offers a more straightforward, shorter, and less expensive one without the additional feature of a choked throat needed in the convergent–divergent diffuser design. The weak normal shock at the inlet of the diffuser in the second design is nearly isentropic. It effectively converts the dynamic pressure into a corresponding rise in the static pressure downstream of the shock over a negligible thickness.

Example 5.15

From the normal shock Table A.6 of Sultanian (2023), we obtain $M_2 = 0.57735$ and $p_{02} / p_{01} = 0.72088$ for $M_1 = 2.0$. Verify these values separately using Fanno flow Table A.3 and Rayleigh flow Table A.4.

Solution

(A) FANNO FLOW TABLE A.3

For $M_1 = 2.0$, we obtain the following values from the Fanno flow Table A.3 of Sultanian (2023):

$$\frac{p_{i1}}{p_i^*} = 1.12268 \text{ and } \frac{p_{01}}{p_0^*} = 1.68750$$

As $p_{i2} / p_i^* = p_{i1} / p_i^*$ across a normal shock, we search for $p_{i2} / p_i^* = 1.12268$ in the subsonic section of the table and obtain by linear interpolation the following values:

$$M_2 = 0.57740 \text{ and } \frac{p_{02}}{p_0^*} = 1.21647$$

which further yields:

$$\frac{p_{02}}{p_{01}} = \frac{p_{02} / p_0^*}{p_{01} / p_0^*} = \frac{1.21647}{1.68750} = 0.72087$$

Thus, for $M_1 = 2.0$, we obtain practically identical values of M_2 and p_{02} / p_{01} from the Fanno flow and the normal shock tables.

(B) RAYLEIGH FLOW TABLE A.4 OF SULTANIAN (2023)

For $M_1 = 2.0$, we obtain the following values from the Rayleigh flow Table A.4:

$$\frac{T_{01}}{T_0^*} = 0.79339 \text{ and } \frac{p_{01}}{p_0^*} = 1.50309$$

As $T_{02} / T_0^* = T_{01} / T_0^*$ across a normal shock, we search for $T_{02} / T_0^* = 0.79339$ in the subsonic section of the table and obtain by linear interpolation the following values:

$$M_2 = 0.57738 \text{ and } \frac{p_{02}}{p_0^*} = 1.08352$$

from which we obtain:

$$\frac{p_{02}}{p_{01}} = \frac{p_{02} / p_0^*}{p_{01} / p_0^*} = \frac{1.50309}{1.08352} = 0.72086$$

Again, for $M_1 = 2.0$, we obtain practically identical values of M_2 and p_{02} / p_{01} from the Rayleigh flow and the normal shock tables.

Besides illustrating the techniques to obtain normal shock solutions from either the Fanno flow table or the Rayleigh flow table, this example validates that the intersection points of Fanno and Rayleigh lines constitute a graphical method to obtain the normal shock solutions.

Example 5.16

A uniform air flow at Mach number 1.8 goes past a wedge of half-angle $15°$. Assuming $\gamma = 1.4$ and $R = 287\ \text{J}/(\text{kg K})$ for air, compute:

a. Wave angles of attached oblique shocks that correspond to weak and strong shock solutions,
b. Maximum deflection angle and the corresponding wave angle,
c. Deflection angle and the corresponding wave angle that renders the downstream flow sonic.

Solution

In this example, we have $M_1 = 1.8$ and $\theta = 15°$.

(A) WAVE ANGLES OF ATTACHED OBLIQUE SHOCKS THAT CORRESPOND TO WEAK AND STRONG SHOCK SOLUTIONS

For a strong shock solution with $\delta = 0$, Equation 5.163 reduces to,

$$\tan\beta = \frac{M_1^2 - 1 + 2\lambda\cos\left(\dfrac{\cos^{-1}\chi}{3}\right)}{3\left(1 + \dfrac{\gamma - 1}{2}M_1^2\right)\tan\theta}$$

where we use Equation 5.164 to evaluate λ and Equation 5.165 to evaluate χ, giving $\beta_{\text{strong}} = 76.757°$.

For the weak shock solution with $\delta = 1$, Equation 5.163 becomes:

$$\tan\beta = \frac{M_1^2 - 1 + 2\lambda\cos\left(\dfrac{4\pi + \cos^{-1}\chi}{3}\right)}{3\left(1 + \dfrac{\gamma - 1}{2}M_1^2\right)\tan\theta}$$

where we again use Equation 5.164 to evaluate λ and Equation 5.165 to evaluate χ, giving $\beta_{weak} = 51.336°$. One can easily check these solutions against the $\theta - \beta$ plot shown in Figure 5.46 for $M_1 = 1.8$.

(B) MAXIMUM DEFLECTION ANGLE AND THE CORRESPONDING WAVE ANGLE

In this case, we first use Equation 5.162 to compute $\beta_{max} = 64.987°$ and then use Equation 5.161 to obtain $\theta_{max} = 19.183°$. These results compare well with the results plotted in Figure 5.46 for $M_1 = 1.8$.

(C) DEFLECTION ANGLE AND THE CORRESPONDING WAVE ANGLE THAT RENDERS THE DOWNSTREAM FLOW SONIC

In this case, we use the fact that the total temperature remains constant across the shock. Using the isentropic relations for the flow upstream and downstream of the shock, we obtain:

$$\frac{T_0}{T_1} = 1 + \frac{\gamma - 1}{2} M^2 = 1 + \left(\frac{1.4 - 1}{2}\right) \times 1.8 \times 1.8 = 1.648$$

$$\frac{T_0}{T_2} = \frac{T_0}{T_2^*} = \frac{\gamma + 1}{2} = \frac{1.4 + 1}{2} = 1.200$$

$$\frac{T_2}{T_1} = \frac{\left(\dfrac{T_0}{T_1}\right)}{\left(\dfrac{T_0}{T_2}\right)} = \frac{1.648}{1.200} = 1.373$$

Through an iterative solution (e.g., "Goal Seek" in Excel) of Equation 5.158, we first obtain $\beta^* = 61.302°$ and then $\theta^* = 18.841°$ from Equation 5.161. As is evident from Figure 5.46, the computed values of β^* and θ^*, which correspond to $M_2 = 1$, are very close to the values of β_{max} and θ_{max} computed in (b).

Example 5.17

A flat plate is placed at an angle of attack of $\theta = 10°$ to an oncoming supersonic uniform air flow at Mach number 1.8 and total pressure 1 bar. Assuming $\gamma = 1.4$ for air, compute the static pressure on the suction surface (top side) and pressure surface (bottom side) of the plate.

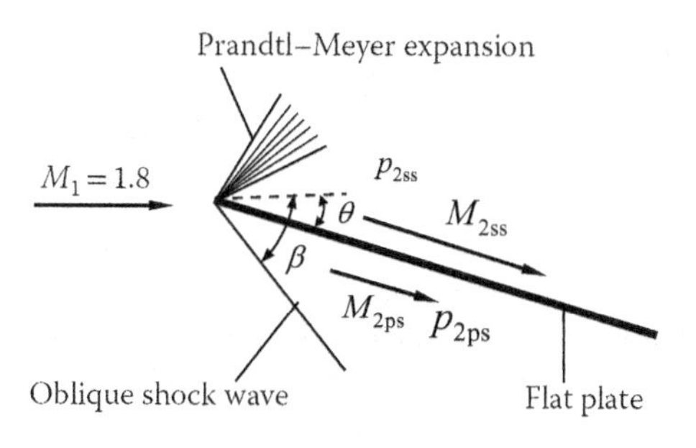

FIGURE 5.58 A flat plate at an angle of attack to an oncoming supersonic flow (Example 5.17).

Solution

In this example, as shown in Figure 5.58, the oncoming supersonic airflow undergoes an expansion wave (Prandtl–Meyer) on the top side (suction surface) of the flat plate and a compression wave (oblique shock) on the bottom side (pressure surface). The angle of attack equals the turning angle for the oblique shock. For the Prandtl–Meyer expansion, we can write:

$$\theta = \phi_2\left(M_{2\_ss}\right) - \phi_1\left(M_1\right)$$

where ϕ_1 and ϕ_2 are the Prandtl–Meyer functions computed using Equation 5.170. Note that the flow is isentropic on the suction surface and nonisentropic (due to the oblique shock) on the pressure surface.

For the oncoming airflow, we can compute static pressure using the isentropic relation as:

$$p_1 = \frac{p_{01}}{\left(1 + \dfrac{\gamma - 1}{2} M_1^2\right)^{\frac{\gamma}{\gamma-1}}} = \frac{1.0}{(1 + 0.2 \times 1.8 \times 1.8)^{3.5}} = 0.174 \text{ bar}$$

Static pressure on the suction surface: Using Equation 5.170, we obtain:

$$\phi_1 = \sqrt{\frac{1.4+1}{1.4-1}} \tan^{-1} \sqrt{\frac{1.4-1}{1.4+1}(1.8^2 - 1)} - \tan^{-1} \sqrt{1.8^2 - 1} = 20.725°$$

which yields:

$$\phi_2\left(M_{2ss}\right) = 20.725° + 10° = 30.725°$$

Now, we use an iterative method (e.g., "Goal Seek" in Excel) to solve Equation 5.170 and obtain $M_{2ss} = 2.161$. Using isentropic relation, we further obtain $p_{2ss} = 0.0993$ bar.

Static pressure on the pressure surface: For the oblique shock, the turning (deflection) angle is 10°. Equation 5.163 yields the weak shock solution as $\beta = 44.057°$. Using Equation 5.157, we obtain:

$$\frac{p_{2ps}}{p_1} = \frac{2\gamma M_1^2 \sin^2\beta - (\gamma - 1)}{(\gamma + 1)} = \frac{2 \times 1.4 \times (1.8 \times \sin(0.769))^2 - (1.4 - 1)}{(1.4 + 1)} = 1.661$$

which further yields $p_{2ps} = 0.289$ bar.

Therefore, the computed static pressure on the suction surface and pressure surface of the flat plate are 0.0993 bar and 0.289 bar, respectively. From these pressures, we can easily compute a lift force of 18687 N and a drag force of 3295 N acting per square meter of the plate. Further, the flow Mach number on the suction surface is 2.161 and that on the pressure surface, computed using Equation 5.155, is 1.449. As the computed static pressure and Mach number on the pressure surface are so different from their values on the suction surface, they result in a very complex flow field when the two flows rejoin at the plate trailing edge.

Example 5.18

For a converging–diverging nozzle, the exit-to-throat area ratio (A_e / A_1^*) is 12. For the steady airflow through the nozzle with exit Mach number 0.15, find the ratio of the area where the normal shock occurs to the nozzle throat area. Assume $\gamma = 1.4$ for air.

Solution

In this example, the normal shock in the diverging section of the C–D nozzle divides the flow field into two isentropic flows, one on either side of the shock. The critical area of the upstream isentropic supersonic flow equals the nozzle throat area. The critical area must be larger for the downstream isentropic subsonic flow because of a reduction in the total pressure across the normal shock. In fact, from the mass conservation across a normal shock with constant total temperature, we have:

$$\frac{A_1^*}{A_2^*} = \frac{p_{02}}{p_{01}}$$

From Equation 5.66, we write:

$$\frac{A_e}{A_2^*} = \frac{1}{M_e}\sqrt{\left(\frac{2+(\gamma-1)M_e^2}{\gamma+1}\right)^{\frac{\gamma+1}{\gamma-1}}} = \frac{1}{0.15}\sqrt{\left(\frac{2+(1.4-1)\times 0.15\times 0.15}{1.4+1}\right)^{\frac{1.4+1}{1.4-1}}}$$

$$\frac{A_e}{A_2^*} = 3.910$$

which yields:

$$\frac{A_1^*}{A_2^*} = \left(\frac{A_e}{A_2^*}\right)\left(\frac{A_1^*}{A_e}\right) = \frac{3.910}{12} = 0.3259 = \frac{p_{02}}{p_{01}}$$

Using an iterative method such as "Goal Seek" in MS Excel to solve Equation 5.148 yields the upstream Mach number as:

$$\frac{p_{02}}{p_{01}} = 0.3259 = \left\{\frac{(1.4+1)}{2\times 1.4\times M_1^2-(1.4-1)}\right\}^{\frac{1}{1.4-1}}\left[\frac{(1.4+1)M_1^2}{\left\{2+(1.4-1)M_1^2\right\}}\right]^{\frac{1.4}{1.4-1}}$$

$$M_1 = 3.0$$

Finally, we use Equation 5.66 to calculate:

$$\frac{A_{ns}}{A_1^*} = \frac{1}{M_1}\sqrt{\left(\frac{2+(\gamma-1)M_1^2}{\gamma+1}\right)^{\frac{\gamma+1}{\gamma-1}}} = \frac{1}{3.0}\sqrt{\left(\frac{2+(1.4-1)\times 3.0\times 3.0}{1.4+1}\right)^{\frac{1.4+1}{1.4-1}}}$$

$$\frac{A_{ns}}{A_1^*} = 4.270$$

Therefore, for the given operating conditions, we calculate the area ratio $A_{ns} / A_1^* = 4.270$ for the normal shock location in the diverging section of the C–D nozzle. The present calculation method for locating a normal shock in a C–D nozzle involves an iterative solution technique to compute M_1 from the computed total pressure ratio. By using a normal shock table, for example, Table A.6 in Sultanian (2015, 2023), we can avoid iteration and instead use interpolation between adjacent tabular values.

Example 5.19

The test section of a supersonic wind tunnel with a variable-area diffuser is to operate at a Mach number of 2.75. Find the maximum ratio between the nozzle throat area and the diffuser throat area needed to swallow the normal shock that appears in the test section during start-up.

Solution

For upstream Mach number 2.75, we use Equation 5.148 to compute the total pressure ratio across the normal shock in the test section as:

$$\frac{p_{02}}{p_{01}} = \left\{ \frac{(1.4+1)}{2\times1.4\times2.75\times2.75-(1.4-1)} \right\}^{\frac{1}{1.4-1}} \left[\frac{(1.4+1)\times2.75\times2.75}{\{2+(1.4-1)\times2.75\times2.75\}} \right]^{\frac{1.4}{1.4-1}}$$

$$\frac{p_{02}}{p_{01}} = 0.4062$$

As $\frac{A_1^*}{A_2^*} = \frac{p_{02}}{p_{01}}$ for a normal shock, we obtain:

$$\frac{A_1^*}{A_2^*} = 0.4062$$

Therefore, we calculate the maximum ratio of the nozzle throat area to the diffuser throat area as 0.4062. With a slight increase in the diffuser area from its sonic value to satisfy mass conservation, the normal shock in the test section will jump to the diverging section of the diffuser.

Problems

5.1 As shown in Figure 5.59, compressed air at a total pressure of 8 bar and a total temperature of 300 K enters the plenum through the inlet pipe at A and leaves the outlet pipe at D to the ambient pressure of 1 bar. Both pipes are identical in length and diameter. Going from inlet to outlet, including a sudden expansion at B and a sudden contraction at C (neglect any vena contracta effect in the flow entering at C), the flow suffers an overall loss of 0.5 bar in total pressure. The entire flow system is adiabatic. Based on your understanding of a compressible flow, choose a section (A, B, C, or D) at which the air flow will choke ($M = 1$). Give the reason for your choice.

5.2 A pipe with constant area ($A = 0.0025\,\text{m}^2$) at inlet and outlet with steady adiabatic air flow is shown in Figure 5.60. The pipe wall is rough. The inlet total pressure p_{01} and total temperature T_{01} are 1.5 bar and 500 K, respectively. The flow exits the pipe at the static pressure of $p_2 = p_{amb} = 1.0$ bar. Under these boundary conditions, the measured loss in total pressure from pipe inlet to exit ($\Delta p_{01\to2}$) is 0.15 bar. Calculate the mass flow rate $\dot{m}$ through the pipe. Assume air as a perfect gas with $\gamma = 1.4$ and $R = 287$ J/(kg K).

5.3 As shown in Figure 5.61, a short conical diffuser of area ratio 1.1 with negligible friction and heat transfer is appended at the exit of the pipe of Problem 5.2, with identical inlet boundary conditions. The exit boundary conditions at section 2 in Problem 5.2 now prevail at section 3. The measured loss in total pressure from inlet to exit ($\Delta p_{01\to3}$) is 0.15

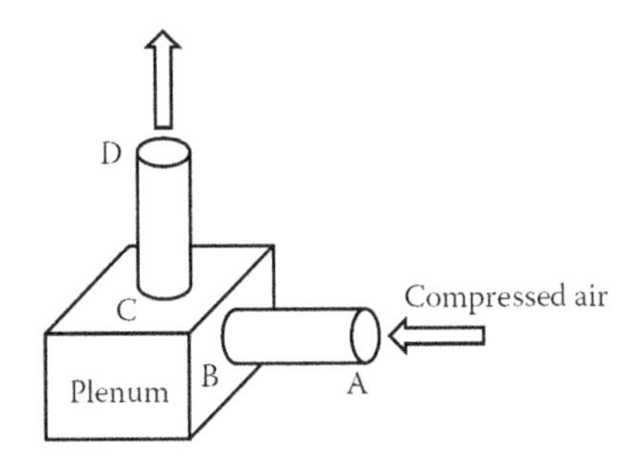

FIGURE 5.59 A compressed air flow system (Problem 5.1).

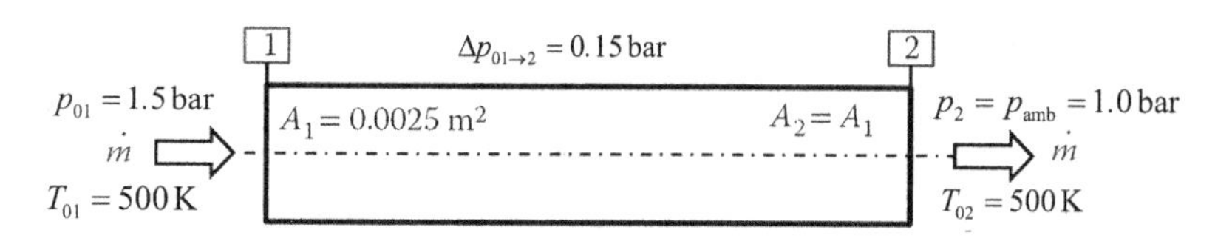

FIGURE 5.60 Adiabatic air flow through a rough pipe (Problem 5.2).

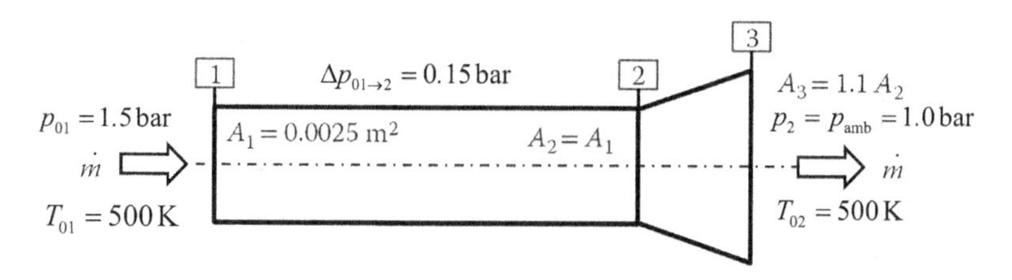

FIGURE 5.61 Adiabatic air flow through a rough pipe extended with a short conical diffuser (Problem 5.3).

bar. Calculate the mass flow $\dot{m}$ rate through the pipe and total-to-static pressure ratio $\frac{p_{01}}{p_2}$ between sections 1 and 2, and compare these values with the corresponding values obtained for Problem 5.2. From these results, comment on the role a diffuser plays in land-based gas turbines for power generation. Assume air as a perfect gas with $\gamma = 1.4$ and $R = 287$ J/(kg K).

5.4 As shown in Figure 5.62, air flows through two identical convergent nozzles into a large plenum and exits from it sideways through a choked divergent nozzle. The throat area of the divergent nozzle equals twice the throat area of each convergent nozzle. For the supply total pressure and total temperature of 8 bar and 436.5 K, respectively, for each convergent nozzle, find the mass flow rate through the choked divergent nozzle. All walls are adiabatic and frictionless. Assume air as a perfect gas with $\gamma = 1.4$ and $R = 287$ J/(kg K).

5.5 The pressure and temperature of air in an automobile tire are 2.4 bar and 298 K, respectively. A hole of 1.0-mm diameter is accidentally punched in the tire. The hole assumes the shape of a convergent nozzle. Neglecting frictional effects, find the air mass flow rate through the hole. What would the air velocity and Mach number be if the hole were correctly shaped for an isentropic expansion to the ambient pressure of 1 bar? Assume air as a perfect gas with $\gamma = 1.4$ and $R = 287$ J/(kg K).

5.6 Figure 5.63 shows an isentropic flow of air in a rubber pipe of constant diameter 0.1 m with inlet total pressure and total temperature of 1.2 bar and 300 K, respectively. The

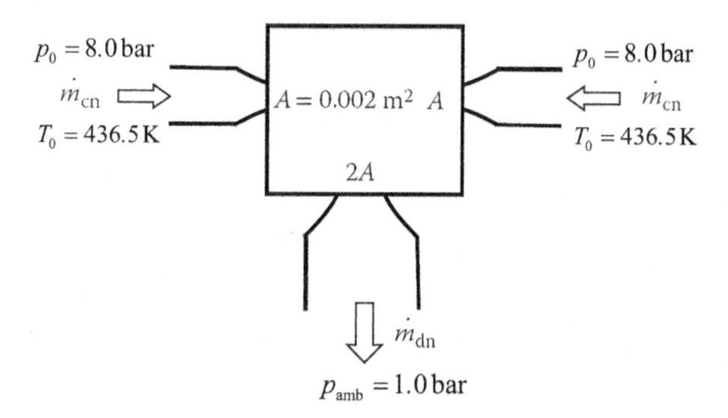

FIGURE 5.62 Adiabatic air flow through two convergent nozzles and one divergent nozzle connected to a plenum (Problem 5.4).

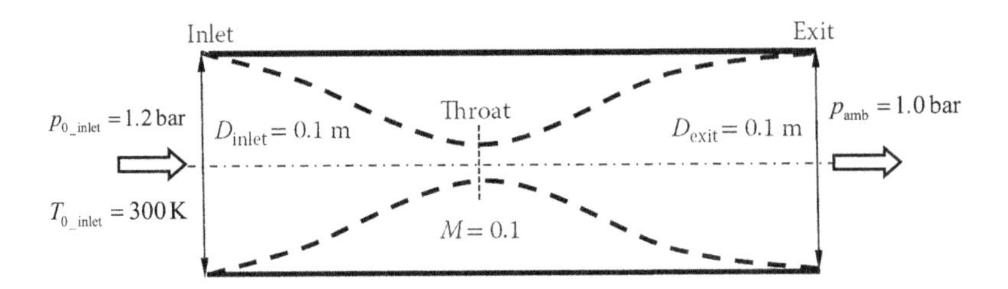

FIGURE 5.63 Isentropic air flow through a deformable rubber pipe (Problem 5.6).

ambient pressure is 1.0 bar. The pipe is slowly deformed into a convergent–divergent nozzle, as shown by the dotted line, until the flow just chokes ($M = 1.0$) at the throat. For the given boundary conditions, calculate the throat diameter. Assume air as a perfect gas with $\gamma = 1.4$ and $R = 287$ J/(kg K).

5.7 For the air flow in a convergent–divergent nozzle shown in Figure 5.64, a normal shock stands in the divergent section. The exit-to-throat area ratio (A_{exit}/A_{throat}) of the C–D nozzle is known. For the given exit Mach number (M_{exit}), write a step-by-step nongraphical and noniterative procedure to determine the ratio of nozzle area (A_{ns}), where the normal shock is located, to the throat area (A^*).

5.8 Figure 5.65 shows a high-pressure inlet bleed heat (IBH) system of a land-based gas turbine used for power generation. When the ambient air is cold, the IBH system is used to raise its temperature to prevent ice formation in the compressor IGVs (inlet guide vanes). In this system, the hot air is bled from an intermediate compressor stage and mixed uniformly with air flow at the engine inlet. Let us consider such a system. The ambient pressure and temperature are 1 bar and 20°C, respectively. The Mach number of the air flow entering the IGV is 0.6. To prevent ice formation, the static temperature at this section is

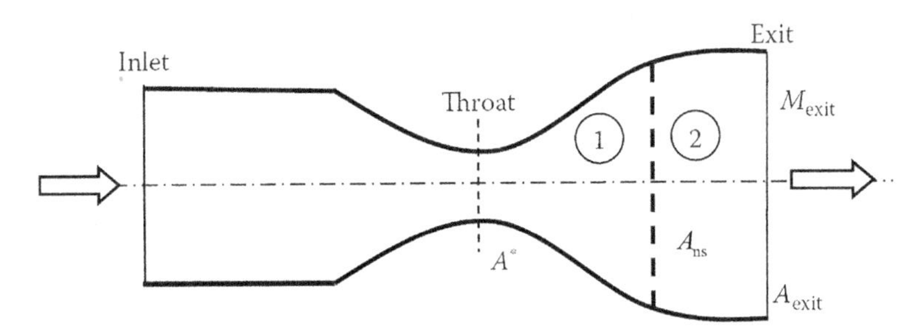

FIGURE 5.64 Normal shock located in a convergent–divergent nozzle (Problem 5.7).

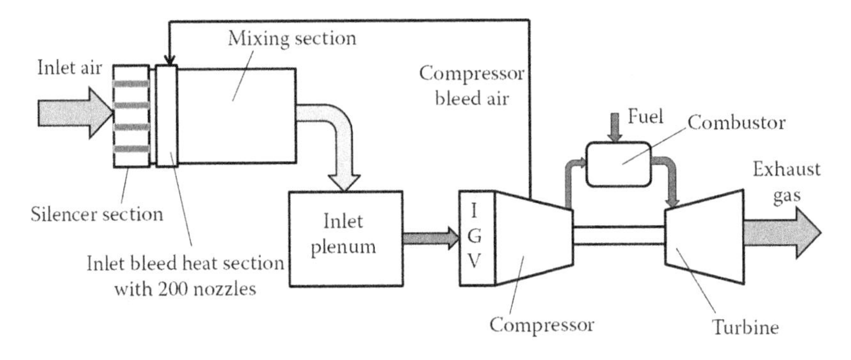

FIGURE 5.65 High-pressure inlet bleed heat (IBH) system of a land-based gas turbine for power generation (Problem 5.8).

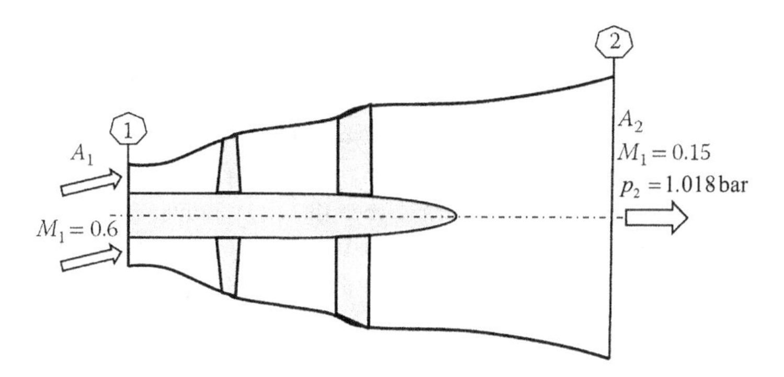

FIGURE 5.66 Exhaust diffuser of a land-based gas turbine for power generation (Problem 5.9).

required to be 2°C. The air mass flow rate entering the engine inlet system (before the high-pressure bleed heat section) at low Mach number ($M<0.2$) is 275 kg/s. The total temperature and pressure of the air bled from the compressor are 269°C and 8 bar, respectively. Neglect any changes in the pressure and temperature in the bleed air supply system. Assume air as a perfect gas with $\gamma = 1.4$ and $R=287$ J/(kg K).

a. Calculate the mass flow rate of the compressor bleed air to meet the design objective at the compressor IGV inlet.
b. To promote uniform mixing of the hot compressor bleed air with the cold inlet ambient air, 200 nozzles for bleed air injection are uniformly placed in the inlet duct cross section. Find the effective flow area of each nozzle.

5.9 In a land-based gas turbine used for power generation, the turbine exhaust enters an annular diffuser, shown in Figure 5.66, at a total velocity Mach number of 0.60 and a total temperature of 450°C. The swirl velocity (tangential velocity) at the diffuser inlet equals 15% of the total velocity. The flow exits the diffuser fully axially at a Mach number of 0.15 with no change in total temperature. The design calls for the static pressure at the diffuser exit to be 1.018 bar to allow the exhaust gases to discharge into the ambient air via a downstream duct system (not shown in the figure). The exit-to-inlet flow area ratio for the annular diffuser is 3.45. Find the pressure rise coefficient (C_p) for the annular diffuser. If the last-stage turbine is redesigned for zero swirl velocity at its exit, how will the pressure rise coefficient (C_p) of the annular diffuser change? Assume exhaust gases with $\gamma = 1.4$ and $R=287$ J/(kg K).

5.10 Air flows through a frictionless adiabatic convergent–divergent nozzle. The stagnation pressure and temperature at the nozzle inlet are 6.0×10^5 Pa and 600 K, respectively. The diverging section of the nozzle has an area ratio $A_{\text{exit}}/A_{\text{throat}}=12$. A normal shock wave stands in the diverging section of the nozzle where the Mach number is 3.3. Determine the Mach number, static pressure, and static temperature at the nozzle exit plane. Assume air as a perfect gas with $\gamma = 1.4$ and $R=287$ J/(kg K).

5.11 For $\gamma = 1.4$, the maximum value of the dimensionless total pressure mass flow function is 0.6847. Use the isentropic flow Table A.2 of Sultanian (2015, 2023) to find (a) the dimensionless total pressure mass flow function at $M=0.5$ and (b) supersonic Mach number for which the value of the dimensionless total pressure mass flow function equals that for $M=0.5$. Verify your answers from Table A.1 of Sultanian (2015, 2023) on compressible flow functions.

5.12 Use Table A.1 on compressible flow functions for $\gamma = 1.4$ to find the Mach number (M_2) downstream of a normal shock when the upstream Mach number is $M_1=3.5$: first, by using the Prandtl–Meyer equation, and second, by using the tabulated normal shock function N. Verify your answer from Table A.6 on normal shock.

5.13 From the normal shock Table A.6 of Sultanian (2015, 2023), we obtain $M_2=0.5774$ and $\frac{p_{01}}{p_{02}}=0.7209$ for $M_1=2.0$. Use the Fanno flow Table A.3 of Sultanian (2015, 2023) and Rayleigh flow Table A.4 of Sultanian (2015, 2023) to show that the points of intersection of the Fanno and Rayleigh flow lines have the same total pressure ratio ($\frac{p_{01}}{p_{02}}=0.7209$). Also, compute the entropy change $\Delta s/R$ between these two points.

5.14 A convergent–divergent nozzle exhausts into a diffuser through a normal shock. Show that the ratio $\frac{p_{02}}{p_{01}}$ across the shock is equal to the ratio of the first throat area A_1^* to that of the second throat area A_2^*.

5.15 A convergent–divergent diffuser is to be used at $M_2=2.3$. The diffuser has to use a variable throat area so as to swallow the starting shock. What percentage increase in throat area will be necessary?

5.16 A straight pipe of 0.05 m diameter is attached to a large air reservoir at pressure 14×10^5 Pa and 300 K. The pipe exit is open to atmosphere. Assuming adiabatic flow with an average Darcy friction factor of 0.02, calculate the maximum pipe length to deliver a mass flow rate of 2.5 kg/s. Assume air as a perfect gas with $\gamma=1.4$ and $R=287$ J/(kg K).

5.17 Air at static pressure 3.5×10^5 Pa and total temperature 300 K is to be transported at the rate of 0.11 kg/s over a distance of 500 m through a constant-diameter pipe. The static pressure at the pipe exit is 1.50×10^5 Pa. Assuming adiabatic flow and average Darcy friction factor of 0.015, determine the pipe diameter. Assume air as a perfect gas with $\gamma=1.4$ and $R=287$ J/(kg K).

5.18 As shown in Figure 5.67, an isentropic convergent–divergent nozzle having an area ratio (exit area/throat area) of 2 discharges air into an insulated pipe of length L and diameter D. The nozzle inlet air has a stagnation pressure of 7×10^5 N/m^2 and a stagnation temperature of 300 K, and the pipe discharges into a space where the static pressure is 2.8×10^5N/m^2. Calculate fL/D of the pipe and the mass flow rate per unit area in the pipe for the cases where a normal shock stands (a) at the nozzle throat, (b) at the nozzle exit, and (c) at the pipe exit.

5.19 A C–D nozzle is connected upstream to an air tank at pressure 6×10^5 Pa and downstream to a reservoir via a short converging duct whose starting cross section area is twice the nozzle throat area. The duct area reduces by 25% along its length. As the back pressure in the downstream reservoir is lowered, the normal shock in the divergence section moves downstream. Show that the normal shock cannot be drawn to the end of the divergent section. Find the maximum Mach number upstream of the normal shock. Neglect friction and heat transfer and assume air as a perfect gas with $\gamma=1.4$ and $R=287$ J/(kg K).

5.20 As shown in Figure 5.68, a thin wedge with a blunt base is placed in a uniform supersonic air flow with $M_1=2.0$. When the sharp end of the wedge is forward, the pressure at point A is measured to be 75 kPa. With the wedge turned blunt end forward in the same flow, a

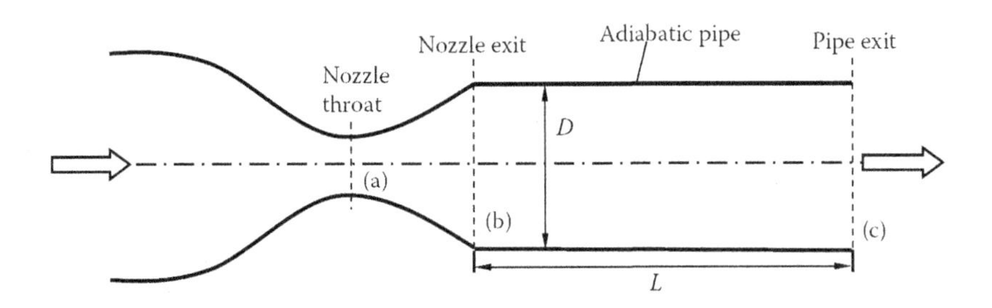

FIGURE 5.67 A convergent–divergent diffuser exhausting into an adiabatic pipe with friction (Problem 5.18).

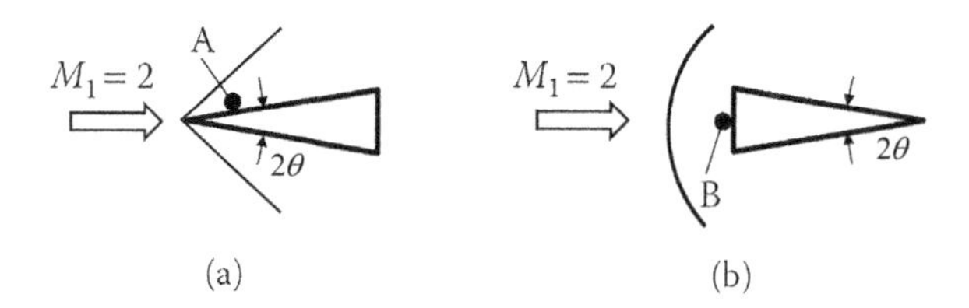

FIGURE 5.68 A wedge in an oncoming supersonic flow. (a) Attached oblique shock and (b) detached bow shock (Problem 5.20).

detached bow shock is formed, and the pressure recorded at point B is 350 kPa. Find the (a) free stream pressure and (b) wedge angle 2θ. Use for air $\gamma = 1.4$ and $R = 287$ J/(kg K).

5.21 A wedge of angle $2\theta = 30°$ is used to measure the Mach number of an oncoming supersonic air flow. If the observed wave angle is $\beta = 39°$, find $M_1, \frac{p_2}{p_1}, \frac{T_2}{T_1}, M_2$ and the entropy change across the oblique wave.

5.22 In Problem 5.21, suppose that the pressure ratio $\frac{p_2}{p_1}$ is measured instead of the wave angle β. Describe a solution method to determine M_1 and the remaining unknowns.

5.23 Consider a C–D nozzle with an area ratio $A_{exit}/A_{throat} = 3.6$. Find the Mach number immediately upstream of a normal shock standing at the exit of this nozzle. Find the back-pressure ratio $\frac{p_b}{p_{01}}$ needed to establish this flow condition. Neglect friction and heat transfer. The working fluid is air, assumed to be a perfect gas with $\gamma = 1.4$ and $R = 287$ J/(kg K).

5.24 In the C–D nozzle flow of Problem 5.23, suppose that the back-pressure ratio is $\frac{p_b}{p_{01}} = 0.1$. Find the wave angle of the oblique shock at the nozzle exit, the flow deflection angle, and the Mach number downstream of the shock.

5.25 Repeat Problem 5.24 but with a back-pressure ratio of 0.02. Hint: The given back-pressure ratio results in Prandtl–Meyer expansion flow at the nozzle exit.

5.26 A gas turbine for aircraft propulsion is mounted on a test bed. Ambient air at pressure 1 bar and temperature 293 K enters the compressor at low velocity and is compressed through a pressure ratio of 4. The air then passes to a combustion chamber where it is heated to 1175 K. The hot gas then expands through a turbine, which drives the compressor. The gas is then further expanded isentropically through a nozzle, leaving at the speed of sound. The exit area of the nozzle is 0.1 m^2. Assume $\gamma = 1.4$ for the compressor air and turbine exhaust gases in the nozzle and $R = 287$ J/(kg K) for the entire flow. Neglect the increase in mass due to the addition of fuel in the combustion chamber. Determine the gas turbine mass flow rate and the thrust produced on the engine mountings.

5.27 Compare the actual pressure rise coefficient C_p with the so-called theoretical incompressible C_{pi} as defined here for the row of axial air flow compressor vanes sketched in Figure 5.69. Comment on any difference. $C_p = \frac{p_2 - p_1}{p_{01} - p_1}$ and $C_{pi} = 1 - \left(\frac{V_2}{V_1}\right)^2$ where V_1 and V_2 are mean velocities at inlet and outlet. The loss in total pressure $(p_{01} - p_{02})$ between Planes 1 and 2 is 15% of the inlet dynamic pressure. Assume air as a perfect gas with $\gamma = 1.4$ and $R = 287$ J/(kg K).

5.28 Total pressure, static pressure, and static temperature immediately upstream of a normal shock in a compressible air flow are given. Using Table A.1 in Sultanian (2015, 2023) on compressible flow functions and, if needed, Table A.2 in Sultanian (2015, 2023) on isentropic compressible flow, write a step-by-step procedure to determine the static pressure

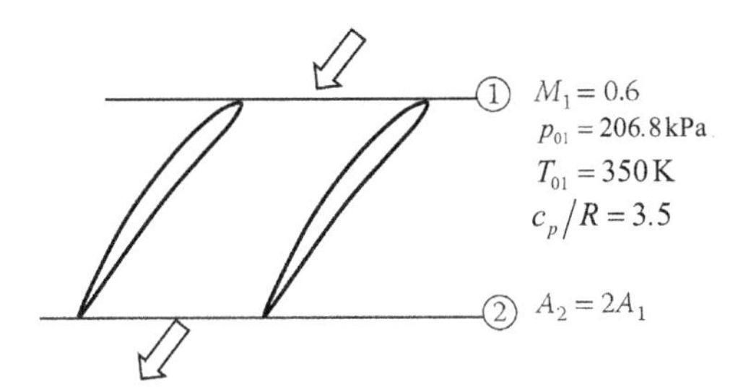

FIGURE 5.69 Diffusion in axial air flow compressor vanes (Problem 5.27).

and the static temperature immediately downstream of the normal shock. Assume air as a perfect gas with $\gamma = 1.4$ and $R = 287$ J/(kg K).

5.29 A normal shock is detected in a convergent–divergent nozzle with known exit-to-throat area ratio, inlet total pressure, and inlet total temperature. The air flow is isentropic on either side of the normal shock. For the given exit static temperature, write a noniterative, nongraphical step-by-step procedure to determine the ratio of the nozzle flow area, where the normal shock is located, to the nozzle throat area. For air, assume $\gamma = 1.4$ and $R = 287$ J/(kg K).

5.30 A supersonic air flow with known total pressure, total temperature, and static pressure enters a pipe of given constant diameter and constant friction factor. The pipe wall is adiabatic and stationary. The pipe is 15% longer than the maximum needed to achieve the choked flow conditions. While the flow remains choked, a normal shock occurs within the pipe. Write a step-by-step procedure to determine the location (the distance from the pipe inlet) of the normal shock. Assume air as a perfect gas with $\gamma = 1.4$ and $R = 287$ J/(kg K).

5.31 Total pressure, static pressure, and static temperature immediately upstream of a normal shock in a compressible air flow are given. Using Table A.3 in Sultanian (2015, 2023) on Fanno flow and, if needed, Table A.2 in Sultanian (2015, 2023) on isentropic compressible flow, write a step-by-step procedure to determine the static pressure and the static temperature immediately downstream of the normal shock. For air, assume $\gamma = 1.4$ and $R = 287$ J/(kg K).

5.32 Total pressure, static pressure, and static temperature immediately upstream of a normal shock in a compressible air flow are given. Using Table A.4 in Sultanian (2015, 2023) on Rayleigh flow and, if needed, Table A.2 in Sultanian (2015, 2013) on isentropic compressible flow, write a step-by-step procedure to determine the static pressure and the static temperature immediately downstream of the normal shock. For air, assume $\gamma = 1.4$ and $R = 287$ J/(kg K).

5.33 Air enters a duct with frictional wall at Mach 0.4 and leaves the duct at Mach 0.6. At what fraction of the duct length will the Mach number be 0.5?

5.34 Air flows adiabatically in a duct with a mean hydraulic diameter of 110 mm and a friction factor of 0.025. The entry conditions are 200 m/s, 50°C, and 2 atm. What is the maximum duct length under these conditions? With the duct exit choked, what are the pressure and temperature there?

5.35 What is the mass flow rate of air from a reservoir at $T_0 = 1.5$°C through a 6 m of insulated 25-mm-diameter pipe with wall friction $f = 0.02$? The exit pressure is the same as that of the surroundings (1 atm). Calculate the corresponding reservoir stagnation pressure.

5.36 In a frictionless air flow through a 120-mm duct, 0.15 kg/s enters at 0°C and 7 kPa. How much heat in kcal/kg can we add without choking the flow.

5.37 In a Rayleigh flow with reversible heat transfer, how much heat transfer is necessary for a flow of air at 500 m/s to increase its Mach number from 2.0 to 2.8? Is the heat transfer to or from the flow? What would be the effect of reversing the direction of heat transfer?

5.38 Heat extraction is used to increase the Mach number of a supersonic flow in a constant-area duct with no wall friction. The initial conditions are $M_1 = 2.0, T_{01} = 1111\text{ K}, p_{01} = 34.47\text{ bar}$. If the downstream Mach number is to be $M_2 = 3.0$, find the downstream pressure, temperature, stagnation pressure, and stagnation temperature. If the duct has a diameter of 15.24 cm, find the mass flow rate and the rate of heat removals in kilowatts.

5.39 In an isentropic flow, show that:

$$\frac{d\rho}{\rho} = -\frac{MdM}{1+\left[(\gamma-1)/2\right]M^2}$$

5.40 Show that at small Mach numbers, changes in Mach number lead to changes in density given by:

$$\frac{d\rho}{\rho} = -\left(\frac{2}{\gamma-1}\right)\frac{dM}{M}$$

What would be the percent change in density if the Mach number change of 10% occurs in an airflow at Mach 0.3?

5.41 Show that as the Mach number in a flow becomes very large, changes in density become proportional to the square of the Mach number.

5.42 A wedge of half-angle $\theta = 15°$ is used to measure the Mach number of a supersonic flow. If the observed wave angle is $\beta = 15°$, find $M_1, \frac{p_2}{p_1}, \frac{T_2}{T_1}, M_2$, and the entropy change across the wave.

5.43 In Problem 5.42, suppose that the pressure ratio $\frac{p_2}{p_1}$ is measured rather than β. Describe a solution method to determine M_1 and the remaining unknowns.

5.44 Consider a C–D nozzle with an area ratio of $\frac{A_{exit}}{A_{throat}} = 3.0$. Find the Mach number immediately upstream of a normal shock standing exit of this nozzle. Find the back-pressure ratio $\left(\frac{p_b}{p_{01}}\right)$ necessary to establish this condition.

5.45 In the nozzle of Problem 5.44, suppose that the back-pressure ratio is $\frac{p_b}{p_{01}} = 0.1$. Find the angle of the oblique shock at the nozzle exit and the Mach number downstream of this shock. What is the angle of the flow downstream of the shock. (A trial-and-error solution will be necessary.)

5.46 Solve Problem 5.7 when we know the nozzle exit static pressure instead of the exit Mach number given in the problem.

5.47 Using a physics-based argument from first principles, explain why the Mach number increases downstream in a Fanno flow with subsonic flow at the inlet.

5.48 Using a physics-based argument from first principles, explain why the Mach number decreases downstream in a Fanno flow with supersonic flow at the inlet.

5.49 We define unit Reynolds number as the Reynolds number per unit length. Using Sutherland's equation to compute dynamic viscosity in a gas, express the local unit Reynolds number in terms of local total pressure, total temperature, and Mach number.

REFERENCES

Anderson, J.D. 2021. *Modern Compressible Flow with Historical Perspective*, 4th Edition. New York: McGraw-Hill.

Oosthuizen, P.H. and W.E. Carscallen. 2013. *Introduction to Compressible Fluid Flow*, 2nd Edition (Heat Transfer). Boca Raton, FL: Taylor & Francis.

Shapiro, A.H. 1953. *The Dynamics and Thermodynamics of Compressible Fluid Flow*, Vols. 1 and 2. New York: Ronald Press.

Sultanian, B.K. 2015. *Fluid Mechanics: An Intermediate Approach*. Boca Raton, FL: Taylor & Francis.

Sultanian, B.K. 2021. *Fluid Mechanics and Turbomachinery: Problems and Solutions*. Boca Raton, FL: Taylor & Francis.

Sultanian, B.K. 2022. *Thermal-Fluids Engineering: Problems with Solutions*. Independently published by Kindle Direct Publishing (Amazon.com).

Sultanian, B.K. 2023. *One-Dimensional Compressible Flow: Physics-Based Modeling for Design Engineering*. Independently published by Kindle Direct Publishing (Amazon.com).

BIBLIOGRAPHY

Balachandran, P. 2012. *Fundamentals of Compressible Fluid Dynamics*. New Delhi, India: PHI.

Chen, R.-H. 2017. *Foundations of Gas Dynamics*. Cambridge: Cambridge University Press.

Daneshyar, H. 1976. *One-Dimensional Compressible Flow*. New York: Pergamon Press.

Durst, F. 2022. *Fluid Mechanics: An Introduction to the Theory of Fluid Flows*, 2nd edition. Heidelberg: Springer Verlag GmbH.

Farokhi, S. 2009. *Aircraft Propulsion*. New York: John Wiley.

Greitzer, E.M., C.S. Tan, and M.B. Graf. 2004. *Internal Flow Concepts and Applications*. Cambridge, UK: Cambridge University Press.

Korpela, S.A. 2011. *Principles of Turbomachinery*. New York: John Wiley.

Kundu, P.K., I.M. Cohen, D.R. Dowling, J. Capecilatro. 2024. *Fluid Mechanics*, 7th Edition. Burlington: Academic Press.

Liepmann, H.W. and A. Roshko. 1958. *Elements of Gas Dynamics*. New York: John Wiley. (Reprinted in 2001 by Dover Publications, Mineola, NY.)

Mattingly, J.D. 2006. *Elements of Propulsion Gas Turbines and Rockets*, 2nd Edition. Reston, VA: AIAA.

Nunn R.H. 1989. *Intermediate Fluid Mechanics*. Boca Raton, FL: Taylor & Francis.

Oates, G.C. 1988. *Aerothermodynamics of Gas Turbine and Rocket Propulsion* (Revised and Enlarged). Washington, DC: AIAA.

Panton, R.L. 2024. *Incompressible Flow*, 5th Edition. New York: Wiley.

Powers, J.M. 2023. *Mechanics of Fluids*. Cambridge: Cambridge University Press.

Rathakrishnan, E. 1995. *Gas Dynamics*. New Delhi, India: PHI.

Saad, M.A. 1992. *Compressible Fluid Flow*, 2nd Edition. Englewood Cliffs, NJ: Prentice-Hall.

Saravanamutto, H.I.H., G.F.C. Rogers, H. Cohen., P.V. Straznicky, and A.C. Nix. 2017. *Gas Turbine Theory*, 7th Edition. New York: Pearson.

Sultanian, B.K. 2018. *Gas Turbines: Internal Flow Systems Modeling* (Cambridge Aerospace Series #44). Cambridge: Cambridge University Press.

Sultanian, B.K. 2019. Logan's Turbomachinery: Flowpath Design and Performance Fundamentals, 3rd Edition. Boca Raton, FL: Taylor & Francis.

Sultanian, B.K. 2022. *Fluid Mechanics:* An Intermediate Approach: Errata for the First Edition Published in 2015. Independently published on KDP (Amazon.com).

Sultanian, B.K. 2022a. *Gas Turbines* Internal Flow Systems Modeling: Errata for the First Edition Published in 2018. USA: Independently published on KDP (Amazon.com).

Sultanian, B.K. 2022b. Axial-Flow Compressors and Turbines: How to Quickly Draw Dimensionless Velocity Diagrams. USA: Independently published on KDP (Amazon.com).

Sultanian, B.K. 2022c. Power Generation Gas Turbines: High-Performance Exhaust Diffuser Design. USA: Independently published on KDP (Amazon.com).

Sutton, G.P. and O. Biblarz. 2010. Rocket Propulsion Elements, 8th Edition. Hoboken, NJ: John Wiley.

Thompson, P.A. 1972. Compressible-Fluid Dynamics. New York: McGraw-Hill.

White, F.M and J. Majdalani. 2022. *Viscous Fluid Flow*, 4th Edition. New York: McGraw-Hill.

Wilson, D.G. and T. Korakianitis. 1998. The Design of High-Efficiency Turbomachinery and Gas Turbines, 2nd edition. Upper Saddle River, NJ: Prentice-Hall.
Zucrow, M.J. and J.D. Hoffman. 1976. Fundamentals of Gas Dynamics. New York: John Wiley.

NOMENCLATURE

A	Area
A^*	Critical throat area with $M = 1$
c_p	Specific heat at constant pressure
c_v	Specific heat at constant volume
C	Speed of sound
C^*	Characteristic speed of sound at T^*
C_V	Velocity coefficient ($C_V = V_x / V$)
C_f	Shear coefficient or Fanning friction factor
D	Nozzle or pipe diameter
D_h	Mean hydraulic diameter
f	Moody or Darcy friction factor
F	Force
F_f	Static pressure mass flow function
$\hat{F}_f$	Dimensionless static pressure mass flow function
F_{f0}	Total pressure mass flow function
$\hat{F}_{f0}$	Dimensionless total pressure mass flow function
g	Acceleration due to gravity
h	Specific enthalpy; heat transfer coefficient
I_f	Static pressure impulse function
I_{f0}	Total pressure impulse function
K_R	Total pressure loss coefficient in a Rayleigh flow with heating
L	Length
$\dot{m}$	Mass flow rate
M	Mach number ($M = V / C$)
M^*	Characteristic Mach number ($M^* = V / C^*$)
M_θ	Rotational Mach number ($M_\theta = r\Omega / \sqrt{\gamma RT}$)
N	Normal shock function
N^∞	Asymptotic value of N as $M \to \infty$
p	Static pressure
p_i	Impulse pressure
P_w	Wetted perimeter of a duct
$\dot{Q}$	Heat transfer rate
$\delta\dot{Q}$	Rate of heat transfer into a differential control volume
R	Gas constant; pipe radius
Re	Reynolds number
s	Specific entropy
S	Point of strong shock solution on a hodograph plane
$\hat{s}$	Dimensionless change in specific entropy
S_T	Stream thrust
t	Time
T	Temperature
V	Velocity; through-flow velocity relative to a duct
V_x^*	Dimensionless x velocity as a coordinate in the hodograph plot ($V_x^* = V_x / C^*$)
V_y^*	Dimensionless y velocity as a coordinate in the hodograph plot ($V_y^* = V_y / C^*$)
W	Relative velocity; point of weak shock solution on a hodograph plane

W_j Relative jet velocity
$\dot{W}$ Work transfer rate
$\delta\dot{W}$ Rate of work transfer into a differential control volume

SUBSCRIPTS AND SUPERSCRIPTS

0 Origin, total (stagnation)
1 Location 1; Section 1
2 Location 2; Section 2
e Exit
max Maximum
min Minimum
n Velocity component normal to shock wave
ns Normal shock
Ps Pressure surface
ref Reference state
rot Rotation
rot _ x Component in the x direction
sh Shear
S Sound-source
ss Suction surface
w Wall
x Component along x coordinate direction; axial direction
β Velocity component along the shock wave
$(^*)$ Properties at $M = 1$; characteristic values; critical value $M = 1/\sqrt{\gamma}$ for an isothermal flow in a constant-area duct with friction

GREEK SYMBOLS

β Wave angle of an oblique shock
β_S Wave angle of an oblique shock corresponding to the strong solution point S in the hodograph plane
β_W Wave angle of an oblique shock corresponding to the weak solution point W in the hodograph plane
δ Parameter in the equation to calculate wave angle from known upstream Mach number and deflection angle: $\delta = 0$ for a strong shock and $\delta = 1$ for a weak shock
η Number of transfer units (NTU)
θ Wedge half-angle or deflection angle
γ Ratio of specific heats ($\gamma = c_P / c_V$)
λ Parameter in the equation to calculate wave angle from known upstream Mach number and deflection angle
μ Dynamic viscosity; Mach wave angle { $\mu = sin^{-1}(1/M)$ }
ρ Density
τ_w Wall shear stress
χ Parameter in the equation to calculate wave angle from known upstream Mach number and deflection angle
ϕ Prandtl–Meyer function
Ω Rotational (angular) speed

6 Computational Fluid Dynamics of Turbulent Flows

An Overview of Theory and Applications

6.1 INTRODUCTION

Computational fluid dynamics (CFD) is a powerful tool offering numerical predictions of velocity, pressure, temperature, and other pertinent properties across a computational domain. Since its inception in the 1960s, CFD has undergone remarkable development, driven by concurrent advancements in computing power and technology. It stands as the singular method capable of computing time-dependent three-dimensional flows and heat transfer phenomena. Despite advancements, the efficacy of CFD results hinges on the predictive capacity of chosen turbulence models, posing a notable challenge in contemporary CFD technology. As aptly noted by Peter Bradshaw (1997), "While progress in turbulence modeling is evident, industry still awaits a model that balances reliability with cost-effectiveness."

The fundamental conservation equations governing mass, momentum, and energy utilized in computational fluid dynamics (CFD) are universally recognized and require no further validation. However, the same cannot be said for turbulence models, which are developed based on various approximations, including data derived from simple shear flows like equilibrium wall boundary layers and free shear flows. The successful application of CFD hinges on comprehensive validation tailored to specific applications. Yet, this endeavor encounters three significant challenges. Firstly, there is a scarcity of benchmark-quality data representative of modern gas turbine internal flow systems. Secondly, many existing datasets lack crucial measurements at key boundaries, such as inflow, outflow, and walls, essential for meaningful CFD analysis and comparison. Thirdly, certain experimental measurements, often considered as gold standards, may contain notable errors, as demonstrated by Sultanian, Neitzel, and Metzger (1987) invalidating detailed Laser Doppler Velocimetry (LDV) measurements published in the *AIAA* journal. These observations underscore the challenges in conducting reliable experiments and accessing accurate data in the literature. Despite these hurdles, CFD technology remains the most effective analytical method for comprehensively assessing existing designs within available datasets through validation, thus enabling informed exploration of new operating regimes.

In modern design engineering, computational fluid dynamics (CFD) technology assumes several pivotal roles. Firstly, CFD enables comprehensive numerical flow visualization, illuminating intricate flow features otherwise obscured. This insight aids in the development of reduced-order models, such as flow network models, optimizing design efficiency. Secondly, in scenarios where empirical flow data are lacking, CFD offers a quantitative assessment of critical parameters like friction factor, discharge coefficient, and loss coefficient. These values subsequently serve as inputs for reduced-order models. The accuracy of these numerical assessments hinges on the turbulence model's ability to faithfully predict the entire flow field, spanning from near-wall to far-wall regions. Thirdly, CFD facilitates the generation of heat transfer coefficients, crucial for thermal analysis.

The greatest strength of CFD technology is to provide quick and inexpensive flow visualization, which is often better than an unrealistic guess. Just knowing the flow behavior in a design can be

DOI: 10.1201/9781003325192-6

extremely valuable to perform a reduced-order analysis with higher speed and direct use of adjusted empirical correlations to yield results in better agreement with the test results.

The original version of this chapter appears as Chapter 10 in the first edition of this book by Sultanian (2015).

6.2 INDUSTRIAL ANALYSIS AND DESIGN SYSTEMS

In the analysis of industrial equipment for flow and heat transfer, intricate velocity, pressure, and temperature distributions within the equipment aren't always requisite. Typically, performance criteria revolve around permissible overall pressure loss and temperature increases at specific flow rates. While maximizing efficiency remains paramount, it often contends with cost-effectiveness and swift design cycles. Hence, the analytical system employed in industrial design must exhibit reliability, precision, swiftness, and cost-efficiency. Furthermore, to accommodate uncertainties in boundary conditions and empirical correlations, contemporary design practices advocate for robust designs incorporating probabilistic analyses entailing multiple deterministic simulations.

The era of back-of-the-envelope design calculations is rapidly fading in today's design landscape. As illustrated in Figure 6.1, a cohesive analysis system has become indispensable for modern product design, prioritizing accuracy, reliability, speed, and cost-efficiency. This figure underscores the iterative nature of the entire design process. The key to streamlining design cycles and minimizing costs lies in error prevention during the early stages of design, thereby averting extensive iterations downstream—a principle akin to the adage "measure twice and cut once."

6.3 PHYSICS OF TURBULENCE

Turbulent flows are ubiquitous in industrial equipment and processes. A qualitative understanding of turbulence is essential before we undertake the task of its quantitative predictions for engineering purposes. Recall the famous dye experiment of Osborne Reynolds (1883), demonstrating the transition of laminar flow in a pipe into a fully developed turbulent flow. While this transition occurs over a finite pipe length and the Reynolds number ranges from 2300 to 4000, the experiment presented in

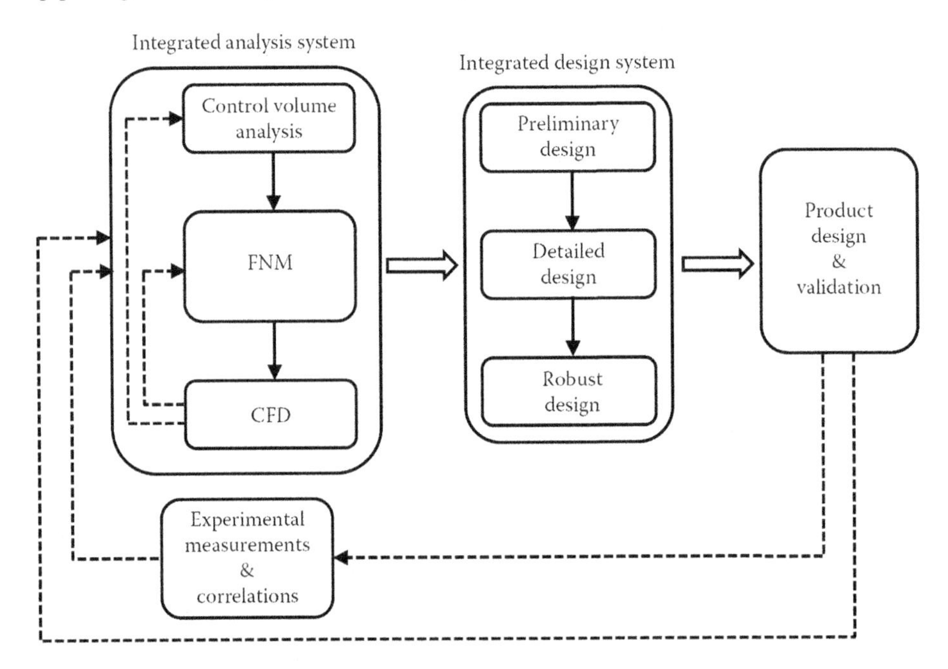

FIGURE 6.1 Typical industrial analysis and design systems.

Reynolds (1894) shows how the dye streakline along the pipe axis in the laminar flow regime mixes thoroughly when the flow becomes turbulent. Thus, unlike a laminar flow, turbulent flows feature vigorous mixing of flow properties, somewhat independent of the fluid viscosity. In his physics-rich little book, Bradshaw (1971) defined turbulence as follows:

> Turbulence is a three-dimensional time-dependent motion in which vortex stretching causes velocity fluctuations to spread to all wavelengths between a minimum determined by viscous forces and maximum determined by the boundary conditions of the flow. It is the usual state of fluid motion except at low Reynolds numbers.

In the above statement, vortex stands for a vortex tube or filament made of vortex lines, which are locally tangent vorticity vectors. It does not stand for a line vortex representing the bulk rotational motion of the flow.

Recall the vorticity transport equation:

$$\frac{D\zeta}{Dt} = (\zeta \bullet \nabla)V + \nu\nabla^2\zeta$$

where ζ and V are three-dimensional vorticity and velocity vectors, respectively. Unlike the Navier–Stokes equations in primitive variables (velocities and static pressure), this equation has no dependence on pressure gradients. The first term on the right-hand side of this equation accounts for turning and stretching of the vorticity vector by velocity gradients. For example, we can write the transport equation for ζ_z (component of vorticity in the z direction) as:

$$\frac{D\zeta_z}{Dt} = \left(\zeta_x \frac{\partial W}{\partial x} + \zeta_y \frac{\partial W}{\partial y} + \zeta_z \frac{\partial W}{\partial z}\right) + \nu\nabla^2\zeta_z$$

On the right-hand side of this equation, the first two terms within the parentheses represent the turning of ζ_x and ζ_y, respectively, while the third term represents the stretching of ζ_z in the z direction. Similarly, the other components of the deformation rate tensor cause turning and stretching of the vorticity vector. In a shear flow, this constitutes the primary mechanism for turbulence production.

When a vortex filament stretches, its rotational speed increases to conserve its angular momentum without any viscous effect, and so does its rotational kinetic energy, extracted from the mean flow. These vortex filaments are also known as turbulent eddies, which characterize a turbulent flow's random and chaotic nature. Interestingly, the turbulent eddies have a continuous spectrum of wavelengths and frequencies. The kinetic energy from the mean flow enters the large-scale eddies and gets cascaded down to smaller and smaller scales until it is dissipated by viscosity at the smallest scale eddies, which are isotropic and nearly universal. This mechanism is known as the energy cascade of turbulence, aptly summarized by Richardson (1922) as:

Big whorls have little whorls,
Which feed on their velocity;
And little whorls have lesser whorls,
And so on to viscosity.

It should be apparent from the preceding discussion that eddies, which are not precisely defined, are responsible for the enhanced transport properties of a turbulent flow. Although these eddies are random, they differ from the gas molecules whose random motion is responsible for gas transport properties (viscosity and thermal conductivity). First, the turbulent eddies are continuous and contiguous, unlike gas molecules, which are discrete and collide at random intervals. Second, molecular mean free paths are small compared to the characteristic dimension of the mean flow. However, the turbulent eddies span from the large scale responsible for energy extraction from the mean flow to the smallest scale at which viscous dissipation takes place.

6.4 PHYSICS-BASED MODELING

Physics is the foundation of engineering; mathematics is the language of physics. All engineering solutions must not only be mathematically sound, but they must also satisfy all the applicable laws of physics to be realized in the physical world. Fortuitously, all the flow and energy transfer processes in a gas turbine are governed by the conservation laws of mass, momentum, and energy (the first law of thermodynamics), while satisfying the entropy constraint as dictated by the second law of thermodynamics.

In thermofluids design engineering, 1D, 2D, and 3D modeling are also called 1D CFD, 2D CFD, and 3D CFD, respectively, as shown schematically in Figure 6.2. The only requirement for a prediction method, be it 1D, 2D, or 3D, to be physics-based is that it must not violate any of the conservation laws. Although each method must validate with the product performance data obtained from in-house testing and field operation, a method that is not physics-based and based entirely on previous empirical data and arbitrary correction factors is undesirable and short-lived as a reliable predictive tool. Physics-based methods tend to make more consistent predictions and, at times, may need only some minor corrections for an acceptable validation with the actual design. These methods are continuously validated using data from gas turbine thermal surveys, and the discrepancies between pre-test predictions and test data are meticulously resolved by improved physical modeling and adjustment of correction factors.

In 1D CFD, shown schematically in Figure 6.2a, the flow domain is divided into large control volumes. Each control volume contains a part of the bounding walls. The prediction yields one-dimensional, usually along the flow direction, variation of various flow properties. The modeling is based on empirical correlations to compute the discharge coefficient, friction factor, loss coefficient, and heat transfer coefficient. Like all CFD analyses, boundary conditions are specified at the inlet, outlet, and walls. The 1D CFD offers designers maximum flexibility to adjust correction factors to the empirical correlations to improve validation with the component test data.

A 2D CFD analysis, shown schematically in Figure 6.2b with a two-dimensional computational grid, becomes necessary when the flow properties vary both in the flow direction and in one more direction normal to the flow direction. A three-dimensional axisymmetric flow is commonly handled by 2D CFD in cylindrical polar coordinates. In this case, for predicting turbulent flows, one of

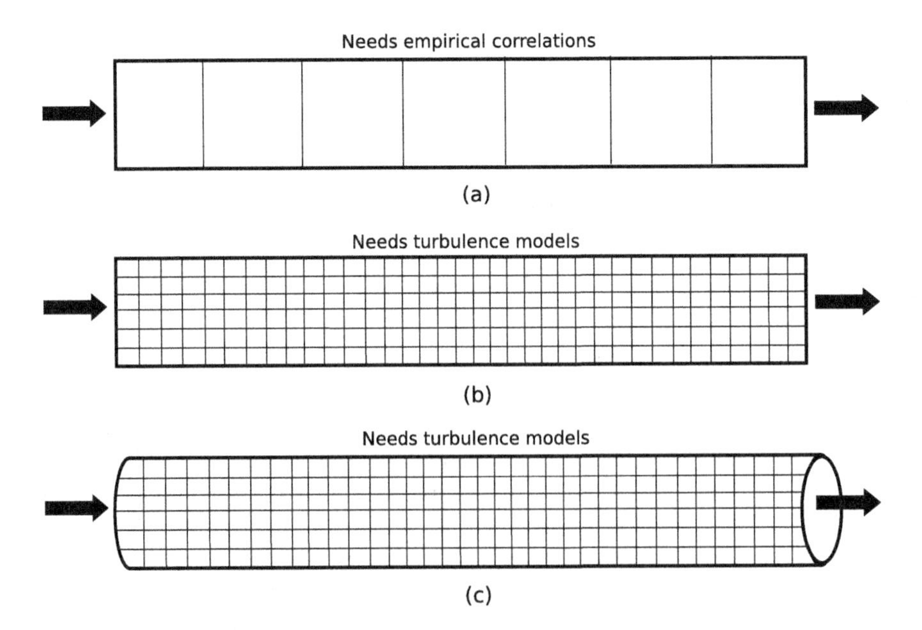

FIGURE 6.2 Physics-based modeling: (1) 1D CFD, (b) 2D CFD, and (c) 3D CFD.

the turbulence models is used instead of the empirical correlations used in 1D CFD. As shown in Figure 6.2b, to resolve local variations in flow properties, we use very small control volumes, making the final results grid-independent, that is to say that further refinement in grid will have little effect on the solution. Except those near a wall, all other control volumes used in a 2D CFD are without a wall. Knowing the flow properties at many locations (grid points) in a 2D CFD has the distinct advantage of providing two-dimensional flow visualization, delineating regions of internal flow recirculation and wall boundary layer separation. This information is certainly missing in a 1D CFD. For computing integral quantities such as the loss in total pressure from inlet to outlet, we need to compute section-average values from the 2D CFD results. These integral quantities may then be compared with the corresponding 1D CFD results or test data. In Section 6.5.5, we present a physics-based method to postprocess CFD results. It is important to note that when using 2D CFD, changing the turbulence model is the only option available to a designer to improve the validation of the CFD results with the test data. The flow visualization aspect of a 2D CFD analysis often proves useful in design engineering for reinforcing 1D CFD modeling. It may come as a surprise to some that the integral results obtained from a 2D CFD analysis may not often be more accurate than those from a 1D CFD, which directly uses the applicable empirical correlations. In this respect, 1D CFD tends to be more postdictive, that is, using the empirical correlations obtained from the test data to predict these data, than predictive.

The general methodology to carry out a 3D CFD analysis, which becomes necessary to understand the three-dimensional behavior of a flow field, is like that used for a 2D CFD. In this case, as shown schematically in Figure 6.2c, we use a three-dimensional grid system with a number of interconnected three-dimensional control volumes. In each of these control volumes, all the governing conservation equations must be satisfied, albeit approximately due to the numerical nature of the solution obtained.

6.5 CFD METHODOLOGY

This section provides a concise outline of the computational fluid dynamics (CFD) methodology, encompassing the governing conservation equations, which entail nonlinearly coupled partial differential equations. Additionally, the overview includes commonly employed turbulence models and the physics-based postprocessing of 3D CFD results for their insightful interpretation, design applications, and comparison with outcomes derived from a reduced-order model. For in-depth discussions on turbulent flow simulation techniques like direct numerical simulation (DNS) and large eddy simulation (LES), readers are directed to the bibliography listed at the conclusion of this chapter.

As illustrated in Figure 6.3, the flow and heat transfer physics of the design are mathematically represented through the conservation equations of mass, momentum, and energy. These equations, along with relevant boundary conditions, are solved numerically employing computational fluid

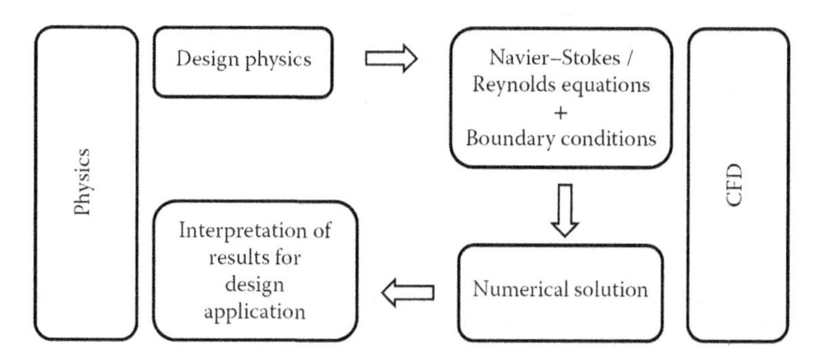

FIGURE 6.3 CFD methodology

dynamics (CFD) methods. Subsequently, the numerical results undergo postprocessing, ensuring comprehensive interpretation and facilitating informed adjustments to enhance design performance.

Until the mid-1980s, most industries and universities independently developed and maintained their own in-house computational fluid dynamics (CFD) computer codes. This effort was driven by a need for tailored solutions and advancements in numerical methods and turbulence models. However, the landscape shifted with the advent of several commercially available general-purpose CFD codes, leading to a significant decline in independent development activities over the subsequent three decades.

The adoption of leading commercial CFD codes has made the investigation of industrial designs routine, as depicted in Figure 6.4. CFD analysts primarily focus on generating high-quality grids, defining boundary conditions, and selecting suitable turbulence models aligned with the design's physics. The commercial CFD software then handles the computational intricacies, leaving analysts to concentrate on postprocessing CFD results for their design applications, a critical endeavor.

Engineers interested in delving deeper into the numerical aspects of CFD can refer to works such as Patankar (1980) and Anderson et al. (2021), which cover the derivation of discretization equations and iterative solution methods. Thompson, Soni, and Weatherill (1998) offered comprehensive insights into CFD grid generation technology. Additionally, detailed discussions on turbulence modeling are available in works like Wilcox (1998, 2006) and Leschziner (2016), while Durbin and Shih (2005) provide a state-of-the-art review of turbulence modeling techniques.

6.5.1 Governing Conservation Equations

6.5.1.1 Continuity Equation

Equation in vector form:

$$\frac{\partial \rho}{\partial t} + \text{div}(\rho \boldsymbol{V}) = 0 \tag{6.1}$$

Equation in tensor notation:

$$\frac{\partial \rho}{\partial t} + \frac{\partial}{\partial x_j}(\rho U_j) = 0 \tag{6.2}$$

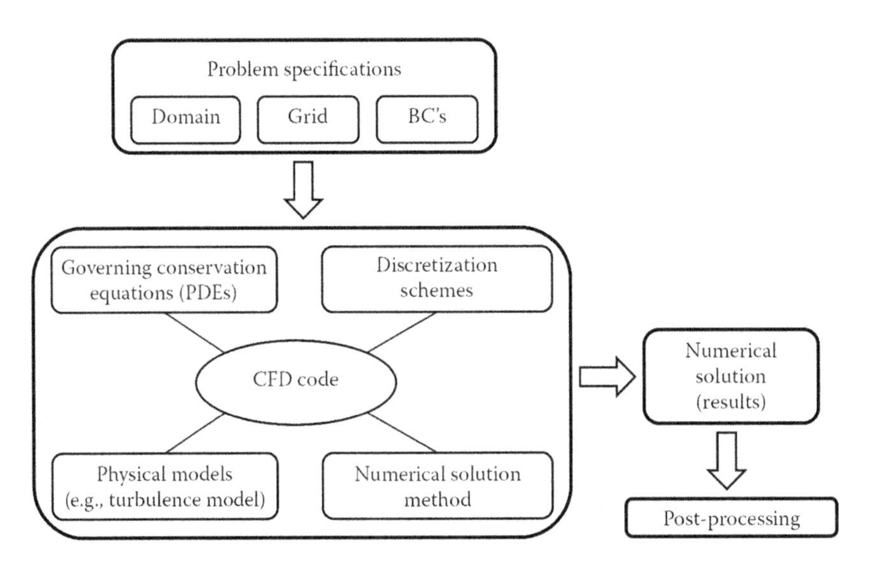

FIGURE 6.4 Solution with a commercial CFD code.

6.5.1.2 Momentum Equations

The x component of momentum equation in vector form:

$$\frac{\partial}{\partial t}(\rho U)+\text{div}(\rho \boldsymbol{V} U)=\text{div}(\mu \text{grad} U)-\frac{\partial p}{\partial x}+\tilde{S}_U \tag{6.3}$$

The y component of momentum equation in vector form:

$$\frac{\partial}{\partial t}(\rho V)+\text{div}(\rho \boldsymbol{V} V)=\text{div}(\mu \text{grad} V)-\frac{\partial p}{\partial x}+\tilde{S}_V \tag{6.4}$$

The z component momentum equation in vector form:

$$\frac{\partial}{\partial t}(\rho W)+\text{div}(\rho \boldsymbol{V} W)=\text{div}(\mu \text{grad} W)-\frac{\partial p}{\partial x}+\tilde{S}_W \tag{6.5}$$

Momentum equations in tensor notation:

$$\frac{\partial}{\partial t}(\rho U_i)+\frac{\partial}{\partial x_j}(\rho U_j U_i)=\frac{\partial}{\partial x_j}\left(\mu \frac{\partial U_i}{\partial x_j}\right)+S_{U_i} \tag{6.6}$$

6.5.1.3 Energy Equation

In terms of static enthalpy (h) in vector form:

$$\frac{\partial}{\partial t}(\rho h)+\text{div}(\rho \boldsymbol{V} h)=\text{div}(\tilde{k}\,\text{grad} T)+S_h \tag{6.7}$$

Using $h=c_p T$ to replace T on the right-hand side of this equation yields:

$$\frac{\partial}{\partial t}(\rho h)+\text{div}\,(\rho \boldsymbol{V} h)=\text{div}\left(\frac{\tilde{k}}{c_p}\,\text{grad}\, h\right)+S_h \tag{6.8}$$

which we write in tensor notation as:

$$\frac{\partial}{\partial t}(\rho h)+\frac{\partial}{\partial x_j}(\rho U_j h)=\frac{\partial}{\partial x_j}\left(\frac{\tilde{k}}{c_p}\frac{\partial h}{\partial x_j}\right)+S_h \tag{6.9}$$

Substituting $h=c_p T$ in Equation 6.8 yields the energy equation in vector form in terms of the static temperature as follows:

$$\frac{\partial}{\partial t}(\rho T)+\text{div}\,(\rho \boldsymbol{V} T)=\text{div}\left(\frac{\tilde{k}}{c_p}\,\text{grad}\, T\right)+\frac{S_h}{c_p} \tag{6.10}$$

which in tensor notation becomes:

$$\frac{\partial}{\partial t}(\rho T)+\frac{\partial}{\partial x_j}(\rho U_j T)=\frac{\partial}{\partial x_j}\left(\frac{\tilde{k}}{c_p}\frac{\partial T}{\partial x_j}\right)+\frac{S_h}{c_p} \tag{6.11}$$

6.5.1.4 Chemical Species Conservation Equation

In vector form:

$$\frac{\partial}{\partial t}(\rho m_\ell)+\text{div}\,(\rho V m_\ell)=\text{div}\,(\Gamma_\ell\,\text{grad}\,m_\ell)+\Re_\ell \tag{6.12}$$

where

$m_\ell \equiv$ Mass fraction of chemical species ℓ

$\Gamma_\ell \equiv$ Diffusion coefficient for chemical species ℓ

$\Re_\ell \equiv$ Generation rate per unit volume (from chemical reaction)

In tensor notation:

$$\frac{\partial}{\partial t}(\rho m_\ell)+\frac{\partial}{\partial x_j}(\rho U_j m_\ell)=\frac{\partial}{\partial x_j}\left(\Gamma_\ell\frac{\partial m_\ell}{\partial x_j}\right)+\Re_\ell \tag{6.13}$$

6.5.1.5 The Common Equation Form

In tensor notations, we can express the common form of the conservation equations suitable for a common numerical solution method as:

$$\frac{\partial}{\partial t}(\rho\Phi)+\frac{\partial}{\partial x_j}(\rho U_j\Phi)=\frac{\partial}{\partial x_j}\left(\Gamma_\Phi\frac{\partial\Phi}{\partial x_j}\right)+S_\Phi \tag{6.14}$$

where $\Phi = 1, U_i, h$, and m_ℓ yield Equations 6.2, 6.6, 6.9, and 6.13, respectively. The interpretation of each term in Equation 6.14 is as follows:

$\frac{\partial}{\partial t}(\rho\Phi) \equiv$ Transient term

$\frac{\partial}{\partial x_j}(\rho U_j\Phi) \equiv$ Convection term

$\frac{\partial}{\partial x_j}\left(\Gamma_\Phi\frac{\partial\Phi}{\partial x_j}\right) \equiv$ Diffusion term

$S_\Phi \equiv$ Source term

Note that Equation 6.14 is valid whether the flow is laminar or turbulent, incompressible or compressible, or steady or unsteady.

6.5.2 Turbulence Modeling

There are two ways of numerically predicting turbulent flows: one by simulation, either direct numerical simulation (DNS) or large eddy simulation (LES), and another by statistical modeling. The first is too expensive and time-consuming for most practical designs, and the second is not universally accurate and reliable. We thus remain trapped between the rock and the hard place.

For statistical modeling of a turbulent flow, we decompose all its randomly varying properties into their statistically average values and their fluctuating parts. Figure 6.5 shows one such

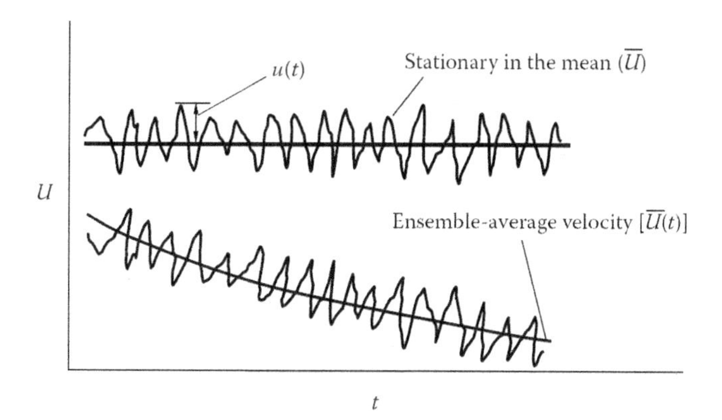

FIGURE 6.5 Velocity at a point in a turbulent flow.

decomposition for $U = \overline{U} + u(t)$. In the top velocity plot, the mean velocity $\overline{U}$ is time-independent, and the flow is considered to be stationary in the mean. In the bottom plot, the mean velocity obtained from ensemble averaging varies with time.

6.5.2.1 Reynolds Equations: The Closure Problem

Reynolds averaging. Let us decompose $U_i(t)$, $p(t)$, and the general flow property $\Phi(t)$ as $U_i = \overline{U}_i + u_i$, $p = \overline{p} + p'$, $\Phi = \overline{\Phi} + \varphi$, respectively. The time-averaging (also called Reynolds averaging) of these quantities yields their mean values as:

$$\overline{U}_i = \frac{1}{t_2 - t_1}\int_{t_1}^{t_2} U_i \mathrm{d}t;\ \overline{p} = \frac{1}{t_2 - t_1}\int_{t_1}^{t_2} p\,\mathrm{d}t;\ \overline{\Phi} = \frac{1}{t_2 - t_1}\int_{t_1}^{t_2} \Phi\,\mathrm{d}t$$

The closure problem. We restrict the rest of our discussion of turbulence modeling to the Navier–Stokes equations governing statistically stationary incompressible flows. The Reynolds averaging of Equation 6.6 with the source term S_{U_i} replaced by the pressure–gradient term, and dropping bars on the time-averaged quantities ($\overline{U}_i$, $\overline{p}$, $\overline{\Phi}$, etc.) in the following, yields:

$$\begin{aligned} &\frac{\partial}{\partial x_j}(\rho U_j U_i) + \frac{\partial}{\partial x_j}\left(\rho \overline{u_i u_j}\right) = -\frac{\partial p}{\partial x_i} + \frac{\partial}{\partial x_j}\left(\mu \frac{\partial U_i}{\partial x_j}\right) \\ &\frac{\partial}{\partial x_j}(\rho U_j U_i) = -\frac{\partial p}{\partial x_i} + \frac{\partial}{\partial x_j}\left(\mu \frac{\partial U_i}{\partial x_j} - \rho \overline{u_i u_j}\right) \end{aligned} \tag{6.15}$$

where U_i and u_i are the mean and fluctuating parts, respectively, of the local instantaneous velocity and $-\rho\, u_j u_i$ are the Reynolds stresses, which result from the time-averaging of the nonlinear convection terms in the Navier–Stokes equations. Clearly, additional equations are required to determine these stresses and thereby have a closed system. This is the task of turbulence modeling. Thus, the turbulence models essentially model the Reynolds stresses in terms of the mean flow quantities.

Boussinesq hypothesis. According to the Boussinesq hypothesis, which invokes the gradient transport model as in a laminar flow, the Reynolds stresses are related to the mean velocity gradients by the following equation:

$$-\rho \overline{u_i u_j} = \mu_t \left(\frac{\partial U_i}{\partial x_j} + \frac{\partial U_j}{\partial x_i}\right) - \frac{2}{3}\rho k \delta_{ij} \tag{6.16}$$

where

$\mu_t \equiv$ Turbulent viscosity

$k \equiv$ Turbulent kinetic energy

$$\delta_{ij} \equiv \text{Kronecker delta} = \begin{cases} 1 \text{ for } i = j \\ 0 \text{ for } i \neq j \end{cases}$$

Note that the turbulent viscosity used in Equation 6.16, unlike a laminar (molecular) viscosity, is not a fluid property but that of a flow. However, by analogy with the molecular viscosity used in a laminar flow, we write:

$$\mu_t \sim \rho \upsilon \ell \tag{6.17}$$

where

$\upsilon \equiv$ Turbulent velocity scale

$\ell \equiv$ Turbulent length scale

According to Equation 6.17, the challenge of turbulence modeling to solve the second-order closure problem, discussed in the preceding, boils down to finding the turbulent velocity and length scales. Although seldom used in current industrial design applications, for historical importance, two early turbulence models, namely, the Prandtl mixing length model and the one-equation $k - L$ model are briefly in the following sections.

6.5.2.2 Prandtl Mixing Length Model

Prandtl proposed the following mixing length hypothesis:

Turbulent length scale:

$$\ell = \ell_m$$

Turbulent velocity scale:

$$\upsilon = \ell_m \left| \frac{\partial U}{\partial y} \right|$$

Hence,

$$\mu_t = \rho \ell_m^2 \left| \frac{\partial U}{\partial y} \right| \tag{6.18}$$

For a wall boundary layer, we specify the mixing length ℓ_m in Equation 6.18 in the inner and outer regions as follows:

Inner region: $\ell_m = K_c y \left(1 - e^{-y^+/A^+}\right)$

Outer region: $\ell_m = 0.09\,\delta$

where

$\delta \equiv$ Velocity boundary layer thickness

$K_c \equiv$ von Karman constant

$$A^+ = 26$$

$$y^+ = \frac{y\sqrt{\tau_w / \rho_w}}{\nu_w}$$

Note that y^+ is a Reynolds number based on shear velocity ($\sqrt{\tau_w / \rho_w}$) and the transverse distance from the wall.

6.5.2.3 One-Equation $k - L$ Model

In this model, while we obtain the turbulent velocity scale from the local value of the turbulent kinetic energy k by solving the corresponding differential transport equation, we directly specify the turbulent length.

Turbulent velocity scale: $\upsilon = k^{\frac{1}{2}}$
Turbulent length scale: $\ell = L$

$$\mu_t = C'_\mu \rho k^{\frac{1}{2}} L \tag{6.19}$$

where $C'_\mu = 0.548$. The transport equation for k is as follows:

$$\frac{\partial}{\partial x_j}(\rho U_j k) = \frac{\partial}{\partial x_j}\left(\frac{\mu_t}{\sigma_k}\frac{\partial k}{\partial x_j}\right) + \rho(P_k - \varepsilon) \tag{6.20}$$

where σ_k (typically assigned a value 1.0) is the turbulent Prandtl number for k, P_k the production rate of k, and ε the dissipation rate of k. We evaluate P_k and ε as follows:

$$P_k = \mu_t \left(\frac{\partial U_i}{\partial x_j} + \frac{\partial U_j}{\partial x_i}\right)\frac{\partial U_i}{\partial x_j} \tag{6.21}$$

and

$$\varepsilon = C_D \frac{k^{\frac{3}{2}}}{L} \tag{6.22}$$

where $C_D = 0.164$. Note that we obtain $C'_\mu C_D = C_\mu = 0.09$ where the coefficient C_μ is used in the two-equation k–ε turbulence model (discussed in the next section) to define the turbulent viscosity.

For a turbulent boundary layer in local equilibrium, we can show that:

$$(C'_\mu / C_D)^{\frac{1}{4}} L = \ell_m \tag{6.23}$$

Specifying the mixing length ℓ_m as in the Prandtl mixing length model, we use Equation 6.23 to specify L in the one-equation model.

6.5.2.4 Two-Equation k–ε Model (BVM)

The conventional high Reynolds number two-equation k–ε model embodies the Boussinesq eddy viscosity hypothesis, which relates Reynolds stresses to mean velocity gradients. This model, hereafter called the Boussinesq viscosity model (BVM), has been in widespread use for more than four decades and is generally now considered the starting point for most complex shear flow predictions in industrial design applications. Even when using a more advanced turbulence model in a design application, the baseline CFD predictions are often carried out using BVM. In this model, we compute the turbulent velocity scale, turbulent length scale, and eddy viscosity using the following equations:

Turbulent velocity scale: $\upsilon = k^{\frac{1}{2}}$

Turbulent length scale: $\ell = \frac{k^{\frac{3}{2}}}{\varepsilon}$

$$\mu_t = C_\mu \rho \frac{k^2}{\varepsilon} \tag{6.24}$$

where
$\varepsilon \equiv$ Dissipation rate of turbulent kinetic energy (k)
$C_\mu \equiv$ Proportionality constant$=0.09$

Equation 6.24 is based on the Kolmogorov–Prandtl relation (see Launder and Spalding (1974)) and the fact that in a region of high turbulent Reynolds number, the dissipation (ε) is essentially controlled by large-scale turbulent motions through an energy cascade. Note that the transport equation for k in this model is the same as Equation 6.20 used in the one-equation k–L model.

While it is possible to derive a transport equation for ε, the resulting form requires several modeling assumptions. Consequently, the common approach has been to utilize an ε equation structured similarly to the k equation, taking the following form:

$$\frac{\partial}{\partial x_j}(\rho U_j \varepsilon) = \frac{\partial}{\partial x_j}\left(\frac{\mu_t}{\sigma_\varepsilon}\frac{\partial \varepsilon}{\partial x_j}\right) + \rho \frac{\varepsilon}{k}(C_{\varepsilon_1} P_k - C_{\varepsilon_2}\varepsilon) \tag{6.25}$$

where σ_ε is the turbulent Prandtl number for ε and C_{ε_1} and C_{ε_2} are additional model constants.

The choice of turbulence model constants greatly influences flow predictions. Rather than allowing them to fit data arbitrarily, we select them with the hope of having universality. The model constants in the BVM typically used by different investigators are summarized in Table 6.1. This table shows a good agreement among different investigators on the values for these constants. The BVM model constants used in most industrial applications are those of Launder and Spalding (1974); a shortcoming of this choice, however, is the failure of the model to predict the growth of an axisymmetric free jet correctly.

6.5.2.5 Reynolds Stress Transport Model (RSTM)

Rodi (1980) presented a convenient method to derive the transport equations for Reynolds stresses. These equations are cast into the following form:

$$C_{\overline{u_i u_j}} - D_{\overline{u_i u_j}} = P_{ij} + \Phi_{ij} - \varepsilon_{ij} \tag{6.26}$$

where

Convection: $C_{\overline{u_i u_j}} = \frac{\partial}{\partial x_\ell}\left(U_\ell \overline{u_i u_j}\right)$

Diffusion: $D_{\overline{u_i u_j}} = -\frac{\partial}{\partial x_\ell}\left(\overline{u_i u_j u_\ell} + \frac{\overline{p u_i}}{\rho}\delta_{j\ell} + \frac{\overline{p u_j}}{\rho}\delta_{i\ell}\right)$

TABLE 6.1
Values of model constants used in BVM

Investigators					
Jones and Launder (1972)	0.09	1.45	1.90	1.0	1.30
Hanjalic and Launder (1972)	0.09	1.45	2.00	1.0	1.30
Wyngaard, Arya and Cote (1974)	0.09	1.50	2.00	1.0	1.30
Launder and Spalding (1974)	0.09	1.44	1.92	1.0	1.30
Gibson and Launder (1976)	0.09	1.45	1.90	1.0	1.30
Pope and Whitelaw (1976)	0.09	1.45	1.90	1.0	1.30
Moon and Rudinger (1977)	0.09	1.44	1.70	1.0	1.30
Gosman et al. (1969)	0.09	1.42	1.92	1.0	1.22
Lilley and Rhode (1982)	0.09	1.44	1.92	1.0	1.23

$$\text{Production: } P_{ij} = -\left(\overline{u_i u_\ell}\frac{\partial U_j}{\partial x_\ell} + \overline{u_j u_\ell}\frac{\partial U_i}{\partial x_\ell}\right)$$

$$\text{Pressure–strain: } \Phi_{ij} = \overline{\frac{p}{\rho}\left(\frac{\partial u_i}{\partial x_j} + \frac{\partial u_j}{\partial x_i}\right)}$$

$$\text{Dissipation: } \varepsilon_{ij} = 2\overline{\frac{\mu}{\rho}\left(\frac{\partial u_i}{\partial x_\ell}\frac{\partial u_j}{\partial x_\ell}\right)}$$

The diffusion, dissipation, and pressure–strain terms in Equation 6.26 require modeling for the second-order closure. In regions of high turbulent Reynolds numbers, the viscous diffusion term is negligible compared to the remaining diffusion terms. In addition, the tiny scales responsible for energy dissipation are essentially isotropic. Thus, Equation 6.26 assumes the following modeled form:

$$\begin{aligned} C_{\overline{u_i u_j}} = {} & C_s \frac{\partial}{\partial x_\ell}\left(\frac{k}{\varepsilon}\overline{u_\ell u_m}\frac{\partial \overline{u_i u_j}}{\partial x_m}\right) \\ & -\left(\overline{u_i u_\ell}\frac{\partial U_j}{\partial x_\ell} + \overline{u_j u_\ell}\frac{\partial U_i}{\partial x_\ell}\right) \\ & \underbrace{-C_1\frac{\varepsilon}{k}\left(\overline{u_i u_j} - \frac{2}{3}k\delta_{ij}\right)}_{(\Phi_{ij})_1} \\ & \underbrace{-C_2\left(P_{ij} - \frac{2}{3}P_k\delta_{ij}\right)}_{(\Phi_{ij})_2} \\ & -\frac{2}{3}\varepsilon\delta_{ij} \end{aligned} \tag{6.27}$$

where $P_k = 0.5P_{ii}$. For the remaining diffusion terms use the gradient transport model proposed by Daly and Harlow (1970). The pressure–strain term is split into two components such that $(\Phi_{ij})_1$, which represents the contribution of the turbulence field alone, is modeled using the "return-to-isotropy" term of Rotta (1951) and $(\Phi_{ij})_2$, which results from the interaction of the mean flow and the fluctuating field, represents the "rapid" part. This part is modeled according to the proposal of Naot, Shavit, and Wolfshtein (1973). Launder, Reece, and Rodi (1975) proposed a more elaborate model for $(\Phi_{ij})_2$ whose leading term is identical to the one given here.

The turbulent kinetic energy k and its isotropic dissipation rate ε still appear as unknowns in Equation 6.27. Because $k = 0.5\overline{u_i u_i}$, Equation 6.27 already contains its transport equation, which we write as:

$$\underbrace{\frac{\partial}{\partial x_j}(U_j k)}_{C_k} = \underbrace{C_s \frac{\partial}{\partial x_\ell}\left(\frac{k}{\varepsilon}\overline{u_\ell u_m}\frac{\partial k}{\partial x_m}\right)}_{D_k} \underbrace{- \overline{u_\ell u_m}\frac{\partial U_\ell}{\partial x_m}}_{P_k = 0.5 P_{ii}} - \varepsilon \tag{6.28}$$

As mentioned earlier, the transport equation for ε is simply a modeled equation assuming the form of the k equation. Accordingly, we write the ε equation as:

$$\frac{\partial}{\partial x_j}(U_j \varepsilon) = C_\varepsilon \frac{\partial}{\partial x_\ell}\left(\frac{k}{\varepsilon}\overline{u_\ell u_m}\frac{\partial \varepsilon}{\partial x_m}\right) - C_{\varepsilon 1}\frac{\varepsilon}{k}P_k - C_{\varepsilon 2}\frac{\varepsilon^2}{k} \tag{6.29}$$

We now have a closed system of equations for the Reynolds stress transport model (RSTM) where the model constants in Equation 6.27 are typically assigned the following values:

$$C_s = 0.22, C_\varepsilon = 0.15, C_1 = 2.2, \text{ and } C_2 = 0.55$$

6.5.2.6 Algebraic Stress Model (ASM)

In the preceding, we have briefly outlined the RSTM in which we model the Reynolds stresses using six partial differential equations. Even though the mathematical system contains the most desirable flow physics, being closely related to the Navier–Stokes equations, it still lacks the necessary appeal for industrial flow computations. One must solve eleven partial differential equations for a three-dimensional flow: four for the mean flow, six for the Reynolds stresses, and one for obtaining the length scale distribution. Because the turbulent field is always three-dimensional, even for a two-dimensional mean flow, there is no substantial saving in the total number of partial differential equations that need to be solved.

The convection and diffusion terms in Equation 6.26 make it a differential equation. Three main hypotheses have been proposed to replace these terms so that Equation 6.26 assumes an algebraic form. Launder (1971) proposed the following hypothesis:

$$C_{\overline{u_i u_j}} - D_{\overline{u_i u_j}} = 0$$

which implies an assumption of local equilibrium for the Reynolds stresses. Later, Mellor and Yamada (1974) assumed that:

$$C_{\overline{u_i u_j}} - D_{\overline{u_i u_j}} = \frac{2}{3}(C_k - D_k)\delta_{ij}$$

where the transport of Reynolds shear stresses is neglected. For the case of non-equilibrium shear flows, Rodi (1976) postulated that the net transport of stress component $\overline{u_i u_j}$ is proportional to the net transport of the turbulent kinetic energy k using $\overline{u_i u_j}/k$ as the constant of proportionality. Accordingly, we obtain:

$$C_{\overline{u_iu_j}} - D_{\overline{u_iu_j}} = \frac{\overline{u_iu_j}}{k}(C_k - D_k) = \frac{\overline{u_iu_j}}{k}(P_k - \varepsilon)$$

Using this hypothesis in Equation 6.46, we obtain the following system of algebraic equations for Reynolds stresses:

$$\frac{\overline{u_iu_j} - \frac{2}{3}k\,\delta_{ij}}{k} = \left(\frac{1 - C_2}{C_1 - 1 + \lambda}\right)\left(\frac{P_{ij} - \frac{2}{3}k\,\delta_{ij}}{\varepsilon}\right) \tag{6.30}$$

where $\lambda = P_k / \varepsilon$, the ratio of local production of turbulent kinetic energy to its dissipation. While a consensus on the model constants used in BVM prevails (see Table 6.1), no such agreement for C_1 and C_2 currently exists. By way of ASM calibration, one may consider variations around the nominal values of $C_1 = 2.2$ and $C_2 = 0.55$ proposed by Launder (1975). Sultanian (1984) and Sultanian, Neitzel, and Metzger (1986 and 1987) found significantly superior performance of ASM over BVM in sudden expansion pipe flows with and without swirl.

6.5.3 Boundary Conditions

All solutions of a given set of governing equations for the primitive and the turbulence model variables are subject to the specified boundary conditions at inflow, outflow, and wall boundaries. We may need to learn the detailed boundary conditions for a high-fidelity CFD solution in a design environment. In such cases, we must perform sensitivity analyses to understand the variations in the computed CFD results due to uncertainties in critical boundary conditions.

In the following discussion, we will consider the BVM (the high Reynolds number k–ε turbulence model), which is widely used as the baseline turbulence model in most industrial CFD applications.

6.5.3.1 Inlet and Outlet Boundary Conditions

As a matter of CFD best practice in industrial design, inlets and outlets of the CFD calculation domain where the flow field is expected to be parabolic are free from any reverse flow. Sometimes, we modify the calculation domain with artificial extensions to achieve desirable inflow and outflow boundaries.

At inlets, we specify uniform or non-uniform profiles of all dependent variables of the mean flow, either from available measurements or from other related analyses. For the k–ε turbulence model, assuming isotropic turbulence at the inlet, we compute k_{in} using the equation:

$$k_{\text{in}} = 1.5(Tu)^2 U_{\text{in}}^2 \tag{6.31}$$

where Tu is the average inlet turbulence intensity and U_{in} the mean inlet velocity. At each inlet, we specify ε_{in} in one of two ways:

Method 1:

$$\varepsilon_{\text{in}} = C_\mu^{3/4} k_{\text{in}}^{3/2} / \ell_{\text{m}} \tag{6.32}$$

Method 2:

$$\varepsilon_{\text{in}} = \rho C_\mu k_{\text{in}}^2 / \mu_t \tag{6.33}$$

In Equation 6.32, ℓ_m is the mixing length, which is determined from the inlet dimensions. In Equation 6.33, μ_t is the assumed inlet turbulent viscosity, that is, a multiple of the fluid dynamic viscosity μ. In both equations, k_{in} is determined from Equation 6.31.

If we place an outlet far enough downstream, we may assign fully developed flow boundary conditions at this boundary. Otherwise, all dependent variables are to be specified at the outlet. For computing compressible flows, specifying the total pressure at inlet and static pressure at outlet, if the outlet is not choked, works better than specifying mass flow rate through predetermined velocity and density distributions.

6.5.3.2 Wall Boundary Conditions: The Wall-Function Treatment

Although steep gradients characterize the near-wall region in mean flow variables and turbulence quantities, these quantities vanish right at the wall. The conservation equations must be integrated into the wall to incorporate this simple wall boundary condition directly. This requirement poses two main difficulties: first, the BVM and ASM, in the form presented here, need to be validated in the region of low turbulent Reynolds numbers that prevail near a wall. Second, a fine grid is required near the wall so that the assumption of a linear profile for each quantity between grid points is valid for proper numerical integration. Using a wall function overcomes both these difficulties since it directly links the near-wall equilibrium region (characteristic of all turbulent wall boundary layers where local production of turbulent kinetic energy balances its dissipation) with the wall.

When a turbulent boundary layer separates from the wall, either under an adverse pressure gradient or due to a step change in wall geometry, a stalled region of flow recirculation occurs at the wall. In this region, the turbulence energy production near the wall is negligible, and turbulence energy diffusion toward the wall nearly balances its local dissipation. Despite some areas of local non-equilibrium, perhaps for simplicity or in the absence of better information, the wall-function approach, as recommended by Launder and Spalding (1974), is widely used to simulate most industrial flows.

Logarithmic law of the wall. In a boundary layer, we define shear velocity U^* as $U^* = \sqrt{\tau_w / \rho}$ where τ_w is the wall shear stress. In terms of this shear velocity, we further define U^+ and y^+ as follows:

$$U^+ = \frac{U}{U^*}$$

and

$$y^+ = \frac{yU^*}{\nu}$$

Note that y^+ is a Reynolds number based on the shear velocity U^* and the distance y from the wall. Plotted in terms U^+ and y^+, Figure 6.6 shows the overall structure of a flat plate turbulent boundary layer, which is devoid of any influence from stream-wise pressure–gradient and streamline curvature. This boundary layer consists of two regions: inner region and outer region, which interface with the nearly potential outer flow. For the wall-function treatment, the inner region, which is further divided into three zones, is of interest. As shown in the figure, the innermost zone in direct contact with the wall is the viscous sublayer, which in the old literature on fluid mechanics was also called the laminar sublayer. The log-law zone at its outer edge interfaces with the outer region and at its inner edge connects with the viscous sublayer through the buffer zone. The log-law zone is considered somewhat universal in nature with the corresponding logarithmic law of the wall given by:

$$U^+ = 5.5\log_{10} y^+ + 5.45 = 2.388\ln y^+ + 5.45$$

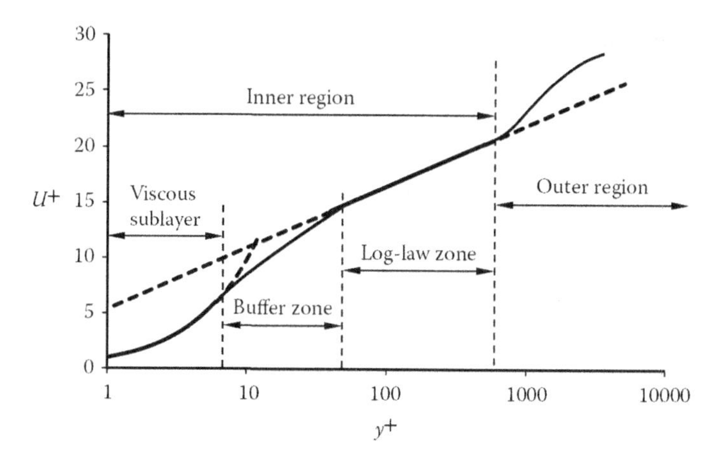

FIGURE 6.6 Velocity variation in the turbulent boundary layer on a flat plate.

$$U^+ = \frac{1}{K_c}\ln(E_c y^+) \tag{6.34}$$

where $K_c = 0.4187$ and $E_c = 9.793$.

Modified log-law for the velocity parallel to the wall. While solving for the velocity component parallel to the wall using a wall-function approach, one essentially applies the shear stress (or momentum flux) boundary condition for the near-wall control volume. Because the wall-function treatment in a turbulent flow CFD is meant for all boundary layers, including those under nonzero stream-wise pressure gradients and with streamline curvature, the logarithmic law of the wall given by Equation 6.34 is modified using the turbulence structure parameter, which in a turbulent boundary layer in local equilibrium is given by (based on measurements):

$$\frac{-\overline{uv}}{k} = C_\mu^{\frac{1}{2}} = 0.3 \tag{6.35}$$

We can express the shear velocity as:

$$U^* = \sqrt{\tau_w / \rho_w} = \sqrt{-\overline{uv}} = C_\mu^{\frac{1}{4}} k^{\frac{1}{2}} \tag{6.36}$$

For the near-wall node shown in Figure 6.7a, we can write:

$$U^+ = \frac{(U_P - U_W)}{U^*} = \frac{1}{K_c}\ln(E_c y^+)$$

$$\frac{(U_P - U_W)U^* \rho}{\tau_w} = \frac{1}{K_c}\ln(E_c y^+)$$

In the aforementioned equation, substituting U^* from Equation 6.36 and solving τ_w yield:

$$\tau_w \frac{(U_P - U_W)K_c \rho C_\mu^{\frac{1}{4}} k_P^{\frac{1}{2}}}{\ln\left(E_c y^+\right)} \tag{6.37}$$

We discuss the method to obtain k_P used in Equation 6.37 in the following.

Specifications of k_P and ε_P. From an equilibrium boundary layer consideration, ε_P is fixed at the near-wall node P shown in Figure 6.7b using the following relation:

$$\varepsilon_P = \frac{C_\mu^{\frac{3}{4}} k_P^{\frac{3}{2}}}{K_c y_P} = \frac{U^{*3}}{K_c y_P} \tag{6.38}$$

To obtain k_P, we solve the k equation assuming $k = 0$ at the wall, approximating the dissipation source term as:

$$\int_0^{y_P} \varepsilon \, dy = \frac{C_\mu k_P^{\frac{3}{2}} \ln(E_c y_P^+)}{K_c} \tag{6.39}$$

Temperature. In the wall-function approach for the energy equation, we calculate the wall heat flux using the equation:

$$\dot{q}_w = \frac{(T_w - T_P)\rho c_p C_\mu^{\frac{1}{4}} k_P^{\frac{1}{2}}}{Pr_t \left(U_P^+ + P_f\right)} \tag{6.40}$$

where

$Pr_t \equiv$ Turbulent Prandtl number ≈ 0.9

$P_f \equiv$ Jaytilleke's P function given by:

$$P_f = 9.24\left\{\left(\frac{Pr}{Pr_t}\right)^{\frac{3}{4}} - 1\right\}\left\{1 + 0.28\exp\left(-0.007\frac{Pr}{Pr_t}\right)\right\}$$

6.5.3.3 Alternative Near-Wall Treatments

The wall-function treatment discussed in the preceding section has its limitations. The main drawback of the approach is that it is only valid for wall boundary layers that satisfy the local equilibrium conditions; that is, the production rate of turbulent kinetic energy is equal to its dissipation rate.

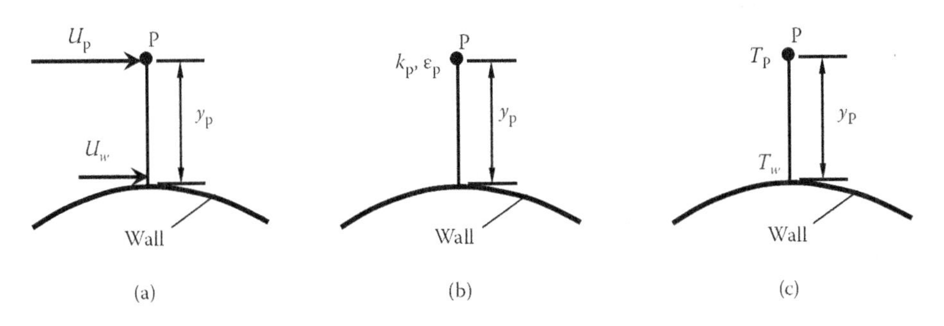

FIGURE 6.7 Wall boundary conditions: (a) velocity parallel to wall, (b) specifications of k_P and ε_P, and (c) temperature.

Complex geometries associated with industrial flow and heat transfer situations do not generally meet the restrictions of a wall-function treatment, whose continued use is primarily driven by a reduced model size as the first near-wall is placed in the y^+ range from 40 to 400 (see the log-law zone in Figure 6.6). Any attempt to resolve the near-wall region with lower y^+ values for the node next to the wall will be counterproductive.

To overcome the limitations of wall functions, alternative near-wall treatments have been developed and used, such as two- and three-layer models and low Reynolds number models. In the latter, we integrate the governing equations for mean flow and turbulence model variables up to the wall with the first layer of near-wall nodes placed at a y^+ of around unity.

Luo, Razinsky, and Moon (2012) compared the performance of several commonly used turbulence models using both wall functions and low Reynolds modeling at the wall. It is important to note that, in addition to packing many more grid points near a wall to resolve the boundary layer correctly, the turbulence models are modified to model the low Reynolds number region and the near-wall non-isotropic turbulence field. Therefore, the standard BVM discussed earlier needs to be revised to model the low Reynolds number wall region on a wall integration grid. For a turbulent heat transfer prediction using a low Reynolds number turbulence model with wall integration, the challenge of modeling a variable turbulent Prandtl number in the boundary layer remains.

6.5.4 Selection of a Turbulence Model

The accuracy of a turbulent flow CFD for a given set of boundary conditions depends mainly upon three factors: grid quality, accuracy of the numerical scheme, and the turbulence model used. While most CFD analysts in an industry rightly focus on generating a high-quality grid in their computation domain and spend a good deal of time obtaining a fully converged solution with a high-order numerical scheme, their choice of turbulence model is limited to the options available in the commercial CFD code used and the design cycle time available to conclude the analysis. If the CFD predictions fail to validate against the available test data, they repeat calculations with another turbulence model available in the CFD code. They can save a lot of time and computing resources if they short-list the available turbulence models based on their design's flow and heat transfer physics. Here, we present a few examples illustrating the proper choice of a turbulence model, depending on the flow and heat transfer physics we are trying to predict.

6.5.5 Physics-Based Postprocessing of CFD Results

CFD uses many small control volumes and generates a detailed description of velocities, pressure, temperature, and other flow and heat transfer properties in each control volume. The last step, and perhaps the most important and nontrivial step, in a 3D CFD analysis is to postprocess the computed results for design applications. This postprocessing is carried out with two initial objectives in mind. First, we use the results to get a clear qualitative understanding of the key features of the computed flow field by generating, for example, a plot of streamlines. Second, we obtain various integral quantities such as section-average values of static pressure, total pressure, static temperature, total temperature, shear force on the bounding walls, and the drag force on an internal design feature. For those who want to use the ultimate power of CFD analysis in a design, generating an entropy map from the computed results will clearly identify local areas of high entropy production. Improving these areas in design will certainly improve the component aerodynamic performance efficiency, which is difficult to achieve using other means.

6.5.5.1 Large Control Volume Analysis of CFD Results

Figure 6.8 shows large parallelepiped control volume drawn around a solid object around which a detailed 3D CFD analysis has been performed. The simple geometry of the solid object shown in

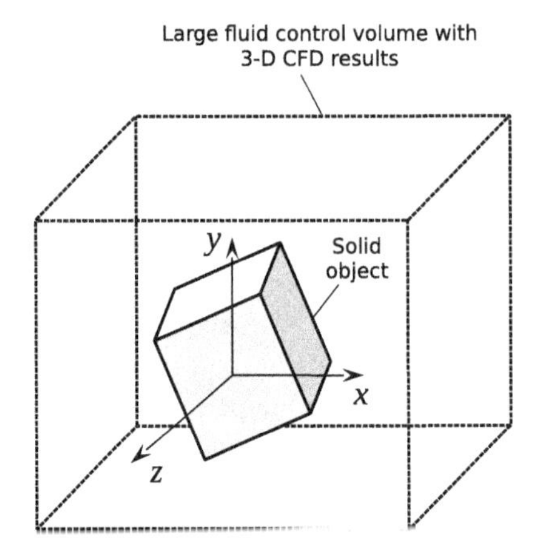

FIGURE 6.8 A large control volume representation of 3D CFD results.

the figure is just for illustration. In an actual design, the object shape may be more complex. Let us assume that the main design objective of the CFD analysis is to evaluate the drag force acting on the solid object in the x direction. A direct method to determine the drag force will be to integrate the x component of the forces from both pressure and shear stress distributions on the surface of the solid object. We can also use an indirect method of first postprocessing the CFD results to determine the x-momentum flux and surface force on each of the six faces of the large control volume, and then perform the x-momentum control volume analysis to find the drag force on the solid object. Except for some small numerical error introduced in the postprocessing of the CFD results, both methods should essentially yield the same result for the drag force. The second method, however, can also yield section-average values of the 3D distribution of various quantities available from the CFD results. These section-average values are extremely useful in design applications to determine the overall performance of a flow component.

As an illustrative example of physics-based postprocessing of 3D CFD results, let us consider a gas turbine exhaust diffuser for which we have completed the CFD analysis, discussed in Section 6.5.8. The key aerodynamic performance parameter for a diffuser is its pressure recovery coefficient C_p, which is defined by the equation:

$$C_P = \frac{\bar{p}_{\text{out}} - \bar{p}_{\text{in}}}{\bar{p}_{0\text{in}} - \bar{p}_{\text{in}}} \tag{6.41}$$

where

$\bar{p}_{\text{in}} \equiv$ Section-average value of static pressure at diffuser inlet

$\bar{p}_{\text{out}} \equiv$ Section-average value of static pressure at diffuser outlet

$\bar{p}_{0\text{in}} \equiv$ Section-average value of total pressure at diffuser inlet

Although we know the detailed distributions of all computed mean flow quantities such as velocities, static pressure, and static temperature from the CFD results at each section, it is not obvious whether we should perform an area-weighted averaging or mass-weighted averaging to compute their section-average values. For a physics-based postprocessing of the CFD results to compute a section-average value, we must satisfy the following two criteria:

1. Averaging must preserve the total flow of mass, momentum, angular momentum, energy, and entropy through the section, including the total surface force and torque at the section used in the CFD solution.
2. The computed section-average quantity obtained either by area-weighted averaging or by mass-weighted averaging of the CFD results must be meaningful in an integral control volume analysis in which the section under consideration is a control surface.

Based on the second criterion, we see that the area-weighted averaging of the static pressure at a section will be a meaningful quantity, representing the total pressure force acting on the surface in a related control volume analysis. The section-average value of the total pressure obtained using either the area-weighted averaging or mass-weighted averaging of the CFD results will not be meaningful. To compute the section-average total pressure at a section, the following step-by-step physics-based method of postprocessing of CFD results is recommended. Without any loss of generality, let us assume that the section normal is along the x direction (the main flow direction). Let us also assume that the thermophysical properties are constant over the section.

Step 1: Calculate the cross-sectional area and the mass flow rate through the section:

$$A = \iint_A \mathrm{d}A \text{ and } \dot{m} = \iint_A \rho V_x \,\mathrm{d}A$$

Step 2: Calculate the section-average static pressure by area-weighted averaging of the static pressure distribution from the CFD results:

$$\overline{p} = \frac{1}{A}\iint_A p\,\mathrm{d}A \tag{6.42}$$

Step 3: Calculate the section-average static temperature by mass-weighted averaging of the static temperature distribution from the CFD results:

$$\overline{T} = \frac{1}{\dot{m}}\iint_A T\rho V_x \,\mathrm{d}A \tag{6.43}$$

Step 4: Calculate the section-average specific kinetic energy by mass-weighted averaging of the kinetic energy flow through the section:

$$\frac{\overline{V}^2}{2} = \frac{1}{\dot{m}}\iint_A \left(\frac{V^2}{2}\right)\rho V_x \,\mathrm{d}A \tag{6.44}$$

Step 5: Calculate the section-average total temperature as:

$$\overline{T}_0 = \overline{T} + \frac{\overline{V}^2}{2c_p} \tag{6.45}$$

where $\overline{T}$ and $\overline{V}^2$ are from Equations 6.43 and 6.44, respectively.

Step 6: Calculate the section-average total pressure as:

$$\overline{p}_0 = \overline{p}\left(\frac{\overline{T}_0}{\overline{T}}\right)^{\frac{\gamma}{\gamma-1}} \tag{6.46}$$

where $\overline{p}$, $\overline{T}$, and $\overline{T}_0$ are from Equations 6.42, 6.43, and 6.45, respectively.

Using this postprocessing method at diffuser inlet and outlet, we can easily compute the pressure recovery coefficient C_P from Equation 6.41. This method may similarly be used in other design applications involving 3D CFD analysis.

6.5.6 Entropy Map Generation for Identifying Lossy Regions

The preceding section presents the postprocessing of 3D CFD results to obtain section-average values to determine the overall equipment performance and efficiency. The knowledge of local areas of significant irreversibility present in the design, as obtained from the detailed CFD results, could help us redesign these areas, leading to an improved overall design.

The conventional approach to assessing the performance of a flow system is to find the loss of total pressure in the system. This approach serves well when dealing with incompressible flows because the total pressure in such flows represents mechanical energy per unit volume, and its loss appears as an increase in fluid internal energy, loosely called thermal energy, which we generally do not track in these flows. For a compressible flow system, the loss in total pressure is used to denote performance loss under adiabatic conditions with constant total temperature.

Entropy, being a scalar quantity and grounded in the second law of thermodynamics, is more beneficial to track irreversibility in a flow system. In terms of changes in total pressure and total temperature between two points in a flow system, we can compute entropy change by the equation:

$$s_2 - s_1 = c_p \ln\left(\frac{T_{02}}{T_{01}}\right) - R\left(\frac{p_{02}}{p_{01}}\right) \tag{6.47}$$

For an adiabatic flow with $T_{02} = T_{01}$, Equation 6.66 yields:

$$\frac{p_{02}}{p_{01}} - 1 = \frac{\Delta p_0}{p_{01}} = \mathrm{e}^{-\left(\frac{\Delta s}{R}\right)} - 1 \tag{6.48}$$

which relates the increase in entropy to the loss in total pressure between any two points. Using static pressure and static temperature, the quantities directly available in the CFD results, we can easily compute the entropy at a point using the equation:

$$s^* = \frac{s}{R} = \frac{c_p}{R} \ln\left(\frac{T}{T_{\mathrm{ref}}}\right) - \left(\frac{p}{p_{\mathrm{ref}}}\right) \tag{6.49}$$

where we have arbitrarily assumed zero entropy at the reference pressure p_{ref} and reference temperature T_{ref}.

We use the following steps to generate an entropy map in a calculation domain to identify lossy regions:

Step 1: Obtain 3D CFD results in the calculation domain earmarked for design improvement.

Step 2: Postprocess the 3D CFD results to compute s^* with Equation 6.68 throughout the calculation domain.

Step 3: Calculate the section-average entropy $\overline{s}_{\text{in}}^*$ at the inlet by mass-weighted averaging of the inlet entropy distribution:

$$\overline{s}_{\text{in}}^* = \frac{1}{\dot{m}} \iint_A T\rho V_x \, \mathrm{d}A \tag{6.50}$$

Step 4: Calculate $\overline{s}^* - \overline{s}_{\text{in}}^*$ and plot their contours in the calculation domain.

Step 5: The regions of high entropy production are the regions to be reduced or eliminated for the next design iteration.

Thus, the entropy map provides an invaluable insight into the design space for local improvements in the regions of excessive entropy production. Naterer and Camberos (2008) and Sciubba (1997) provided further details in this promising area of CFD application to design optimization.

6.6 CFD APPLICATIONS

6.6.1 CFD Technology Used in Various Industries

CFD technology has many advantages and remains highly promising. Some of these advantages are:

- Physics-based (satisfies conservation equations of mass, momentum, and energy)
- A powerful tool for 3D flow and heat transfer analysis in many design applications of various industries (see Table 6.2)
- Provides low-cost numerical flow visualization.
- Provides microanalysis with detailed results of all flow properties.
- Quick turn-around for reduced design cycle time.
- Synergy with experimental testing and instrumentation.
- Synergy with flow network modeling (macro-analysis).
- Conjugate heat transfer (solution of conduction in solid and convection in fluid in one domain).

Table 6.2 summarizes typical applications of CFD technology in a few energy conversion and process industries, partly adapted from Engelman (1993). Other areas of CFD applications include, but are not limited to, HVAC design, Formula 1, and yacht design (America's Cup), where this technology helps make rapid, continuous improvements.

6.6.2 Secondary Flows

Secondary flows are flows normal to the main flow direction. In a pipe flow, for example, the streamlines in the pipe cross section (normal to the main flow direction) indicate secondary flows. Although the velocities in secondary flows are generally an order of magnitude smaller than those along the main flow direction, they play a significant role in augmenting the flow's heat, mass, and momentum transport properties. As discussed below, there are two kinds of secondary flows: the secondary flow of the first kind and the secondary flow of the second kind.

TABLE 6.2
Industrial applications of CFD technology

Industry	Design Application
Turbomachinery Industry for Aircraft Propulsion and Power Generation	• Combustor modeling • Inlet and exhaust system • Nozzles and diffusers • 3D aerodynamic design of blades and vanes • Rotor–rotor and rotor–stator cavities • Hot gas ingestion • Seals • Internal and film cooling of gas turbine airfoils • ATEX certification of gas turbine enclosures • Generator cooling
Automotive Industry	• Vehicle aerodynamics for drag reduction • Cooling of engine block • Climate control • Flow through valves and filters
Chemical/Process Industry	• Mixing and reacting flows • Plastic flows • Glass flows • Die and extrusion flows • Filling of molds
Electronics Industry	• Component cooling • Air flow in disk drives • Controlled thermal environment in electronic packaging • Conjugate heat transfer
Thin Film/Coating Industry	• Coating of photographic film • Coating of magnetic tape • Conjugate heat transfer
Biomedical/Pharmaceutical Industry	• Blood flow in veins and arteries • Heart flow system • Injection delivery system • Centrifugal processing
Nuclear Industry	• Reactor cooling • Flow through piping systems • Reactor flows • Cooling towers and thermal plumes
Food/Beverage Industry	• Convection ovens • Pasteurization processes • Extrusion of dough-like fluids • Velocity, temperature, and concentration distribution in process equipment

6.6.2.1 Secondary Flow of the First Kind

The generation of the secondary flow of the first kind occurs when the primary flow follows a curved path with a finite radius of curvature, such as in a bent pipe. As illustrated in Figure 6.9, the flow enters a circular pipe with uniform velocity and pressure at its lower end and then exits a 180-degree bend at its upper end. As the primary flow traverses the bend, centrifugal force propels the fluid from the pipe's inner wall (lower radius of curvature) toward the outer wall (higher radius of curvature) in the central region. Subsequently, the fluid returns to its original position along the

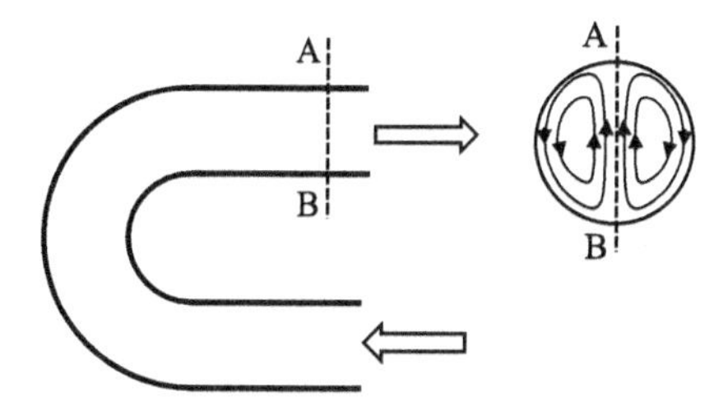

FIGURE 6.9 Secondary flow of the first kind.

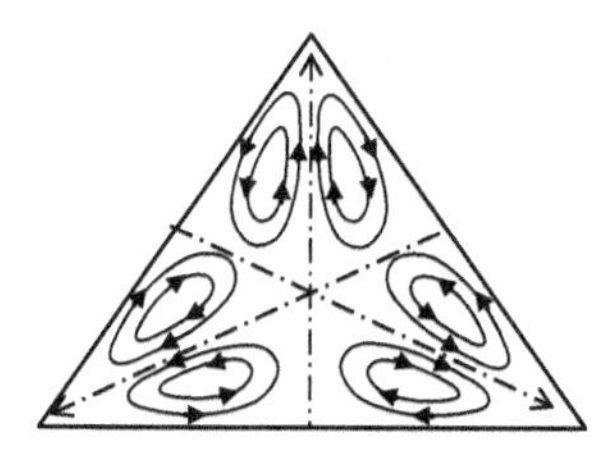

FIGURE 6.10 Secondary flow of the second kind.

side wall due to the pressure gradient induced by the centrifugal force in the pipe cross section. This mechanism gives rise to counter-rotating vortices in the pipe cross section at AB.

6.6.2.2 Secondary Flow of the Second Kind

A turbulent flow in straight noncircular duct features the second kind of secondary flow, which is not found in constant-viscosity laminar flow in circular or noncircular ducts or in turbulent flow in a circular duct. These secondary flows are driven primarily by the gradient in turbulence pressure along the duct wall. For example, the secondary flows shown in Figure 6.10 in a duct of triangular cross-sectional result from the variation of turbulence normal intensity around the triangle. It may be conjectured that these secondary flows cannot be predicted by using an isotropic turbulence model such as the standard high Reynolds number $k-\varepsilon$ turbulence model.

6.6.3 Turbulent Flow in a Noncircular Duct

As discussed in the preceding section, the turbulent flow in a noncircular duct features secondary flows of the second kind, which are nonexistent in a laminar flow in the duct. While the secondary flows are weak in magnitude, they significantly affect the wall shear stress and heat transfer. Because the gradients in the normal turbulent stresses at the wall drive these secondary flows, an isotropic turbulence model will fail to predict them, regardless of the quality of the CFD grid or the accuracy of the numerical scheme used. For example, the widely used two-equation, high Reynolds number k–ε turbulence model will fail to predict these secondary flows in a noncircular duct. We must, therefore, choose a turbulence model, such as the Reynolds stress model or its variants, which can model anisotropic turbulence.

6.6.4 A Sudden Expansion Pipe Flow without Swirl

The gross features of separation and redevelopment found in an axisymmetric sudden expansion flow are depicted in Figure 6.11. The separation consists of detachment (the flow leaving the inlet pipe wall), recirculation, growth of the free shear layer, and reattachment to the larger pipe wall. The

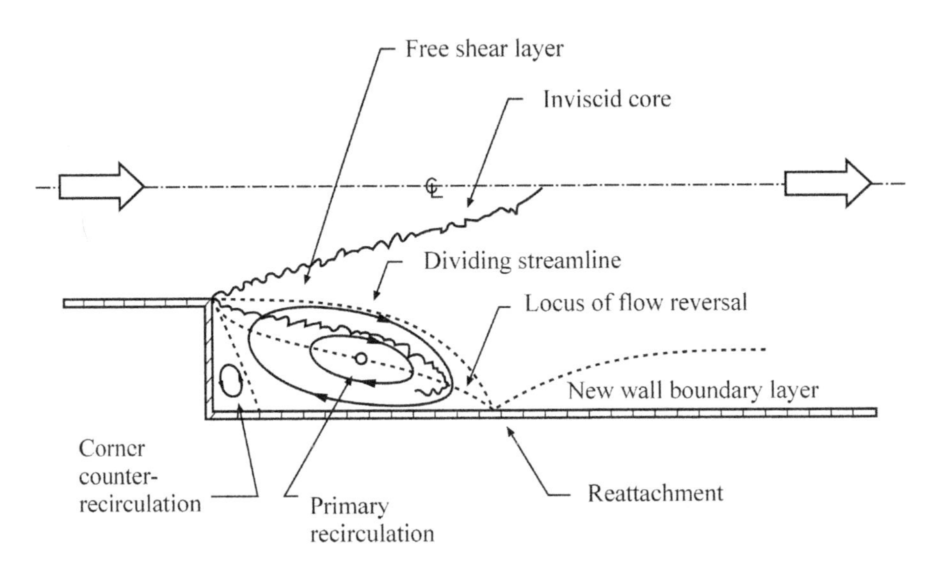

FIGURE 6.11 A sudden expansion pipe flow without swirl.

redevelopment region has a pseudo-boundary-layer character where the flow eventually becomes fully developed if the pipe is long enough. The early part of the redevelopment region in this flow is found to be quite different from the usual entrance region in a developing pipe flow.

Following the sudden expansion, the flow enters the larger pipe in the form of a circular jet. The detachment of flow from the wall converts what had been the wall boundary layer into a complex free shear layer that emanates from the edge of the expansion and envelops the core flow in the jet. This shear layer grows both inward toward the pipe axis and outward toward the pipe wall. While the inward growth of the shear layer occurs at the expense of the potential (inviscid) core flow, which decays downstream until the shear layers merge at the pipe axis, its outward growth happens through the process of entrainment until it finally reattaches to the pipe wall. The fluid entrained into this shear layer is continuously replenished from downstream through a favorable reverse pressure gradient. This gives rise to a primary recirculation region such that the flow in its forward branch is driven by the central jet momentum and, in its reverse branch, by an adverse pressure gradient.

Engineering consideration of a sudden expansion flow geometry dates back to the original analysis of Borda (1766). The existence of a large wall-bounded recirculation region in this flow makes it predominantly elliptic in character. Drewry (1978) reported an improved flow visualization study of the recirculation region using a surface oil-film technique. His results show that the reattachment length ranges between 7.9 and 9.2 step (difference between the radii of two pipes) heights for Reynolds numbers (based on the smaller pipe diameter) in the range of 1.3×10^{6}–2.2×10^{6}. Sultanian, Neitzel, and Metzger (1986) presented accurate CFD predictions of this flow field using an advanced algebraic stress model (ASM) of turbulence and demonstrated that the standard high Reynolds number $\kappa-\varepsilon$ turbulence model significantly under-predicts the experimentally found reattachment lengths.

6.6.5 A Sudden Expansion Pipe Flow with Swirl

Figure 6.12 shows the main features of a swirling flow in a sudden pipe expansion. The wall-bounded primary recirculation zone in this flow field compared to the case without swirl, shown in Figure 6.11 and described above, shrinks with increasing swirl. When the swirl exceeds a critical value (greater than 45-degree swirl angle), an on-axis recirculation region appears. Sultanian (1984) expounded

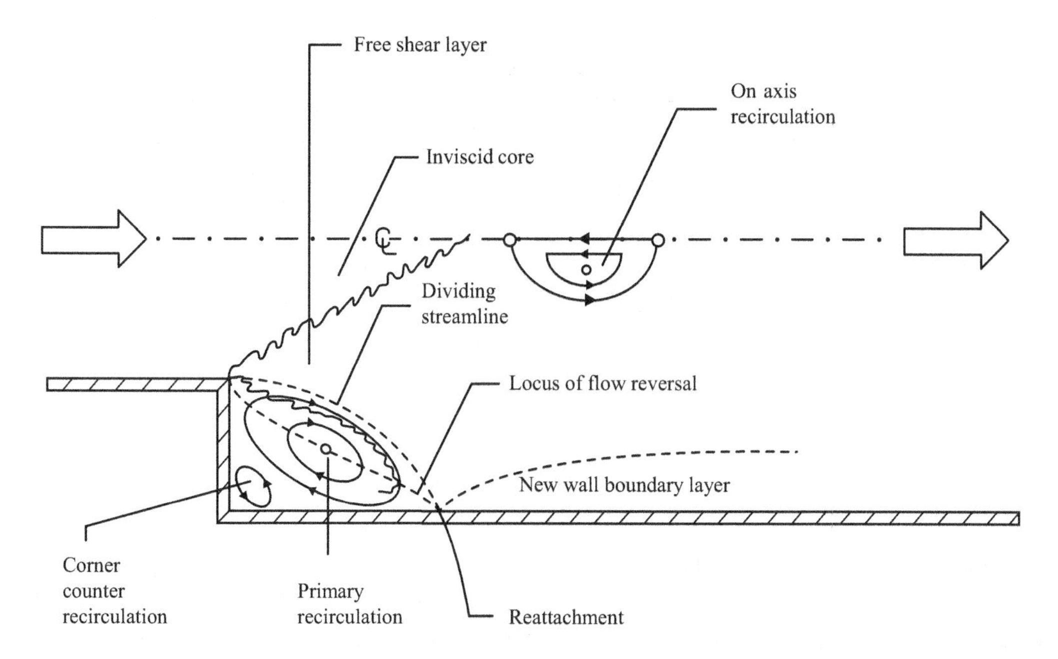

FIGURE 6.12 A sudden expansion pipe flow with swirl.

on the overwhelming complexity of this flow field from a computational viewpoint. The major difficulties are associated with predicting both the mean flow, which is dominated by all three velocity components, and the turbulence field, which is highly anisotropic.

The most distinguishing feature of a swirling flow in a sudden pipe expansion shown in Figure 6.12 is the appearance of an on-axis recirculation zone caused by vortex breakdown. Such a feature is widely used in modern gas turbine combustors as an aerodynamic flame holder, which in early designs corresponded to the wake region of a bluff body, to ignite the fresh charge entering the combustor. Here is a simple explanation behind the formation of the central recirculation zone in a strong swirling flow in a sudden pipe expansion. A swirl in a free or forced vortex causes radial pressure gradient; pressure always increases radially outward—the higher the swirl, the higher this pressure gradient becomes. Near the sudden expansion, due to high swirl, the pressure near the downstream pipe axis is much smaller than the average pressure. Further downstream, swirl decays due to wall shear stress and the pressure distribution, tends to become more uniform at the pipe cross section. As a result, the static pressure on the pipe axis is higher downstream than upstream, causing on-axis flow reversal.

6.6.6 Turbulent Flow and Heat Transfer in a Rotor Cavity

Figure 6.13 shows streamlines of a complex shear flow in a cavity between two rotating disks. The axial flow entering the cavity through the upstream disk undergoes a sudden geometric expansion. The growth of the outer shear layer of the annular jet occurs through entrainment of the pressure–gradient-driven backflow from the downstream stagnation region. This creates the primary recirculation region shown in the figure. The size and strength of this recirculation region are found to depend mainly on the flow rate and rotational speed as discussed in Sultanian and Nealy (1987). The entering axial flow turns 90 degrees over the concave corner and flows radially outward, aided in part by frictional pumping over the downstream disk induced by its rotation. A part of the flow (almost half in this case!) turns back toward the upstream disk and moves radially outward as a result of similar pumping action over that disk. The development of boundary layers on the two

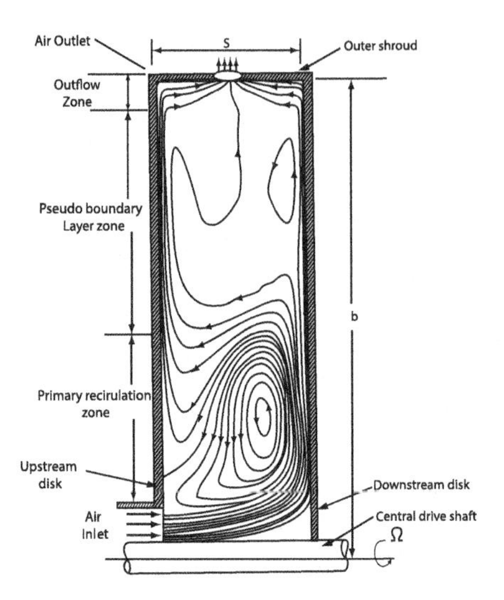

FIGURE 6.13 Rotor cavity flow field.

disks shares many of the features of a regular Ekman layer (see Greenspan (1968) and Lugt (1995)) where the inertia force is negligible, and the viscous and Coriolis forces nearly balance each other in a reference frame rotating with the cavity. Unlike regular wall boundary layers, the Ekman boundary layers are nearly non-entraining, maintaining a constant mass flow rate within them.

When it comes to modeling heat transfer in a complex shear flow, the near-wall modeling of turbulence becomes most important and challenging. The flow and heat transfer near a wall is dominated by low Reynolds numbers and variable turbulent Prandtl numbers for which we lack sufficient empirical data and correlations. The wall-function treatment presented in the foregoing section is valid only for a boundary layer in local equilibrium where the rate of turbulence production equals its rate of dissipation. In addition, the turbulent Prandtl number is assumed constant. In any other boundary layer, the application of wall functions to link the wall conditions to the near-wall nodes is not accurate. To circumvent the shortcomings of the wall-function treatment, alternate methods of near-wall turbulence modeling such as two- and three-layer modeling have been proposed. Another approach is to use a low Reynolds number turbulence model to integrate the governing mean flow and turbulence model equations right up to the wall. Sultanian and Nealy (1987) used such an approach to predict heat transfer in a rotor cavity where the rotor walls are dominated by Ekman boundary layers (see Lugt (1995)).

6.6.7 Cooling Flow in a Compressor Rotor Drum Cavity

Figure 6.14 schematically shows a rotor drum cavity of an axial flow compressor, where some cooling flow is bled from the primary flow path. This cooling flow, which is intended for cooling turbine parts operating at higher temperatures, has a mass flow rate of $\dot{m}$, which flows radially inward and passes through the bore of the right rotor disk. The quantities of design interest are the drop in static pressure from point A to point B and the swirl strength of the cooling flow at point B. The higher the swirl velocity at B, the higher will be the strength of the "vortex whistle" that should be eliminated in a good design. The static pressure at point B should be large enough to prevent any backflow from the downstream flow path at a very high temperature.

In the absence of a microanalysis or a flow visualization experiment, one would expect the cooling flow in the rotor cavity of Figure 6.14 to flow from point A to point B via a complex flow shear flow dominated by a free vortex. A CFD analysis of the flow, however, shows an unexpectedly interesting and simple behavior. The coolant essentially flows along two paths hugging the rotor walls,

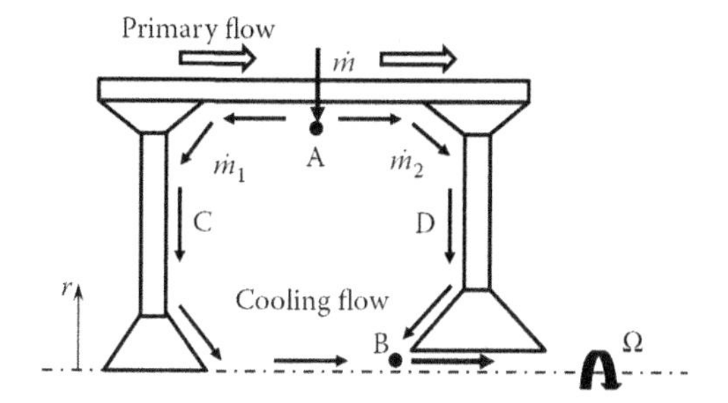

FIGURE 6.14 Radially inward cooling flow through a compressor rotor drum cavity.

one along ACB and another along ADB. The total mass flow rate $\dot{m}$ is split at point A into $\dot{m}_1$ and $\dot{m}_2$ flowing along these paths and becoming wholesome again at point B. The CFD analysis features little flow activity in the core of the rotor drum cavity. This understanding of the flow field obtained from CFD can be used to alternatively model the flow as two one-dimensional control volume analyses (along ACB and ADB) with the common static pressure conditions at points A and B. In fact, such a constraint can be used to establish the flow split along two paths, obtaining mixed-mean swirl and total temperature at point B. This design example demonstrates a synergistic application of CFD and flow network modeling (FNM) methods for higher speed and accuracy of results.

6.6.8 Aerodynamic Design of a Gas Turbine Exhaust Diffuser

A gas turbine exhaust diffuser, schematically depicted in Figure 6.15, plays an important design role in improving the turbine efficiency by setting up sub-ambient static pressure conditions at the last stage turbine exit. The ducting downstream of the diffuser exhausts the hot gases to ambient under a simple cycle operation or to a heat recovery steam generator (HRSG) system in a combined-cycle operation. The CFD technology in concert with the flow network modeling method (FNM) offers a powerful tool for developing a gas turbine exhaust diffuser of high aerodynamic performance. The key to improved diffuser performance is to eliminate all regions of flow separation and recirculation that cause excessive entropy production and loss in total pressure.

Sultanian, Nagao, and Sakamoto (1999) presented the results of both experimental and three-dimensional CFD investigation in a scale model of an industrial gas turbine exhaust system (GE-MS9001E type) to better understand its complex flow field and to validate CFD prediction capabilities. The CFD predictions were based on the standard high Reynolds number $k-\varepsilon$ turbulence model with the wall-function treatment at the diffuser walls. The investigation concludes that, overall, the applied CFD method offers a useful design engineering tool capable of predicting

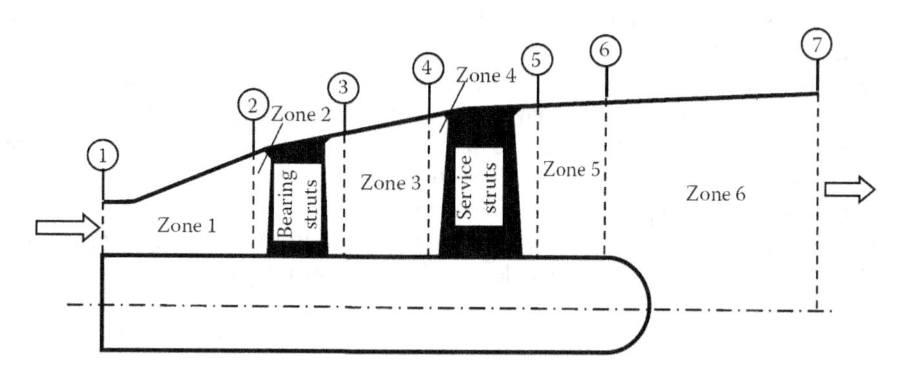

FIGURE 6.15 A typical gas turbine exhaust diffuser.

complex gas turbine exhaust system flows, including the quantitative prediction of the total pressure loss and static pressure recovery.

For purposes of analysis, we divide the exhaust diffuser into multiple zones, as shown in Figure 6.15. The overall aerodynamic design objective in each zone is to reduce the loss in total pressure. Additionally, one should aim for the highest recovery in static pressure in zone 1, which also helps to reduce loss in total pressure in each downstream zone due to reduction in available dynamic pressure. The design objective for zone 6 shifts to attaining nearly uniform flow velocities at the diffuser exit. Three-dimensional CFD analysis in each zone plays important design roles: first, to understand the flow behavior, and second, to reinforce one-dimensional modeling and performance prediction for the diffuser to validate with the data obtained under full-scale testing or sub-scale model testing. Additionally, knowing the regions of flow separation at the wall provides a quick way to reshape the diffuser wall for improved aerodynamic performance and pressure recovery. Sultanian (2019) presented additional details on gas turbine exhaust diffusers.

6.6.9 High-Pressure Inlet Bleed Heat (IBH) System Design

Figure 5.65 schematically shows a high-pressure inlet bleed heat system (IBH) of a land-based gas turbine used for power generation. When the ambient air is cold, IBH system is used to raise its temperature to prevent ice formation in the compressor inlet guide vanes. In this system, a small amount of hot air from an intermediate compressor stage is bled and mixed uniformly with the ambient air flow at the engine inlet. The key design objective in this case is to achieve complete mixing of the hot compressor bleed air and the incoming cold ambient air within the short mixing section. A hot streak resulting from a non-uniform mixing in the inlet flow would adversely affect the compressor performance.

Uniform mixing and slightly elevated temperature in the inlet system of a gas turbine are typically achieved by using a number of bleed air injection nozzles. These nozzles are distributed over the large inlet area and operate under choked-flow conditions. In this design, one can leverage 3D CFD technology to achieve optimum aerodynamic mixing with minimum use of the bleed air injection hardware and significantly reduce equipment cost.

6.6.10 Gas Turbine Enclosure Ventilation Design: ATEX Certification

Modern gas turbine enclosures must be adequately ventilated to achieve the following two key objectives: (1) to maintain acceptable enclosure air temperature distribution (for the safety of internal devices) and effective dilution to ensure the absence of dangerous gas pockets within the enclosure, and (2) to maintain acceptable variations in compressor and turbine casing temperatures to prevent casing distortion and maintain designed blade tip clearances for required operational efficiencies. Full compliance with ATEX (atmosphere explosive) requirements is presented in Santon, Kidger, and Lea (2002). It is interesting to note that a good ventilation design should not be based on a common misconception that "more is better." An effective distribution of ventilation, which restricts any flammable gas concentration to the mixing zone and its immediate vicinity, is more important than its quantity. High ventilation rates will tend to mask the detection of small leaks, which may result in potentially explosive larger flammable volumes. An optimum ventilation design, therefore, calls for good flow distribution and low gas detection settings.

CFD technology as an acceptable three-dimensional analysis method has proven to be very successful in acquiring ATEX certification for a gas turbine enclosure. Gilham, Cowan, and Kaufman (1999) and Ponnuraj et al. (2003) presented details of CFD investigations in a gas turbine enclosure ventilation system.

6.6.11 Turbomachinery Flowpath Design

6.6.11.1 3D Flow Field

All turbomachines feature complex turbulent shear flows, which are inherently unsteady in the mean with unsteady wake interactions due to the proximity of blade rows moving relative to each other. Additional complexities arise from buildup of boundary layers on the casing and hub, leakage flows around blade tip and stator shroud, and the low-momentum secondary flows mixing with the high-momentum primary flow. In early years, during the pre-CFD era, Wu (1952) proposed to analyze the complex 3D turbomachinery flow by analyzing it on 2D intersecting stream surfaces called S_1 (blade-to-blade) and S_2 (hub-to-tip). Even with Euler's momentum equations, which neglect the effects of viscosity, obtaining the iteratively converged S_1/S_2 solutions was not trivial, and therefore, the approach could not become a part of the prevailing turbomachinery design process. Currently, S_1/S_2 solutions are largely handled by fully 3D CFD solutions with mixing planes. Some of the basic ideas of Wu's proposal, however, remain integral to the state-of-the-art turbomachinery aerodynamic design, particularly in the detailed design phase, which involves 2D axisymmetric throughflow analysis in the meridional (axial–radial) plane and 2D blade-to-blade analyses.

Figure 6.16 shows a highly iterative process for the aerodynamic design of turbomachines. Overall, the design is carried out in two phases: preliminary design and detailed design. We will briefly discuss each design phase here.

6.6.11.2 Preliminary Design

The preliminary design's goal, essentially a meanline design, is to produce a safe, reliable, efficient, and economically competitive turbomachinery system. The meanline design, depicted in Figure 6.17a, involves solutions of one-dimensional forms of the conservation equations of mass, momentum, and energy at locations halfway between hub and casing, incorporating empirical correlations and prior product experience to account for various aerodynamic losses. This design phase ensures that the specified requirements of the engine cycle in terms of overall pressure ratio, mass flow rate, target efficiency, etc., are fully met by the final design. The key outputs of the preliminary design include:

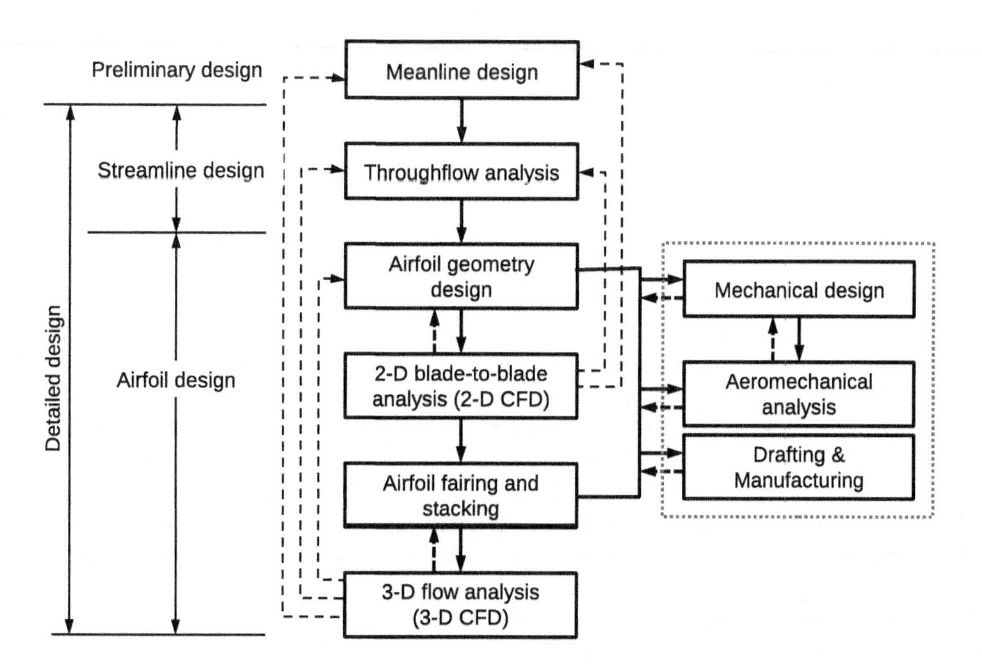

FIGURE 6.16 Turbomachinery aerodynamic design process.

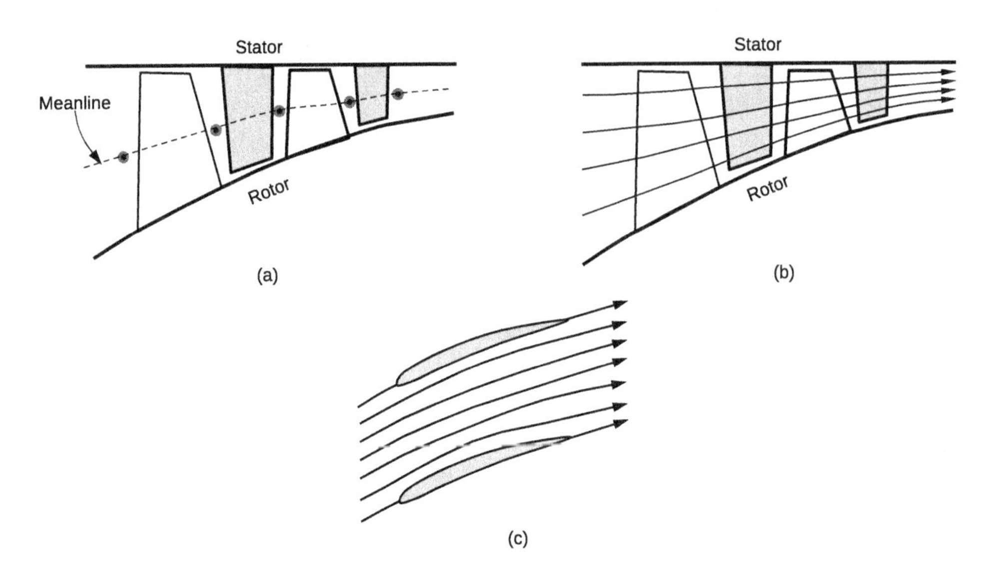

FIGURE 6.17 (a) Meanline design, (b) throughflow analysis, and (c) blade-to-blade analysis.

- Flowpath shape.
- Number of stages and airfoils in each stage.
- Meanline vector diagrams.
- Meanline 2D geometry of airfoils.

The initial configuration produced in a preliminary design is refined in a detailed design, which is discussed next.

6.6.11.3 Detailed Design

As shown in Figure 6.16, with inputs from the preliminary design, the detailed design mainly consists of the throughflow design, which embodies a 2D axisymmetric throughflow analysis along streamlines in the meridional (axial–radial) plane depicted in Figure 6.17b, and the airfoil design. Some design features evaluated in the detailed design phase include blade lean, bow, scallop, sweep, fillet radius, midspan damper position, impeller splitter, blade position, and tip clearance. In addition to meeting the required aerodynamic efficiency, turbomachines are designed to have no significant aerodynamically excited vibrations, have good mechanical integrity and acceptable life, and must be manufactured within acceptable cost and turn-around time. However, these design aspects shown within the dotted box in Figure 6.16 are handled by other specialists.

6.6.11.4 Throughflow Design

The throughflow design uses 2D axisymmetric analysis in the radial–axial (meridional) plane (S2 mean) and provides circumferentially averaged values of pressure, temperature, density, velocity vector diagrams, and loading parameters. As shown in Figure 6.16, throughflow design is iterated with multiple 2D blade-to-blade analyses at spanwise locations to capture the 3D variations and factor in the smeared airfoil solidity (blockage) and blade forces in the throughflow analyses. Denton and Dawes (1999) pointed out that, even though 3D calculations are used in the design procedure, the throughflow analysis remains an essential tool in designing turbomachines. Cumptsy (1989) and Lakshminarayana (1996) provided comprehensive details on the throughflow design.

The most widely used method for throughflow design is the streamline curvature method (SCM), which is based on the radial equilibrium equation and a space-marching procedure coupled to boundary layer calculations along the annulus end walls. Smith (1966) derived the most general form of the radial equilibrium equation for turbomachinery design applications, essentially the linear momentum equation in the radial direction. Novak (1967) and Marsh (1968) presented the complete procedure and the related computer codes for performing a throughflow analysis. Spur (1980) presented a method that combines throughflow and blade-to-blade analyses to predict 3D transonic flow in a turbomachine.

Streamline curvature (SCM) method. Although the actual flow field in the turbomachine is unsteady with rotating airfoils moving past stationary ones, SCM assumes the flow is steady, adiabatic, compressible, and axisymmetric. This method solves the discrete equations of continuity, momentum, and energy along with the equation of state on the computational grid constructed in the meridional plane. Computation nodes are defined at the intersections of streamlines and quasi-orthogonal computation stations located at blade edges and inside blade passages. The throughflow analysis is carried out in concentric annuli (stream tubes) bounded by meridional streamlines, thus allowing no spanwise mixing. As the physical presence of the airfoils will change the slopes and curvatures of the streamlines in the axisymmetric throughflow analysis, this analysis is re-run but now with airfoil geometry present. The process shown in Figure 6.16 is then repeated until the designer is satisfied with everything. Given multiple iterations needed to obtain a converged design solution, the throughflow analysis in an actual turbomachinery design uses an in-house computer code.

6.6.11.5 Airfoil Design

As shown in Figure 6.16, the airfoil design is an iterative process to achieve the specified work transfer and vector diagrams obtained from the throughflow analysis while ensuring the lowest aerodynamic loss at any given radius. Although the airfoils are designed once the thermodynamic and aerodynamic properties are established, their presence is recognized by work input/output, flow input, flow-turning ability, blockage, aerodynamic loss, and force terms in the throughflow analysis. Both 2D blade-to-blade analyses and 3D flow analysis play a vital role in establishing the final design of 3D airfoils.

2D blade-to-blade analysis. This analysis, shown in Figure 6.17c, is performed at multiple radial locations as a 2D channel flow with periodic boundary conditions on lateral surfaces, including airfoil pressure and suction surfaces. Considering advances in grid generation technology, the overall capabilities of a commercial CFD code, and available high-performance computing power, an aero design engineer can now perform a 2D blade-to-blade analysis in minutes. The results of this analysis identify regions of excess entropy production in the flow, including the areas of flow separation, if present, on the airfoil surfaces. Aero design engineers use these data to improve the 2D airfoil design further, which initially created from the velocity vectors obtained from the throughflow analysis.

3D flow analysis. When the 2D airfoil sections are stitched (stacking) together, the airfoil becomes 3D. The throughflow design does not permit spanwise mixing, as the streamtube (concentric annuli) is bounded by meridional streamlines. One must consider the spanwise (radial) mixing in turbomachinery flows to design a realistic airfoil. The following are various mechanisms that cause spanwise mixing in turbomachines:

- Stream-wise vorticity off the blades (stream surface twist), rendering the blade-to-blade stream surfaces non-axisymmetric
- Secondary flows in the end-wall boundary layers and in the blade boundary layers
- Wake momentum transport downstream of blade rows
- Tip clearance flows with tip clearance vortices
- Turbulent diffusion
- Shock–boundary-layer interactions
- Flow separation
- Any other regions of high loss (entropy production)

In the state-of-the-art turbomachinery aero design system, all 3D flow analyses (steady and unsteady) are carried out using a 3D CFD tool, most likely commercial software, with an advanced turbulence model and an accurate near-wall treatment.

6.6.11.6 Role of 3D CFD

Based on the preceding brief overview of the state-of-the-art turbomachinery aero design system, it is inconceivable to improve further the machine efficiency, which is already around 90%, with the help of 3D CFD technology, which quickly (with high-performance computing resources) provides a detailed distribution of all flow properties throughout the flowpath, highlighting areas of further design improvement. Some empirical correlations that have served the past designs well are being replaced by the ability to compute flow fields with improved accuracy using CFD methodologies. Two-dimensional solutions of the flow through a blade row are gradually being supplemented and even replaced by three-dimensional ones.

Denton (2014) presented various loss mechanisms found in turbomachines. These mechanisms are related to excess entropy production. 3D CFD results are ideal for generating an entropy map of the complete flowpath toward making the design as isentropic as possible.

6.7 CONCLUDING REMARKS

In the past half-century, computational fluid dynamics (CFD) has undergone remarkable advancements, paralleling the surge in computing capabilities and establishing itself as a staple in industrial fluid engineering designs. This chapter has offered a nuanced appraisal of CFD, acknowledging its strengths and weaknesses as a versatile, swift, and cost-effective design tool. Within the realm of turbulent flow prediction, two predominant methodologies exist simulation-based approaches like direct numerical simulation (DNS) and large eddy simulation (LES), and statistical modeling through turbulence models. While the former is often prohibitively resource-intensive for industrial applications, the latter, albeit widely employed, lacks universal accuracy and reliability, with turbulence models serving as engineering approximations rather than steadfast scientific laws.

Despite the inherent limitations of turbulence models, this chapter offers pragmatic insights to harness the inherent strengths of CFD technology in industrial design endeavors. Firstly, 3D CFD serves as a potent tool for rapid flow visualization, enhancing understanding of flow dynamics to complement flow network modeling (FNM). Unlike CFD, FNM allows for easy calibration of empirical correlations, facilitating swift product validation. Given stringent design timelines, engineers often prefer the efficiency of finely tuned FNM over extensive 3D CFD modeling, particularly for robust designs necessitating probabilistic analyses. Secondly, 3D CFD finds utility in generating entropy maps, invaluable for identifying and mitigating areas of unacceptable irreversibility, thereby enhancing overall design performance.

In sum, this chapter offers a concise overview of state-of-the-art CFD-based turbulent flow predictions, indispensable across diverse high-tech industries. The field of CFD remains vibrant, continuously evolving through ongoing research and development efforts. For those inclined to delve deeper into this domain, the provided references and bibliography serve as invaluable resources for further exploration.

REFERENCES

Anderson, D.A., J.C. Tannehill, R.H. Pletcher, R. Munipalli, and V. Shankar. 2021. *Computational Fluid Dynamics and Heat Transfer.* 4th edition. Boca Raton, Florida: CRC Press.

Borda, Jeane-Charles de. 1766. Memoire sur l'ecoulement des fluids par les orifices des vases. In *Memoire de l'academie royal des sciences*.

Bradshaw, P. 1971. *An Introduction to Turbulence*. New York: Pergamon Press.

Bradshaw, P. 1997. Understanding and prediction of turbulent flow – 1996. *Int. J Heat Fluid Flow*. 18:45–54.

Cumpsty, N.A. 1989. *Compressor Aerodynamics*. London: Longman.

Daly, B.J. and Harlow, F.H. 1970. Transport equations of turbulence. *Phys. Fluids*. 13: 2634–2649.

Denton, J.D. 2014. Loss mechanisms in turbomachines. *ASME J. Turbomach.*115: 621–656.

Denton, J.D. and W.N. Dawes. 1999. Computational fluid dynamics for turbomachinery design, *Proc Inst. Mechanical Engineers*. 213 (Part C): 107–124.

Drewry, J.E. 1978. Fluid dynamic characterization of sudden expansion ramjet combustor flow fields. *AIAA J.* 16:313–319.

Durbin, P.A. and T.I-P. Shih. 2005. An overview of turbulence modeling. *Modeling and Simulation of Turbulent Heat Transfer.* Editors: B. Sunden and M. Faghri. Ashurst (Southampton): WIT Press

Engelman, M.S. 1993. CFD – an industrial perspective. *Incompressible Computational Fluid Dynamics.* Editors: M.D. Gunzburger and R.A. Nicolaides. Cambridge: Cambridge University Press.

Gosman, A.D., W.M. Pun, A.K. Runchal, D.B. Spalding, and M. Wolfshtein. 1969. *Heat and Mass Transfer in Recirculating Flows*. London: Academic Press.

Gibson, M.M. and B.E. Launder. 1976. On the calculation of horizontal turbulent shear flows under gravitational influence. *Trans. ASME, J. Heat Transfer.* 98: 81–87.

Gilham, S., I.R. Cowan, and E.S. Kaufman. 1999. Improving gas turbine power plant safety: the application of computational fluid dynamics to gas leaks. *Proceed. Inst. Mech. Eng. Part A: J. Power Energy.* 213 (6): 475–489.

Greenspan, H.P. 1968. *The Theory of Rotating Fluids*. Cambridge: Cambridge University Press.

Hanjalic, K. and B.E. Launder. 1972. A Reynolds stress model of turbulence and its application to thin shear flows. *J. Fluid Mech.* 52: 609–638.

Jones, W.P. and B.E. Launder. 1972. The prediction of laminarization with a two-equation model of turbulence. *Int. J. Heat Mass Transfer.* 15: 301–314.

Lakshminarayana, B. 1996. *Fluid Dynamics and Heat Transfer of Turbomachinery.* New York: John Wiley & Sons.

Launder, B.E. 1971. *An Improved Algebraic Stress Model of Turbulence.* Mechanical Engineering Department, Imperial College, London, Report TM/TN/A8.

Launder, B.E. 1975. On the effects of gravitational field on turbulent transport of heat and momentum. *J. Fluid Mech.* 67: 569–581.

Launder, B.E. and D.B Spalding. 1974. The numerical computation of turbulent flows. *Comp. Methods Appl. Mech. Eng.* 3: 269–289.

Launder, B.E., G.J. Reece, and W. Rodi. 1975. Progress in the development of a Reynolds-stress turbulence closure. *J. Fluid Mech.* 68: 537–566.

Leschziner, M. 2016. *Statistical Turbulence Modeling for Fluid Dynamics—Demystified: An Introductory Text for Graduate Engineering Students*. London: Imperial College Press.

Lilley, D.B. and D.L. Rhode. 1982. *STARPIC: A Computer Code for Swirling Turbulent Axisymmetric Recirculating Flows in Practical Isothermal Combustor Geometries.* Cleveland, OH: NASA CR-3442.

Lugt, H.J. 1995. *Vortex Flow in Nature and Technology.* Malabar: Krieger Publishing Company.

Luo, J., E.H. Razinsky, and H-K. Moon. 2012. Three-Dimensional RANS prediction of gas-side heat transfer coefficients on turbine blade and end-wall. *J. Turbomachinery.* 135 (2): 1 11.

Marsh, H. 1968. *A Computer Program for the Through Flow Fluid Mechanics in an Arbitrary Turbomachine Using a Matrix Method.* Aeronautical Research Council R. & M. No. 3509, Cranfield University.

Mellor, G. and T. Yamada. 1974. A hierarchy of turbulence models for planetary boundary layers. *J. Atmos. Sci.* 31: 1971–1982.

Moon, L.F. and G. Rudinger. 1977. Velocity distribution in an abruptly expanding circular duct. *Trans. ASME, J. Fluids Eng.* 99: 226–230.

Naot, D., A. Shavit, and M. Wolfshtein. 1973. Two-point correlation model and redistribution of Reynolds stresses. *Phys. Fluids*. 16: 738–743.

Naterer, G.F. and J.A. Camberos. 2008. *Entropy-Based Design and Analysis of Fluids Engineering Systems.* Boca Raton, FL: CRC Press.

Novak, R.A. 1967. Streamline curvature computing procedures for fluid-flow problems. *ASME J. Eng. Power.* 89 (4): 478–490.

Patankar, S.V. 1980. *Numerical Heat Transfer and Fluid Flow*. Boca Raton, FL: CRC Press.

Ponnuraj, B., B.K. Sultanian, A. Novori, and P. Pecchi. 2003. 3D CFD analysis of an industrial gas turbine compartment ventilation system. Proceedings of ASME IMECE, Washington, DC.

Pope, S.B. and J.H. Whitelaw. 1976. The calculation of near-wake flows. *J. Fluid Mech.* 73: 9–32.

Reynolds, O. 1883. An experimental investigation of the circumstances which determine whether the motion of water shall be direct or sinuous, and the law of resistance in parallel channels. *Philos. Trans. R. Soc. London Ser. A* 174: 935–982.
Reynolds, O. 1894. On the dynamical theory of incompressible viscous flows and the determination of the criterion. *Philos. Trans. R. Soc. London Ser. A* 186: 123–161.
Richardson, L.F. 1922. *Weather Prediction by Numerical Process.* Cambridge: Cambridge University Press.
Rodi, W. 1976. A new algebraic relation for calculating the Reynolds stresses. *ZAMM* 56: 219–221.
Rodi, W. 1980. *Turbulence Models and Their Applications in Hydraulics—A State of the Art Review.* Inst. Assoc. for Hydraulic Research, Deft, the Netherlands.
Rotta, J.C. 1951. Statistische theorie nchthomogener turbulence. *Z.F. Physiks.* 129: 547–572. 131: 51–77.
Santon, R.C., J.W. Kidger, and C.J. Lea. 2002. Safety developments in gas turbine power applications. Proceedings of ASME Turbo Expo 2002. ASME Paper GT2002–30469. Amsterdam, Netherlands.
Sciubba, E. 1997. Calculating entropy with CFD. *Mech. Eng.* 119(10): 86–88.
Smith, L.H. 1966. The radial-equilibrium equation of turbomachinery. *ASME J. Eng. Power.* 88:1–12.
Spur, A. 1980. The prediction of 3D transonic flow in turbomachinery using a combined through-flow and blade-to-blade time marching method. *Int. J. Heat Fluid Flow.* 2 (4): 189–199.
Sultanian, B.K. 1984. Numerical modeling of turbulent swirling flow downstream of an abrupt pipe expansion. Ph.D. diss., Arizona State University.
Sultanian, B.K., and D.A. Nealy. 1987. Numerical modeling of heat transfer in the flow through a rotor cavity. *Heat Transfer in Gas Turbines*, HTD-Vol. 87, ed. D.E. Metzger, 11–24, New York: The ASME.
Sultanian, B.K., G.P. Neitzel, and D.E. Metzger. 1986. A study of sudden expansion pipe flow using an algebraic stress transport model. Paper presented at the *AIAA/ASME 4th Fluid Mechanics, Plasma Dynamics and Lasers Conference*, Atlanta.
Sultanian, B.K., S. Nagao, and T. Sakamoto. 1999. Experimental and three-dimensional CFD investigation in a gas turbine exhaust system. *Trans. ASME, J. Eng. Gas Turbines Power.* 121: 364–374.
Sultanian, B.K., G.P. Neitzel, and D.E. Metzger. 1987. Comment on the flow field in a suddenly enlarged combustion chamber. *AIAA J.* 25(6): 893–895.
Sultanian, B.K. 2015. *Fluid Mechanics: An Intermediate Approach.* Boca Raton, FL: Taylor & Francis.
Sultanian, B.K. 2019. *Logan's Turbomachinery: Flowpath Design and Performance Fundamentals,* Third Edition (Mechanical Engineering). Boca Raton, FL: Taylor & Francis.
Thompson, J.F., B.K. Soni, and N.P. Weatherill (editors). 1998. *Handbook of Grid Generation.* Boca Raton, FL: CRC Press.
Wilcox, D.C. 1998. *Turbulence Modeling for CFD*, 2nd Edition. La Canada: DCW Industries.
Wilcox, D.C. 2006. *Turbulence Modeling for CFD*, 3rd Edition. La Canada: DCW Industries.
Wu, C.H. 1952. *A General Theory of Three-Dimensional Flow in Subsonic und Supersonic Turbomachines of Axial-, Radial- and Mixed-flow Types.* NACA Report, TN 2604.
Wyngaard, J.C., S.P. Arya, and O.R. Cote. 1974. Some aspects of the structure of convective planetary boundary layers. *J. Atmos. Sci.* 31: 747–754.

BIBLIOGRAPHY

Baldwin, B.S. and H. Lomax. 1978. Thin Layer Approximation of Algebraic Model for Separated Turbulent Flows. AIAA Paper No. 78–257.
Barselli, L.C. 2021. *Three-Dimensional Navier–Stokes Equations for Turbulence.* New York: Academic Press.
Bernard, P.S. 2019. *Turbulent Fluid Flow.* Hoboken, NJ: John Wiley.
Billard, F. (2011). Development of robust elliptic-blending turbulence model for near-wall separated and buoyant flows. *Ph.D. Thesis.* University of Manchester.
Bradshaw, P. 1997. Understanding and prediction of turbulent flow – 1996. *Int. J. Heat Fluid Flow.* 18:45–54.
Bradshaw, P., D.H. Ferriss, and N.D. Atwell. 1967. Calculation of boundary layer development using the turbulent energy equation. *J. Fluid Mech.* 28: 593–616.
Brunton, S.L., B.R. Noack, and P. Koumoutsakos. 2019. Machine learning for fluid mechanics. *Ann. Rev. Fluid Mech.* 52: 477–508.
Cebeci, T. and P. Bradshaw. 1984. *Physical and Computational Aspects of Convective Heat Transfer.* New York: Springer.
Cebeci, T. and A.M.O. Smith. 1974. *Analysis of Turbulent Boundary Layers.* New York: Academic Press.
Craft, T.J., N.Z. Ince, and B.E. Launder. 1996. Recent developments in second-moment closure for buoyancy-affected flows. *Dynam. Atmos. Ocean* 23: 99–114.

Durbin, P.A. 1991. Near-wall turbulence modeling without damping functions. *Theor. Comput. Fluid Dyn.* 3: 1–13.

Durbin, P.A. 1993. A Reynolds-stress model for near-wall turbulence. *J. Fluid Mech.* 249: 465- 498.

Durbin, P.A. 1995. Separated flow computations with $k - \varepsilon - \overline{v^2}$ model. *AIAA J.* 33: 659–664.

Durbin, P.A. and B.A. Peterson Reif. 2001. *Statistical Theory and Modeling for Turbulent Flows*, 2nd Edition. Chichester, England: Wiley.

Durbin, P.A. and G. Medic. 2007. *Fluid Dynamics with a Computational Perspective*. Cambridge: Cambridge University Press.

Ferziger, J.H. and M. Peric. 2013. *Computational Methods for Fluid Dynamics*, 3rd Edition. New York: Springer.

Hinch, E.J. 2020. *Think Before You Compute: A Prelude to Computational Fluid Dynamics*. Cambridge: Cambridge University Press.

Hinze, J.O. 1975. *Turbulence*, 2nd Edition. New York: McGraw-Hill.

Hunt, J.C.R. and Vassilicos, J.C. 1991. Kolmogorov's contributions to the physical and geometrical understanding of small-scale turbulence and recent developments. *Turbulence and Stochastic Processes: Kolmogorov's Ideas 50 Years On*, pp. 183–210. Editors: J.C.R. Hunt, O.M. Phillips, and D. Williams. Cambridge: Cambridge University. Press.

Hussain, F. and Melander, M.V. 1992. Understanding turbulence via vortex dynamics. *Studies in Turbulence*, pp. 157–178. Editors: T.B. Gatski, S. Sarkar, and C.G. Speziale. New York: Springer-Verlag.

Iannelli, J. 2013. An exact non-linear Navier–Stokes compressible-flow solution for CFD code verification. *Int. J. Numer. Methods Fluids*. 72(2): 157–176.

Kolmogorov, A.N. 1942. Equations of turbulent motion of an incompressible fluid. *Izvestiya Akademiya Nauk SSSR*, Seria Fizicheska VI (1–2): 56–58. (Translated from the original Russian by D.B. Spalding).

Menter, F.R., M. Kuntz, and R. Langtry. 2003. Ten years of industrial experience with the SST turbulence model. *4th Internal Symposium, Turbulence, Heat and Mass Transfer*, Antalya, Turkey, 4: 625–632.

Myong, H.K. and N. Kasagi. 1990. A new approach to the improvement of $k - \varepsilon$ turbulence model for wall bounded shear flows. *JSME Int. J. Series II*. 33: 63–72.

Nisizima, S. and A. Yoshizawa. 1987. Turbulent channel and Couette flow using an anisotropic $k - \varepsilon$ model. *J. AIAA*. 25: 414–420.

Piomelli, U. 1993. Application of large eddy simulations in engineering: an overview. *Large Eddy Simulation of Complex Engineering and Geophysical Flows*, pp. 119–137. Editors: B. Galperin and S.A. Orszag. Cambridge: Cambridge University Press.

Pope, S.B. 2000. *Turbulent Flows*. Cambridge: Cambridge University Press.

Landahl, M.T. and E. Mollo-Christensen. 1992. *Turbulence and Random Process in Fluid Mechanics*, 2nd Edition. Cambridge: Cambridge University Press.

Lilley, D.B. and D.L. Rhode. 1982. *STARPIC: A Computer Code for Swirling Turbulent Axisymmetric Recirculating Flows in Practical Isothermal Combustor Geometries*. NASA CR-3442.

Rodriguez, S. 2011. Swirling jets for the mitigation of hot spots and thermal stratification in the VHTR lower plenum. *Ph.D. Thesis*. University of New Mexico.

Rodriguez, S. 2019. *Applied Computational Fluid Dynamics and Turbulence Modeling: Practical Tools, Tips and Techniques*. New York: Springer.

Rubinstein R. and J.M. Barton. 1990. Nonlinear Reynolds stress models and the renormalization group. *Phys. Fluids A*. 2: 1472–1476.

Saffman, P.G. 1970. A model for inhomogeneous turbulent flow. *Proc. R. Soc. A*. 317: 417–433.

Shih, T.I-P. and P. Durbin. 2014. Modeling and simulation of turbine cooling. *Turbine Aerodynamics, Heat Transfer, Materials, and Mechanics*, pp. 389–438, Editors: T. Shih and V. Yang. AIAA Progress Series, AIAA, Vol. 243.

Shih, T.I-P. and B.K. Sultanian. 2001. Computations of internal and film cooling. *Heat Transfer in Gas Turbines*, pp. 175–225. Editors: B. Sunden and M. Faghri. Ashurst, Southampton: WIT Press.

Shih, T.-H, W.W. Liou, and J.L. Lumley (1995). A realizable Reynolds stress algebraic equation model. *Comp. Methods Appl. Mech. Engrg*. 125: 287–302.

Spalart, P.R. and S.R. Allmaras (1992). *A One-Equation Turbulence Model for Aerodynamic Flows*. AIAA Paper 92-439, Reno, NV.

Speziale, C.G. 1987. On nonlinear $k - \ell$ and $k - \varepsilon$ models of turbulence. *J. Fluid Mech.* 178: 459–475.

Sultanian, B.K. 2018. *Gas Turbines: Internal Flow Systems Modeling* (Cambridge Aerospace Series). Cambridge: Cambridge University Press.

Sultanian, B.K. 2021. *Fluid Mechanics and Turbomachinery: Problems and Solutions*. Boca Raton, FL: Taylor & Francis.
Sultanian, B.K. 2022a. *Fluid Mechanics: An Intermediate Approach*: Errata for the First Edition Published in 2015. USA: Independently published on KDP (Amazon.com).
Sultanian, B.K. 2022b. *Gas Turbines Internal Flow Systems Modeling*: Errata for the First Edition Published in 2018. USA: Independently published on KDP (Amazon.com).
Sultanian, B.K. and H.C. Mongia. 1986. Fuel nozzle air flow modeling. *AIAA/ASME/SAE/ASEE 22nd Joint Propulsion Conference*, Huntsville, Alabama. Paper No. AIAA-86-1667.
Sultanian, B.K., G.P. Neitzel, and D.E. Metzger. 1987. Turbulent flow prediction in a sudden axisymmetric expansion. *Turbulence measurements and Flow Modeling*, pp. 655–674. Editors: C.J. Chen, L.D. Chen, and F.M. Holly. New York: Hemisphere.
Tennekes, H. and J.L. Lumley. 1972. *A First Course in Turbulence*. Cambridge, MA: MIT Press.
Townsend, A.A. 1976. *The Structure of Turbulent Shear Flow*. Cambridge: Cambridge University Press.
Tucker, P.G. 2016. *Advanced Computational Fluid and Aerodynamics*. Cambridge: Cambridge University Press.
Van Driest, E. 1956. On turbulent flow near a wall. *J. Aeronaut. Sci.* 23: 1007–1011.
Wilcox, D.C. 1988. Multiscale models for turbulent flows. *AIAA J.* 26 (11): 1311–1320.
Wilcox, D.C. 1993a. *Turbulence Modeling for CFD,* First Edition. La Canada: DCW Industries.
Wilcox, D.C. 1993b. Comparison of two-equation turbulence models for boundary layers with pressure gradient. *AIAA J.* 31: 1414–1421.
Wilcox, D.C. 1994. Simulation of transition with a two-equation turbulence model. *AIAA J.* 32(2): 247–255.
Wilcox, D.C. 2007. Formulation of the $k-\omega$ turbulence model revisited. 45th AIAA Aerospace Sciences Meeting and Exhibit. *AIAA J.* 46(11): 2823–2838.

NOMENCLATURE

A	Section area
A^+	Prandtl Mixing length model constant
c_p	Specific heat at constant pressure
$C_\mathrm{D}, C_{\mu'}$	Model constants in the one-equation $(k-L)$ turbulence model
$C_\mu, C_{\varepsilon_1}, C_{\varepsilon_2}$	Model constants in the two-equation $(k-\varepsilon)$ turbulence model
$C_\mathrm{s}, C_\varepsilon, C_1, C_2$	Model constants in turbulence models RSTM and ASM
C_k	Convection of turbulent kinetic energy k
C_p	Pressure recovery coefficient
$C_{\overline{u_iu_j}}$	Convection of $\overline{u_iu_j}$
D_k	Diffusion of turbulent kinetic energy k
$D_{\overline{u_iu_j}}$	Diffusion of $\overline{u_iu_j}$
E	Constant in the logarithmic law of the wall
h	Specific enthalpy
k	Turbulent kinetic energy
$\tilde{k}$	Thermal conductivity
K	Von Karman constant
l	Turbulent length scale
l_m	Mixing length

L	Mixing length in the one-equation $k-L$ turbulence model
$\dot{m}$	Mass flow rate
m_ℓ	Mass fraction of chemical species ℓ
p'	Fluctuating part of P
p	Pressure
P_f	Jaytilleke's P function
P_k	Production rate of k
P_{ij}	Production of $\overline{u_i u_j}$
Pr	Prandtl number
q	Heat flux
Pr_t	Turbulent Prandtl number
R	Gas constant
$\Re$	Generation rate per unit volume (from chemical reaction)
s	Specific entropy
S	Source term in the common-form transport equation
$\tilde{S}$	Source terms other than pressure gradient terms in the momentum equation
ta	Time
T	Temperature
Tu	Turbulence intensity
u	Fluctuating part of U
u_i	Fluctuating velocities in tensor notations
U	Velocity component along x direction
U^+	Dimensionless velocity in a wall boundary layer
U^*	Shear velocity
U_i	Velocities in tensor notation
v	Fluctuating part of velocity along y direction
V	Total velocity; velocity component along y direction
$\boldsymbol{V}$	Velocity vector
W	Velocity component along z direction
x	Cartesian coordinate x
x_i	Coordinates in tensor notation
y	Cartesian coordinate y
y^+	Dimensionless transverse distance in a wall boundary layer
z	Cartesian coordinate z

SUBSCRIPTS AND SUPERSCRIPT

1	Location 1; Section 1
2	Location 2; Section 2
h	For enthalpy in Inlet
ℓ	Chemical species ℓ
m_ℓ	For mass fraction of chemical species ℓ

P	Point P
ref	Reference state
0	Total (stagnation)
U_i	For U_i
w	Wall
x	Component in x coordinate direction
y	Component in y coordinate direction
z	Component in z coordinate direction
$(\bar{\ })$	Time average or ensemble average
$(^{*})$	Dimensionless

GREEK SYMBOLS

Γ	Diffusion coefficient
δ_{ij}	Kronecker delta=1 for $i = j$ and=0 for $i \neq j$
ε	Dissipation rate of k
ε_{ij}	Dissipation rate of $\overline{u_i u_j}$
ζ	Vorticity
$\boldsymbol{\zeta}$	Vorticity vector
γ	Ratio of specific heats
λ	Ratio P_k / ε
μ	Dynamic viscosity
μ_t	Turbulent (or eddy) viscosity
ν	Kinematic viscosity ($\nu = \mu / \rho$)
ν_t	Kinematic eddy viscosity ($\nu_t = \mu_t / \rho$)
σ_k	Turbulent Prandtl number for k
σ_ε	Turbulent Prandtl number for ε
ρ	Density
τ	Shear stress
υ	Turbulent velocity scale
φ	Fluctuating part of Φ
φ_{ij}	Pressure–strain term in RSTM
Φ	General transport variable of the common-form governing transport equation

Appendix A
Review of Necessary Mathematics

A.1 INTRODUCTION

This appendix reviews necessary mathematics used in the main text, including tensor algebra; gradient, divergence, curl, and Laplacian; dyad in total derivative, total derivative; vector identities; Gauss's divergence theorem; and Leibniz formula for differentiating an integral.

A.2 SUFFIX NOTATION AND TENSOR ALGEBRA

In Cartesian coordinates, the unit vectors along the x, y, and z coordinate directions are denoted by $\hat{\boldsymbol{i}}$, $\hat{\boldsymbol{j}}$, and $\hat{\boldsymbol{k}}$, respectively. In Cartesian tensor, we name the x,y, and z coordinates as x_1, x_2, and x_3, respectively, with the corresponding unit vectors denoted by $\hat{\boldsymbol{e}}_1$, $\hat{\boldsymbol{e}}_2$, and $\hat{\boldsymbol{e}}_3$. A typical vector $\boldsymbol{a}$ with components a_1, a_2, and a_3 in Cartesian coordinates can be expressed as:

$$\boldsymbol{a} = a_1\hat{\boldsymbol{i}} + a_2\hat{\boldsymbol{j}} + a_3\hat{\boldsymbol{k}} = a_1\hat{\boldsymbol{e}}_1 + a_2\hat{\boldsymbol{e}}_2 + a_3\hat{\boldsymbol{e}}_3$$

A.2.1 Summation Convention

When an index is repeated precisely twice in a term, it implies summation over all possible values of the index, which has values 1, 2, and 3 in Cartesian coordinates. Accordingly, we can express our vector $\boldsymbol{a}$ in the compact tensor notation as: $\boldsymbol{a} = a_i\hat{\boldsymbol{e}}_i$.

Note that the summation convention possesses the commutative and distributive properties as shown below.

Commutative:

$$a_i b_i = b_i a_i$$

Distributive:

$$a_i(b_i + c_i) = a_i b_i + a_i c_i$$

A.2.2 Free and Dummy Indices

In the term on the left-hand side of the following equation, the index j is repeated twice, implying summation. This index is called the dummy index. On the other hand, the index m can have any value and is called the free index:

$$a_{mj} x_j = c_m$$

This equation can be alternatively written as:

$$a_{kq} x_q = c_k$$

Both these equations are a compact form (in tensor notation) of the following there equations:

$$a_{11}x_1 + a_{12}x_2 + a_{13}x_3 = c_1$$

$$a_{21}x_1 + a_{22}x_2 + a_{23}x_3 = c_2$$

$$a_{31}x_1 + a_{32}x_2 + a_{33}x_3 = c_3$$

A.2.3 Two Special Symbols

Kronecker delta:

$$\delta_{ij} = \hat{e}_i \cdot \hat{e}_j = \begin{cases} 1 & \text{if } i = j \\ 0 & \text{if } i \neq j \end{cases}$$

$$\delta_{ij} = \begin{bmatrix} \delta_{11} & \delta_{12} & \delta_{13} \\ \delta_{21} & \delta_{22} & \delta_{23} \\ \delta_{31} & \delta_{32} & \delta_{33} \end{bmatrix} = \begin{bmatrix} 1 & 0 & 0 \\ 0 & 1 & 0 \\ 0 & 0 & 1 \end{bmatrix}$$

The alternating symbol:

$$\varepsilon_{ijk} = \hat{e}_i \bullet (\hat{e}_j \times \hat{e}_k) = \begin{cases} 1 & \text{if } i,j,k \text{ are a cyclic permutation of } 1,2, \text{ and } 3 \\ -1 & \text{if } i,j,k \text{ are a anticyclic permutation of } 1,2, \text{ and } 3 \\ 0 & \text{if any index is repeated} \end{cases}$$

This definition of ε_{ijk} yields the following identities:

$$\varepsilon_{ijk} = \varepsilon_{kij} = \varepsilon_{jki} \quad \text{and} \quad \varepsilon_{jik} = -\varepsilon_{ijk}$$

A.3 GRADIENT, DIVERGENCE, CURL, AND LAPLACIAN

A.3.1 Gradient

Define:

$$\nabla = \frac{\partial}{\partial x}\hat{i} + \frac{\partial}{\partial y}\hat{j} + \frac{\partial}{\partial z}\hat{k} = \hat{e}_i \frac{\partial}{\partial x_i}$$

For example, we can write the gradient of a scalar Φ as:

$$\nabla\Phi = \frac{\partial\Phi}{\partial x}\hat{i} + \frac{\partial\Phi}{\partial y}\hat{j} + \frac{\partial\Phi}{\partial z}\hat{k} = \frac{\partial\Phi}{\partial x_1}\hat{e}_1 + \frac{\partial\Phi}{\partial x_2}\hat{e}_2 + \frac{\partial\Phi}{\partial x_3}\hat{e}_3 = \hat{e}_i \frac{\partial\Phi}{\partial x_i}$$

In cylindrical coordinates, this equation becomes:

$$\nabla\Phi = \frac{\partial\Phi}{\partial r}\hat{e}_r + \frac{1}{r}\frac{\partial\Phi}{\partial\theta}\hat{e}_\theta + \frac{\partial\Phi}{\partial z}\hat{e}_z$$

where $\hat{e}_r$, $\hat{e}_\theta$, and $\hat{e}_z$ are the unit vectors along the r, θ, and z coordinate directions, respectively.

A.3.2 Divergence

We can write the divergence of a vector $\boldsymbol{V} = u\hat{\boldsymbol{i}} + v\hat{\boldsymbol{j}} + w\hat{\boldsymbol{k}} = \hat{\boldsymbol{e}}_i u_i$ as:

$$\nabla \bullet \boldsymbol{V} = \frac{\partial u}{\partial x} + \frac{\partial v}{\partial y} + \frac{\partial w}{\partial z} = \frac{\partial u_1}{\partial x_1} + \frac{\partial u_2}{\partial x_2} + \frac{\partial u_3}{\partial x_3} = \frac{\partial u_i}{\partial x_i}$$

In cylindrical coordinates, this equation becomes:

$$\nabla \bullet \boldsymbol{V} = \frac{1}{r}\frac{\partial (r u_r)}{\partial r} + \frac{1}{r}\frac{\partial u_\theta}{\partial \theta} + \frac{\partial u_z}{\partial z}$$

A.3.3 Curl

We can write the curl of a vector $\boldsymbol{V} = u\hat{\boldsymbol{i}} + v\hat{\boldsymbol{j}} + w\hat{\boldsymbol{k}} = \hat{\boldsymbol{e}}_i u_i$

$$\nabla \times \boldsymbol{V} = \begin{bmatrix} \hat{\boldsymbol{i}} & \hat{\boldsymbol{j}} & \hat{\boldsymbol{k}} \\ \frac{\partial}{\partial x} & \frac{\partial}{\partial y} & \frac{\partial}{\partial z} \\ u & v & w \end{bmatrix} = \begin{bmatrix} \hat{\boldsymbol{e}}_1 & \hat{\boldsymbol{e}}_2 & \hat{\boldsymbol{e}}_3 \\ \frac{\partial}{\partial x_1} & \frac{\partial}{\partial x_2} & \frac{\partial}{\partial x_3} \\ u_1 & u_2 & u_3 \end{bmatrix}$$

$$\nabla \times \boldsymbol{V} = \left(\frac{\partial w}{\partial y} - \frac{\partial v}{\partial z}\right)\hat{\boldsymbol{i}} + \left(\frac{\partial u}{\partial z} - \frac{\partial w}{\partial x}\right)\hat{\boldsymbol{j}} + \left(\frac{\partial v}{\partial x} - \frac{\partial u}{\partial y}\right)\hat{\boldsymbol{k}}$$

$$\nabla \times \boldsymbol{V} = \left(\frac{\partial u_3}{\partial x_2} - \frac{\partial u_2}{\partial x_3}\right)\hat{\boldsymbol{e}}_1 + \left(\frac{\partial u_1}{\partial x_3} - \frac{\partial u_3}{\partial x_1}\right)\hat{\boldsymbol{e}}_2 + \left(\frac{\partial u_2}{\partial x_1} - \frac{\partial u_1}{\partial x_2}\right)\hat{\boldsymbol{e}}_3 = \hat{\boldsymbol{e}}_i \varepsilon_{ijk} \frac{\partial u_k}{\partial x_j}$$

In cylindrical coordinates, this equation becomes:

$$\nabla \times \boldsymbol{V} = \left(\frac{1}{r}\frac{\partial u_z}{\partial \theta} - \frac{\partial u_\theta}{\partial z}\right)\hat{\boldsymbol{e}}_r + \left(\frac{\partial u_r}{\partial z} - \frac{\partial u_z}{\partial r}\right)\hat{\boldsymbol{e}}_\theta + \left(\frac{1}{r}\frac{\partial (r u_\theta)}{\partial r} - \frac{1}{r}\frac{\partial u_r}{\partial u_\theta}\right)\hat{\boldsymbol{e}}_z$$

A.3.4 Laplacian

We can write the Laplacian of a scalar Φ as:

$$\nabla^2 \Phi = \nabla \bullet \nabla \Phi = \frac{\partial^2 \Phi}{\partial x^2} + \frac{\partial^2 \Phi}{\partial y^2} + \frac{\partial^2 \Phi}{\partial z^2} = \frac{\partial^2 \Phi}{\partial x_1^2} + \frac{\partial^2 \Phi}{\partial x_2^2} + \frac{\partial^2 \Phi}{\partial x_3^2} = \frac{\partial}{\partial x_i}\frac{\partial \Phi}{\partial x_i}$$

In cylindrical coordinates, this equation becomes:

$$\nabla^2 \Phi = \frac{\partial^2 \Phi}{\partial r^2} + \frac{1}{r}\frac{\partial \Phi}{\partial r} + \frac{1}{r^2}\frac{\partial^2 \Phi}{\partial \theta^2} + \frac{\partial^2 \Phi}{\partial z^2} = \frac{1}{r}\frac{\partial}{\partial r}\left(r\frac{\partial \Phi}{\partial r}\right) + \frac{1}{r^2}\frac{\partial^2 \Phi}{\partial \theta^2} + \frac{\partial^2 \Phi}{\partial z^2}$$

A.4 DYAD IN TOTAL DERIVATIVE

The total or substantial derivative of the velocity vector $\boldsymbol{V}$ can be written in vector form as:

$$\frac{\mathrm{D}\boldsymbol{V}}{\mathrm{D}t} = \frac{\partial \boldsymbol{V}}{\partial t} + (\boldsymbol{V} \bullet \nabla)\boldsymbol{V}$$

which makes it independent of the coordinate system. In this equation, $\boldsymbol{V} \bullet \nabla$ is called a dyad, which is easily handled in the tensor notation. Using the vector identity, we can express the dyad in this equation as:

$$(\boldsymbol{V} \bullet \nabla)\boldsymbol{V} = \nabla\left(\frac{1}{2}V^2\right) - \boldsymbol{V} \times (\nabla \times \boldsymbol{V})$$

A.5 TOTAL DERIVATIVE

The total or substantial or material derivative following a particle (Lagrangian viewpoint) in a fluid flow can be written in various forms (vector, differential, and tensor) as given in the following:

Cartesian coordinates:

$$\frac{\mathrm{D}\boldsymbol{V}}{\mathrm{D}t} = \frac{\partial \boldsymbol{V}}{\partial t} + (\boldsymbol{V} \bullet \nabla)\boldsymbol{V}$$

$$\frac{\mathrm{D}\boldsymbol{V}}{\mathrm{D}t} = \left(\frac{\partial u}{\partial t} + u\frac{\partial u}{\partial x} + v\frac{\partial u}{\partial y} + w\frac{\partial u}{\partial z}\right)\hat{\boldsymbol{i}} + \left(\frac{\partial v}{\partial t} + u\frac{\partial v}{\partial x} + v\frac{\partial v}{\partial y} + w\frac{\partial v}{\partial z}\right)\hat{\boldsymbol{j}}$$
$$+ \left(\frac{\partial w}{\partial t} + u\frac{\partial w}{\partial x} + v\frac{\partial w}{\partial y} + w\frac{\partial w}{\partial z}\right)\hat{\boldsymbol{k}}$$

Cartesian tensor notation:

$$\frac{\mathrm{D}\boldsymbol{V}}{\mathrm{D}t} = \frac{\partial u_i}{\partial t} + u_j\frac{\partial u_i}{\partial x_j}$$

Cylindrical coordinates:

$$\frac{\mathrm{D}\boldsymbol{V}}{\mathrm{D}t} = \left(\frac{\partial u_r}{\partial t} + u_r\frac{\partial u_r}{\partial r} + \frac{u_\theta}{r}\frac{\partial u_r}{\partial \theta} + u_z\frac{\partial u_r}{\partial z} - \frac{u_\theta^2}{r}\right)\hat{\boldsymbol{e}}_r$$
$$+ \left(\frac{\partial u_\theta}{\partial t} + u_r\frac{\partial u_\theta}{\partial r} + \frac{u_\theta}{r}\frac{\partial u_\theta}{\partial \theta} + u_z\frac{\partial u_\theta}{\partial z} + \frac{u_r u_\theta}{r}\right)\hat{\boldsymbol{e}}_\theta$$
$$+ \left(\frac{\partial u_z}{\partial t} + u_r\frac{\partial u_z}{\partial r} + \frac{u_\theta}{r}\frac{\partial u_z}{\partial \theta} + u_z\frac{\partial u_z}{\partial z}\right)\hat{\boldsymbol{e}}_z$$

A.6 VECTOR IDENTITIES

In the following vector identities, Φ is a scalar and $\boldsymbol{A}$, $\boldsymbol{B}$, and $\boldsymbol{C}$ are vectors:

$$\nabla \times \nabla\Phi = 0$$

$$\nabla \bullet (\nabla \times \boldsymbol{A}) = 0$$

$$\nabla \bullet (\Phi\boldsymbol{A}) = \Phi(\nabla \bullet \boldsymbol{A}) + \boldsymbol{A} \bullet \nabla\Phi$$

$$\nabla \times (\Phi\boldsymbol{A}) = \nabla\Phi \times \boldsymbol{A} + \Phi(\nabla \times \boldsymbol{A})$$

$$(\boldsymbol{A} \bullet \nabla)\boldsymbol{A} = \frac{1}{2}\nabla(\boldsymbol{A} \bullet \boldsymbol{A}) - \boldsymbol{A} \times (\nabla \times \boldsymbol{A})$$

$$\nabla \times (\nabla \times \boldsymbol{A}) = \nabla(\nabla \bullet \boldsymbol{A}) - \nabla^2 \boldsymbol{A}$$

$$\nabla \bullet (\boldsymbol{A} \times \boldsymbol{B}) = \boldsymbol{B} \bullet (\nabla \times \boldsymbol{A}) - \boldsymbol{A} \bullet (\nabla \times \boldsymbol{B})$$

$$\nabla \times (\boldsymbol{A} \times \boldsymbol{B}) = \boldsymbol{A}(\nabla \bullet \boldsymbol{B}) - \boldsymbol{B}(\nabla \bullet \boldsymbol{A}) - (\boldsymbol{A} \bullet \nabla)\boldsymbol{B} + (\boldsymbol{B} \bullet \nabla)\boldsymbol{A}$$

A.7 GAUSS'S DIVERGENCE THEOREM

Gauss's divergence theorem allows us to convert a volume integral into a surface integral or vice versa:

$$\iiint_{\mathcal{V}} \nabla \bullet \boldsymbol{V} \mathrm{d}\mathcal{V} = \iint_{S} \boldsymbol{V} \bullet \boldsymbol{n} \mathrm{d}s$$

A.8 LEIBNITZ FORMULA FOR DIFFERENTIATING AN INTEGRAL

A.8.1 Integral in One Dimension

$$\frac{\mathrm{d}}{\mathrm{d}t}\int_{\alpha_1(t)}^{\alpha_2(t)} f(x,t)\mathrm{d}x = \int_{\alpha_1(t)}^{\alpha_2(t)} \frac{\partial f}{\partial t}\mathrm{d}x + \left[f(\alpha_2,t)\frac{\mathrm{d}\alpha_2}{\mathrm{d}t} - f(\alpha_1,t)\frac{\mathrm{d}\alpha_1}{\mathrm{d}t} \right]$$

A.8.2 Integral in Three Dimensions

For a deformable volume $\mathcal{V}(t)$ having time-dependent surface area $S(t)$ with its surface element moving with an arbitrary velocity $\boldsymbol{V}_s(t)$, if $f(x,y,z,t)$ is a scalar function of space and time, the Leibnitz formula yields:

$$\frac{\mathrm{d}}{\mathrm{d}t}\iiint_{\mathcal{V}(t)} f \mathrm{d}\mathcal{V} = \iiint_{\mathcal{V}(t)} \frac{\partial f}{\partial t}\mathrm{d}\mathcal{V} + \iint_{S(t)} f(\boldsymbol{V}_s \bullet \boldsymbol{n})\mathrm{d}S$$

For further details on the topics presented in this appendix, and for related additional topics, interested readers may refer to the references listed in the Bibliography section.

BIBLIOGRAPHY

Aris, R. 1962. *Vectors, Tensors, and the Basic Equations of Fluid Mechanics.* Englewood Cliffs: Prentice-Hall.
Hughes, W.A. and Gaylord, E.W. 1964. *Basic Equations of Engineering Science.* New York: McGraw-Hill.

Appendix B
Vorticity, Vortex, and Circulation

B.1 INTRODUCTION

In this appendix, we present additional insightful discussion on vorticity, vortex, and circulation to reinforce the material presented in Chapter 2. We discuss here the key characteristics of forced and free vortexes, the analogy between a stream tube and a vortex tube, Helmholtz theorems of vorticity, and Kelvin's circulation theorem in an inviscid flow. For further details on the topics presented here, interested readers may refer to the references listed in the bibliography at the end of this appendix.

The original version of this appendix appears as Appendix E Vorticity and Circulation in the first edition of this book by Sultanian (2015).

B.2 PROPERTIES OF VELOCITY FIELD

The velocity field is a vector field where the velocity at a point may have components in all three coordinate directions. In an unsteady flow, the velocity components are functions of spatial coordinates (x, y, and z) and time (t); in a steady flow, they do not depend on time. Mathematically, we can express velocity components in unsteady and steady flows as follows:

Unsteady flow:

$$V_x = f_1(x,y,z,t), V_y = f_2\,(x,y,z,t), \text{and } V_z = f_3(x,y,z,t)$$

Steady flow:

$$V_x = f_1(x,y,z), V_y = f_2\,(x,y,z), \text{and } V_z = f_3(x,y,z)$$

Note that a turbulent flow is inherently always unsteady. If, however, it is stationary in the mean (i.e., the ensemble average value of each velocity component at a point does not change with time), we call the flow "steady."

All vector fields do not represent a flow field; they must satisfy the local mass conservation (the continuity equation).

B.2.1 VORTICITY

Like velocity, the vorticity is a kinematic vector property of a flow field and equals the curl of the velocity vector at each point in the flow. The component of the vorticity vector along a coordinate direction represents twice the rate of local counterclockwise rotation of the fluid particles about the coordinate direction.

As shown in Figure B.1, let us consider a tiny fluid element ABCD in a two-dimensional flow at time t. In time interval dt, the element occupies a new position A′B′C′D′, the side AB rotates counterclockwise by $\mathrm{d}\alpha$ into its new position A′B′, and the side AD rotates clockwise by $\mathrm{d}\beta$ into its new position A′D′. The net counterclockwise rotation of the diagonal element A′C′ about the z axis becomes $(\mathrm{d}\alpha - \mathrm{d}\beta)/2$. Let us determine the value of this rotation in terms of velocities and their spatial gradients.

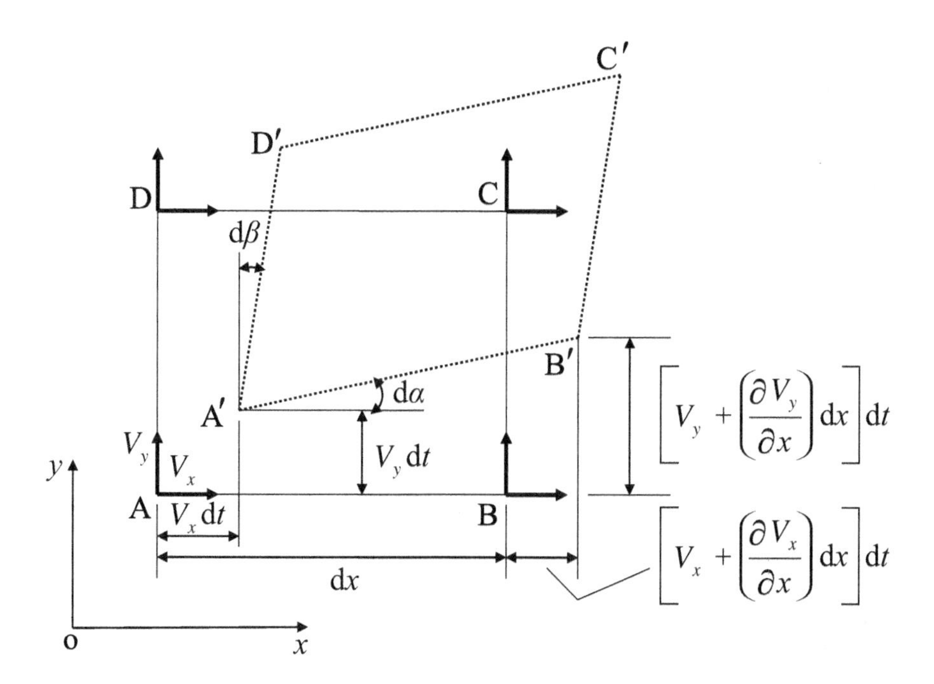

FIGURE B.1 Rotation in a two-dimensional flow.

From Figure B.1, we can write the difference between the y coordinates of A′ and B′ as:

$$\Delta y = \left[V_y + \left(\frac{\partial V_y}{\partial x}\right)\mathrm{d}x\right]\mathrm{d}t - V_y\mathrm{d}t = \left(\frac{\partial V_y}{\partial x}\right)\mathrm{d}x\,\mathrm{d}t$$

Similarly, we can write the difference between the x coordinates of A' and B' as:

$$\Delta x = \left[V_x + \left(\frac{\partial V_x}{\partial x}\right)\mathrm{d}x\right]\mathrm{d}t + \mathrm{d}x - V_x\mathrm{d}t = \mathrm{d}x + \left(\frac{\partial V_x}{\partial x}\right)\mathrm{d}x\,\mathrm{d}t$$

As $\left(\frac{\partial V_x}{\partial x}\right)\mathrm{d}x\,\mathrm{d}t \ll \mathrm{d}x$, the above equation reduces to $\Delta x = \mathrm{d}x$. We can now write:

$$\tan(\mathrm{d}\alpha) = \frac{\Delta y}{\Delta x} = \frac{\left(\partial V_y / \partial x\right)\mathrm{d}x\,\mathrm{d}t}{\mathrm{d}x} = \left(\frac{\partial V_y}{\partial x}\right)\mathrm{d}t$$

Similarly, we obtain:

$$\tan(\mathrm{d}\beta) = \left(\frac{\partial V_x}{\partial y}\right)\mathrm{d}t$$

We can write the z component of rotation vector at a point in this flow field as:

$$\omega_z = \lim_{\mathrm{d}t\to 0}\frac{(\mathrm{d}\alpha - \mathrm{d}\beta)/2}{\mathrm{d}t} = \frac{1}{2}\lim_{\mathrm{d}t\to 0}\frac{\tan(\mathrm{d}\alpha) - \tan(\mathrm{d}\beta)}{\mathrm{d}t}$$

where we have used the identity $\mathrm{d}\alpha \cong \tan(\mathrm{d}\alpha)$ and $\mathrm{d}\beta \cong \tan(\mathrm{d}\beta)$ for small $\mathrm{d}\alpha$ and $\mathrm{d}\beta$, respectively. Substituting for $\tan(\mathrm{d}\alpha)$ and $\tan(\mathrm{d}\beta)$ from above in this equation yields:

$$\omega_z = \frac{1}{2}\left(\frac{\partial V_y}{\partial x} - \frac{\partial V_x}{\partial y}\right) \quad \text{(B.1)}$$

The rotation vector component ω_z given by Equation B.1 represents the average angular velocity of two orthogonal line segments, dx and dy, which are vanishingly small, about the z axis. Being the average angular velocity of all line segments in a vanishingly small fluid volume at a point, we can interpret ω_z as the local rigid-body angular velocity of a fluid particle about the z axis. In a three-dimensional flow field, we can determine the remaining two components of the local rotation vector as:

$$\omega_x = \frac{1}{2}\left(\frac{\partial V_z}{\partial y} - \frac{\partial V_y}{\partial z}\right) \quad \text{(B.2)}$$

$$\omega_y = \frac{1}{2}\left(\frac{\partial V_x}{\partial z} - \frac{\partial V_z}{\partial x}\right) \quad \text{(B.3)}$$

The vorticity vector in a flow field is related to the velocity vector through the curl operation, that is:

$$\zeta = \text{curl}\,\boldsymbol{V} = \nabla \times \boldsymbol{V} = \left(\frac{\partial V_z}{\partial y} - \frac{\partial V_y}{\partial z}\right)\hat{i} + \left(\frac{\partial V_x}{\partial z} - \frac{\partial V_z}{\partial x}\right)\hat{j} + \left(\frac{\partial V_y}{\partial x} - \frac{\partial V_x}{\partial y}\right)\hat{k} \quad \text{(B.4)}$$

From above, we see that the vorticity vector is twice the rotation vector, that is:

$$\zeta = 2\boldsymbol{\omega} \quad \text{(B.5)}$$

where $\boldsymbol{\omega}$ represents the local fluid rotation vector.

B.2.2 Vortex

While vorticity, which is twice the rotation vector, characterizes the local rigid-body rotation of a fluid particle, the vortex pertains to the bulk circular motion of the entire flow field. The streamlines in a vortex flow are concentric circles. A forced vortex features solid-body rotation with constant angular velocity (radians per second) for all streamlines. An outer streamline in this vortex features higher tangential or swirl velocity than the inner one. In a free vortex, free from any external torque, the angular momentum (the product of radius and tangential velocity), not the angular velocity, remains constant everywhere in the flow. An outer streamline in a free vortex features a lower tangential velocity than the inner one. Readers may find an excellent discussion of vortex flows in nature and technology in Lugt (1995).

B.2.2.1 Forced Vortex

Figure B.2b shows the velocity distribution in a two-dimensional forced vortex. Let us first verify if the velocity field satisfies the continuity equation (divergence-free). We will then determine its vorticity distribution using the Cartesian and cylindrical coordinates shown in Figure B.2a.

Forced vortex in Cartesian coordinates:

Velocity:

$$\boldsymbol{V} = -\Omega y\hat{\boldsymbol{i}} + \Omega x\hat{\boldsymbol{j}} \quad \text{(B.6)}$$

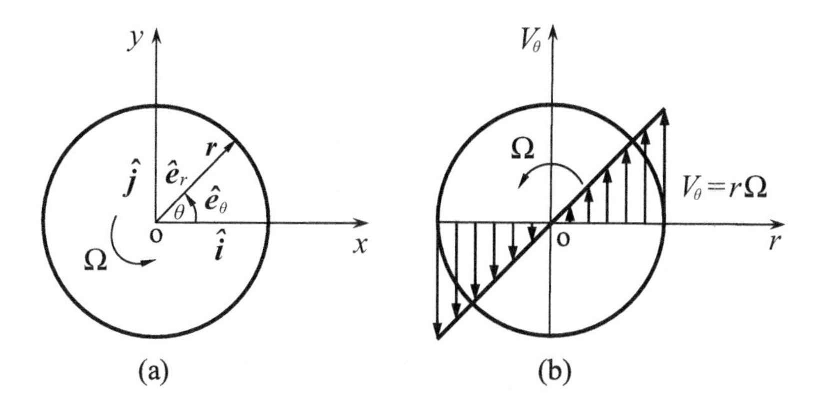

FIGURE B.2 (a) Cartesian and cylindrical coordinates and (b) forced vortex.

Divergence:

$$\operatorname{div}\boldsymbol{V} = \nabla \bullet \boldsymbol{V} = \frac{\partial(-\Omega y)}{\partial x} + \frac{\partial(\Omega x)}{\partial y} = 0$$

Vorticity:

$$\operatorname{curl}\boldsymbol{V} = \nabla \times \boldsymbol{V} = \left(\frac{\partial V_y}{\partial x} - \frac{\partial V_x}{\partial y}\right)\hat{\boldsymbol{k}}$$

$$\operatorname{curl}\boldsymbol{V} = \xi = \left(\frac{\partial(\Omega x)}{\partial x} - \frac{\partial(-\Omega y)}{\partial y}\right)\hat{\boldsymbol{k}} = 2\Omega\,\hat{\boldsymbol{k}} \tag{B.7}$$

Forced vortex in cylindrical coordinates:
Velocity:

$$V = r\Omega\,\hat{\boldsymbol{e}}_\theta \tag{B.8}$$

Divergence:

$$\operatorname{div}\boldsymbol{V} = \nabla \bullet \boldsymbol{V} = \frac{1}{r}\frac{\partial(rV_r)}{\partial r} + \frac{1}{r}\frac{\partial V_\theta}{\partial \theta}$$

which for $V_r = 0$ and $V_\theta = r\Omega$ yields $\operatorname{div}\boldsymbol{V} = 0$.
Vorticity:

$$\operatorname{curl}\boldsymbol{V} = \nabla \times \boldsymbol{V} = \left(\frac{1}{r}\frac{\partial(rV_\theta)}{\partial r} - \frac{1}{r}\frac{\partial V_r}{\partial \theta}\right)\hat{\boldsymbol{k}}$$

which for $V_r = 0$ and $V_\theta = r\Omega$ yields:

$$\operatorname{curl}V = \zeta = \left(\frac{1}{r}\frac{\partial(r^2\Omega)}{\partial r} - \frac{1}{r}\frac{\partial(0)}{\partial \theta}\right)\hat{k} = 2\Omega\,\hat{k} \tag{B.9}$$

In a two-dimensional forced vortex, Equations B.7 and B.9 show that the vorticity equals twice the rate of solid body rotation about the z axis

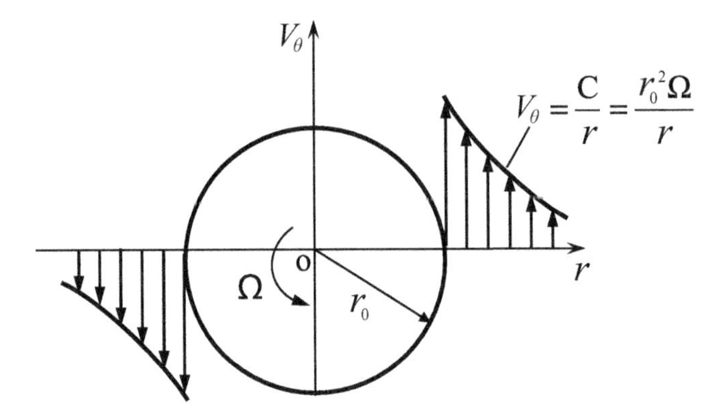

FIGURE B.3 Free vortex.

B.2.2.2 Free Vortex

Figure B.3 shows the velocity distribution in a two-dimensional free vortex. Before determining its vorticity distribution using Cartesian and cylindrical coordinates, shown in Figure B.2a, let us verify if the velocity field is divergence-free (satisfies the continuity equation).

Free vortex in Cartesian coordinates:

Velocity:

$$V = -\frac{C_1 y}{x^2 + y^2}\hat{\boldsymbol{i}} + \frac{C_1 x}{x^2 + y^2}\hat{\boldsymbol{j}} \tag{B.10}$$

Divergence:

$$\text{div}\boldsymbol{V} = \nabla \bullet \boldsymbol{V} = \frac{\partial}{\partial x}\left(-\frac{C_1 y}{x^2 + y^2}\right) + \frac{\partial}{\partial y}\left(\frac{C_1 x}{x^2 + y^2}\right) = 0$$

Vorticity:

$$\text{curl}V = \nabla \times V = \left(\frac{\partial V_y}{\partial x} - \frac{\partial V_x}{\partial y}\right)\hat{\boldsymbol{k}}$$

$$\text{curl}V = \left\{\frac{\partial}{\partial x}\left(\frac{C_1 x}{x^2 + y^2}\right) - \frac{\partial}{\partial y}\left(-\frac{C_1 y}{x^2 + y^2}\right)\right\}\hat{\boldsymbol{k}} = 0 \tag{B.11}$$

Free vortex in cylindrical coordinates:

Velocity:

$$\boldsymbol{V} = \frac{C_1}{r}\hat{\boldsymbol{e}}_\theta \tag{B.12}$$

Divergence:

$$\text{div}\boldsymbol{V} = \nabla \bullet \boldsymbol{V} = \frac{1}{r}\frac{\partial(rV_r)}{\partial r} + \frac{1}{r}\frac{\partial V_\theta}{\partial \theta}$$

which for $V_r = 0$ and $V_\theta = C_1 / r$ yields:

$$\text{div}V = 0$$

Vorticity:

$$\text{curl}\boldsymbol{V} = \nabla \times \boldsymbol{V} = \left(\frac{1}{r} \frac{\partial (rV_\theta)}{\partial r} - \frac{1}{r} \frac{\partial V_r}{\partial \theta} \right) \hat{\boldsymbol{k}}$$

which for $V_r = 0$ and $V_\theta = C_1 / r$ yields:

$$\text{curl}\boldsymbol{V} = \left(\frac{1}{r} \frac{\partial (C_1)}{\partial r} - \frac{1}{r} \frac{\partial (0)}{\partial \theta} \right) \hat{\boldsymbol{k}} = 0 \tag{B.13}$$

Equations B.11 and B.13 demonstrate that the free vortex possesses zero vorticity. This finding may seem counterintuitive to some, as it suggests that despite bulk rotation within the flow, individual fluid particles experience no local rotation. Figure B.4a illustrates this concept by depicting a small floating rod undergoing continuous orientation changes (rotation) along a circular streamline within a forced vortex. Conversely, in a free vortex, as depicted in Figure B.4b, the rod travels along a circular streamline without altering its orientation (exhibiting no rotation).

B.2.2.3 Rankine Vortex

Upon examination of Equations B.10 and B.12, it is evident that a free vortex displays a singularity at its origin (zero radius), theoretically resulting in infinite tangential velocity. Nevertheless, in reality, such a scenario is impossible for a fluid with nonzero viscosity. Instead, a genuine free vortex comprises a forced vortex core with zero tangential velocity at its origin. This structure, illustrated in Figure B.5, is termed a Rankine vortex, characterized by its flow field devoid of singularities.

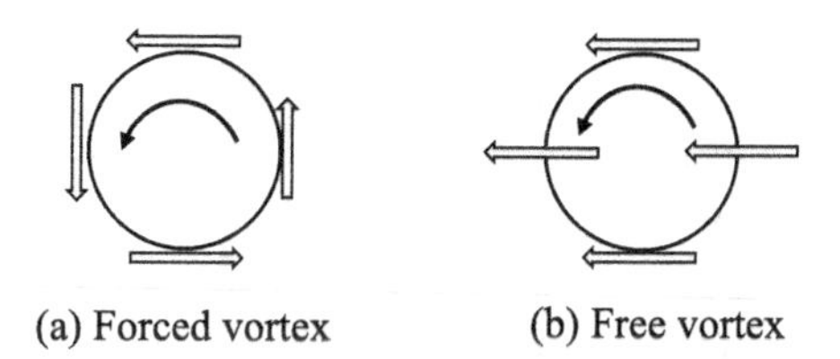

FIGURE B.4 Movement of a tiny floating rod in (a) forced vortex and (b) free vortex.

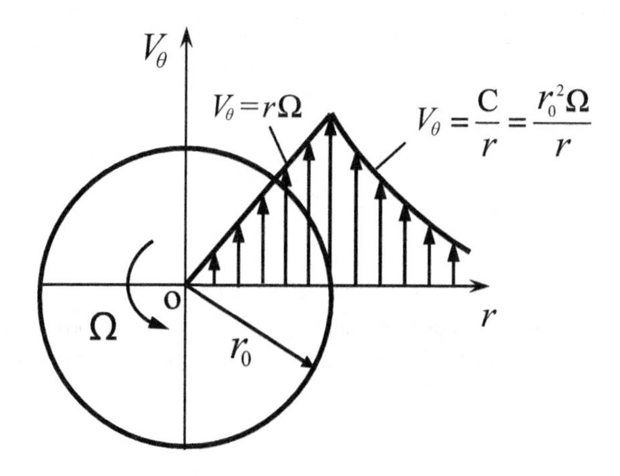

FIGURE B.5 Rankine vortex.

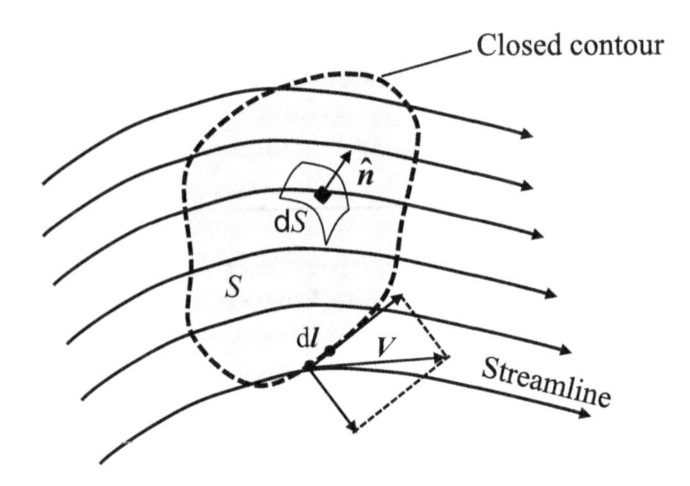

FIGURE B.6 Circulation around a closed contour in a flow field.

B.2.3 Circulation

Circulation is another property of a velocity field. For a closed contour shown in Figure B.6, the circulation measures the total vorticity contained within the contour. To compute the circulation around a positively oriented closed contour, we perform the line integral of the velocity vector using the equation:

$$\Gamma = \oint_C \boldsymbol{V} \cdot \mathrm{d}\boldsymbol{l} \tag{B.14}$$

Note that the velocity vector $\boldsymbol{V}$ is always tangent to the streamline but need not be tangent to the chosen closed contour at any of its points.

Applying Stokes theorem to Equation B.14 yields:

$$\Gamma = \oint V \cdot \mathrm{d}l = \iint_S (\nabla \times V) \cdot \hat{n} \mathrm{d}S = \iint_S \zeta \cdot \hat{n} \mathrm{d}S \tag{B.15}$$

which shows that the vorticity contained in a fluid element is related to the circulation associated with a closed contour around the element. For arbitrary choices of closed contours and enclosed areas, if $\zeta = 2\omega = 0$, then Equation B.15 yields zero circulation ($\Gamma = 0$).

B.3 STREAM TUBE VERSUS VORTEX TUBE

B.3.1 Stream Tube

Consider a closed contour C_1 with area A_1 shown in Figure B.7a. A unique streamline passes through each point on this contour. All the streamlines going through C_1 form the lateral surface of the stream tube shown in the figure. The contour C_2 with area A_2 is at the other end of the stream tube. By definition of a streamline, no flow can cross the lateral surface of a stream tube.

For an incompressible flow, the continuity equation implies that the flow field must be divergence-free, giving:

$$\nabla \cdot \boldsymbol{V} = 0$$

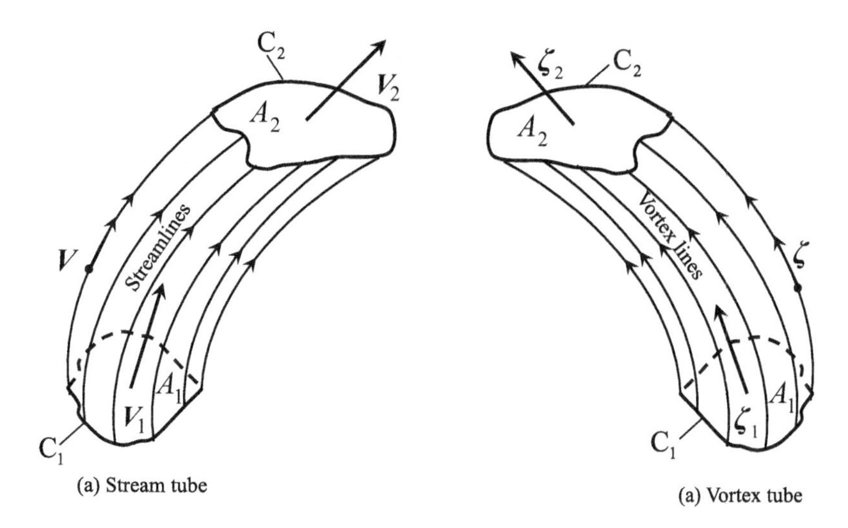

FIGURE B.7 (a) Stream tube and (b) vortex tube.

Integrating the above equation over the entire fluid control volume of the stream tube shown in Figure B.7a yields:

$$\iiint_{\forall} (\nabla \bullet V)\,d\forall = 0$$

Using Gauss' divergence theorem, we convert the above volume integral to the following surface integral:

$$\iiint_{\forall} (\nabla \bullet V)\,d\forall = \iint_{S} V \bullet \hat{n}\, dS = 0$$

where S is the surface area of the control volume and $\hat{n}$ is its outward-pointing unit normal vector. For the stream tube, the nonzero contributions to the above surface integral are the inflow at area A_1 and the outflow at A_2. If V_1 is the average velocity at A_1 and V_2 at A_2, the above equation yields:

$$Q = A_1 V_1 = A_2 V_2$$

where Q is the constant volumetric flow rate through the stream tube.

B.3.2 Vortex Tube

A unique vortex line passes through each point on a closed contour C_1 with area A_1 shown in Figure B.7b. All the vortex lines going through C_1 form the vortex tube shown in the figure. The contour C_2 with area A_2 is at the other end of the vortex tube.

The vorticity vector at any point in the flow is given by the curl of the local velocity vector. As the divergence of the curl of a vector is identically zero, we can write:

$$\nabla \bullet \zeta = 0$$

Integrating the above equation over the entire fluid control volume of the vortex tube shown in Figure B.7b yields:

$$\iiint_{\forall} (\nabla \bullet \zeta) \mathrm{d}\forall = 0$$

Using Gauss' divergence theorem, we convert the above volume integral to the following surface integral:

$$\iiint_{\forall} (\nabla \bullet \zeta) \mathrm{d}\forall = \iint_{S} \zeta \bullet \hat{n} \, \mathrm{d}S = 0$$

where S is the surface area of the control volume and $\hat{\boldsymbol{n}}$ is its outward-pointing unit normal vector. By definition, the vortex lines form the lateral surface of the vortex tube. Accordingly, $\zeta \bullet \hat{n} = 0$, making zero contribution to the surface integral. The nonzero contributions to the above surface integral correspond to areas A_1 and A_2. If V_1 is the average vorticity at A_1 and ζ_2 at A_2, the above equation yields:

$$\Gamma = A_1\zeta_1 = A_2\zeta_2$$

where Γ is the constant circulation at each cross-section of the vortex tube.

B.3.3 Analogy between Stream Tube and Vortex Tube

Based on the deductions presented in the foregoing, Table B.1 summarizes analogous properties between a stream tube and a vortex tube.

B.4 HELMHOLTZ THEOREMS OF VORTICITY

In 1858, Hermann von Helmholtz published the following three theorems that govern the behavior of vorticity in an inviscid, incompressible flow under the influence of a conservative force field (e.g., gravity):

B.4.1 Theorem 1

No fluid element that was not originally in rotation will rotate. This theorem implies that the vorticity of fluid particles in an inviscid flow remains unchanged.

B.4.2 Theorem 2

The elements that belong to one vortex line at any time remain on that line, regardless of their translation. This theorem implies that a vortex line moving in an inviscid flow constitutes identical fluid particles.

TABLE B.1
Analogous Properties between a Stream Tube and a Vortex Tube

Stream Tube	Vortex Tube
Velocity vector ($\boldsymbol{V}$)	Vorticity vector ($\boldsymbol{\zeta}$)
Streamline	Vortex line
Divergence-free velocity field: $\nabla \bullet \boldsymbol{V} = 0$	Divergence-free vorticity field: $\nabla \bullet \zeta = 0$
Constant volumetric flow rate: $Q = A\bar{V}$	Constant circulation: $\Gamma = A\bar{\zeta}$

B.4.3 Theorem 3

The product of the section area and the angular velocity of a vortex filament (an infinitely thin vortex tube) is constant throughout its whole length and retains the same value during all filament displacements. From this theorem, we deduce that a vortex tube must either terminate on itself (e.g., a smoke ring), a solid boundary, or a free surface. For example, a vortex tube in a deep river may have one end touching the river bottom and the other at the free surface.

B.5 KELVIN'S CIRCULATION THEOREM

In 1867, Lord Kelvin published the following circulation theorem:

In an inviscid, barotropic (fluid pressure is a function of density only) flow subject to conservative body forces (e.g., gravity), the circulation around a closed curve moving with the fluid remains constant with time. According to this theorem, which is closely related to Helmholtz's third theorem, if we observe a closed contour in a fluid flow at one instant and follow it (Lagrangian viewpoint) over time, the circulation at the two locations of this contour remains unchanged.

In an equation form, we can express Kelvin's circulation theorem as:

$$\frac{\mathrm{D}\Gamma}{\mathrm{D}t} = 0 \tag{B.16}$$

To derive this theorem, let us substitute the definition of circulation in Equation B.6, giving:

$$\frac{\mathrm{D}\Gamma}{\mathrm{D}t} = \frac{\mathrm{D}}{\mathrm{D}t}\oint_{\mathrm{C}} u_i \mathrm{d}x_i$$

where the integration is to be performed around a closed contour C in the flow. Taking the total derivative of the terms before the line integration yields:

$$\frac{\mathrm{D}\Gamma}{\mathrm{D}t} = \oint_{\mathrm{C}} \frac{\mathrm{D}}{\mathrm{D}t}(u_i \mathrm{d}x_i)$$

$$\frac{\mathrm{D}\Gamma}{\mathrm{D}t} = \oint_{\mathrm{C}} \left(\frac{\mathrm{D}u_i}{\mathrm{D}t} dx_i + u_i \frac{\mathrm{D}(\mathrm{d}x_i)}{\mathrm{D}t} \right) \tag{B.17}$$

Let us now evaluate $\mathrm{D}(\mathrm{d}x_i) / \mathrm{D}t$ in Equation B.17:

$$\frac{\mathrm{D}(\mathrm{d}x_i)}{\mathrm{D}t} = \mathrm{d}\left(\frac{\mathrm{D}x_i}{\mathrm{D}t} \right) = \mathrm{d}\left(\frac{\partial x_i}{\partial t} + u_k \frac{\partial x_i}{\partial x_k} \right)$$

Noting that $\partial x_i / \partial t = 0$ and $\partial x_i / \partial x_k = \delta_{ik}$, the above equation reduces to:

$$\frac{\mathrm{D}(\mathrm{d}x_i)}{\mathrm{D}t} = \mathrm{d}u_i$$

whose substitution in Equation B.17 yields:

$$\frac{\mathrm{D}\Gamma}{\mathrm{D}t} = \oint_{\mathrm{C}} \left(\frac{\mathrm{D}u_i}{\mathrm{D}t} \mathrm{d}x_i + u_i \mathrm{d}u_i \right) \tag{B.18}$$

To replace the total derivate of u_i in Equation B.18, let us invoke the three-dimensional Navier–Stokes equation discussed in Chapter 3. Dropping the viscous terms in Equation 3.29, we obtain the following equation for an inviscid flow with the inclusion of gravitational body forces. This equation is also known as the three-dimensional Euler's momentum equation:

$$\rho\frac{\mathrm{D}u_i}{\mathrm{D}t} = -\rho g\frac{\partial h}{\partial x_i} - \frac{\partial p}{\partial x_i}$$

Substituting this equation in Equation B.18, we obtain:

$$\frac{\mathrm{D}\Gamma}{\mathrm{D}t} = \oint_C \left(-g\frac{\partial h}{\partial x_i}\mathrm{d}x_i - \frac{\partial p}{\rho\,\partial x_i}\mathrm{d}x_i + u_i\mathrm{d}u_i\right)$$

$$\frac{\mathrm{D}\Gamma}{\mathrm{D}t} = \oint_C \left(-g\,\mathrm{d}h - \frac{\mathrm{d}p}{\rho} + \frac{1}{2}\mathrm{d}(u_iu_i)\right)$$

As h (the height measured from an arbitrary datum) and u_i are single valued in the flow field, we obtain:

$$\oint_C (-g)\,\mathrm{d}h = 0 \text{ and } \frac{\mathrm{D}\Gamma}{\mathrm{D}t} = \oint_C \frac{1}{2}\mathrm{d}(u_iu_i) = 0$$

giving

$$\frac{\mathrm{D}\Gamma}{\mathrm{D}t} = \oint_C \left(-\frac{\mathrm{d}p}{\rho}\right)$$

For an incompressible flow with constant density and for a barotropic flow where static pressure is a function of density only, the integral on right-hand side of the above equation becomes zero, giving:

$$\frac{\mathrm{D}\Gamma}{\mathrm{D}t} = 0$$

which proves Kelvin's circulation theorem. Note that the circulation of a given closed contour is also a measure of the total vorticity enclosed within the contour. In this way, Kevin's theorem is also proof of Helmholtz's third theorem presented in the foregoing.

B.6 PRESSURE AND TEMPERATURE CHANGES IN AN ISENTROPIC FREE VORTEX AND FORCED VORTEX

Mathematically, we can express free and forced vortices as follows:

Free vortex:

$$rV_\theta = C_1 \tag{B.19}$$

Forced vortex:

$$\frac{V_\theta}{r} = \Omega_\mathrm{f} \tag{B.20}$$

where C_1 is a constant and Ω_f is fluid angular velocity, which remains constant for a forced vortex.

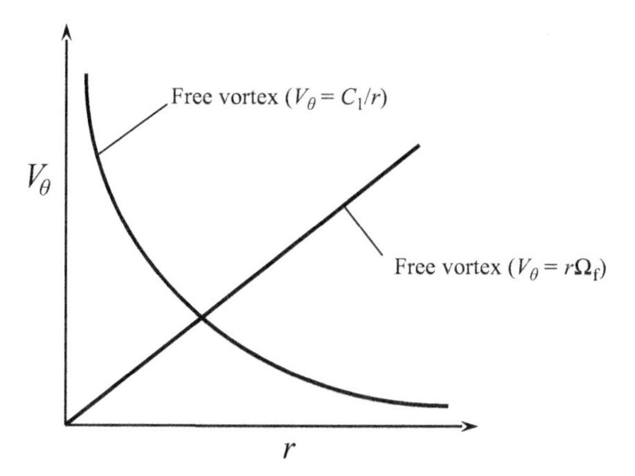

FIGURE B.8 Radial variation of tangential velocity in free and forced vortices.

Figure B.8 shows the variation of tangential velocity with radius in both free and forced vortices. We may combine Equations B.19 and B.20 to represent both free and forced vortices:

$$V_\theta r^n = C_2 \tag{B.21}$$

where we have $n = 1$ and $C_2 = C_1$ for a free vortex and $n = -1$ and $C_2 = \Omega_f$ for a forced vortex. Note that n has no physical significance, being merely a mathematical artifice with binary values to combine Equations B.19 and B.20 into Equation B.21.

To compute changes in the static pressure and temperature in isentropic free and forced vortices, setting $ds = 0$ in Equation 1.37 yields:

$$\frac{dp}{p} = \frac{c_p}{R}\frac{dT}{T} \tag{B.22}$$

Substituting for density from the equation of state ($p/\rho = RT$) into the radial equilibrium equation ($dp/dr = \rho V_\theta^2/r$) yields

$$\frac{dp}{p} = \frac{V_\theta^2}{RT}\frac{dr}{r} \tag{B.23}$$

Using Equation B.22 in this equation, we obtain:

$$dT = \frac{V_\theta^2 dr}{c_p r}$$

which with the substitution for V_θ from Equation B.21 yields:

$$dT = \frac{C_2^2}{c_p r^{2n}}\frac{dr}{r} = \frac{C_2^2}{c_p}\frac{dr}{r^{2n+1}} \tag{B.24}$$

Integrating Equation B.24 between any two points in the vortex results in:

$$\int_1^2 dT = \frac{C_2^2}{c_p}\int_{r_1}^{r_2}\frac{dr}{r^{2n+1}}$$

$$T_2 - T_1 = \frac{C_2^2}{c_p(-2n)}\left(r_2^{-2n} - r_1^{-2n}\right) \tag{B.25}$$

which using Equation B.21 yields:

$$T_2 - T_1 = \Delta T = \frac{1}{(-n)}\left(\frac{V_{\theta 2}^2}{2c_p} - \frac{V_{\theta 1}^2}{2c_p}\right) \tag{B.26}$$

where $V_{\theta 1}^2/2c_p$ and $V_{\theta 2}^2/2c_p$ are the dynamic temperatures associated with the tangential velocities at points 1 and 2, respectively. This equation shows that, in a free vortex ($n = 1$), the change in static temperature is equal to and opposite of the change in the corresponding dynamic temperature. In contrast, these changes are equal for a forced vortex ($n = -1$).

Let us suppose that we are given the vortex properties—namely, T_1, p_1, and $V_{\theta 1}$ at point 1 where $r = r_1$. For both free and forced vortices, we present here a simple step-wise procedure to compute T_2 and p_2 at point 2, which corresponds to $r = r_2$.

B.6.1 Free Vortex ($n = 1$)

Step 1: Compute $V_{\theta 2}$ from the equation:

$$V_{\theta 2} = \frac{r_1 V_{\theta 1}}{r_2} \tag{B.27}$$

Step 2: Use Equation B.26 to compute $(\Delta T)_{\text{free vortex}}$

$$(\Delta T)_{\text{free vortex}} = \frac{V_{\theta 1}^2 - V_{\theta 2}^2}{2c_p} = \frac{V_{\theta 1}^2}{2c_p}\left\{1 - \left(\frac{r_1}{r_2}\right)^2\right\} = \frac{V_{\theta 1}^2}{2c_p}\left\{\left(\frac{r_2}{r_1}\right)^2 - 1\right\}\left(\frac{r_1}{r_2}\right)^2 \tag{B.28}$$

Step 3: Compute T_2

$$T_2 = T_1 + (\Delta T)_{\text{free vortex}} \tag{B.29}$$

Step 4: Compute p_2

$$p_2 = p_1\left(\frac{T_2}{T_1}\right)^{\frac{\gamma}{\gamma-1}} \tag{B.30}$$

B.6.2 Forced Vortex ($n = -1$)

Step 1: Compute $V_{\theta 2}$ from the equation:

$$V_{\theta 2} = \frac{r_2 V_{\theta 1}}{r_1} \tag{B.31}$$

Step 2: Use Equation B.26 to compute $(\Delta T)_{\text{forced vortex}}$

$$(\Delta T)_{\text{forced vortex}} = \frac{V_{\theta 2}^2 - V_{\theta 1}^2}{2c_p} = \frac{V_{\theta 1}^2}{2c_p}\left\{\left(\frac{r_2}{r_1}\right)^2 - 1\right\} \tag{B.32}$$

which using Equation B.28 may be written as:

$$(\Delta T)_{\text{forced vortex}} = (\Delta T)_{\text{free vortex}} \left(\frac{r_2}{r_1}\right)^2 \tag{B.33}$$

Step 3: Compute T_2

$$T_2 = T_1 + (\Delta T)_{\text{forced vortex}} \tag{B.34}$$

Step 4: Compute P_2

$$p_2 = p_1 \left(\frac{T_2}{T_1}\right)^{\frac{\gamma}{\gamma-1}} \tag{B.35}$$

Equation B.33 reveals that, for a radially outward flow ($r_2 > r_1$), the increase in static temperature for a forced vortex will be higher than that for a free vortex. For a radially inward flow ($r_2 < r_1$), the decrease in static temperature for a forced vortex will be lower than that for a free vortex. In both cases, using the foregoing equations, we can show that T_2 and p_2 for a forced vortex will be higher than those for a free vortex.

At any point in a vortex with axial, radial, and tangential velocity components, total and static temperatures are related by the equation:

$$T_0 = T + \frac{V_x^2}{2c_p} + \frac{V_r^2}{2c_p} + \frac{V_\theta^2}{2c_p} = T + \frac{V_{xr}^2}{2c_p} + \frac{V_\theta^2}{2c_p} \tag{B.36}$$

where V_{xr} is the meridional velocity. For no change in V_{xr} between two points in a vortex, Equation B.36 yields:

$$T_{02} - T_{01} = (T_2 - T_1) + \left(\frac{V_{\theta 2}^2}{2c_p} - \frac{V_{\theta 1}^2}{2c_p}\right) \tag{B.37}$$

in which the first parenthetical term on the right-hand side represents the change in static temperature and the second parenthetical term the change in dynamic temperature. In an isentropic free vortex ($n = 1$), these two changes negate each other, giving $T_{02} = T_{01}$ and $p_{02} = p_{01}$; that is, both total temperature and total pressure remain constant in this vortex.

In an isentropic forced vortex ($n = -1$), both parenthetical terms in Equation 2.71 are equal. The change in total temperature in this vortex is therefore twice the difference in static temperatures between two radial locations. Thus, for this vortex, we can write:

$$T_{02} - T_{01} = (\Delta T_0)_{\text{forced vortex}} = 2(T_2 - T_1) = 2(\Delta T)_{\text{forced vortex}}$$

$$T_{02} = T_{01} + 2(\Delta T)_{\text{forced vortex}} \tag{B.38}$$

where $(\Delta T)_{\text{forced vortex}}$ is given by Equation B.32.

After computing T_{01} and T_{02} using Equation B.33, we can compute p_{01} from the equation:

$$p_{01} = p_1 \left(\frac{T_{01}}{T_1}\right)^{\frac{\gamma}{\gamma-1}} \tag{B.39}$$

and then p_{02} either from the equation:

$$p_{02} = p_2 \left(\frac{T_{02}}{T_2} \right)^{\frac{\gamma}{\gamma-1}} \tag{B.40}$$

or from the equation:

$$p_{02} = p_{01} \left(\frac{T_{02}}{T_{01}} \right)^{\frac{\gamma}{\gamma-1}} \tag{B.41}$$

Sultanian (2018) presents equations to compute pressure and temperature changes in a nonisentropic generalized vortex.

B.7 CONCLUDING REMARKS

Earlier, we discussed two basic types of vortices: free and forced. Except at the origin (zero radius), the angular momentum remains constant in a free vortex. The vorticity remains constant everywhere in a forced vortex. All the vorticities in a free vortex are confined to the singularity at the origin. Away from it, the flow is irrotational (zero vorticity). An atmospheric tornado is a good example of a free vortex. The line passing through the eye of a three-dimensional tornado (or a free vortex) is known as a line vortex with nonzero vorticity. A vortex line, however, is a line that is tangent to the local vorticity vector, just as a streamline in a flow is a line that is tangent to the local velocity vector. Analogous properties between a stream tube and a vortex tube presented in this appendix are worth noting.

WORKED EXAMPLE

Example B.1

The streamlines in a two-dimensional vortex are circles. Calculate the circulation around a circle of radius R from the origin for each vortex type: (a) forced vortex (Figure B.2), (b) free vortex (Figure B.3), and (c) Rankine vortex (Figure B.5).

Solution

a. **Circulation in a Forced Vortex**
   Because the tangential flow velocity is tangent to the circle at each point, we can use Equation B.14 to calculate the circulation in a forced vortex for the circle of radius R as:

$$\Gamma_{\text{force vortex}} = \oint V \bullet \mathrm{d}l = \int_0^{2\pi} R\Omega(R\mathrm{d}\theta) = \pi R^2 (2\Omega)$$

   For a forced vortex, this result shows that the circulation around a circle of radius R is equal to the product of the area of the circle and 2Ω. Recall from Equations B.7 and B.9 that 2Ω is the constant vorticity of a forced vortex.

b. **Circulation in a Free Vortex**
   In a free vortex given by Equation B.12, we can write the circulation around a circle of radius R as:

$$\Gamma_{\text{force vortex}} = \oint V \bullet \mathrm{d}l = \int_0^{2\pi} \frac{C}{R}(R\mathrm{d}\theta) = 2\pi C$$

which shows that the circulation is constant and independent of the radius of the circle. The annular region between two circles around the origin will have zero circulation (vorticity), the constant circulation being confined within the inner circle. Thus, in an ideal free vortex, we can associate the constant circulation and the related vorticity to the singularity at the origin.

c. **Circulation in a Rankine Vortex**

Because the Rankine vortex is a combination of a forced vortex and a free vortex, from the results obtained in (a) and (b), we can conclude that, for a circle inside the forced vortex core ($R \le r_o$), the circulation will be equal to $2\pi R^2\Omega$. For all circles with ($R \ge r_o$), the circulation will remain constant at $2\pi r_o^2\Omega$.

REFERENCES

Lugt, H.J. 1995. *Vortex Flow in Nature and Technology.* Malabar: Krieger Publishing Company.

Sultanian, B.K. 2015. *Fluid Mechanics: An Intermediate Approach*, 1st Edition. Boca Raton, FL: Taylor & Francis.

Sultanian, B.K. 2018. *Gas Turbines: Internal Flow Systems Modeling* (Cambridge Aerospace Series #44). UK, Cambridge: Cambridge University Press.

BIBLIOGRAPHY

Currie, I.G. 2013. *Fundamental Mechanics of Fluids*, 4th edition. Boca Raton: CRC Press.

Kundu, P.K., I.M. Cohen, D.R. Dowling, J. Capecilatro. 2024. *Fluid Mechanics*, 7th edition. Burlington: Academic Press.

Panton, R.L. 2024. *Incompressible Flow*, 5th edition. New York: Wiley.

Sultanian, B.K. 2022a. *FLUID MECHANICS An Intermediate Approach*: Errata for the First Edition Published in 2015. Independently published on KDP (Amazon.com, USA).

Sultanian, B.K. 2022b. *GAS TURBINES Internal Flow Systems Modeling*: Errata for the First Edition Published in 2018. Independently published on KDP (Amazon.com, USA).

Truesdell, C. 2018. *The Kinematics of Vorticity.* Mineola, NY: Dover.

Appendix C
Euler's Turbomachinery Equation

C.1 INTRODUCTION

Euler's turbomachinery equation, widely used to determine power transfer between the fluid and the rotor blades in all types of turbomachines, is based on the angular momentum equation discussed in Chapter 1. The aerodynamic power transfer to or from the fluid is the product of the torque and the rotor angular velocity in radians per second. In pumps, fans, and compressors, the power transfer occurs into the fluid to increase its rate of angular momentum outflow over its rate of inflow. In turbines, the power transfer occurs from the fluid to the rotor, which decreases the rate of fluid angular momentum outflow over its rate of inflow.

Euler's turbomachinery equation leads to the concept of rothalpy and equations to compute it in both stator and rotor reference frames. These equations provide the easiest way to convert total temperature and pressure from the rotor reference frame to the stator reference frame and vice versa.

This appendix includes a significantly improved version of the material presented in Appendix D: Converting Quantities between Stator and Rotor Reference Frames in the first edition of this book by Sultanian (2015).

C.2 DERIVATION OF EULER'S TURBOMACHINERY EQUATION

Let us consider a steady adiabatic flow in a rotating passage formed between adjacent blades of the rotor, as shown in Figure C.1. The velocity vectors V_1 at the inlet 1 and V_2 at outlet 2 have components in the axial, radial, and tangential directions. The meridional velocity V_{xr}, which is the resultant of the axial and the radial velocities, is the mass velocity at sections 1 and 2. Let $V_{\theta 1}$ be the tangential velocity at section 1 (inlet) and $V_{\theta 2}$ the tangential velocity at section 2 (outlet).

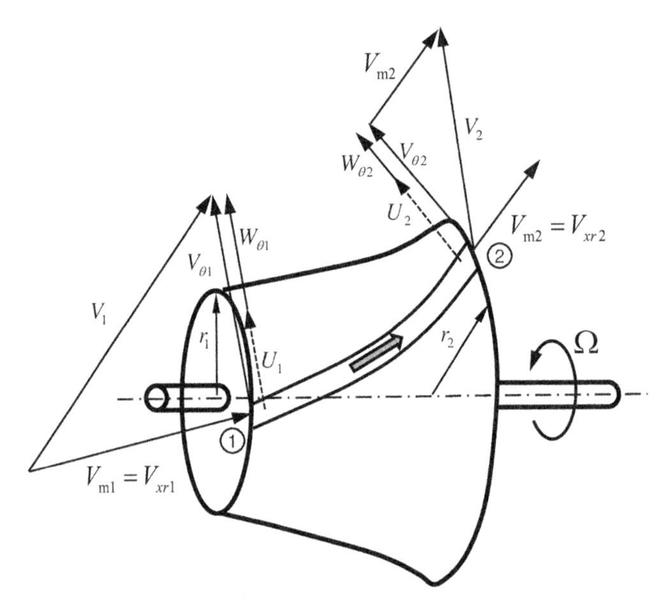

FIGURE C.1 Flow through an axial–radial turbomachinery passage between adjacent blades.

Using the angular momentum equation, Equation 1.23 presented in Chapter 1, we write for a constant mass flow rate $\dot{m}$ through the blade passage:

$$\Gamma = \dot{m}(r_2 V_{\theta 2} - r_1 V_{\theta 1}) \tag{C.1}$$

We can write the aerodynamic power transfer, which is the rate of work transfer due to the aerodynamic torque acting on the fluid control volume as:

$$P = \Gamma \Omega = \dot{m}\left(r_2 V_{\theta 2} - r_1 V_{\theta 1}\right)\Omega = \dot{m}\left(U_2 V_{\theta 2} - U_1 V_{\theta 1}\right) \tag{C.2}$$

where U_1 and U_2 are rotor tangential velocities at section 1 (inlet) and section 2 (outlet), respectively.

Using steady flow energy equation in terms of total enthalpy at these sections, we can also write:

$$P = \dot{m}(h_{02} - h_{01}) \tag{C.3}$$

which together with Equation C.2 yields:

$$h_{02} - h_{01} = U_2 V_{\theta 2} - U_1 V_{\theta 1} \tag{C.4}$$

which is known as Euler's turbomachinery equation. This equation states that, under adiabatic condition, the change in specific total enthalpy of the fluid flow between any two sections of a rotor equals the difference in the products of the rotor tangential velocity and the flow tangential velocity at these sections. This equation further reveals that, for turbines, where the work transfer occurs from the fluid to the rotor, we have $h_{02} < h_{01}$ and $U_2 V_{\theta 2} < U_1 V_{\theta 1}$; for compressors, fans, and pumps, where the work transfer occurs from the rotor to the fluid, we have $h_{02} > h_{01}$ and $U_2 V_{\theta 2} > U_1 V_{\theta 1}$.

We note here that Equation C.4 allows us to compute energy transfer in an inertial reference frame, although the coordinate system attached to the turbomachinery is a noninertial one, featuring the centrifugal and Coriolis forces in this reference frame. In support of this equation, Sultanian (2015) presents a mathematical derivation to convert the integral angular momentum equation written in the noninertial reference frame to that in the inertial reference frame.

For axial flow machines, where $r_1 \approx r_2$ and $U_1 \approx U_2$, Equation C.4 reveals that the change in total enthalpy is entirely due to the change in flow tangential velocity—i.e., $\Delta h_0 = \bar{U}\Delta V_\theta$—requiring blades with camber (bow). For radial-flow machines, however, the change in total enthalpy results largely from the change in rotor tangential velocity due to the change in radius—i.e., $\Delta h_0 = \Delta U \bar{V}_\theta$.

Substituting $V_\theta = W_\theta + r\,\Omega$ in Equation C.4 yields:

$$h_{02} - h_{01} = (U_2 W_{\theta 2} - U_1 W_{\theta 1}) + (r_2^2 - r_1^2)\Omega^2$$

where $(U_2 W_{\theta 2} - U_1 W_{\theta 1})$ represents the specific work transfer due to aerodynamic forces and $(r_2^2 - r_1^2)\Omega^2$ represents the specific work transfer due to Coriolis forces, see Lewis (1996).

C.2.1 Rothalpy

The concept of rothalpy is grounded in Euler's turbomachinery equation presented in the preceding section. Rearranging Equation C.4 yields:

$$h_{01} - U_1 V_{\theta 1} = U_2 V_{\theta 2} \tag{C.5}$$

which reveals that the quantity $(h_0 - UV_\theta)$ at any point in a rotor flow remains constant under adiabatic condition (no heat transfer). We call this quantity rothalpy and express it as:

$$I = h_0 - UV_\theta = h + \frac{V^2}{2} - UV_\theta \tag{C.6}$$

where both h_0 and V_θ are in the stator reference frame.

$$V^2 = V_x^2 + V_r^2 + V_\theta^2 \tag{C.7}$$

$$W^2 = W_x^2 + W_r^2 + W_\theta^2 \tag{C.8}$$

$$V_\theta = W_\theta + U \tag{C.9}$$

where U is the local tangential velocity of the rotor.

Substituting for V_θ from Equation C.9 into Equation C.7 and noting that $W_x = V_x$ and $W_r = V_r$, we obtain:

$$V^2 = W_x^2 + W_r^2 + W_\theta^2 + 2W_\theta U + U^2 \tag{C.10}$$

Using Equations C.9 and C.10 in Equation C.6 yields:

$$I = h + \frac{W_x^2 + W_r^2 + W_\theta^2 + 2W_\theta U + U^2}{2} - U(W_\theta + U)$$

which reduces to:

$$I = h + \frac{W^2}{2} - \frac{U^2}{2} = h_{0R} - \frac{U^2}{2} \tag{C.11}$$

where h_{0R} is the specific total enthalpy in the rotor reference frame. For a calorically perfect gas with constant c_p, we can write this equation as:

$$I = c_p T_{0R} - \frac{U^2}{2} \tag{C.12}$$

where T_{0R} is the fluid total temperature in the rotor reference frame. According to Equation C.11, the rothalpy of a flow at a point in a rotor tells us as to how much higher its relative total enthalpy is compared to its corresponding dynamic enthalpy of solid-body rotation.

For an isentropic flow, we can write:

$$\mathrm{d}h = \frac{\mathrm{d}p}{\rho} \tag{C.13}$$

which, when combined with Equation C.11, yields for an isentropic incompressible flow in a rotor the following relation between points 1 and 2:

$$\begin{gathered}
\frac{p_1}{\rho} + \frac{W_1^2}{2} - \frac{U_1^2}{2} = \frac{p_2}{\rho} + \frac{W_2^2}{2} - \frac{U_2^2}{2} \\
\frac{p_{0R1}}{\rho} - \frac{U_1^2}{2} = \frac{p_{0R2}}{\rho} - \frac{U_2^2}{2} \\
p_{0R2} - p_{0R1} = \frac{\rho U_2^2}{2} - \frac{\rho U_1^2}{2}
\end{gathered} \tag{C.14}$$

where p_{0R1} and p_{0R2} are relative total pressures at points 1 and 2, respectively.

C.2.2 An Alternate Form of Euler's Turbomachinery Equation

For the adiabatic flow in a rotor, the rothalpy remains constant between two points—i.e., $I_1 = I_2$. We can write from Equation C.11:

$$h_1 + \frac{W_1^2}{2} - \frac{U_1^2}{2} = h_2 + \frac{W_2^2}{2} - \frac{U_2^2}{2}$$

$$h_2 - h_1 + \left(\frac{W_1^2}{2} - \frac{W_2^2}{2}\right) + \left(\frac{U_2^2}{2} - \frac{U_1^2}{2}\right) \tag{C.15}$$

which expresses the change in specific static enthalpy in a rotor in terms of changes in the specific kinetic energy of the flow relative velocity and that of the rotor tangential velocity.

From the definition of specific total enthalpy, we can write its change between locations 1 and 2 as:

$$h_{02} - h_{01} = (h_2 - h_1) + \left(\frac{V_2^2}{2} - \frac{V_1^2}{2}\right) \tag{C.16}$$

which upon substituting for $(h_2 - h_1)$ from Equation C.15 yields the following alternate form of the Euler's turbomachinery equation—earlier derived as Equation C.4:

$$h_{02} - h_{01} = \left(\frac{W_1^2}{2} - \frac{W_2^2}{2}\right) + \left(\frac{U_2^2}{2} - \frac{U_1^2}{2}\right) + \left(\frac{V_2^2}{2} - \frac{V_1^2}{2}\right) \tag{C.17}$$

C.3 CONVERTING TOTAL PRESSURE AND TEMPERATURE BETWEEN STATOR AND ROTOR REFERENCE FRAMES

Static properties, such as static temperature and static pressure, are independent of the reference frame, be it inertial (stator) or noninertial (rotor). However, the total properties, such as total pressure and total temperature, generally change from one reference frame to the other due to the change in the flow velocity relative to the reference frame. In a turbomachine with a single axis of rotation, only the tangential (swirl) velocity changes between the stator and rotor reference frames; the absolute and relative tangential velocities are related by Equation C.9. In turbomachinery design applications, we often need to convert total temperature and total pressure from the stator reference frame to the rotor reference frame. Let us use the definitions of rothalpy in the two reference frames to perform this task.

To convert total temperature from the rotor reference frame to the stator reference frame, we use Equations C.6 and C.11 to write:

$$h_0 - UV_\theta = h_{0R} - \frac{U^2}{2}$$

$$c_pT_0 - UV_\theta = c_pT_{0R} - \frac{U^2}{2}$$

which, with the substitution of V_θ from Equation C.9, yields:

$$T_0 = T_{0R} + \frac{U}{2c_p}(U + 2W_\theta) \tag{C.18}$$

giving $T_0 = T_{0R}$ for $W_\theta = -U/2$.

For an isentropic compressible flow, we can compute the total pressure in the stator reference frame with the equation:

$$\frac{p_0}{p_{0R}} = \left(\frac{T_0}{T_{0R}}\right)^{\frac{\gamma}{\gamma-1}} = \left\{1 + \frac{U(U + 2W_\theta)}{2c_p T_{0R}}\right\}^{\frac{\gamma}{\gamma-1}} \tag{C.19}$$

where we have used the fact that both static pressure and static temperature, being fluid properties, are independent of the reference frame.

For converting total temperature from stator reference frame to rotor reference frame, we rewrite Equation C.18 as:

$$T_{0R} = T_0 - \frac{U}{2c_p}(U + 2W_\theta)$$

which upon substitution for W_θ from Equation C.9 and further simplification of the resulting expression yields:

$$T_{0R} = T_0 + \frac{U}{2c_p}(U - 2V_\theta) \tag{C.20}$$

giving $T_{0R} = T_0$ for $V_\theta = U/2$.

For an isentropic compressible flow, we can compute the total pressure in the rotor reference frame using the equation:

$$\frac{p_{0R}}{p_0} = \left(\frac{T_{0R}}{T_0}\right)^{\frac{\gamma}{\gamma-1}} = \left\{1 + \frac{U(U - 2V_\theta)}{2c_p T_0}\right\}^{\frac{\gamma}{\gamma-1}} \tag{C.21}$$

C.4 CONCLUDING REMARKS

Euler's turbomachinery equation, derived from the angular momentum equation in this appendix, is fundamental in determining power transfer between fluid and rotor blades in turbomachines. It calculates the aerodynamic power transfer by multiplying torque and rotor angular velocity. In devices like pumps and compressors, power transfer increases fluid momentum outflow rate over inflow, while in turbines, it decreases fluid momentum outflow rate. Euler's equation introduces the concept of rothalpy and provides equations to compute it in stator and rotor reference frames, facilitating the conversion of total temperature and pressure between the two frames.

REFERENCES

Lewis, R.I. 1996. *Turbomachinery Performance Analysis.* New York: John Wiley and Sons.

Sultanian, B.K. 2015. *Fluid Mechanics: An Intermediate Approach*, 1st Edition. Boca Raton, FL: Taylor & Francis.

BIBLIOGRAPHY

Sultanian, B.K. 2018. *Gas Turbines: Internal Flow Systems Modeling* (Cambridge Aerospace Series). Cambridge: Cambridge University Press.

Sultanian, B.K. 2021. *Fluid Mechanics and Turbomachinery: Problems and Solutions.* Boca Raton, FL: Taylor & Francis.

Sultanian, B.K. 2019. *Logan's Turbomachinery: Flowpath Design and Performance Fundamentals*, 3rd Edition. Boca Raton, FL: Taylor & Francis.

Sultanian, B.K. 2022a. *Thermal-Fluids Engineering*: *Problems with Solutions*. Independently published on KDP (Amazon.com).

Sultanian, B.K. 2022b. *Fluid Mechanics: An Intermediate Approach*: Errata for *the First Edition Published in 2015*. Independently published on KDP (Amazon.com).

Index

Printed in the United States
by Baker & Taylor Publisher Services